Classics in Mathematics

André Weil Basic Number Theory

André Weil was born May 6, 1906 in Paris. After
studying mathematics at the Ecole Normale
Supérieure and receiving a doctoral degree from
the University of Paris in 1928, he held professorial
positions in India, France, the United States and
Brazil before being appointed to the Institute
for Advanced Study, Princeton in 1958 where he
is now Professor Emeritus.

André Weil's work laid the foundation for abstract
algebraic geometry and the modern theory of
abelian varieties. A great deal of his work was
directed towards establishing the links between
number theory and algebraic geometry and
devising modern methods in analytic number
theory. Weil was one of the founders, around 1934,
of the group that published, under the collective
name of N. Bourbaki, the highly influential multi-
volume treatise *Eléments de mathématique.*

André Weil

Basic Number Theory

Reprint of the 1974 Edition

Springer

André Weil
Institute for Advanced Study
Princeton, NJ 08540
USA

Originally published as Vol. 144 of the
Grundlehren der mathematischen Wissenschaften

Mathematics Subject Classification (1991): 11R

ISBN 3-540-58655-5 Springer-Verlag Berlin Heidelberg New York

Photograph by kind permission of Lucien Gillet

CIP data applied for

This work is subject to copyright. All rights are reserved, whether the whole or part of the material is concerned, specifically the rights of translation, reprinting, reuse of illustration, recitation, broadcasting, reproduction on microfilm or in any other way, and storage in data banks. Duplication of this publication or parts thereof is permitted only under the provision of the German Copyright Law of September 9, 1965, in its current version, and permission for use must always be obtained from Springer-Verlag. Violations are liable for prosecution under the German Copyright Law.

© Springer-Verlag Berlin Heidelberg 1995
Printed in Germany

The use of general descriptive names, registered names, trademarks, etc. in this publication does not imply, even in the absence of a specific statement, that such names are exempt from the relevant protective laws and regulations and therefore free for general use.

SPIN 10701941 41/3111 - 5 4 3 2 1 - Printed on acid-free paper

André Weil

Basic Number Theory

Third Edition 1974

Springer-Verlag Berlin • Heidelberg • New York

1974

Professor André Weil

The Institute for Advanced Study, Princeton, N. J. 08540

ISBN 0-387-06177-0 Springer-Verlag New York · Heidelberg · Berlin
ISBN 3-540-06177-0 Springer-Verlag Berlin · Heidelberg · New York

All rights, especially that of translation into foreign languages, reserved. It is also
forbidden to reproduce this book, either whole or in part, by photomechanical
means (photostat, microfilm and/or microcard) or by other procedure without
written permission from Springer-Verlag

© 1973 by Springer-Verlag New York Inc.

Library of Congress Catalog Card Number 73-75018

PRINTED IN THE UNITED STATES OF AMERICA

Foreword

Ἀριθμόν, ἔξοχον σοφισμάτων

Αἰσχ., Προμ. Δεσμ.

The first part of this volume is based on a course taught at Princeton University in 1961–62; at that time, an excellent set of notes was prepared by David Cantor, and it was originally my intention to make these notes available to the mathematical public with only quite minor changes. Then, among some old papers of mine, I accidentally came across a long-forgotten manuscript by Chevalley, of pre-war vintage (forgotten, that is to say, both by me and by its author) which, to my taste at least, seemed to have aged very well. It contained a brief but essentially complete account of the main features of classfield theory, both local and global; and it soon became obvious that the usefulness of the intended volume would be greatly enhanced if I included such a treatment of this topic. It had to be expanded, in accordance with my own plans, but its outline could be preserved without much change. In fact, I have adhered to it rather closely at some critical points.

To improve upon Hecke, in a treatment along classical lines of the theory of algebraic numbers, would be a futile and impossible task. As will become apparent from the first pages of this book, I have rather tried to draw the conclusions from the developments of the last thirty years, whereby locally compact groups, measure and integration have been seen to play an increasingly important role in classical number-theory. In the days of Dirichlet and Hermite, and even of Minkowski, the appeal to "continuous variables" in arithmetical questions may well have seemed to come out of some magician's bag of tricks. In retrospect, we see now that the real numbers appear there as one of the infinitely many completions of the prime field, one which is neither more nor less interesting to the arithmetician than its p-adic companions, and that there is at least one language and one technique, that of the adeles, for bringing them all together under one roof and making them cooperate for a common purpose. It is needless here to go into the history of these developments; suffice it to mention such names as Hensel, Hasse, Chevalley, Artin; every one of these, and more recently Iwasawa, Tate, Tamagawa, helped to make some significant step forward along this road. Once the presence of the real field, albeit at infinite distance, ceases to be regarded as a necessary ingredient in the arithmetician's brew, it

goes without saying that the function-fields over finite fields must be granted a fully simultaneous treatment with number-fields, instead of the segregated status, and at best the separate but equal facilities, which hitherto have been their lot. That, far from losing by such treatment, both races stand to gain by it, is one fact which will, I hope, clearly emerge from this book.

It will be pointed out to me that many important facts and valuable results about local fields can be proved in a fully algebraic context, without any use being made of local compacity, and can thus be shown to preserve their validity under far more general conditions. May I be allowed to suggest that I am not unaware of this circumstance, nor of the possibility of similarly extending the scope of even such global results as the theorem of Riemann-Roch? We are dealing here with mathematics, not with theology. Some mathematicians may think that they can gain full insight into God's own way of viewing their favorite topic; to me, this has always seemed a fruitless and a frivolous approach. My intentions in this book are more modest. I have tried to show that, from the point of view which I have adopted, one could give a coherent treatment, logically and aesthetically satisfying, of the topics I was dealing with. I shall be amply rewarded if I am found to have been even moderately successful in this attempt.

Some of my readers may be surprised to find no explicit mention of cohomology in my account of classfield theory. In this sense, while my approach to number-theory may be called a "modern" one in the first half of this book, it may well be described as thoroughly "unmodern" in the second part. The sophisticated reader will of course perceive that a certain amount of cohomology, and in fact no more and no less than is required for the purposes of classfield theory, hides itself in the theory of simple algebras. For anyone familiar with the language of "Galois cohomology", it will be an easy and not unprofitable exercise to translate into it some of the definitions and results of our Chapters IX, XII and XIII; in one or two places (the most conspicuous case being that of the "transfer theorem" in Chapter XII, § 5), this even makes it possible to substitute more satisfactory proofs for ours. For me to develop such an approach systematically would have meant loading a great deal of unnecessary machinery on a ship which seemed well equipped for this particular voyage; instead of making it more seaworthy, it might have sunk it.

In charting my course, I have been careful to steer clear of the arithmetical theory of algebraic groups; this is a topic of deep interest, but obviously not yet ripe for book treatment. Partly for this reason, I have refrained from discussing zeta-functions of simple algebras beyond what was needed for the sake of classfield theory. Artin's non-abelian L-func-

tions have also been excluded; the reader of this book will find it easy to proceed to the study of Artin's beautiful papers on this subject and will find himself well prepared to enjoy them, provided he has some knowledge of the representation theory of finite groups.

It remains for me to discharge the pleasant duty of expressing my thanks to David Cantor, who prepared from my lectures at Princeton University the set of notes which reappears here as Chapters I to VII of this book (in many places with no change at all), and to Chevalley, who generously allowed me to make use of the above-mentioned manuscript and expand it into Chapters XII and XIII. My thanks are also due to Iwasawa and Lazard, who read the book in manuscript and offered many suggestions for its improvement; to H. Pogorzelski, for his assistance in proofreading; to B. Eckmann, for the interest he took in its publication; and to the staff of the Springer Verlag, and that of the Zechnersche Buchdruckerei, for their expert cooperation and their invaluable help in the process of bringing out this volume.

Princeton, May 1967. ANDRÉ WEIL

Foreword to the second edition

The text of the first edition has been left unchanged. A few corrections, references, and some brief remarks, have been added as Notes at the end of the book; the corresponding places in the text have been marked by a * in the margin. Somewhat more substantial additions will be found in the Appendices, originally prepared for the Russian edition (M. I. R., Moscow 1971). The reader's attention should be drawn to the collective volume: J. W. S. Cassels and A. Fröhlich (edd.), *Algebraic Number Theory*, Acad. Press 1967, which covers roughly the same ground as the present book, but with far greater emphasis on the cohomological aspects.

Princeton, December 1971. ANDRÉ WEIL

Contents

Chronological table

(In imitation of Hecke's "*Zeittafel*" at the end of his "*Theorie der algebraischen Zahlen*", and as a partial substitute for a historical survey, we give here a chronological list of the mathematicians who seem to have made the most significant contributions to the topics treated in this volume.)

Fermat (1601–1665)
Euler (1707–1783)
Lagrange (1736–1813)
Legendre (1752–1833)
Gauss (1777–1855)
Dirichlet (1805–1859)
Kummer (1810–1893)
Hermite (1822–1901)
Eisenstein (1823–1852)
Kronecker (1823–1891)

Riemann (1826–1866)
Dedekind (1831–1916)
H. Weber (1842–1913)
Hensel (1861–1941)
Hilbert (1862–1943)
Takagi (1875–1960)
Hecke (1887–1947)
Artin (1898–1962)
Hasse (1898–)
Chevalley (1909–)

Prerequisites and notations

No knowledge of number-theory is presupposed in this book, except for the most elementary facts about rational integers; it is useful but not necessary to have some superficial acquaintance with the p-adic valuations of the field \mathbf{Q} of rational numbers and with the completions \mathbf{Q}_p of \mathbf{Q} defined by these valuations. On the other hand, the reader who wishes to acquire some historical perspective on the topics treated in the first part of this volume cannot do better than take up Hecke's unsurpassed *Theorie der algebraischen Zahlen*, and, if he wishes to go further back, the *Zahlentheorie* of Dirichlet-Dedekind (either in its 4th and final edition of 1894, or in the 3rd edition of 1879), with special reference to Dedekind's famous "eleventh Supplement". For similar purposes, the student of the second part of this volume may be referred to Hasse's *Klassenkörperbericht* (J. D. M. V., Part I, 1926; Part II, 1930).

The reader is expected to possess the basic vocabulary of algebra (groups, rings, fields) and of linear algebra (vector-spaces, tensor-products). Except at a few specific places, which may be skipped in a first reading, Galois theory plays no role in the first part (Chapters I to VIII). A knowledge of the main facts of Galois theory for finite and for infinite extensions is an indispensable requirement in the second part (Chapters IX to XIII).

Already in Chapter I, and throughout the book, essential use is made of the basic properties of locally compact commutative groups, including the existence and unicity of the Haar measure; the reader is expected to have acquired some familiarity with this topic before taking up the present book. The Haar measure for non-commutative locally compact groups is used in Chapters X and XI (but nowhere else). The basic facts from the duality theory of locally compact commutative groups are briefly recalled in Chapter II, § 5, and those about Fourier transforms in Chapter VII, § 2, and play an essential role thereafter.

As to our basic vocabulary and notations, they usually agree with the usage of Bourbaki. In particular, this applies to \mathbf{N} (the set of the "finite cardinals" or "natural integers" $0, 1, 2, \ldots$), \mathbf{Z} (the ring of rational integers), \mathbf{Q} (the field of rational numbers), \mathbf{R} (the field of real numbers), \mathbf{C} (the field of complex numbers), \mathbf{H} (the field of "classical", "ordinary" or "Hamiltonian" quaternions). If p is any rational prime, we write \mathbf{F}_p for the prime field with p elements, \mathbf{Q}_p for the field of p-adic numbers (the completion of \mathbf{Q} with respect to the p-adic valuation; cf. Chapter I,

§ 3), \mathbf{Z}_p for the ring of p-adic integers (i.e. the closure of \mathbf{Z} in \mathbf{Q}_p). The fields \mathbf{R}, \mathbf{C}, \mathbf{H}, \mathbf{Q}_p are always understood to be provided with their usual (or "natural") topology; so are all finite-dimensional vector-spaces over these fields. By \mathbf{F}_q we understand the finite field with q elements when there is one, i.e. when q is of the form p^n, p being a rational prime and n an integer $\geqslant 1$ (cf. Chapter I, § 1). We write \mathbf{R}_+ for the set of all real numbers $\geqslant 0$.

All rings are assumed to have a unit. If R is a ring, its unit is written 1_R, or 1 when there is no risk of confusion; we write R^\times for the multiplicative group of the invertible elements of R; in particular, when K is a field (commutative or not), K^\times denotes the multiplicative group of the non-zero elements of K. We write \mathbf{R}_+^\times for the multiplicative group of real numbers > 0. If R is any ring, we write $M_n(R)$ for the ring of matrices with n rows and n columns whose elements belong to R, and we write 1_n for the unit in this ring, i.e. the matrix (δ_{ij}) with $\delta_{ij} = 1_R$ or 0 according as $i = j$ or $i \neq j$. We write tX for the transpose of any matrix $X \in M_n(R)$, and $\mathrm{tr}(X)$ for its trace, i.e. the sum of its diagonal elements; if R is commutative, we write $\det(X)$ for its determinant. Occasionally we write $M_{m,n}(R)$ for the set of the matrices over R with m rows and n columns.

If R is a commutative ring, and T is an indeterminate, we write $R[T]$ for the ring of polynomials in T with coefficients in R; such a polynomial is called *monic* if its highest coefficient is 1. If S is a ring containing R, and x an element of S commuting with all elements of R, we write $R[x]$ for the subring of S generated by R and x; it consists of the elements of S of the form $F(x)$, with $F \in R[T]$. If K is a commutative field, L a field (commutative or not) containing K, and x an element of L commuting with all elements of K, we write $K(x)$ for the subfield of L generated by K and x; it is commutative. We do not speak of a field L as being an "extension" of a field K unless both are commutative; usually this word is reserved for the case when L is of finite degree over K, and then we write $[L:K]$ for this degree, i.e. for the dimension of L when L is regarded as a vector-space over K (the index of a group g' in a group g is also denoted by $[g:g']$ when it is finite; this causes no confusion).

All topologies should be understood to be Hausdorff topologies, i.e. satisfying the Hausdorff "separation" axiom ("separated" in the sense of Bourbaki). The word "homomorphism", for groups, rings, modules, vector-spaces, should be understood with the following restrictions: (a) when topologies are involved, *all homomorphisms are understood to be continuous*; (b) homomorphisms of rings are understood to be "unitary"; this means that a homomorphism of a ring R into a ring S is assumed to map 1_R onto 1_S. On the other hand, in the case of groups, homomorphisms are *not* assumed to be open mappings (i.e. to map open sets

onto open sets); when necessary, one will speak of an "open homomorphism". The word "morphism" is used as a shorter synonym for "homomorphism"; the word "representation" is used occasionally, as a synonym for "homomorphism", in certain situations, e.g. when the homomorphism is one of a group into C^\times, or for certain homomorphisms of simple algebras (cf. Chapter IX, § 2). By a *character* of a group G, commutative or not, we understand as usual a homomorphism (or "representation") of G into the subgroup of C^\times defined by $z\bar{z}=1$; as explained above, this should be understood to be continuous when G is given as a topological group. The words "endomorphism", "automorphism", "isomorphism" are subject to the same restrictions (a), (b) as "homomorphism"; for "automorphism" and "isomorphism", this implies, in the topological case, that the mapping in question is bijective and bicontinuous. Occasionally, when a mapping f of a set A into a set B, both with certain structures (usually fields), determines an isomorphism of A onto its image in B, we speak of it by "abuse of language" as an "isomorphism" of A into B.

In a group G, an element x is said to be of *order n* if n is the smallest integer $\geqslant 1$ such that $x^n = e$, e being the neutral element of G. If K is a field, an element of K^\times of finite order is called a *root of* 1 *in* K; in accordance with a long-standing tradition, any root of 1 of order dividing n is called an *n-th root of* 1 *in* K; it is called a *primitive n-th root of* 1 if its order is n. Thus the n-th roots of 1 in K are the roots of the equation $X^n = 1$ in K.

If a, b are in Z, (a, b) denotes their g.c.d., i.e. the element d of N such that $dZ = aZ + bZ$. If R is any ring, the mapping $n \to n \cdot 1_R$ of Z into R maps Z onto the subring $Z \cdot 1_R$ of R, known as "the prime ring" in R; the kernel of the morphism $n \to n \cdot 1_R$ of Z onto $Z \cdot 1_R$ is a subgroup of Z, hence of the form $m \cdot Z$ with $m \in N$; if R is not $\{0\}$ and has no zero-divisor, m is either 0 or a rational prime and is known as the *characteristic* of R. If $m = 0$, $n \to n \cdot 1_R$ is an isomorphism of Z onto $Z \cdot 1_R$, by means of which $Z \cdot 1_R$ will frequently be identified with Z. If the characteristic of R is a prime $p > 1$, the prime ring $Z \cdot 1_R$ is isomorphic to the prime field F_p.

We shall consider left modules and right modules over non-commutative rings, and fix notations as follows. Let R be a ring; let M and N be two left modules over R. Then morphisms of M into N, for their structures as left R-modules, will be written as *right operators* on M; in other words, if α is such a morphism, we write it as $m \to m\alpha$, where $m \in M$; thus the property of being a morphism, apart from the additivity, is expressed by $r(m\alpha) = (rm)\alpha$ for all $r \in R$ and all $m \in M$. This applies in particular to endomorphisms of M. Morphisms of right R-modules are similarly written as left operators. This notation will be consistently used, in particular in Chapter IX.

As morphisms of fields into one another are assumed to be "unitary" (as explained above), such morphisms are always injective; as we have said, we sometimes refer to a morphism of a field K into a field L as an "isomorphism", or also as an *embedding*, of K into L. In part of this book, we use for such mappings the "functional" notation; beginning with Chapter VIII, § 3, where the role of Galois theory becomes essential, we shall use for them the "exponential" notation. This means that such a mapping λ is written in the former case as $x \to \lambda(x)$ and in the latter case as $x \to x^\lambda$. If L is a Galois extension of K, and λ, μ are two automorphisms of L over K, we define the law of composition $(\lambda, \mu) \to \lambda\mu$ in the Galois group \mathfrak{g} of L over K as being identical with the law $(\lambda, \mu) \to \lambda \circ \mu$ in the former case, and as its opposite in the latter case; in other words, it is defined in the former case by $(\lambda\mu)x = \lambda(\mu x)$, and in the latter case by $x^{\lambda\mu} = (x^\lambda)^\mu$. For instance, if K' is a field between K and L, and \mathfrak{h} is the corresponding subgroup of \mathfrak{g}, consisting of the automorphisms which leave fixed all the elements of K', the automorphisms of L over K which coincide on K' with a given one λ make up the right coset $\lambda\mathfrak{h}$ when the functional notation is used, and the left coset $\mathfrak{h}\lambda$ when the exponential notation is used.

When A, B, C are three additively written commutative groups (usually with some additional structures) and a "distributive" (or "bi-additive", or "bilinear") mapping $(a, b) \to ab$ of $A \times B$ into C is given, and when X, Y are respectively subgroups of A and of B, it is customary to denote by $X \cdot Y$, not the image of $X \times Y$ under that mapping, but the subgroup of C generated by that image, i.e. the group consisting of the finite sums $\sum x_i y_i$ with $x_i \in X$ and $y_i \in Y$ for all i. This notation will be used occasionally, e.g. in Chapter V.

For typographical reasons, we frequently write $\exp(z)$ instead of e^z, and $\mathbf{e}(z)$ instead of $\exp(2\pi i z) = e^{2\pi i z}$, for $z \in \mathbf{C}$; ordinarily $\mathbf{e}(z)$ occurs only for $z \in \mathbf{R}$.

Finally we must explain the method followed for cross-references; these have been inserted quite generously, with a view to helping the inexperienced reader; the reader is advised to follow them up only when the argument is not otherwise clear. Theorems have been numbered continuously throughout each chapter; the same is true for propositions, for lemmas, for definitions, for the numbered formulas. Each theorem and each proposition may be followed by one or several corollaries. Generally speaking, theorems are to be regarded as more important than propositions, but the distinction between them would hardly stand a close scrutiny. Lemmas are merely auxiliary results. Not all new concepts are the object of a numbered definition; all concepts, except those which are assumed to be known, are listed in the index at the end of the book, with proper references. Formulas are numbered only for purposes

of quotation, and not as an indication of their importance. When a reference is given thus: "by prop. 2", "by corollary 1 of th. 3", etc., it refers to a result in the same §; when thus: "by prop. 2 of § 2", "by th. 3 of § 3", etc., it refers to another § of the same chapter; when thus: "by prop. 2 of Chap. IV-2", it refers to proposition 2 of Chapter IV, § 2. Numbers of Chapter and § are given at the top of every page. A table of the most frequently used notations is given below, in the order of their first appearance.

Table of notations

Chapter I.

§ 2: \mod_G, \mod_V, \mod_K.

§ 3: $|x|_p$, $|x|_\infty$, $\mathbf{Q}_\infty = \mathbf{R}$, $|x|_v$, \mathbf{Q}_v ($v = $ rational prime or ∞).

§ 4: K (any p-field), R, P, π, q, ord_K, ord, M^\times, M.

Chapter II.

§ 3: $1 + P^n$ (as subgroup of K^\times for $n \geqslant 1$).

§ 5: $\langle g, g^* \rangle_G$, $\langle g, g^* \rangle$, G^*, H_*, V^*, L_*, V', $[v, v']_V$, $[v, v']$, χ, $\text{ord}(\chi)$.

Chapter III.

§ 1: (for a place v of an A-field k) $|x|_v$, k_v, r_v, p_v (for q_v, see Chap. VII-1); ∞ (as a place of \mathbf{Q}), $w|v$, $E_v = E \otimes_k k_v$, ε_v, \mathscr{A}_v, α_v.

§ 3: $\text{End}(E)$, $Tr_{\mathscr{A}/k}$, $N_{\mathscr{A}/k}$, $Tr_{k'/k}$, $N_{k'/k}$.

Chapter IV.

§ 1: P, P_∞, $k_\mathbf{A}(P)$, $k_\mathbf{A}$, χ, χ_v, $E_\mathbf{A}(P, \varepsilon)$, $E_\mathbf{A}$, $\mathscr{A}_\mathbf{A}$, $\mathscr{A}_\mathbf{A}(P, \alpha)$, $(k'/k)_\mathbf{A}$, $(E/k)_\mathbf{A}$.

§ 3: $\text{Aut}(E)$, $\mathscr{A}_\mathbf{A}^\times$, $\mathscr{A}_\mathbf{A}(P, \alpha)^\times$, $|a|_\mathbf{A}$.

§ 4: $k_\mathbf{A}^1$, M, $\Omega(P) = k_\mathbf{A}(P)^\times$, $\Omega_1(P)$, $E(P)$.

Chapter V.

§ 2: k_∞, E_∞, \mathfrak{r}, L_v.

§ 3: \mathfrak{p}_v, $I(k)$, $\text{id}(a)$, $P(k)$, h, $\mathfrak{N}(\mathfrak{a})$.

§ 4: $|dx \wedge d\bar{x}|$, R, c_k.

Chapter VI.

$\deg(\mathfrak{a})$, $\mathfrak{a} \succ \mathfrak{b}$, $\text{div}(a)$, $D(k)$, $P(k)$, $D_0(k)$, g, $\text{div}(\chi)$.

Chapter I

Locally compact fields

§ 1. Finite fields. Let F be a finite field (commutative or not) with the unit-element 1. Its characteristic must clearly be a prime $p>1$, and the prime ring in F is isomorphic to the prime field $\mathbf{F}_p = \mathbf{Z}/p\mathbf{Z}$, with which we may identify it. Then F may be regarded as a vector-space over \mathbf{F}_p; as such, it has an obviously finite dimension f, and the number of its elements is $q = p^f$. If F is a subfield of a field F' with $q' = p^{f'}$ elements, F' may also be regarded e.g. as a left vector-space over F; if its dimension as such is d, we have $f' = df$ and $q' = q^d = p^{df}$.

THEOREM 1. *All finite fields are commutative.*

This theorem is due to Wedderburn, and we will reproduce Witt's modification of Wedderburn's original proof. Let F be a finite field of characteristic p, Z its center, $q = p^f$ the number of elements of Z; if n is the dimension of F as a vector-space over Z, F has q^n elements. The multiplicative group F^\times of the non-zero elements of F can be partitioned into classes of "conjugate" elements, two elements x, x' of F^\times being called conjugate if there is $y \in F^\times$ such that $x' = y^{-1}xy$. For each $x \in F^\times$, call $N(x)$ the set of the elements of F which commute with x; this is a subfield of F containing Z; if $\delta(x)$ is its dimension over Z, it has $q^{\delta(x)}$ elements. As we have seen above, n is a multiple of $\delta(x)$, and we have $\delta(x) < n$ unless $x \in Z$. As the number of elements of F^\times conjugate to x is clearly the index of $N(x)^\times$ in F^\times, i.e. $(q^n - 1)/(q^{\delta(x)} - 1)$, we have

$$(1) \qquad q^n - 1 = q - 1 + \sum_x \frac{q^n - 1}{q^{\delta(x)} - 1},$$

where the sum is taken over a full set of representatives of the classes of non-central conjugate elements of F^\times. Now assume that $n > 1$, and call P the "cyclotomic" polynomial $\prod(T - \zeta)$, where the product is taken over all the primitive n-th roots of 1 in the field \mathbf{C} of complex numbers. By a well-known elementary theorem (easily proved by induction on n), this has integral rational coefficients; clearly it divides $(T^n - 1)/(T^\delta - 1)$ whenever δ is a divisor of n other than n. Therefore, in (1), all the terms except $q - 1$ are multiples of $P(q)$, so that $P(q)$ must divide $q - 1$. On the other hand, each factor in the product $P(q) = \prod(q - \zeta)$ has an absolute value $> q - 1$. This is a contradiction, so that we must have $n = 1$ and $F = Z$.

We can now apply to every finite field the following elementary result:

LEMMA 1. *If K is a commutative field, every finite subgroup of K^{\times} is cyclic.*

In fact, let Γ be such a group, or, what amounts to the same, a finite subgroup of the group of all roots of 1 in K. For every $n \geqslant 1$, there are at most n roots of $X^n = 1$ in K, hence in Γ; we will show that every finite commutative group with that property is cyclic. Let α be an element of Γ of maximal order N. Let β be any element of Γ, and call n its order. If n does not divide N, there is a prime p and a power $q = p^{\nu}$ of p such that q divides n and not N. Then one verifies at once that the order of $\alpha \beta^{n/q}$ is the l.c.m. of N and q, so that it is $> N$, which contradicts the definition of N. Therefore n divides N. Now $X^n = 1$ has the n distinct roots $\alpha^{iN/n}$ in Γ, with $0 \leqslant i < n$; as β is a root of $X^n = 1$, it must be one of these. This shows that α generates Γ.

THEOREM 2. *Let K be an algebraically closed field of characteristic $p > 1$. Then, for every $f \geqslant 1$, K contains one and only one field $F = \mathbf{F}_q$ with $q = p^f$ elements; F consists of the roots of $X^q = X$ in K; F^{\times} consists of the roots of $X^{q-1} = 1$ in K and is a cyclic group of order $q - 1$.*

If F is any field with q elements, lemma 1 shows that F^{\times} is a cyclic group of order $q - 1$. Thus, if K contains such a field F, F^{\times} must consist of the roots of $X^{q-1} = 1$, hence F of the roots of $X^q - X = 0$, so that both are uniquely determined. Conversely, if $q = p^f$, $x \to x^q$ is an automorphism of K, so that the elements of K which are fixed under it make up a field F consisting of the roots of $X^q - X = 0$; as it is clear that $X^q - X$ has only simple roots in K, F is a field with q elements.

COROLLARY 1. *Up to isomorphisms, there is one and only one field with $q = p^f$ elements.*

This follows at once from theorem 2 and the fact that all algebraic closures of the prime field \mathbf{F}_p are isomorphic. It justifies the notation \mathbf{F}_q for the field in question.

COROLLARY 2. *Put $q = p^f$, $q' = p^{f'}$, with $f \geqslant 1$, $f' \geqslant 1$. Then $\mathbf{F}_{q'}$ contains a field \mathbf{F}_q with q elements if and only if f divides f'; when that is so, $\mathbf{F}_{q'}$ is a cyclic extension of \mathbf{F}_q of degree f'/f, and its Galois group over \mathbf{F}_q is generated by the automorphism $x \to x^q$.*

We have already said that, if $\mathbf{F}_{q'}$ contains \mathbf{F}_q, it must have a finite degree d over \mathbf{F}_q, and then $q' = q^d$ and $f' = df$. Conversely, assume that $f' = df$, hence $q' = q^d$, and call K an algebraic closure of $\mathbf{F}_{q'}$; by theorem 2, the fields \mathbf{F}_q, $\mathbf{F}_{q'}$, contained in K, consist of the elements of K respectively

invariant under the automorphisms α, β of K given by $x \to x^q$, $x \to x^{q'}$; as $\beta = \alpha^d$, $\mathbf{F}_{q'}$ contains \mathbf{F}_q. Clearly α maps $\mathbf{F}_{q'}$ onto itself; if φ is the automorphism of $\mathbf{F}_{q'}$ induced by α, \mathbf{F}_q consists of the elements of $\mathbf{F}_{q'}$ invariant under φ, hence under the group of automorphisms of $\mathbf{F}_{q'}$ generated by φ; this group is finite, since φ^d is the identity; therefore, by Galois theory, it is the Galois group of $\mathbf{F}_{q'}$ over \mathbf{F}_q and is of order d.

COROLLARY 3. *Notations being as in corollary 2, assume that* $f' = df$. *Then, for every* $n \geqslant 1$, *the elements of* $\mathbf{F}_{q'}$, *invariant under* $x \to x^{q^n}$, *make up the subfield of* $\mathbf{F}_{q'}$ *with* q^r *elements, where* $r = (d, n)$.

Let K be as in the proof of corollary 2; the elements of K, invariant under $x \to x^{q^n}$, make up the subfield F' of K with q^n elements; then $F' \cap \mathbf{F}_{q'}$ is the largest field contained both in F' and $\mathbf{F}_{q'}$; as it contains \mathbf{F}_q, the number of its elements must be of the form q^r, and corollary 2 shows that r must be (d, n).

§ 2. **The module in a locally compact field.** An arbitrary field, provided with the discrete topology, becomes locally compact; thus the question of determining and studying locally compact fields becomes significant only if one adds the condition that the field should not be discrete.

We recall the definition of the "module" of an automorphism, which is basic in what follows. For our purposes, it will be enough to consider automorphisms of locally compact commutative groups. Let G be such a group (written additively), λ an automorphism of G, and α a Haar measure on G. As the Haar measure is unique up to a constant factor, λ transforms α into $c\alpha$, with $c \in \mathbf{R}_+^\times$; the constant factor c, which is clearly independent of the choice of α, is called the *module* of λ and is denoted by $\mathrm{mod}_G(\lambda)$. In other words, this is defined by one of the equivalent formulas

(2) $\alpha(\lambda(X)) = \mathrm{mod}_G(\lambda)\alpha(X)$, $\int f(\lambda^{-1}(x))\,d\alpha(x) = \mathrm{mod}_G(\lambda) \int f(x)\,d\alpha(x)$,

where X is any measurable set, f any integrable function, and $0 < \alpha(X) < +\infty$, $\int f d\alpha \neq 0$; the second formula may be written symbolically as $d\alpha(\lambda(x)) = \mathrm{mod}_G(\lambda)\,d\alpha(x)$. If G is discrete or compact, the first formula (applied to $X = \{0\}$, $X = G$, respectively) shows that the module is 1. Obviously, if λ, λ' are two automorphisms of G, the module of $\lambda \circ \lambda'$ is the product of those of λ and λ'. We shall need the following lemma:

LEMMA 2. *Let* G' *be a closed subgroup of* G, *and* λ *an automorphism of* G *which induces on* G' *an automorphism* λ' *of* G'. *Put* $G'' = G/G'$, *and call* λ'' *the automorphism of* G'' *determined by* λ *modulo* G'. *Then*:

$$\mathrm{mod}_G(\lambda) = \mathrm{mod}_{G'}(\lambda')\mathrm{mod}_{G''}(\lambda'').$$

In fact, it is well-known that one can choose Haar measures $\alpha, \alpha', \alpha''$ on G, G', G'' so as to have, for every continuous function f with compact support on G:

$$\int\limits_G f(x)\,d\alpha(x) = \int\limits_{G''} \left(\int\limits_{G'} f(x+y)\,d\alpha'(y) \right) d\alpha''(\dot{x});$$

here \dot{x} denotes the image of x in G'', and the function $\int f(x+y)\,d\alpha'(y)$, which is written as a function of $x \in G$, but is constant on the classes modulo G' in G, is to be understood as a function of \dot{x} on G'' in the obvious manner. Applying λ to both sides, one gets the conclusion of the lemma.

Now, if K is any topological field, and $a \in K^\times$, $x \to ax$ and $x \to xa$ are automorphisms of the additive group of K; if K is locally compact, we may consider their modules. Similarly, if V is a topological left vector-space over K, $v \to av$ is an automorphism of V for every $a \in K^\times$; if V is locally compact, we may consider the module of this automorphism; this will be denoted by $\mathrm{mod}_V(a)$; we also define $\mathrm{mod}_V(0)$ to be 0. In other words, if μ is a Haar measure on V and X any measurable subset of V with $0 < \mu(X) < +\infty$ (e.g. any compact neighborhood of 0 in V), $\mathrm{mod}_V(a)$ is defined, for all $a \in K$, by

$$\mathrm{mod}_V(a) = \frac{\mu(aX)}{\mu(X)}.$$

In particular, for any locally compact field K, we define $\mathrm{mod}_K(a)$ to be the module of $x \to ax$ in K if $a \neq 0$, and 0 if $a = 0$. It will be seen later that the module of $x \to xa$ is always the same as that of $x \to ax$. Clearly, if $K = \mathbf{R}, \mathbf{C}$ or \mathbf{H}, $\mathrm{mod}_K(a)$ is equal, respectively, to $|a|$, $|a|^2 = a\bar{a}$ or $|a|^4 = (a\bar{a})^2$.

In the rest of this section, K will denote, once for all, a nondiscrete locally compact field (commutative or not), and α a Haar measure on the additive group of K.

PROPOSITION 1. *The function* mod_K *is continuous on* K, *and* $\mathrm{mod}_K(ab) = \mathrm{mod}_K(a)\mathrm{mod}_K(b)$ *for all* $a \in K$, $b \in K$.

The latter assertion is obvious. Now let X be a compact neighborhood of 0 in K. For any $a \in K$ and any $\varepsilon > 0$, there is an open neighborhood U of the compact set aX such that $\alpha(U) \leqslant \alpha(aX) + \varepsilon$; let W be a neighborhood of a such that $WX \subset U$. Then, for all $x \in W$, we have

$$\mathrm{mod}_K(x) \leqslant \mathrm{mod}_K(a) + \alpha(X)^{-1}\varepsilon.$$

This shows that mod_K is upper semicontinuous. In particular, it is continuous at 0. As $\mathrm{mod}_K(x) = \mathrm{mod}_K(x^{-1})^{-1}$ for $x \neq 0$, it is also lower semicontinuous everywhere on K^\times, hence continuous on K^\times.

As K is not discrete, prop. 1 shows that there is, for every $\varepsilon > 0$, $a \in K$ such that $0 < \mathrm{mod}_K(a) \leqslant \varepsilon$, hence also, for every $M > 0$, $b \in K$ such that $\mathrm{mod}_K(b) \geqslant M$. As mod_K is not bounded, K cannot be compact.

PROPOSITION 2. *For all* $m > 0$, *the set* B_m *of the elements* x *of* K *such that* $\mathrm{mod}_K(x) \leqslant m$ *is compact.*

Let V be a compact neighborhood of 0 in K; let W be a neighborhood of 0 such that $WV \subset V$. As above, we can choose $r \in V \cap W$ such that $0 < \mathrm{mod}_K(r) < 1$; by induction on n, we have $r^n \in V$ for all $n \geqslant 1$. If r' is any limit point of the sequence $\{r^n\}_{n \geqslant 1}$, $\mathrm{mod}_K(r')$ must be 0, since $\mathrm{mod}_K(r^n)$ has the limit 0 for $n \to +\infty$. Therefore that sequence can have no other limit point than 0; as it is contained in the compact set V, it has the limit 0. Now take $m > 0$ and $a \in B_m$; as $r^n a$ tends to 0, there is a smallest integer $\nu \geqslant 0$ such that $r^\nu a \in V$; if a is not in V, then $r^{\nu-1} a \notin V$, hence $r^\nu a \in V - (rV)$. Call X the closure of $V - (rV)$; clearly X is compact, and 0 is not in X; therefore, if we put $\mu = \inf_{x \in X} \mathrm{mod}_K(x)$, we have $\mu > 0$. Let N be an integer such that $\mathrm{mod}_K(r)^N \leqslant \mu/m$. Then, if $a \in B_m$, $a \notin V$, and ν is defined as above, we have

$$\mathrm{mod}_K(r)^N m \leqslant \mu \leqslant \mathrm{mod}_K(r^\nu a) = \mathrm{mod}_K(r)^\nu \mathrm{mod}_K(a) \leqslant \mathrm{mod}_K(r)^\nu m,$$

hence $\nu \leqslant N$. This proves that B_m is contained in the union of the compact sets $V, r^{-1}V, \dots, r^{-N}V$. As prop. 1 shows that B_m is closed, this completes the proof.

COROLLARY 1. *The sets B_m, for $m > 0$, make up a fundamental system of neighborhoods of 0 in K.*

Let V be any compact neighborhood of 0 in K; take $m > \sup_{x \in V} \mathrm{mod}_K(x)$, so that $B_m \supset V$; call X the closure of $B_m - V$, and put $m' = \inf_{x \in X} \mathrm{mod}_K(x)$. Then $0 \notin X$ and $X \subset B_m$, so that, by prop. 2, X is compact; therefore $0 < m' \leqslant m$. Take $0 < \mu < m'$; then $B_\mu \subset B_m$, $B_\mu \cap X = \emptyset$, hence $B_\mu \subset V$.

COROLLARY 2. *For $a \in K$, $\lim_{n \to +\infty} a^n = 0$ if and only if $\mathrm{mod}_K(a) < 1$.*

COROLLARY 3. *A discrete subfield of K is finite.*

Let L be such a field. If $a \in L$, we must have $\mathrm{mod}_K(a) \leqslant 1$, since otherwise, by corollary 2, the sequence $\{a^{-n}\}_{n \geqslant 0}$ would be contained in L and not discrete. Therefore L is a discrete subset of the compact set B_1, hence finite. Of course this cannot happen if K is of characteristic 0.

THEOREM 3. *Let V be a topological left vector-space over K, and let V' be a finite-dimensional subspace of V, with a basis $\{v_1, \dots, v_n\}$. Then the mapping*

$$(x_1, \dots, x_n) \to \sum_{i=1}^n x_i v_i$$

of K^n onto V' is an isomorphism for the structures of K^n and V' as topological left vector-spaces; V' is closed in V and locally compact.

Let f be the mapping defined above; it is bijective, K-linear, and continuous by the definition of a topological vector-space. In order to

show that it is an isomorphism, it is enough to prove that f^{-1} is conti-nuous, i.e. that f is an open mapping; in view of corollary 1 of prop. 2 and of the linearity of f, we need only show that the image of $(B_m)^n$ by f contains an neighborhood of 0 in V' for every $m>0$. Call S the subset of K^n defined by

$$\sup_i \mathrm{mod}_K(x_i) = 1.$$

Then $0 \notin S$; by prop. 1, S is closed; it is contained in $(B_1)^n$, hence compact by prop. 2. Therefore $0 \notin f(S)$, and $f(S)$ is compact. Hence there exists a neighborhood W of 0 in V, and a neighborhood of 0 in K which we may assume to be of the form B_ε with $\varepsilon > 0$, such that $B_\varepsilon W \subset V - f(S)$, i.e. $yW \cap f(S) = \emptyset$ whenever $\mathrm{mod}_K(y) \leqslant \varepsilon$. Now take $m>0$, and take $a \in K$ such that $0 < \mathrm{mod}_K(a) \leqslant m\varepsilon$. Let $v = \sum x_i v_i$ be any point in $V' \cap aW$, other than 0, and take h such that $\sup_i \mathrm{mod}_K(x_i) = \mathrm{mod}_K(x_h)$; then $x_h \neq 0$. Put $x_i' = x_h^{-1} x_i$ for $1 \leqslant i \leqslant n$, and

$$v' = \sum_1^n x_i' v_i = x_h^{-1} v.$$

As $(x_1', ..., x_n')$ is in S, we have $v' \in f(S)$; as $v \in aW$, we have $v' \in yW$ with $y = x_h^{-1} a$; by the definition of W and ε, this implies $\mathrm{mod}_K(y) > \varepsilon$, hence $\mathrm{mod}_K(x_h) < \varepsilon^{-1} \mathrm{mod}_K(a) \leqslant m$. Therefore $(x_1, ..., x_n)$ is in $(B_m)^n$, and v is in the image of that set by f. We have thus shown that this image contains $V' \cap aW$, which is a neighborhood of 0 in V'. Let now w be in the closure of V' in V, and apply what we have proved to the finite-dimensional subspace V'' of V generated by V' and w; we see then that V' must be closed in V''. As this implies that $w \in V'$, it completes the proof of the theorem.

COROLLARY 1. *Every finite-dimensional left vector-space over K can be provided with one and only one structure of topological left vector-space over K.*

In fact, if V is of dimension n, one can define such a structure on V by means of any K-linear bijective mapping of K^n onto V; the unicity is an immediate consequence of th. 3 applied to V. From now on, every such vector-space will tacitly be assumed to carry the structure defined by this corollary.

COROLLARY 2. *If V is a locally compact topological left vector-space over K, then V has a finite dimension d over K, and $\mathrm{mod}_V(a) = \mathrm{mod}_K(a)^d$ for every $a \in K$.*

The latter assertion, for a space of dimension d, is an immediate consequence of Fubini's theorem and of the fact that such a space is isomorphic to K^d by corollary 1. Now assume merely that V is locally

compact, and take $a \in K$ such that $0 < \mathrm{mod}_K(a) < 1$. Then, by corollary 2 of prop. 2, $\lim a^n = 0$, hence $\mathrm{mod}_V(a) < 1$. Let V' be a subspace of V of finite dimension δ; by th. 3, it is closed in V; put $V'' = V/V'$. By lemma 2 we have then

$$\mathrm{mod}_V(a) = \mathrm{mod}_{V'}(a)\,\mathrm{mod}_{V''}(a) = \mathrm{mod}_K(a)^\delta\,\mathrm{mod}_{V''}(a),$$

and therefore, since $\mathrm{mod}_{V''}(a)$ also must be <1 if $V'' \neq \{0\}$, and is 1 if $V'' = \{0\}$:

$$\mathrm{mod}_V(a) \leqslant \mathrm{mod}_K(a)^\delta.$$

This gives an upper bound for δ, valid for the dimension of all finite-dimensional subspaces of V; therefore V itself has a finite dimension.

If V is a left vector-space over K, of finite dimension n, topologized as we have said above, Fubini's theorem shows at once that every subspace of V of dimension $n' < n$ is of measure 0. Now let A be any K-linear mapping of V into V; if it is of rank n, it is an automorphism of V also in the topological sense, and we may consider its module $\mathrm{mod}_V(A)$. If it is of rank $n' < n$, it maps V onto a subset of V of measure 0, and we define $\mathrm{mod}_V(A)$ to be 0.

COROLLARY 3. *Let A be an endomorphism of a left vector-space V of finite dimension over K. If K is commutative, then $\mathrm{mod}_V(A) = \mathrm{mod}_K(\det A)$.*

Call n the dimension of V. If A is of rank $<n$, the assertion is clear. If not, identify V with K^n by choosing a basis for V. It is well-known that every automorphism of K^n can be written as a product of automorphisms of the following three types: (a) permutations of the coordinates; (b) mappings of the type

$$(x_1, x_2, \ldots, x_n) \to (a x_1, x_2, \ldots, x_n)$$

with $a \in K^\times$; (c) mappings of the type

$$(x_1, x_2, \ldots, x_n) \to \left(x_1 + \sum_{i=2}^{n} a_i x_i, x_2, \ldots, x_n\right).$$

For type (a), the assertion is obvious; for types (b) and (c), it follows from a straightforward application of Fubini's theorem, just as in classical analysis (where one proves the theorem for the case $K = \mathbf{R}$).

PROPOSITION 3. *The function mod_K induces on K^\times an open homomorphism of K^\times onto a closed subgroup Γ of \mathbf{R}_+^\times.*

Call Γ, Γ' the images of K^\times and of K under the mapping mod_K; clearly Γ is a subgroup of \mathbf{R}_+^\times, and $\Gamma' = \Gamma \cup \{0\}$. For every $m > 0$, the intersection of Γ' with the closed interval $[0, m]$ is the image of B_m under mod_K; by prop. 1 and 2, this is compact; therefore Γ' is closed in \mathbf{R}_+,

and Γ is so in \mathbf{R}_+^\times. Now call U the kernel of mod_K in K^\times, i. e. the set $\{x \in K | \mathrm{mod}_K(x) = 1\}$. Let V be any neighborhood of 1 in K^\times, and V' its image under mod_K; in order to prove the openness of the homomorphism mod_K of K^\times onto Γ, we have to show that V' is a neighborhood of 1 in Γ. Assume that this is not so; then there is a sequence (γ_n) in $\Gamma - V'$ such that $\lim \gamma_n = 1$. For each n, let $a_n \in K^\times$ be such that $\gamma_n = \mathrm{mod}_K(a_n)$. By prop. 2, the sequence (a_n) has at least one limit point a; clearly $\mathrm{mod}_K(a) = 1$, i.e. $a \in U$. But UV is a neighborhood of U, and so there must be some n such that $a_n \in UV$, hence $\gamma_n \in V'$. This contradicts the assumption.

THEOREM 4. *There is a constant $A > 0$ such that*

(3) $$\mathrm{mod}_K(x + y) \leqslant A \sup(\mathrm{mod}_K(x), \mathrm{mod}_K(y))$$

for all $x \in K$, $y \in K$. If (3) is valid for $A = 1$, then the image Γ of K^\times under mod_K is discrete in \mathbf{R}_+^\times. Morever, (3) is valid for

$$A = \sup_{x \in K, \, \mathrm{mod}_K(x) \leqslant 1} \mathrm{mod}_K(1 + x),$$

and this is the smallest value of A for which it is valid.

Define A by the last formula; clearly $1 \leqslant A < +\infty$. For $x = y = 0$, (3) is obvious; otherwise we may, after interchanging x and y if necessary, assume that $x \neq 0$ and $\mathrm{mod}_K(y) \leqslant \mathrm{mod}_K(x)$. Put $z = yx^{-1}$; then $\mathrm{mod}_K(z) \leqslant 1$, hence $\mathrm{mod}_K(1 + z) \leqslant A$, and therefore

$$\mathrm{mod}_K(x + y) = \mathrm{mod}_K(1 + z)\mathrm{mod}_K(x) \leqslant A \, \mathrm{mod}_K(x).$$

This proves (3). Also, taking $y = 1$ and $x \in B_1$ in (3), with B_1 as in prop. 2, we see that the value we have chosen for A is the smallest for which (3) can be valid. Now assume $A = 1$. Then the image of $1 + B_1$ by mod_K is contained in the interval $[0, 1]$; as this, by prop. 2 and 3, must contain a neighborhood of 1 in Γ, Γ must be discrete.

COROLLARY. *If (3) is valid with $A = 1$, then $\mathrm{mod}_K(x + y) = \mathrm{mod}_K(x)$ whenever $\mathrm{mod}_K(y) < \mathrm{mod}_K(x)$.*

As $(-1)^2 = 1$, we have $\mathrm{mod}_K(-1) = 1$, hence $\mathrm{mod}_K(-y) = \mathrm{mod}_K(y)$. As $x = (x + y) + (-y)$, our assumptions imply

$$\mathrm{mod}_K(x) \leqslant \sup(\mathrm{mod}_K(x + y), \mathrm{mod}_K(y)) \leqslant \mathrm{mod}_K(x),$$

hence the conclusion.

DEFINITION 1. *The inequality (3) with $A = 1$ is called the ultrametric inequality; if this is valid, then mod_K, and K itself, are said to have the ultrametric property, or to be ultrametric.*

§ 3. **Classification of locally compact fields.** Here we shall need the following elementary lemma:

LEMMA 3. *Let F be a function on the set* \mathbf{N} *of natural integers, with values in* \mathbf{R}_+. *Assume that $F(mn) = F(m)F(n)$ for all m, n, and that there is $A > 0$ such that*

$$F(m+n) \leqslant A \sup(F(m), F(n))$$

for all m, n. Then either $F(m) \leqslant 1$ for all m, or there is $\lambda > 0$ such that $F(m) = m^\lambda$ for all m.

The first assumption on F implies, for $m = 0$, that $F(0) = 0$ unless F is the constant 1, and, for $m = 1$, that $F(1) = 1$ unless F is the constant 0; it also implies that $F(m^k) = F(m)^k$ for all integers $k \geqslant 1$. Leaving aside the trivial cases where F is the constant 0 or 1, we may assume that $F(0) = 0$ and $F(m^k) = F(m)^k$ for all integers $k \geqslant 0$. Put $f(m) = \sup(0, \log F(m))$, this being understood to mean in particular that $f(m) = 0$ whenever $F(m) = 0$. Our lemma amounts now to saying that $f(m) = \lambda \log m$ for all $m \geqslant 2$, with some constant $\lambda \geqslant 0$. Put $a = \sup(0, \log A)$; then we have, for all m, n, k:

$$f(m^k) = k f(m), \quad f(mn) \leqslant f(m) + f(n), \quad f(m+n) \leqslant a + \sup(f(m), f(n)).$$

The last relation gives, by induction on r:

$$(4) \qquad\qquad f\left(\sum_{i=0}^{r} m_i\right) \leqslant r a + \sup_i(f(m_i)).$$

Now let m, n be integers $\geqslant 2$; m may be expressed in the form

$$m = \sum_{i=0}^{r} e_i n^i,$$

with $n^r \leqslant m < n^{r+1}$, and $0 \leqslant e_i < n$ for $0 \leqslant i \leqslant r$. Put

$$b = \sup(f(0), f(1), \dots, f(n-1)).$$

Then we have, for every i:

$$f(e_i n^i) \leqslant b + i f(n),$$

and therefore, in view of (4):

$$f(m) \leqslant r a + b + r f(n).$$

As $n^r \leqslant m$, i.e. $r \log n \leqslant \log m$, this gives

$$\frac{f(m)}{\log m} \leqslant \frac{a + f(n)}{\log n} + \frac{b}{\log m}.$$

In this inequality, replace m by m^k; this does not change the left-hand side, and, for $k \to +\infty$, we get

$$\frac{f(m)}{\log m} \leqslant \frac{a + f(n)}{\log n}.$$

Now replace n by n^k; for $k \to +\infty$, we get

$$\frac{f(m)}{\log m} \leqslant \frac{f(n)}{\log n}.$$

Interchanging m and n, we see that $f(m)/\log m$ is constant for $m \geqslant 2$; as we have observed above, this proves the lemma.

Now we consider again a non-discrete locally compact field K. For greater clarity, in the rest of this section, we shall denote by 1_K (not by 1) the unit-element of K; then the prime ring in K consists of the elements $m \cdot 1_K$ with $m \in \mathbf{Z}$; if K is of characteristic $p > 1$, then $p \cdot 1_K = 0$. For $m \in \mathbf{N}$, we write $F(m) = \mathrm{mod}_K(m \cdot 1_K)$; then, for every $m \in \mathbf{Z}$ and every $x \in K$, we have $\mathrm{mod}_K(mx) = F(|m|)\,\mathrm{mod}_K(x)$.

LEMMA 4. *Assume that F is bounded, i.e. that mod_K is bounded on the prime ring in K. Then $F \leqslant 1$, and mod_K is ultrametric on K.*

Since $F(mn) = F(m)F(n)$, the first assertion is obvious. Now let A be as in th. 4 of § 2; take $n \geqslant 1$, put $N = 2^n$, and let x_1, \ldots, x_N be N elements of K. By induction on n, one gets the inequality

$$\mathrm{mod}_K\left(\sum_{i=1}^{N} x_i\right) \leqslant A^n \sup_i(\mathrm{mod}_K(x_i)).$$

Replacing some of the x_i by 0, one sees that this same inequality remains valid whenever $N \leqslant 2^n$. Applying this to the relation

$$(x+y)^{2^n} = \sum_{i=0}^{2^n} \binom{2^n}{i} x^i y^{2^n-i},$$

we get

$$\mathrm{mod}_K(x+y)^{2^n} \leqslant A^{n+1} \sup_i \left(F\left(\binom{2^n}{i}\right) \mathrm{mod}_K(x)^i \mathrm{mod}_K(y)^{2^n-i} \right).$$

Assume for instance that $\mathrm{mod}_K(y) \leqslant \mathrm{mod}_K(x)$; as $F \leqslant 1$, we get:

$$\mathrm{mod}_K(x+y)^{2^n} \leqslant A^{n+1} \mathrm{mod}_K(x)^{2^n}.$$

This is so for all $n \geqslant 1$; for $n \to +\infty$, we get

$$\mathrm{mod}_K(x+y) \leqslant \mathrm{mod}_K(x),$$

i.e. the ultrametric inequality.

Next we recall the definition of the usual "valuations" on the field \mathbf{Q} of rational numbers. Let first p be a rational prime. Every $x \in \mathbf{Q}^\times$ can be written in one and only one way in the form $x = p^n a/b$, where n, a, b are integers, $b > 0$, and a and b are relatively prime to each other and to p; when that is so, put $|x|_p = p^{-n}$; also, put $|0|_p = 0$. The function $x \to |x|_p$

defined in this way on \mathbf{Q} is known as the *p-adic valuation* on \mathbf{Q}; clearly it satisfies the ultrametric inequality; it determines a topology on \mathbf{Q}, viz. the one defined by the distance function

$$\delta(x,y)=|x-y|_p.$$

The completion of \mathbf{Q} for this metric is *the field of p-adic numbers* and is denoted by \mathbf{Q}_p; the closure of \mathbf{Z} in that field is *the ring of p-adic integers* and is denoted by \mathbf{Z}_p. Clearly the *p*-adic valuation on \mathbf{Q} can be extended by continuity to \mathbf{Q}_p and remains ultrametric on \mathbf{Q}_p; this extension is still denoted by $|x|_p$. It is easily seen that \mathbf{Z}_p is compact (the reason for this may be expressed by saying that \mathbf{Z}_p is the "projective limit" of the finite groups $\mathbf{Z}/p^n\mathbf{Z}$ for $n \to +\infty$); as it is a neighborhood of 0 in \mathbf{Q}_p, \mathbf{Q}_p is locally compact; clearly it is not discrete.

On the other hand, we shall write $|x|_\infty$ whenever convenient, instead of $|x|$, for the "ordinary" absolute value on \mathbf{Q} and on \mathbf{R}. As \mathbf{R} is nothing else than the completion of \mathbf{Q} for the distance function $|x-y|_\infty$, we shall sometimes write \mathbf{Q}_∞ for \mathbf{R}. Thus the symbol \mathbf{Q}_v, where v may be either ∞ or a rational prime, denotes any one of the completions $\mathbf{Q}_\infty = \mathbf{R}$ and \mathbf{Q}_p of \mathbf{Q}.

THEOREM 5. *Let K be a non-discrete locally compact field; put $F(m) = \mathrm{mod}_K(m \cdot 1_K)$ for $m \in \mathbf{N}$. Then: either* (a) *K is of characteristic $p > 1$, and then $F(m) = 0$ for $m \equiv 0 \pmod{.} p$ and $F(m) = 1$ for $(m,p) = 1$; or* (b) *K is a division algebra of finite dimension δ over a field \mathbf{Q}_v, and then $F(m) = |m|_v^\delta$.*

By prop. 1 and th. 4 of § 2, F satisfies the assumptions in lemma 3; hence, by that lemma, it is of the form $m \to m^\lambda$ with $\lambda > 0$, or it is $\leqslant 1$. Assume that we are in the latter case; with B_1 as in prop. 2 of § 2, this means that the sequence $(m \cdot 1_K)$, for $m \in \mathbf{N}$, is contained in B_1; as B_1 is compact, it must have at least one limit point a. Then, by corollary 1 of prop. 2, there are, for every $\varepsilon > 0$, infinitely many $m \in \mathbf{N}$ such that $\mathrm{mod}_K(m \cdot 1_K - a) \leqslant \varepsilon$. Let m, m' be two such integers, with $m < m'$. Since $F \leqslant 1$ implies, by lemma 4, that mod_K is ultrametric, we have then

$$\mathrm{mod}_K(m' \cdot 1_K - m \cdot 1_K) \leqslant \varepsilon,$$

i.e. $F(m' - m) \leqslant \varepsilon$. In particular, this shows that there are integers $n \geqslant 1$ such that $F(n) < 1$; let p be the smallest of such integers. Since $F(mn) = = F(m)F(n)$ for all m, n, clearly p must be a prime. For any $x \in \mathbf{N}$, we have $F(px) < 1$, hence $F(1 + px) = 1$ by the corollary of th. 4, § 2. For any integer $m \geqslant 1$, prime to p, we have $m^{p-1} \equiv 1 \ (p)$, hence $F(m^{p-1}) = 1$ by what we have just proved, and therefore $F(m) = 1$. If K is of characteristic $p' > 1$, then $F(p') = 0$, so that p' can be no other than p; then F is as stated in case (a) of our theorem. If K is of characteristic 0, $F(p)$ cannot be 0,

and we may put $F(p)=p^{-\lambda}$ with $\lambda>0$; then $F(m)=|m|_p^\lambda$ for all m, as one sees at once by writing $m=p^n m'$ with $(m',p)=1$. Accordingly, whenever K is of characteristic 0, F must be of the form $m\to|m|_v^\lambda$ with $\lambda>0$. The mapping $n\to n\cdot 1_K$ of \mathbf{Z} onto the prime ring $\mathbf{Z}\cdot 1_K$ of K is then an algebraic (not necessarily a topological) isomorphism, which can be extended to an isomorphism of \mathbf{Q} onto the prime field in K; to simplify the language, identify the latter with \mathbf{Q} by means of that isomorphism. From what we have found about F, it follows at once that mod_K induces the function $x\to|x|_v^\lambda$ on \mathbf{Q}; therefore, by corollary 1 of prop. 2, § 2, the topological group structure induced on \mathbf{Q} by that of K is the one determined by the distance function $|x-y|_v$. As the closure of \mathbf{Q} in K is locally compact, hence complete for that structure, it follows that this closure is isomorphic to the completion \mathbf{Q}_v of \mathbf{Q} for the valuation v. As the prime ring, hence also the prime field, are clearly contained in the center of K, the same is true of \mathbf{Q}_v. Now K can be regarded as a vector-space over \mathbf{Q}_v; as such, by corollary 2 of th. 3, § 2, it must have a finite dimension δ, and we have, for every $x\in\mathbf{Q}_v$, $\mathrm{mod}_K(x)=\mathrm{mod}_{\mathbf{Q}_v}(x)^\delta$. To complete the proof, it only remains to be shown, in the case $v=\infty$, that $\mathrm{mod}_{\mathbf{R}}(m)=m$ for $m\in\mathbf{N}$, which is clear, and, in the case $v=p$, that $\mathrm{mod}_{\mathbf{Q}_p}(p)=p^{-1}$; this follows at once from the fact that \mathbf{Z}_p is a compact neighborhood of 0 in \mathbf{Q}_p, and that its image $p\cdot\mathbf{Z}_p$, under $x\to px$, is a compact subgroup of \mathbf{Z}_p of index p, so that its measure, for any Haar measure α on \mathbf{Q}_p, is $p^{-1}\alpha(\mathbf{Z}_p)$. It will be convenient to formulate separately what has just been proved:

COROLLARY. *In the case* (b) *of theorem 5,* $\mathrm{mod}_K(x)=|x|_v^\delta$ *for* $x\in\mathbf{Q}_v$.

DEFINITION 2. *A non-discrete locally compact field K will be called a p-field if p is a prime and $\mathrm{mod}_K(p\cdot 1_K)<1$, and an R-field if it is an algebra over* \mathbf{R}.

By lemmas 3 and 4 and th. 4 of § 2, the image Γ of K^\times under mod_K is discrete when K is a p-field, so that such a field cannot be connected; this shows that a topological field is an \mathbf{R}-field if and only if it is connected and locally compact. It is well known that there are no such fields except \mathbf{R}, \mathbf{C} and the field \mathbf{H} of "ordinary" (or "classical") quaternions; a proof for this will be included in Chap. IX-4.

§ 4. **Structure of p-fields.** In this section, p will be a prime and K will be a p-field with the unit element 1.

THEOREM 6. *Let K be a p-field; call R, R^\times and P the subsets of K respectively given by*

$$R=\{x\in K|\mathrm{mod}_K(x)\le 1\}, \quad R^\times=\{x\in K|\mathrm{mod}_K(x)=1\},$$
$$P=\{x\in K|\mathrm{mod}_K(x)<1\}.$$

*Then K is ultrametric; R is the unique maximal compact subring of K;
R^\times is the group of invertible elements of R; P is the unique maximal left,
right or two-sided ideal of R, and there is $\pi \in P$ such that $P = \pi R = R\pi$.
Moreover, the residual field $k = R/P$ is a finite field of characteristic p;
if q is the number of its elements, the image Γ of K^\times in \mathbf{R}_+^\times under mod_K
is the subgroup of \mathbf{R}_+^\times generated by q; and $\mathrm{mod}_K(\pi) = q^{-1}$.*

The set R is the same as the one previously denoted by B_1; it is compact, and so is R^\times. By th. 5 of § 3, mod_K is $\leqslant 1$ on the prime ring of K; therefore, by lemma 4 of § 3, K is ultrametric. This, by th. 4 of § 2, is the same as to say that $R + R = R$; as R is obviously closed under multiplication, it is a ring. Clearly every relatively compact subset of K which is closed under multiplication is contained in R; therefore R is the maximal compact subring of K. The invertible elements of R are those of R^\times. By th. 4 of § 2, Γ is a discrete subgroup of \mathbf{R}_+^\times; let γ be the largest element of Γ which is <1, and let $\pi \in K^\times$ be such that $\mathrm{mod}_K(\pi) = \gamma$. Clearly γ generates Γ; therefore, for every $x \in K^\times$, there is one and only one $n \in \mathbf{Z}$ such that $\mathrm{mod}_K(x) = \gamma^n$; then $x\pi^{-n}$ and $\pi^{-n}x$ are in R^\times. It is clear that $P = \pi R = R\pi$; this implies that P is compact. As $R - P = R^\times$, P has the maximal properties stated in our theorem. As R is a neighborhood of 0, and $R = R + R$, R is open; so is P; as R is compact, $k = R/P$ is finite. As $p \cdot 1 \in P$, the image of $p \cdot 1$ in k is 0, so that k is of characteristic p; if it has q elements, q is the index of $P = \pi R$ in the additive group of R. Therefore, if α is a Haar measure on K, $\alpha(R) = q\alpha(\pi R)$, hence $\mathrm{mod}_K(\pi) = q^{-1}$. This completes the proof.

DEFINITION 3. *With the notations of theorem 6, q will be called the
module of K; any element π of K^\times such that $P = \pi R = R\pi$ will be called
a prime element of K. For any $x \in K^\times$, the integer n such that $\mathrm{mod}_K(x) = q^{-n}$
will be denoted by $\mathrm{ord}_K(x)$. For each $n \in \mathbf{Z}$, one writes $P^n = \pi^n R = R\pi^n$.*

We will write $\mathrm{ord}(x)$, instead of $\mathrm{ord}_K(x)$, when there is no danger of confusion. We also put $\mathrm{ord}(0) = +\infty$; then P^n is the set of the elements x of K such that $\mathrm{ord}(x) \geqslant n$. With these notations, we can state as follows some corollaries of theorem 6:

COROLLARY 1. *Let (x_0, x_1, \ldots) be any sequence with the limit 0 in K.
Then the series $\sum\limits_0^{+\infty} x_i$ is commutatively convergent in K.*

For each $n \in \mathbf{N}$, put

$$\varepsilon_n = \sup_{i>n} \mathrm{mod}_K(x_i).$$

Our assumption means that $\lim \varepsilon_n = 0$. Let now S, S' be two finite sums of terms in the series $\sum x_i$, both containing the terms x_0, x_1, \ldots, x_n and

possibly some others. The ultrametric inequality gives $\mathrm{mod}_K(S - S') \leqslant \varepsilon_n$. The conclusion follows from this at once (the "filter" of finite sums of the series $\sum x_i$ is a "Cauchy filter" for the distance-function $\mathrm{mod}_K(x - y)$).

COROLLARY 2. *Let ξ be an element of P, other than 0; put $n = \mathrm{ord}(\xi)$, and let A be a full set of representatives of the classes modulo P^n in R. Then, for all $\nu \in \mathbf{Z}$, every $x \in P^{n\nu}$ can be expressed in one and only one way in the form*

$$x = \sum_{i = \nu}^{+\infty} a_i \xi^i$$

with $a_i \in A$ for all $i \geqslant \nu$.

Writing $x = x' \xi^\nu$ with $x' \in R$, we see that it is enough to deal with the case $\nu = 0$. Then one sees at once, by using induction on N, that one can determine the $a_i \in A$ in one and only one way by the condition

$$x \equiv \sum_{i = 0}^{N} a_i \xi^i \quad (P^{n(N + 1)})$$

for $N = 0, 1, \ldots$ This is equivalent with the assertion in our corollary.

COROLLARY 3. *Every automorphism of K (as a topological field) maps R onto R, P onto P, and has the module 1 when it is viewed as an automorphism of the additive group of K.*

COROLLARY 4. *For every $a \in K^\times$, the automorphisms $x \to ax$ and $x \to xa$ of the additive group of K have the same module.*

This follows at once from corollary 3, applied to the automorphism $x \to a^{-1}xa$. As the same fact is easily verified for the field \mathbf{H} of "ordinary" quaternions, it holds for all locally compact fields.

COROLLARY 5. *Let K be a commutative p-field, and K' a division algebra over K. Then K' is a p-field; every automorphism of K' over K in the algebraic sense is a topological automorphism; and, if R and R' are the maximal compact subrings of K and of K', and P and P' are the maximal ideals in R and in R', then $R = K \cap R'$ and $P = K \cap P'$.*

Regarding K' as a finite-dimensional vector-space over K, we provide it with its "natural" topology according to corollary 1 of th. 3, § 2. As this is unique, it is invariant under all K-linear mappings of K' onto itself, and in particular under all automorphisms of K' over K. Identifying K, as usual, with the subfield $K \cdot 1_{K'}$ of K', we see that K' is not discrete. By corollary 2 of th. 3, § 2, and th. 5 of § 3, it is a p-field. The rest is obvious.

COROLLARY 6. *Assumptions and notations being as in corollary 5, call q and q' the modules of K and of K', respectively; let π be a prime*

element of K, and put $e = \mathrm{ord}_{K'}(\pi)$. Then $q' = q^f$, where f is an integer $\geqslant 1$, and the dimension of K' over K is ef.

Put $k = R/P$ and $k' = R'/P'$; in view of the last assertion of corollary 5, we may identify k with the image of R in $k' = R'/P'$; if then f is the degree of k' over k, we have $q' = q^f$. Now apply corollary 2 of th. 3, § 2, to $\mathrm{mod}_K(\pi)$ and to $\mathrm{mod}_{K'}(\pi)$; we get the result stated above.

The last corollary shows in particular that $\mathrm{ord}_{K'}(\pi)$ is $\geqslant 1$ and is independent of the choice of the prime element π in K. This justifies the following definition:

DEFINITION 4. *Let assumptions and notations be as in corollaries 5 and 6 of theorem 6. Then e is called the order of ramification of K' over K, and f the modular degree of K' over K; K' is said to be unramified over K if $e = 1$, and to be fully ramified over K if $f = 1$.*

PROPOSITION 4. *Let K be a commutative p-field; let K' be a fully ramified division algebra of finite dimension over K; let R, R' be the maximal compact subrings of K and of K', respectively, and let π' be a prime element of K'. Then $K' = K(\pi')$, $R' = R[\pi']$, and K' is commutative.*

Let P, P' be the maximal ideals in R and in R', respectively, and let A be a full set of representatives of the classes modulo P in R. As K' is fully ramified over K, corollaries 5 and 6 of theorem 6 show at once that A is also a full set of representatives of the classes modulo P' in R'. Applying corollary 2 of th. 6 to K', R', P' and A, and to $\xi = \pi'$, we see that the elements of R' of the form $\sum_{i=0}^{e-1} a_i \pi'^i$, with $a_i \in A$ for $0 \leqslant i \leqslant e-1$, make up a full set of representatives A' of the classes modulo P'^e in R'. Take now a prime element π of K, and put $e = \mathrm{ord}_{K'}(\pi)$; e is the order of ramification of K' over K, hence also the dimension of K' over K, by corollary 6 of th. 6. Applying now corollary 2 of th. 6 to K', R', P'^e, A' and to $\xi = \pi$, we see that every element of P'^{ev} can be written in one and only one way in the form $\sum_{j=v}^{+\infty} a_j' \pi^j$, with $a_j' \in A'$ for all $j \geqslant v$. As K is contained in the center of K', π commutes with π'; therefore, in view of the definition of A', every such element can be written as

$$\sum_{i=0}^{e-1} \left(\sum_{j=v}^{+\infty} a_{ij} \pi^j \right) \pi'^i$$

with $a_{ij} \in A$ for $0 \leqslant i \leqslant e-1$, $j \geqslant v$, or, what amounts to the same in view of corollary 2 of th. 6, as $\sum_{i=0}^{e-1} \alpha_i \pi'^i$ with $\alpha_i \in P^v$ for $0 \leqslant i \leqslant e-1$. This shows that $K' = K(\pi')$, and, for $v = 0$, it shows that $R' = R[\pi']$. As K is contained

in the center of K', π' commutes with all elements of K; therefore K' is commutative.

COROLLARY 1. *Let K be a commutative p-field of characteristic p; call K^p its image under the endomorphism $x \to x^p$, and let π be a prime element of K. Then K is a fully ramified extension of K^p of degree p, and $K = K^p(\pi)$.*

Put $K' = K^p$; $x \to x^p$ is an isomorphism of K onto K', which we may use to transfer to K' the topology of K; K may then be regarded as a topological vector-space over K'; as such, by corollary 2 of th. 3, § 2, it must have a finite dimension. This shows that K is of finite degree over K'. As K and K' are isomorphic, they have the same module, so that the modular degree of K over K' is 1. By proposition 4, this implies that $K = K'(\pi)$; as $\pi^p \in K'$, the degree of K over K' must be p or 1. As $\mathrm{ord}_K(\pi) = 1$, π is not in K^p, so that $K \neq K'$. Therefore K is of degree p over K'.

COROLLARY 2. *Let K be as in corollary 1, and let \bar{K} be an algebraic closure of K. Then, for every $n \geqslant 0$, \bar{K} contains one and only one purely inseparable extension of K of degree p^n; this is the image $K^{p^{-n}}$ of K under the automorphism $x \to x^{p^{-n}}$ of \bar{K}.*

It follows at once from corollary 1 that $K^{p^{-1}}$ is of degree p over K; by induction on n, one sees then that $K^{p^{-n}}$ is of degree p^n over K. On the other hand, it is well-known, and easily proved, that, if K' is purely inseparable of degree $\leqslant p^n$ over K, it must be contained in $K^{p^{-n}}$. Our conclusion follows from this at once.

THEOREM 7. *Let K be a p-field; call q its module, R its maximal compact subring and P the maximal ideal of R. Then K^\times has at least one subgroup of order $q-1$; every such subgroup is cyclic; if M^\times is such a subgroup, the set $M = M^\times \cup \{0\}$ is a full set of representatives of the classes modulo P in R, and there is a prime element π of K such that $\pi M^\times = M^\times \pi$. If K is commutative, there is only one such group M^\times; it is the group of the roots of 1 of order prime to p in K.*

The construction of M^\times depends upon the following lemma:

LEMMA 5. *For all $n \geqslant 0$, $(1+P)^{p^n} \subset 1 + P^{n+1}$.*

This can be immediately verified by induction on n. It amounts to saying that, if $x \equiv 1$ (P), $x^{p^n} \equiv 1$ (P^{n+1}).

Now call ρ the canonical homomorphism of R onto $k = R/P$. By th. 2 of § 1, k^\times is cyclic of order $q-1$. In particular, for all $x \in R^\times$, we have $\rho(x)^{q-1} = 1$, i.e. $x^{q-1} \equiv 1$ (P). If $q = p^f$, lemma 5 shows now that $x^{(q-1)q^n} \equiv 1$ (P^{fn+1}); this can also be written as

$$x^{q^{n+1}} \equiv x^{q^n} \ (P^{fn+1}).$$

Applying now corollary 1 of th. 6 to the series

$$x + (x^q - x) + (x^{q^2} - x^q) + \cdots,$$

we see that it is convergent for all $x \in R^\times$, so that we may write

$$\omega(x) = \lim_{n \to +\infty} x^{q^n}$$

for $x \in R^\times$, and of course also for $x \in P$, hence for all $x \in R$. Clearly $\omega(xy) = \omega(x)\omega(y)$ whenever $xy = yx$; in particular, we have $\omega(x^v) = \omega(x)^v$ for all $x \in R^\times$, $v \in \mathbf{Z}$. As the above series for $\omega(x)$ shows, we have $\omega(x) \equiv x$ (P) for all $x \in R$; obviously $\omega(x) = 0$ for $x \in P$, and lemma 5 shows that $\omega(x) = 1$ for $x \in 1 + P$. Therefore $\omega^{-1}(0) = P$ and $\omega^{-1}(1) = 1 + P$. As $x^{q-1} \in 1 + P$ for all $x \in R^\times$, we have $\omega(x)^{q-1} = 1$ for $x \in R^\times$. Take a representative x_1 in R^\times of a generator of the cyclic group $k^\times = (R/P)^\times$, and put $\mu_1 = \omega(x_1)$; for $n \in \mathbf{Z}$, we have $\mu_1^n = 1$ if and only if $\omega(x_1^n) = 1$; as this is equivalent to $x_1^n \equiv 1$ (P), hence to $n \equiv 0$ $(q-1)$ in view of our choice of x_1, this shows that μ_1 generates a cyclic subgroup of R^\times of order $q-1$. Conversely, let Γ be any finite subgroup of K^\times of order n prime to p; clearly it is a subgroup of R^\times. The image of q, in the multiplicative group $(\mathbf{Z}/n\mathbf{Z})^\times$ of the integers prime to n modulo n, must have a finite order N; then $q^N \equiv 1$ (n). As $z^n = 1$ for every $z \in \Gamma$, we get now $z^{q^{Nv}} = z$ for all $v \geqslant 0$ and all $z \in \Gamma$, hence $\omega(z) = z$, so that $z \equiv 1$ (P) implies $z = 1$. This shows that the morphism of Γ into $k^\times = (R/P)^\times$ induced by ρ is injective, and therefore that Γ is cyclic, that its order divides $q-1$, and that, if it is of order $q-1$, $\Gamma \cup \{0\}$ is a full set of representatives of R/P in R. In particular, if K is commutative, we see that ω induces on R^\times a morphism of R^\times onto the group M^\times of the $(q-1)$-th roots of 1 in K, that it maps R onto $M = M^\times \cup \{0\}$, and that it determines a bijection of R/P onto M; moreover, every subgroup Γ of K^\times of order prime to p is then contained in M^\times; in particular, M^\times contains all the roots of 1 of order prime to p in K. As to the existence of a prime element of K with the property stated in our theorem, it is trivial if K is commutative. Assume that this is not so, and take any prime element π of K. For every $a \in K^\times$, $x \to axa^{-1}$ is an automorphism of K; by corollary 3 of th. 6, it maps R onto R, P onto P, so that it determines an automorphism $\lambda(a)$ of $k = R/P$; clearly $a \to \lambda(a)$ is a homomorphism of K^\times into the group of automorphisms of k. For $a \in R^\times$, $\lambda(a)$ is $\xi \to \rho(a) \xi \rho(a)^{-1}$, which is the identity since k is commutative. Therefore, if a is any element of K^\times, and $\mathrm{ord}(a) = n$, $\lambda(a) = \lambda(\pi)^n$. By corollary 2 of th. 2, § 1, applied to k and to the prime field in k, $\lambda(\pi)$ must be of the form $\xi \to \xi^{p^r}$; this means that we have, for every $x \in R$:

$$\pi x \pi^{-1} \equiv x^{p^r} \quad (P),$$

or, what amounts to the same:

$$\pi x \equiv x^{p^r} \pi \quad (P^2).$$

Take now M^\times as above, and put

$$\pi' = - \sum_{\mu \in M^\times} \mu^{p^r} \pi \mu^{-1}.$$

In view of the above congruences, each one of the $q-1$ terms in the sum in the right-hand side is $\equiv \pi$ modulo P^2. Since $q \cdot 1 \in P$, this gives

$$\pi' \equiv (1-q)\pi \equiv \pi \quad (P^2),$$

which implies that π' is a prime element of K. At the same time, the definition of π' gives

$$\pi' \mu = \mu^{p^r} \pi'$$

for all $\mu \in M^\times$, and therefore $\pi' M^\times = M^\times \pi'$. This completes the proof. One could show, by a similar argument, that, if M^\times and N^\times are two subgroups of K^\times of order $q-1$, there is a prime element π of K such that $\pi M^\times = N^\times \pi$.

COROLLARY 1. *If K and M are as in theorem 7, and K is of characteristic p, then M is a subfield of K. If at the same time K is commutative, M is the algebraic closure of the prime field in K.*

Let k_0 be the prime field in K, and let μ be a generator of the group M^\times. Then $k_0(\mu)$ is a commutative field of characteristic p in which the equation $X^q - X = 0$ has q roots, viz., the elements of M; therefore, by th. 2 of § 1, M is a field. If K is commutative, every element, other than 0, of the algebraic closure of k_0 in K is a root of 1 of order prime to p, again by th. 2 of § 1; therefore, by theorem 7, it must be in M.

COROLLARY 2. *Let K be a commutative p-field, q its module, and K' an extension of K of finite degree, generated by roots of 1 of order-prime to p. Then K' is unramified and cyclic over K, and its Galois group over K is generated by an automorphism φ which induces the permutation $\mu \to \mu^q$ on the group of roots of 1 of order prime to p in K'.*

By corollary 5 of th. 6, K' is a p-field. Let $R, P, q, k, \rho, M^\times$ be as in theorem 7 and its proof, and let $R', P', q', k', \rho', M'^\times$ be similarly defined for K'. By theorem 7, K' is generated over K by M'^\times, i.e. by the roots of $X^{q'-1} = 1$; therefore it is a Galois extension of K, and an automorphism of K' over K is uniquely determined by the permutation it induces on M'^\times. By corollary 5 of th. 6, we have $R = K \cap R'$, $P = K \cap P'$; we may therefore identify k with a subfield of k', and then ρ is the mapping induced by ρ' on R. Let α be an automorphism of K' over K; as it maps R' onto R', P' onto P', and leaves fixed every element of R, it determines an automorphism $\lambda(\alpha)$ of k' over k. Then λ, i.e. the mapping $\alpha \to \lambda(\alpha)$, is a morphism of the Galois group of K' over K into that of k' over k. By

corollary 2 of th. 2, § 1, $\lambda(\alpha)$ must be of the form $\xi \to \xi^{q^s}$. Therefore we have, for all $\mu \in M'^{\times}$:

$$\rho'(\alpha(\mu)) = \rho'(\mu)^{q^s} = \rho'(\mu^{q^s}).$$

As ρ' induces on M'^{\times}, by theorem 7, an isomorphism of M'^{\times} onto k'^{\times}, this implies that $\alpha(\mu) = \mu^{q^s}$. In particular, if $s = 0$, i.e. if $\lambda(\alpha)$ is the identity, α is the identity; this shows that λ is injective; therefore, if n is the degree of K' over K, and f that of k' over k, we have $n \leqslant f$. As $q' = q^f$, corollary 6 of th. 6 shows now that K' is unramified over K and that $n = f$, so that λ is an isomorphism of the Galois group of K' over K onto that of k' over k. In view of corollary 2 of th. 2, § 1, this completes our proof.

COROLLARY 3. *Let K and q be as in corollary 2; then a division algebra of finite dimension over K is unramified if and only if it is commutative and can be generated over K by roots of 1 of order prime to p. For every $f \geqslant 1$, K has one and (up to an isomorphism) only one unramified extension of degree f; this is the extension generated over K by a primitive $(q^f - 1)$-th root of 1.*

Let K' be an unramified division algebra of dimension f over K; let q, q' be the modules of K and of K', respectively; then $q' = q^f$, by corollary 6 of th. 6. Take a subgroup M'^{\times} of K'^{\times} of order $q' - 1$; by theorem 7, it is cyclic; take a generator μ of M'^{\times}, and put $K'' = K(\mu)$. Clearly K'' is commutative; as it contains M'^{\times}, its module is at least q', so that, by corollary 6 of th. 6, its degree over K is at least f; therefore $K'' = K'$, which, together with corollary 2, proves the first part of our corollary. Now take any $f \geqslant 1$; put $q' = q^f$, and call K' the extension of K generated by a primitive $(q' - 1)$-th root of 1, or, what amounts to the same, by the set M'^{\times} of all the roots of $X^{q'-1} = 1$; by theorem 7, its module is at least q', so that, by corollary 6 of th. 6, its degree over K is at least f. On the other hand, by corollary 2, it is unramified and cyclic over K, and its Galois group over K is generated by the automorphism φ defined there; as φ^f induces the identity on M'^{\times}, it is the identity, so that the degree of K' over K is at most f. Therefore it is f. As the foregoing results show that every unramified extension of K of degree f must contain an extension isomorphic to K', this completes our proof.

COROLLARY 4. *Let K' be a finite extension of a commutative p-field K; call f its modular degree over K, and e its order of ramification over K. Then there is a unique maximal unramified extension K_1 of K, contained in K'; it is of degree f over K, and K' is fully ramified of degree e over K_1.*

This follows at once from the foregoing results, K_1 being generated by the roots of 1 of order prime to p in K'.

DEFINITION 5. *Let K be a commutative p-field, and K' an unramified extension of K; the generator φ of the Galois group of K' over K which is defined by corollary 2 of theorem 7 is called the Frobenius automorphism of K' over K.*

In corollary 2 of theorem 6, one can take for ξ a prime element π of K, and then take for A the set M defined in theorem 7. For commutative fields of characteristic p, this gives the following:

THEOREM 8. *Every commutative p-field of characteristic p is isomorphic to a field of formal power-series in one indeterminate with coefficients in a finite field.*

Take notations as in theorem 7; corollary 1 of th. 7 shows that M is a field with q elements. Taking $\xi = \pi$ and $A = M$ in corollary 2 of th. 6, we get for every $x \in K$ with $\mathrm{ord}(x) \geqslant n$ a unique series expansion

$$x = \sum_{i=n}^{+\infty} \mu_i \pi^i,$$

with $\mu_i \in M$ for all $i \geqslant n$. One verifies at once that the rules for the addition and multiplication of such series are the usual ones for formal power-series in algebra (or for convergent power-series in classical analysis). Moreover, this is an isomorphism also in the topological sense if the field of formal power-series is provided with its usual topology, that for which the ring R_0 of "integral" power-series (those containing no power of the indeterminate with an exponent <0), and the ideals generated in it by the powers of the indeterminate, make up a fundamental system of neighborhoods of 0. We recall that, for this topology, the ring R_0 of integral formal power-series in one indeterminate over any finite field F is compact, since the additive group of R_0 is clearly isomorphic to the product of enumerably many groups isomorphic to F; therefore the corresponding field is locally compact. Thus theorem 8 shows that the commutative p-fields of characteristic p are all of that type, so that (up to an isomorphism) they are in a one-to-one correspondence with the finite fields \mathbf{F}_q, with $q = p^n$, $n \geqslant 1$.

By a *local field*, we will understand a commutative non-discrete locally compact field. We have thus obtained a complete list of the local fields of characteristic $p > 1$, while those of characteristic 0 are given by theorem 5 of § 3; they are \mathbf{R}, \mathbf{C} and the finite algebraic extensions of the fields \mathbf{Q}_p, for all p.

Using the same idea as in the proof of theorem 8, we give now one more result for the non-commutative case.

PROPOSITION 5. *Let K be a p-field, commutative or not, with the maximal compact subring R. Then the center K_0 of K is a p-field; if d*

is the modular degree of K *over* K_0, *its order of ramification over* K_0 *is also* d, *and its dimension over* K_0 *is* d^2; *it contains a maximal commutative subfield* K_1 *which is unramified and of degree* d *over* K_0. *Moreover, if* K_1 *is such, and if* R_1 *is its maximal compact subring,* K *has a prime element* π *with the following properties:* (a) π^d *is a prime element of* K_0; (b) $\{1, ..., \pi^{d-1}\}$ *is a basis of* K *as a left vector-space over* K_1, *and generates* R *as a left* R_1*-module;* (c) *the inner automorphism* $x \to \pi^{-1} x \pi$ *of* K *induces on* K_1 *an automorphism* α *which generates the Galois group of* K_1 *over* K_0.

Let notations be as in theorems 6 and 7; choose M and π as in theorem 7, and apply corollary 2 of th. 6 to π and M; this shows that, for every $n \in \mathbf{Z}$, each $x \in P^n$ can be uniquely written as

$$(5) \qquad\qquad x = \sum_{i=n}^{+\infty} \mu_i \pi^i$$

with $\mu_i \in M$ for all $i \geqslant n$. Therefore an element of K is in the center K_0 of K if and only if it commutes with π and with every element of M (or, what amounts to the same, with some generator of the cyclic group M^\times). As $x \to \pi^{-1} x \pi$ induces a permutation on M, some power of it must induce the identity on M; this amounts to saying that there is $\nu > 0$ such that π^ν commutes with every element of M. Then K_0 contains $\pi^{\nu n}$ for all $n \in \mathbf{Z}$; this proves that it is not discrete; as it is clearly closed in K, it is locally compact; if now we consider K as a vector-space, hence an algebra, over K_0, we see, by corollary 2 of th. 3, § 2, that it has a finite dimension over K_0; corollary 5 of th. 6 shows then that K_0 is a p-field. Call q the module of K_0, d the modular degree of K over K_0, and K_1 the field generated over K_0 by M, or, what amounts to the same, by any generator of the cyclic group M^\times; as M^\times is of order $q^d - 1$, such a generator is a primitive $(q^d - 1)$-th root of 1, so that, by corollary 3 of th. 7, K_1 is unramified of degree d over K_0. As $x \to \pi^{-1} x \pi$ induces a permutation on M, and the identity on K_0, it induces on K_1 an automorphism α of K_1 over K_0. An element of K_1 commutes with all the elements of M; it commutes with π if and only if it is invariant under α; in other words, the elements of K_1 which are invariant under α are those of K_0, so that α generates the Galois group of K_1 over K_0; it is therefore of order d, so that, as we have seen above, π^d is in K_0, and π^ν is not in K_0 unless ν is a multiple of d. Now take $x \in K$ and $\mu \in M^\times$; write x in the form (5). Then we have

$$\mu^{-1} x \mu = \sum_{i=n}^{+\infty} \mu_i' \pi^i$$

where we have put

$$\mu_i' = \mu^{-1} \mu_i \cdot (\pi^i \mu \pi^{-i}).$$

In this last formula, the last factor on the right-hand side belongs to M^\times, so that μ'_i is in M. In view of the unicity of the expansion (5) for $x \in K$, this shows that $x = \mu^{-1} x \mu$, i.e. that x commutes with μ, if and only if $\mu'_i = \mu_i$ for all i. Now clearly, for each i, $\mu'_i = \mu_i$ if and only if either $\mu_i = 0$ or π^i commutes with μ. Consequently, x commutes with all elements of M^\times if and only if π^i does so whenever $\mu_i \neq 0$. In view of what has been proved above, this is so if and only if $\mu_i = 0$ whenever i is not a multiple of d; we have then

$$x = \sum_i \mu_{di}(\pi^d)^i.$$

As $\pi^d \in K_0$, x is then in the closure of K_1, hence in K_1 itself, which is therefore a maximal commutative subfield of K. It is also clear now, in view of (5) and of the unicity of (5), that $\{1, \pi, \ldots, \pi^{d-1}\}$ is a basis of K as a left vector-space over K_1, that it generates R as a left R_1-module, and that π^d is a prime element of K_1, hence also of K_0 since it lies in K_0. As this implies that the order of ramification of K over K_0 is d, it completes the proof.

Notations being as in proposition 5, let φ be the Frobenius automorphism of K_1 over K_0; as this also generates the Galois group of K_1 over K_0, we must have $\varphi = \alpha^r$, with r prime to d and uniquely determined modulo d. It will be shown in Chapter XII that, when K_0 is given, d and r may be chosen arbitrarily, subject to these conditions, and characterize the structure of the division algebra K uniquely; in other words, two division algebras of finite dimension over K_0, with the center K_0, are isomorphic if and only if they have the same dimension d^2 over K_0, and the integer r has the same value modulo d for both.

We conclude this Chapter with a result about the maximal compact subrings in p-fields. We recall that, if R is any commutative ring, and x an element of a ring containing R, x is called *integral over R* if and only if it is a root of some monic polynomial over R, i.e. of some polynomial with coefficients in R and the highest coefficient equal to 1.

PROPOSITION 6. *Let K be a p-field and K_0 a p-field contained in the center of K; let R, R_0 be the maximal compact subrings of K and of K_0. Then R consists of the elements of K which are integral over R_0.*

Let x be in K and integral over R_0; this means that it satisfies an equation

$$x^n + a_1 x^{n-1} + \cdots + a_n = 0$$

with $a_i \in R_0$ for $1 \leqslant i \leqslant n$. Assume that x is not in R, i.e. that $\mathrm{ord}_K(x) < 0$. Then $x \neq 0$, and we have

$$1 = -a_1 x^{-1} - \cdots - a_n x^{-n};$$

here all the terms in the right-hand side are in the maximal ideal P of R, so that $1 \in P$, which is absurd. Conversely, let x be any element of R. By corollary 2 of th. 3, § 2, K has a finite dimension over K_0; therefore, if we put $K' = K_0(x)$, this is a commutative field and a finite extension of K_0. Call F the irreducible monic polynomial, with coefficients in K_0, such that $F(x) = 0$; in some algebraic closure of K', call K'' the field generated over K_0 by all the roots of F, so that F splits into linear factors in K''. As K', K'' are finite extensions of K_0, they are p-fields; call R', R'' their maximal compact subrings. Then $R' = K' \cap R = K' \cap R''$; as x is in R, it is in R' and in R''. As F is irreducible, every root x' of F in K'' is the image of x under some automorphism of K'' over K_0; as such an automorphism maps R'' onto R'', all such roots are in R''. Therefore all the coefficients of F are in R''; as they are in K_0, they are in R_0. This completes the proof.

If K is commutative, proposition 6 may be expressed by saying that R is the integral closure of R_0 in K.

Chapter II

Lattices and duality over local fields

§ 1. Norms. In this § and the next one, K will be a p-field, commutative or not. We shall mostly discuss only left vector-spaces over K; everything will apply in an obvious way to right vector-spaces. Only vector-spaces of finite dimension will occur; it is understood that these are always provided with their "natural topology" according to corollary 1 of th. 3, Chap. I-2. By th. 3 of Chap. I-2, every subspace of such a space V is closed in V. Taking coordinates, one sees that all linear mappings of such spaces into one another are continuous; in particular, linear forms are continuous. Similarly, every injective linear mapping of such a space V into another is an isomorphism of V onto its image. As K is not compact, no subspace of V can be compact, except $\{0\}$.

DEFINITION 1. *Let V be a left vector-space over the p-field K. By a K-norm on V, we understand a function N on V, with values in \mathbf{R}_+, such that:* (i) $N(v)=0$ *if and only if* $v=0$; (ii) $N(xv)=\text{mod}_K(x)N(v)$ *for all* $x \in K$ *and all* $v \in V$; (iii) N *satisfies the ultrametric inequality*

$$(1) \qquad\qquad N(v+w) \leqslant \sup(N(v), N(w))$$

for all v, w in V.

On K^n, one defines a K-norm N_0 by putting $N_0(x) = \sup_{1 \leqslant i \leqslant n}(\text{mod}_K(x_i))$ for all $x=(x_1, \ldots, x_n)$ in K^n. As every vector-space of finite dimension over K is isomorphic to a space K^n, this shows that there are K-norms on all such spaces.

One can obviously use any K-norm on V in order to topologize V, by taking $N(v-w)$ as distance-function.

PROPOSITION 1. *Let V be a left vector-space of finite dimension over the p-field K. Then every K-norm N on V defines the natural topology on V. In particular, every such norm N is continuous, and the subsets L_r of V defined by $N(v) \leqslant r$ are compact neighborhoods of 0 for all $r>0$.*

As to the first assertion, in view of corollary 1 of th. 3, Chap. I-2, we need only show that the topology defined by N on V makes V into a topological vector-space over K. This follows at once from the inequality

$$N(x'v' - xv) \leqslant \sup(\text{mod}_K(x')N(v'-v), \text{mod}_K(x'-x)N(v))$$

which is an immediate consequence of def. 1. Therefore N is continuous, and the sets L_r make up a fundamental system of closed neighborhoods

of 0; in particular, L_r must be compact for some $r > 0$. Now, for any $s > 0$, take $a \in K^{\times}$ such that $\mathrm{mod}_K(a) \leqslant r/s$; then, as one sees at once, L_s is contained in $a^{-1} L_r$; therefore it is compact.

COROLLARY 1. *There is a compact subset A of $V - \{0\}$ which contains some scalar multiple of every v in $V - \{0\}$.*

Call q the module of K, and take a K-norm N in V. If π is a prime element of K, we have $\mathrm{mod}_K(\pi) = q^{-1}$, by th. 6 of Chap. I-4, hence $N(\pi^n v) = q^{-n} N(v)$ for all $n \in \mathbf{Z}$ and all $v \in V$. Let A be the subset of V defined by $q^{-1} \leqslant N(v) \leqslant 1$; by proposition 1, it is compact; and, for every $v \neq 0$, one can choose $n \in \mathbf{Z}$ so that $\pi^n v \in A$.

Corollary 1 implies the fact that the "projective space" attached to V is compact.

COROLLARY 2. *Let φ be any continuous function on $V - \{0\}$, with values in \mathbf{R}, such that $\varphi(av) = \varphi(v)$ for all $a \in K^{\times}$ and all $v \in V - \{0\}$. Then φ reaches its maximum at some point v_1 of $V - \{0\}$.*

In fact, this will be so if we take A as in corollary 1 and take for v_1 the point of A where φ reaches its maximum on A.

COROLLARY 3. *Let f be any linear form on V, and N a K-norm on V. Then there is $v_1 \neq 0$ in V, such that*

$$(2) \qquad N(v)^{-1} \mathrm{mod}_K(f(v)) \leqslant N(v_1)^{-1} \mathrm{mod}_K(f(v_1))$$

for all $v \neq 0$ in V.

This is a special case of corollary 2, that corollary being applied to the left-hand side of (2). If one denotes by $N^*(f)$ the right-hand side of (2), then $N^*(f)$ is the smallest positive number such that

$$\mathrm{mod}_K(f(v)) \leqslant N^*(f) \cdot N(v)$$

for all $v \in V$, and $f \to N^*(f)$ is a K-norm on the dual space of V, i.e. on the right vector-space made up of the linear forms on V (where the addition is the obvious one, and the scalar multiplication is defined by putting $(fa)(v) = f(v)a$ when f is such a form, and $a \in K$).

By a hyperplane in V, one understands a subspace of V of codimension 1, i.e. any subset H of V defined by an equation $f(v) = 0$, where f is a linear form other than 0; when H is given, f is uniquely determined up to a scalar factor other than 0. Now, if (2) is valid for all $v \neq 0$, and for a given norm N, a given linear form $f \neq 0$ and a given $v_1 \neq 0$, it remains so if one replaces f by fa, with $a \in K^{\times}$, and v_1 by bv_1 with $b \in K^{\times}$; in other words, its validity for all $v \neq 0$ depends only upon the hyperplane H defined by $f = 0$ and the subspace W of V generated by v_1; when it holds for all $v \neq 0$, we shall say that H and W are *N-orthogonal* to each other.

PROPOSITION 2. *A hyperplane H and a subspace W of V of dimension 1 are N-orthogonal if and only if V is the direct sum of H and W, and $N(h+w)=\sup(N(h),N(w))$ for all $h\in H$ and $w\in W$.*

Let H be defined by $f(v)=0$, and assume first that H and W are N-orthogonal. Then (2) is satisfied if one replaces v_1 in it by any $w\in W$ other than 0. This implies that $f(w)$ is not 0, for otherwise f would be 0; therefore V is the direct sum of H and W. Now replace v in (2) by $h+w$ with $h\in H$; as $f(h+w)=f(w)\neq0$, (2) gives $N(h+w)\geqslant N(w)$. Applying the ultrametric inequality (1) to $h=(h+w)+(-w)$, we get $N(h)\leqslant N(h+w)$; applying it to $h+w$, we get the formula in our proposition, for $w\neq0$; as it is trivial for $w=0$, this proves the necessity of the condition stated there. Now suppose that V is the direct sum of H and W; take any $v\neq0$, and write it as $v=h+w$ with $h\in H$ and $w\in W$, so that $f(v)=f(w)$. If $w\neq0$ and $N(h+w)\geqslant N(w)$, then we have

$$N(v)^{-1}\operatorname{mod}_K(f(v))\leqslant N(w)^{-1}\operatorname{mod}_K(f(w)).$$

As the right-hand side does not change if we replace w by any generator v_1 of W, this shows that (2) holds for any such v_1, and any v not in H. For $v\in H$, i.e. $w=0$, it holds trivially. This completes the proof.

Accordingly, we shall also say that two subspaces V', V'' of V are *N-orthogonal* to each other whenever V is the direct sum of V' and V'', and $N(v'+v'')=\sup(N(v'),N(v''))$ for all $v'\in V'$ and all $v''\in V''$.

PROPOSITION 3. *Let V be of dimension n over K, and let N be a K-norm on V. Then there is a decomposition $V=V_1+\cdots+V_n$ of V into a direct sum of subspaces V_i of dimension 1, such that $N(\sum v_i)=\sup_i N(v_i)$ whenever $v_i\in V_i$ for $1\leqslant i\leqslant n$. Moreover, if $W_1=V$, W_2,\ldots,W_n is a sequence of subspaces of V such that W_i is a subspace of W_{i-1} of codimension 1 for $2\leqslant i\leqslant n$, then the V_i may be so chosen that $W_i=V_i+\cdots+V_n$ for all i.*

This is clear for $n=1$. For $n>1$, use induction on n. By corollary 3 of prop. 1, we may choose v_1 so that the space V_1 generated by v_1 is N-orthogonal to W_2; then, by prop. 2, $N(v'_1+w_2)=\sup(N(v'_1),N(w_2))$ whenever $v'_1\in V_1$, $w_2\in W_2$. Applying the induction assumption to the K-norm induced by N on W_2, and to the sequence W_2,\ldots,W_n, we get our result.

COROLLARY. *To every subspace W of V, there is a subspace W' which is N-orthogonal to W.*

Take a sequence W_1,\ldots,W_n, as in proposition 3, such that W is one of the spaces in that sequence, say W_i. Take the V_i as in proposition 3. Then the space $W'=V_1+\cdots+V_{i-1}$ is N-orthogonal to W.

PROPOSITION 4. *Let N, N' be two K-norms in V. Then there is a decomposition $V=V_1+\cdots+V_n$ of V into a direct sum of subspaces V_i of*

dimension 1, *such that* $N(\sum v_i) = \sup_i N(v_i)$ *and* $N'(\sum v_i) = \sup_i N'(v_i)$ *whenever* $v_i \in V_i$ *for* $1 \leqslant i \leqslant n$.

For $n = 1$, this is clear. For $n > 1$, use induction on n. Applying corollary 2 of prop. 1 to $\varphi = N/N'$, we get a vector $v_1 \neq 0$ such that

$$N(v) N'(v)^{-1} \leqslant N(v_1) N'(v_1)^{-1}$$

for all $v \neq 0$; call V_1 the space generated by v_1. By the corollary of prop. 3, there is a hyperplane W which is N-orthogonal to V_1; then, if $f = 0$ is an equation for W, we have

$$N(v)^{-1} \mathrm{mod}_K(f(v)) \leqslant N(v_1)^{-1} \mathrm{mod}_K(f(v_1))$$

for all $v \neq 0$. Multiplication of these two inequalities gives

$$N'(v)^{-1} \mathrm{mod}_K(f(v)) \leqslant N'(v_1)^{-1} \mathrm{mod}_K(f(v_1)),$$

which means that W is N'-orthogonal to V_1. Applying now prop. 2 to N, V_1 and W, and also to N', V_1 and W, and applying the induction assumption to the norms induced by N and N' on W, we get the announced result.

One should notice the close analogy between propositions 3 and 4, *
and their proofs, and the corresponding results and proofs for norms defined by positive-definite quadratic forms in vector-spaces over **R**, or hermitian forms in vector-spaces over **C** or **H**. For instance, prop. 4 corresponds to the simultaneous reduction of two quadratic or hermitian forms to "diagonal form".

§ 2. **Lattices.** In this section, K will again be a p-field, and we shall use the notations introduced in Chapter I. In particular, we write R for the maximal compact subring of K, P for the maximal ideal in R, q for the module of K, and π for a prime element of K. For $n \in \mathbf{Z}$, we write $P^n = \pi^n R = R \pi^n$.

We shall be concerned with R-modules in left vector-spaces of finite dimension over K; if V is such a space, an R-module in V is a subgroup M of V such that $R \cdot M = M$.

PROPOSITION 5. *Let V be a left vector-space of finite dimension over K. Let M be an R-module in V, and call W the subspace of V generated by M over K. Then M is open and closed in W; it is compact if and only if it is finitely generated as an R-module.*

Let m_1, \ldots, m_r be a maximal set of linearly independent elements over K in M; they make up a basis of W over K. By th. 3 of Chap. I-2, the set $R m_1 + \cdots + R m_r$ is an open subgroup of W; as both M and $W - M$ are unions of cosets with respect to that subgroup, they are both open. If M is compact, it is the union of finitely many such cosets and therefore finitely generated; the converse is obvious.

On the other hand, in view of corollary 2 of th. 6, Chap. I-4, a closed subgroup X of V satisfies $R \cdot X = X$ if and only if $\pi X \subset X$ and $aX \subset X$ for every a in a full set A of representatives of R/P in R. In particular, if $q = p$, i.e. if R/P is the prime field, we may take $A = \{0, 1, \ldots, p-1\}$, and then $aX \subset X$ for all $a \in A$, so that X is then an R-module if and only if $\pi X \subset X$. In the case $K = \mathbf{Q}_p$, we may take $\pi = p$, and then every closed subgroup of V is a \mathbf{Z}_p-module.

In K itself, viewed as a left vector-space over K, every R-module, if not reduced to $\{0\}$, is a union of sets P^n, and thus is either K or one of these sets.

DEFINITION 2. *By a K-lattice in a left vector-space V of finite dimension over K, we understand a compact and open R-module in V.*

When no confusion can occur, we say "lattice" instead of K-lattice. If L is a p-field contained in K, every K-lattice is an L-lattice; the converse is not true unless $L = K$.

Clearly, if L is a lattice in V, and W is a subspace of V, $L \cap W$ is a lattice in W; similarly, if f is an injective linear mapping of a space V' into V, $f^{-1}(L)$ is a lattice in V'; if f' is a surjective linear mapping of V onto a space V'', $f'(L)$ is a lattice in V''.

If N is a K-norm in V, the subset L_r of V defined by $N(v) \leqslant r$ is a K-lattice in V for every $r > 0$. In fact, (iii), in def. 1 of § 1, together with (ii) applied to $x = -1$, shows that it is a subgroup of V; then (ii) shows that it is an R-module, and prop. 1 of § 1 shows that it is a compact neighborhood of 0 in V, hence open since it is a subgroup of V. This has a converse; more generally, we prove:

PROPOSITION 6. *Let M be an open R-module in V; for every $v \in V$, put*

$$N_M(v) = \inf_{x \in K^\times, xv \in M} \mathrm{mod}_K(x)^{-1}.$$

Then the function N_M on V satisfies conditions (ii) and (iii) in definition 1 of § 1, and M is the subset of V defined by $N_M(v) \leqslant 1$; N_M is a K-norm if and only if M is a K-lattice in V.

For $a \in K^\times$, we have $x \cdot av \in M$ if and only if $x = ya^{-1}$ with $yv \in M$; this gives $N_M(av) = \mathrm{mod}_K(a) N_M(v)$; as $N_M(0) = 0$, this is also true for $a = 0$. Therefore N_M satisfies (ii) of def. 1. For each $v \in V$, call M_v the set of the elements x of K such that $xv \in M$; as this is an open R-module in K, it is either K or a set P^n with some $n \in \mathbf{Z}$. If $M_v = K$, $N_M(v) = 0$; if $M_v = P^n$, we have $xv \in M$ if and only if $\mathrm{mod}_K(x) \leqslant q^{-n}$, so that $N_M(v) = q^n$. In particular, we have $N_M(v) \leqslant 1$ if and only if $M_v \supset R$, hence if and only if $v \in M$. Let v, w be in V and such that $N_M(v) \geqslant N_M(w)$; then $M_v \subset M_w$, so that $xv \in M$ implies $xw \in M$, hence also $x(v+w) \in M$; therefore $M_{v+w} \supset M_v$,

hence $N_M(v+w) \leqslant N_M(v)$; this proves (iii) of def. 1. Finally, M is a K-lattice if and only if it is compact, and N_M is a K-norm if and only if $N_M(v)>0$ for all $v \neq 0$, i.e. if and only if $M_v \neq K$ for $v \neq 0$. By prop. 1 of § 1, if N_M is a K-norm, M is compact. Conversely, assume that M is compact, and take $v \neq 0$; then M_v is the subset of K corresponding to $(Kv) \cap M$ under the isomorphism $x \to xv$ of K onto Kv; therefore M_v is compact and cannot be K. This completes our proof.

COROLLARY 1. *An open R-module M in V is a K-lattice if and only if it contains no subspace of V other than* 0.

It has been shown above that, if M is not compact, N_M cannot be a K-norm, so that there is $v \neq 0$ in V such that $N_M(v)=0$, hence $M_v = K$, i.e. $Kv \subset M$. Conversely, as every subspace of V, other than 0, is closed in V and not compact, no such subspace can be contained in M if M is compact.

COROLLARY 2. *Let M be an open R-module in V; let W be a maximal subspace of V contained in M, and let W' be any supplementary subspace to W in V. Then $M \cap W'$ is a K-lattice in W', and $M = (M \cap W') + W$.*

The first assertion is a special case of corollary 1; the second one is obvious.

Proposition 6 shows that every K-lattice in V may be defined by an inequality $N(v) \leqslant 1$, where N is a K-norm; this was our chief motive in discussing norms in § 1. For a given K-lattice M, the norm N_M defined in prop. 6 may be characterized, among all the norms N such that M is the set $N(v) \leqslant 1$, as the one which takes its values in the set of values taken by mod_K on K, i.e. in the set $\{0\} \cup \{q^n\}_{n \in \mathbf{Z}}$.

PROPOSITION 7. *If V has the dimension 1 over K, and if L is a K-lattice in V, then V has a generator v such that $L = Rv$.*

Take any generator w of V; the subset L_w of K defined by $xw \in L$ must be of the form P^n; taking $v = \pi^n w$, we get $L = Rv$.

THEOREM 1. *Let L be a K-lattice in a left vector-space V of dimension n over K. Then there is a basis $\{v_1, \ldots, v_n\}$ of V such that $L = \sum Rv_i$. Moreover, if $W_1 = V, W_2, \ldots, W_n$ is any sequence of subspaces of V such that W_i is a subspace of W_{i-1} of codimension 1 for $2 \leqslant i \leqslant n$, the v_i may be so chosen that, for each i, $\{v_i, \ldots, v_n\}$ is a basis of W_i.*

Take a K-norm N such that L is defined by $N(v) \leqslant 1$. Choose subspaces V_1, \ldots, V_n of V as in prop. 3 of § 1; then $L = \sum (L \cap V_i)$. Applying prop. 7 to V_i and $L \cap V_i$ for each i, we get the basis (v_i).

Theorem 1 may be applied for instance whenever K' is a p-field containing K, and R' is the maximal compact subring of K'. Clearly, if K' is viewed as a left vector-space over K, R' is a K-lattice in K'. Therefore there is a basis $\{y_1,...,y_n\}$ of K' over K such that $R'=\sum R y_i$; then, if we write, for any $y \in R'$, $yy_i = \sum a_{ij} y_j$, with $a_{ij} \in K$ for $1 \leqslant i, j \leqslant n$, all the a_{ij} must be in R. In particular, if K is commutative, these relations, which hold in the commutative field $K(y)$, imply $\det(y \cdot 1_n - A) = 0$, where 1_n is the unit matrix and $A = (a_{ij})$, so that we get an alternative proof for the second part of prop. 6, Chap. I-4.

THEOREM 2. *Let L, L' be two K-lattices in a left vector-space V of finite dimension over K. Then there is a basis $\{v_1,...,v_n\}$ of V, and a sequence of integers $(v_1,...,v_n)$, such that $L = \sum R v_i$ and $L' = \sum P^{v_i} v_i$.*

Take K-norms N, N' such that L is defined by $N(v) \leqslant 1$ and L' by $N'(v) \leqslant 1$. Choose subspaces $V_1,...,V_n$ of V as in prop. 4 of § 1; then $L = \sum(L \cap V_i)$ and $L' = \sum(L' \cap V_i)$. For each i, apply prop. 7 to V_i and $L \cap V_i$, and also to V_i and $L' \cap V_i$; this gives v_i such that $L \cap V_i = R v_i$ and v_i' such that $L' \cap V_i = R v_i'$. Writing $v_i' = x_i v_i$ with $x_i \in K^\times$, and putting $v_i = \mathrm{ord}(x_i)$, we get integers v_i with the required property.

COROLLARY 1. *Let V and L be as in theorems 1 and 2, and let M be an R-module in V. Then there is a basis $\{v_1,...,v_n\}$ of V over K, and there are integers r, s and $v_1,...,v_r$, such that $0 \leqslant r \leqslant s \leqslant n$, $L = \sum R v_i$ and*

$$M = \sum_{j=1}^{r} P^{v_j} v_j + \sum_{h=r+1}^{s} K v_h.$$

Let W be the subspace of V generated by M, and W' the maximal subspace contained in M; call s the dimension of W, and r the codimension of W' in W. In th. 1, choose the sequence $W_1,...,W_n$ so that it includes W and W'. Then th. 1 gives us a basis $\{w_1,...,w_n\}$ of V which generates L as an R-module and contains bases for W and for W'; renumbering this basis in an obvious manner, we may assume that $\{w_1,...,w_s\}$ is a basis for W and that $\{w_{r+1},...,w_s\}$ is one for W'. Call W'' the subspace of V with the basis $\{w_1,...,w_r\}$. By prop. 5, M is open in W; therefore, by corollary 2 of prop. 6, we have $M = M' + W'$, where $M' = M \cap W''$ is a K-lattice in W''. Applying now th. 2 to M' and to $L' = L \cap W''$, we get a basis $\{v_1,...,v_r\}$ for W'', and integers $v_1,...,v_r$, such that $L' = \sum R v_j$ and $M' = \sum P^{v_j} v_j$. Taking $v_i = w_i$ for $i > r$, we get the basis required by our corollary.

COROLLARY 2. *Every finitely generated R-module \mathfrak{M} is the direct sum of finitely many summands, each of which is isomorphic either to R or to a module R/P^v with $v > 0$. Moreover, the number of summands of type R, and, for each v, the number of summands of type R/P^v, are uniquely determined when \mathfrak{M} is given.*

Let \mathfrak{M} be generated by elements $\mathfrak{m}_1,\dots,\mathfrak{m}_n$. Take a vector-space V of dimension n over K, with a basis $\{v_1,\dots,v_n\}$; put $L=\sum Rv_i$. Then the formula

$$\sum x_i v_i \rightarrow \sum x_i \mathfrak{m}_i,$$

where the x_i are taken in R for $1\leqslant i\leqslant n$, defines a morphism of L onto \mathfrak{M}; therefore \mathfrak{M} is isomorphic to L/M, where M is the kernel of that morphism. Apply now corollary 1 to L and M; as $M\subset L$, we have $v_j\geqslant 0$ for $1\leqslant j\leqslant r$, and $r=s$. Our first assertion follows from this at once. As to the second one, put $\mathfrak{M}_i=\pi^i\mathfrak{M}$ for all $i\geqslant 0$; as these are R-modules, their quotients $\mathfrak{N}_i=\mathfrak{M}_i/\mathfrak{M}_{i+1}$ are R-modules; as $\pi\mathfrak{n}=0$ for all $\mathfrak{n}\in\mathfrak{N}_i$, \mathfrak{N}_i may be regarded as a module, i.e. as a vector-space, over the field $k=R/P$; as such, it has a dimension n_i, which depends only upon \mathfrak{M} and i. Write now \mathfrak{M} as a direct sum of modules R and R/P^v, in numbers respectively equal to N_0 and N_v; then one sees at once that $n_i=N_0+\sum_{v>i}N_v$. Therefore $N_0=n_i$ for i large enough, and $N_v=n_{v-1}-n_v$.

COROLLARY 3. *In corollary 1, the integers* r,s,v_1,\dots,v_r *depend only upon L and M.*

As s is the dimension of the subspace W generated by M, and $s-r$ is the dimension of the maximal subspace contained in M, they depend only upon M. Now put $L_1=L\cap W$, and take $i\geqslant 0$ such that $\pi^iL_1\subset M$; our assertion follows now at once from the application of corollary 2 to the R-module $M/(\pi^iL_1)$.

In corollary 2, the number of summands of \mathfrak{M} isomorphic to R is called the *rank* of \mathfrak{M}; with this definition, we have:

COROLLARY 4. *Let \mathfrak{M} be a finitely generated R-module, and \mathfrak{M}' a submodule of \mathfrak{M}. Then the rank of \mathfrak{M} is the sum of those of \mathfrak{M}' and of $\mathfrak{M}/\mathfrak{M}'$.*

As in the proof of corollary 2, write \mathfrak{M} as L/M, where L is the lattice $\sum Rv_i$ in the vector-space V with the basis $\{v_1,\dots,v_n\}$, and M is an R-module. Then the inverse image of \mathfrak{M}' in L is an R-module L_1', and the three modules in our corollary are respectively isomorphic to L/M, L_1'/M and L/L_1'. Let W, V' be the subspaces of V respectively generated by M and by L_1'; then, as corollary 1 shows at once, the ranks of L/M, L_1'/M and L/L_1' are respectively the codimensions of W in V, of W in V' and of V' in V.

§ 3. **Multiplicative structure of local fields.** Let notations be as above; then, for each integer $n\geqslant 1$, the set $1+P^n$ of the elements x of R which are $\equiv 1\ (P^n)$ is clearly an open and compact subgroup of R^\times, and these subgroups make up a fundamental system of neighborhoods of 1 in R^\times.

Moreover, th. 7 of Chap. I-4 shows that $R^\times = M^\times \cdot (1+P)$ if M^\times is any subgroup of order $q-1$ of R^\times, and th. 6 of Chap. I-4 shows that $K^\times = \Pi \cdot R^\times$ if Π is the discrete subgroup of K^\times, isomorphic to \mathbf{Z}, which is generated by any prime element π of K. In these formulas, the products are "semidirect".

From now on, until the end of this §, it will be assumed that K is a commutative p-field; then the above products are direct products, so that we may write $K^\times = \Pi \times R^\times$ and $R^\times = M^\times \times (1+P)$; moreover, by th. 7 of Chap. I-4, M^\times is now the group of roots of 1 of order prime to p in K. Consequently, the investigation of the structure of K^\times amounts to that of $1+P$.

Take any $x \in 1+P$; then, for every $a \in \mathbf{Z}$, x^a is in $1+P$, and the mapping $a \to x^a$ is a homomorphism of the additive group \mathbf{Z} into the multiplicative group $1+P$; as lemma 5 in the proof of th. 7, Chap. I-4, shows that $x^a \in 1+P^{n+1}$ whenever $a \equiv 0 \ (p^n)$, i.e. $|a|_p \leqslant p^{-n}$, this homomorphism is continuous when \mathbf{Z} is provided with the p-adic topology, i.e. that induced on \mathbf{Z} by \mathbf{Q}_p; as $1+P$ is compact, it can therefore be uniquely extended to a continuous homomorphism, which we again denote by $a \to x^a$, of the additive group \mathbf{Z}_p into the multiplicative group $1+P$. If $x \in 1+P^n$, x^a is in $1+P^n$ for all $a \in \mathbf{Z}$, hence for all $a \in \mathbf{Z}_p$. From this, using the formula $y^b(x^a)^{-1} = (yx^{-1})^b x^{b-a}$, one concludes at once, in the usual manner, that the mapping $(a,x) \to x^a$ of $\mathbf{Z}_p \times (1+P)$ into $1+P$ is continuous. One verifies then immediately that this mapping defines, on the group $1+P$, a structure of \mathbf{Z}_p-module (the "addition" of vectors being written multiplicatively, and the "scalar multiplication" by elements of \mathbf{Z}_p being written exponentially).

PROPOSITION 8. *If n is any integer prime to p, and v any integer $\geqslant 1$, $x \to x^n$ induces on $1+P^v$ an automorphism of $1+P^v$; $(K^\times)^n$ is an open subgroup of K^\times, of index $n \cdot (n, q-1)$ in K^\times; if n divides $q-1$, that index is n^2.*

The first assertion is a special case of the fact that $x \to x^a$ is an automorphism of $1+P^v$ whenever a is an invertible element of \mathbf{Z}_p; it implies that $(K^\times)^n$ is open in K^\times. Moreover, as we have seen above, K^\times is the direct product of the group Π, which is isomorphic to \mathbf{Z}, of the cyclic group M^\times of order $q-1$, and of $1+P$; therefore the index of $(K^\times)^n$ in K^\times is the product of the similar indices for Π, M^\times and $1+P$; clearly, these are respectively equal to n, to the g.c.d. $(n, q-1)$ of n and $q-1$, and to 1. This proves our proposition.

We will now determine the structure of the \mathbf{Z}_p-module $1+P$; this depends upon the characteristic of K. If K is of characteristic 0, it is a finite algebraic extension of \mathbf{Q}_p, so that, as we have observed, its maximal compact subring may be regarded as a \mathbf{Q}_p-lattice in K; th. 1

of § 2 shows then that it is the direct product of factors, all isomorphic to \mathbf{Z}_p, whose number is equal to the degree of K over \mathbf{Q}_p.

PROPOSITION 9. *Let K be a commutative p-field of characteristic 0, with the maximal compact subring R. Then there is an integer $m \geqslant 0$ such that $1 + P$, as a (multiplicatively written) \mathbf{Z}_p-module, is isomorphic to the (additively written) \mathbf{Z}_p-module $R \times (\mathbf{Z}_p/p^m \mathbf{Z}_p)$; m is then the largest integer such that K contains a primitive p^m-th root of 1.*

For any $x \in R$ and $a \in \mathbf{N}$, the binomial formula may be written as:

$$(1+x)^a = 1 + ax + ax \sum_{i=2}^{a} \binom{a-1}{i-1} x^{i-1}/i.$$

For $i \geqslant 2$, call p^h the largest power of p dividing i; if $h = 0$, $i - 1 > h$; if $h > 0$, then, as $i \geqslant p^h$, one verifies at once that $i - 1 > h$ except for $i = p = 2$, so that $2(i-1) > h$ in all cases. Therefore, in the above formula, the sum in the last term in the right-hand side is in pR whenever $x \in p^2 R$. This gives, for $x \in p^2 R$, $a \in \mathbf{N}$:

(3) $(1+x)^a \equiv 1 + ax$ $(p\,ax\,R)$,

which must remain valid, by continuity, for all $x \in p^2 R$ and $a \in \mathbf{Z}_p$, since \mathbf{N} is dense in \mathbf{Z}_p. Now call d the degree of K over \mathbf{Q}_p; by th. 1 of § 2, we can find a basis $\{v_1, \ldots, v_d\}$ of K over \mathbf{Q}_p such that $R = \sum \mathbf{Z}_p v_i$. By (3), we have now, for $1 \leqslant i \leqslant d$, $v \geqslant 1$, $a_i \in \mathbf{Z}_p$:

$$(1+p^2 v_i)^{p^{\nu-1} a_i} \equiv 1 + p^{\nu+1} a_i v_i \qquad (p^{\nu+2} R),$$

and therefore:

(4) $\displaystyle\prod_{i=1}^{d} (1+p^2 v_i)^{p^{\nu-1} a_i} \equiv 1 + p^{\nu+1} \sum_{i=1}^{d} a_i v_i \quad (p^{\nu+2} R).$

It follows from this that, if $x_1 \in p^2 R$, we can define by induction a sequence (x_1, x_2, \ldots), with $x_\nu \in p^{\nu+1} R$ for all $\nu \geqslant 1$, by putting, for each ν:

$$x_\nu = p^{\nu+1} \sum_i a_{\nu i} v_i$$

with $a_{\nu i} \in \mathbf{Z}_p$ for $1 \leqslant i \leqslant d$, and then

$$1 + x_{\nu+1} = (1+x_\nu) \prod_i (1+p^2 v_i)^{-p^{\nu-1} a_{\nu i}}.$$

It is now clear that we have

(5) $\displaystyle 1 + x_1 = \prod_i (1 + p^2 v_i)^{b_i}$

where the b_i are given, for $1 \leqslant i \leqslant d$, by

$$b_i = \sum_{v=1}^{+\infty} p^{v-1} a_{vi}.$$

This shows that, as a multiplicative \mathbf{Z}_p-module, the group $1 + p^2 R$ is generated by the d elements $1 + p^2 v_i$; as it is an open subgroup of the compact group $1 + P$, hence of finite index in $1 + P$, and as $1 + P$, as a \mathbf{Z}_p-module, is generated by the elements $1 + p^2 v_i$ and by a full set of representatives of the classes modulo $1 + p^2 R$ in $1 + P$, this implies that $1 + P$ is finitely generated. Now assume that (5) can hold with $x_1 = 0$ while the b_i are not all 0; then, taking for $v - 1$ the smallest of the orders of the b_i in \mathbf{Q}_p, we can write $b_i = p^{v-1} a_i$ with $v \geqslant 1$, $a_i \in \mathbf{Z}_p$ for $1 \leqslant i \leqslant d$, and the a_i not all in $p\mathbf{Z}_p$. Then (4) gives $\sum a_i v_i \equiv 0 \,(pR)$, i.e. $\sum (p^{-1} a_i) v_i \in R$, which contradicts the definition of the v_i. This shows that $1 + p^2 R$, as a \mathbf{Z}_p-module, is the free module generated by the $1 + p^2 v_i$, so that it is isomorphic to $(\mathbf{Z}_p)^d$. We can now apply corollary 4 of th. 2, § 2, to the \mathbf{Z}_p-modules $1 + P$ and $1 + p^2 R$. As their quotient is finite, it is of rank 0; as $1 + p^2 R$ is isomorphic to $(\mathbf{Z}_p)^d$, it is of rank d. Therefore $1 + P$ is of rank d, hence, by corollary 2 of th. 2, § 2, the direct product of d factors isomorphic to \mathbf{Z}_p and of finitely many factors, each isomorphic to a module $\mathbf{Z}_p/p^v \mathbf{Z}_p$. As the latter are finite groups, their product is the group of all elements of finite order in $1 + P$ and is itself a finite group, whose order is a power of p; it is therefore the group of all roots of 1 in $1 + P$; by lemma 1 of Chap. I-1, if p^m is the largest of the orders of its elements, it is cyclic of order p^m, hence, as a \mathbf{Z}_p-module, isomorphic to $\mathbf{Z}_p/p^m \mathbf{Z}_p$. Finally, writing K^\times as the direct product of Π, M^\times and $1 + P$, we see that any root of 1 in K whose order is a power of p must be in $1 + P$. This completes the proof.

COROLLARY. *Let K be as in proposition 9. Then, for every integer $n \geqslant 1$, $(K^\times)^n$ is an open subgroup of K^\times, of finite index in K^\times, and that index is $n \cdot (n,r) \cdot \mathrm{mod}_K(n)^{-1}$ if r is the order of the group of all roots of 1 in K.*

Clearly the latter group is the direct product of M^\times and of the group of roots of 1 in $1 + P$, which is of order p^m; therefore it is cyclic of order $r = (q - 1)p^m$, and K^\times is the direct product of Π, of that group, and of a \mathbf{Z}_p-module isomorphic to R. Now nR is an open subgroup of the additive group R, whose index in R, by the definition of mod_K, is $\mathrm{mod}_K(n)^{-1}$. The conclusion follows from this at once, by the same argument as in the proof of prop. 8.

PROPOSITION 10. *Let K be a commutative p-field of characteristic p. Then $1 + P$, as a \mathbf{Z}_p-module, is the direct product of a countably infinite family of modules isomorphic to \mathbf{Z}_p.*

By th. 8 of Chap. I-4, we may regard K as the field of formal power-series in one indeterminate π, with coefficients in the field \mathbf{F}_q with $q = p^f$ elements. Here it is easy to give explicitly a family of free generators for the \mathbf{Z}_p-module $1 + P$. In fact, take a basis $\{\alpha_1, \ldots, \alpha_f\}$ for \mathbf{F}_q over the prime field \mathbf{F}_p. As generators of $1 + P$, we take the elements $1 + \alpha_i \pi^n$, where $1 \leqslant i \leqslant f$, n running through the set of all integers > 0, prime to p. For any $N > 0$, put $N = n p^\nu$, with $\nu \geqslant 0$ and n prime to p. For any integers $a_i \geqslant 0$ $(1 \leqslant i \leqslant f)$, we have

$$\prod_{i=1}^f (1 + \alpha_i \pi^n)^{a_i p^\nu} = \prod_i (1 + \beta_i \pi^N)^{a_i} \equiv 1 + \left(\sum_i a_i \beta_i \right) \pi^N \quad (P^{N+1})$$

with $\beta_i = \alpha_i^{p^\nu}$. As $x \to x^{p^\nu}$ is an automorphism of \mathbf{F}_q over \mathbf{F}_p, the β_i also make up a basis of \mathbf{F}_q over \mathbf{F}_p; thus, for any given $\alpha \in \mathbf{F}_q$, one may, in one and only one way, choose integers a_i such that $0 \leqslant a_i < p$ and that $\sum a_i \beta_i = \alpha$. Now take any $x_1 \in P$; we define inductively a sequence (x_1, x_2, \ldots), with $x_N \in P^N$ for all $N \geqslant 1$, as follows. For each N, putting $N = n p^\nu$ with n prime to p as above, we choose the integers a_i so that $0 \leqslant a_i < p$ for $1 \leqslant i \leqslant f$ and that

$$y_N = \prod_i (1 + \alpha_i \pi^n)^{a_i p^\nu} \equiv 1 + x_N \quad (P^{N+1}),$$

which can be done in one and only one way in view of the foregoing remarks, and put then

$$1 + x_{N+1} = (1 + x_N) y_N^{-1}.$$

One sees at once, putting these formulas together, that they give for $1 + x_1$ an expression as a convergent infinite product of factors of the form $(1 + \alpha_i \pi^n)^b$, with $1 \leqslant i \leqslant f$, n prime to p, and $b \in \mathbf{Z}_p$. Moreover, the above calculations show also that this expression is unique, which proves our assertions.

§ 4. Lattices over R.

The concept of lattice, as developed for p-fields in §§ 1–2, cannot be applied to **R**-fields. The appropriate concept is here as follows:

DEFINITION 3. *By an **R**-lattice in a vector-space V of finite dimension over an **R**-field, we understand a discrete subgroup L of V such that V/L is compact.*

We have to recall here some elementary facts about discrete subgroups. Let G be a topological group, Γ a discrete subgroup of G, and φ the canonical mapping of G onto G/Γ. Then, if U is a neighborhood of the neutral element e in G, such that $U^{-1} \cdot U$ contains no element

of Γ other than e, φ induces, on each set of the form gU with $g \in G$, a homeomorphism of that set onto its image in G/Γ; one expresses this by saying that φ is a "local homeomorphism"; one may say that it is a "local isomorphism" in U if Γ is normal in G, since in that case it maps the group law in G onto the group law in G/Γ. Assume that G is locally compact, and let a right-invariant measure α be given on G. Then it is easily seen that there is one and only one measure α' on G/Γ such that, whenever X is a measurable subset of G which is mapped by φ in a one-to-one manner onto its image $X' = \varphi(X)$ in G/Γ, $\alpha'(X')$ is equal to $\alpha(X)$; in particular, this will hold for every measurable subset of every set gU, where U is as above. Then, if f is any continuous function with compact support in G, we have

$$(6) \qquad \int_G f(g)\, d\alpha(g) = \int_{G/\Gamma} \left(\sum_{\gamma \in \Gamma} f(g\gamma) \right) d\alpha'(\dot{g});$$

here we have put $\dot{g} = \varphi(g)$, and the integrand in the right-hand side, which is written as a function of g but is constant on cosets $g\Gamma$, is to be understood as a function of \dot{g}. This, in fact, is clear if the support of f is contained in any set gU, and the general case follows from this at once; also, as well known in integration theory, the validity of (6) for continuous functions with compact support implies its validity for all integrable functions, and for all measurable functions with values in \mathbf{R}_+. Clearly, α' is invariant under the action of G on G/Γ if and only if α is left-invariant; this will be so, in particular, whenever G/Γ is compact, since then G/Γ is a set of finite measure which is invariant under the action of G. Then, if at the same time Γ is normal in G, α' is a Haar measure on G/Γ.

Things being as above, α' will be called the *image* of α in G/Γ; we will denote this image simply by α when no confusion is likely. The following lemma (which takes the place of what was known as Minkowski's theorem in classical number-theory) is now obvious:

LEMMA 1. *Let G be a locally compact group with a Haar measure α; let Γ be a discrete subgroup of G, such that G/Γ is compact; let X be a measurable subset of G such that $\alpha(X) > \alpha(G/\Gamma)$. Then there are two distinct elements x, x' of X such that $x^{-1}x' \in \Gamma$.*

One should only note that, since G/Γ is compact, any right-invariant measure on G is also left-invariant; therefore the Haar measure α is bi-invariant, and its image in G/Γ is well-defined.

LEMMA 2. *Let G, α and Γ be as in lemma 1, and let Γ_1 be a discrete subgroup of G, containing Γ. Then Γ has a finite index $[\Gamma_1 : \Gamma]$ in Γ_1, and this is given by*

$$\alpha(G/\Gamma) = [\Gamma_1 : \Gamma]\, \alpha(G/\Gamma_1).$$

As G/Γ is compact, there is a continuous function $f_0 \geqslant 0$ with compact support on G, such that

$$f_1(g) = \sum_{\gamma \in \Gamma} f_0(g\gamma) > 0$$

for all $g \in G$. Then the function $f = f_0/f_1$ is continuous on G, has the same support as f_0, and is such that $\sum f(g\gamma)$, where the sum is extended to all $\gamma \in \Gamma$, is 1 for all g; this implies that the similar sum, extended to all $\gamma \in \Gamma_1$, has the constant value $[\Gamma_1 : \Gamma]$. If now we apply (6) to G, f and Γ, and also to G, f and Γ_1, we get the result in our lemma.

From these facts, we now deduce the following classical result about **R**-lattices:

PROPOSITION 11. *Let L be a subgroup of a vector-space V of dimension n over* **R**. *Then the following three statements are equivalent:* (i) L *is an* **R**-*lattice in V;* (ii) L *is discrete in V, finitely generated, and contains a basis for V over* **R**; (iii) *there is a basis $\{v_1, ..., v_n\}$ of V over* **R** *which generates the group L.*

Assume (iii), and consider the isomorphism

(7) $(x_1, ..., x_n) \to \sum x_i v_i$

of \mathbf{R}^n onto V; L is the image of \mathbf{Z}^n under that isomorphism; therefore it is discrete in V, and V/L is isomorphic to $(\mathbf{R}/\mathbf{Z})^n$, hence compact. Therefore (iii) implies (i) and (ii). Now assume (i); let W be the subspace of V generated by L, and call W' a supplementary subspace to W in V. Then V, as a locally compact group, is the direct product of W and W', and L is a discrete subgroup of W; therefore V/L is isomorphic to the direct product of W/L and W'. This cannot be compact unless W' is so; then W' must be $\{0\}$, and $W = V$, so that L contains a basis of V over **R**. Now let α be the Haar measure on V which is such that $\alpha(V/L) = 1$. For every basis $B = \{v_1, ..., v_n\}$ of V, contained in L, call φ_B the isomorphism of \mathbf{R}^n onto V defined by (7); this maps \mathbf{Z}^n onto the sublattice L_B of L generated by B, and maps the Lebesgue measure λ on \mathbf{R}^n onto some scalar multiple $m_B^{-1}\alpha$ of α. As $\lambda(\mathbf{R}^n/\mathbf{Z}^n) = 1$, we have $m_B^{-1}\alpha(V/L_B) = 1$; by lemma 2, this shows that m_B is the index of L_B in L. Now choose B so that this index has the smallest possible value; it will be shown that then $L_B = L$. In fact, assume that L contains a vector w, not in L_B, and write $w = \sum a_i v_i$ with coefficients a_i in **R**; as w is not in L_B, at least one of the a_i, say a_1, is not in **Z**; replacing then w by $w - mv_1$ with $m \in \mathbf{Z}$, $m < a_1 < m+1$, we may assume that $0 < a_1 < 1$. Now put $v_1' = w$, $v_i' = v_i$ for $2 \leqslant i \leqslant n$, and $B' = \{v_1', ..., v_n'\}$; clearly B' is a basis for V, contained in L. A trivial cal-

culation shows that $\varphi_B^{-1} \circ \varphi_{B'}$ is the automorphism of \mathbf{R}^n given by

$$(x_1, \ldots, x_n) \rightarrow (a_1 x_1, x_2 + a_2 x_1, \ldots, x_n + a_n x_1),$$

whose module is a_1 (cf. corollary 3 of th. 3, Chap. I-2). Take a measurable set X in \mathbf{R}^n; put $Y = \varphi_B(X)$ and $Y' = \varphi_{B'}(X)$. By the definition of m_B, $m_{B'}$, we have $\alpha(Y) = m_B \lambda(X)$, $\alpha(Y') = m_{B'} \lambda(X)$; therefore $\varphi_{B'} \circ \varphi_B^{-1}$, which maps Y onto Y', has the module $m_{B'}/m_B$. Since $\varphi_{B'} \circ \varphi_B^{-1}$ can be written as $\varphi_{B'} \circ (\varphi_B^{-1} \circ \varphi_{B'}) \circ \varphi_{B'}^{-1}$, it has the same module as $\varphi_B^{-1} \circ \varphi_{B'}$. Thus we get $m_{B'}/m_B = a_1 < 1$, which contradicts the definition of B. This completes the proof, as it shows that (i) implies (ii) and (iii).

§ 5. Duality over local fields.

Among the most important properties of commutative locally compact groups are those which make up the content of the "duality theory". We recall that, if G is such a group, a *character* of G (in the sense of that theory) is a continuous representation of G into the multiplicative group of complex numbers of absolute value 1. If g^* is such a character, its value at a point g of G will frequently be written as $\langle g, g^* \rangle_G$, for which we write $\langle g, g^* \rangle$ if there is no danger of confusion. We shall write the group law on G additively; and, on the set G^* of the characters of G, we put a commutative group structure, also written additively, by writing

$$\langle g, g_1^* + g_2^* \rangle_G = \langle g, g_1^* \rangle_G \cdot \langle g, g_2^* \rangle_G;$$

one should note that the neutral element of G^*, which is denoted by 0 in this additive notation, corresponds to the "trivial" character of G with the constant value 1 on G. One topologizes G^* by assigning to it the topology of uniform convergence on compact subsets of G; this makes it into a locally compact group, called *the topological dual* of G, or simply its dual if there is no danger of confusion. Conversely, the characters of G^* are the functions $g^* \rightarrow \langle g, g^* \rangle_G$, for all $g \in G$, and this determines an isomorphism between G and the dual of G^*. In other words, G may be identified with the dual of G^* by writing

$$\langle g^*, g \rangle_{G^*} = \langle g, g^* \rangle_G,$$

and it will always be tacitly assumed that they are so identified. The group G is compact if and only if G^* is discrete; therefore G is discrete if and only if G^* is compact.

If H is any closed subgroup of G, the characters of G which induce the trivial character on H make up a closed subgroup of G^*, which will be denoted by H_* and is said to be *associated with H by duality*; it is isomorphic to the dual of G/H. When G is regarded as the dual of G^*, the subgroup of G associated with H_* is then H itself, which is therefore

isomorphic to the dual of G^*/H_*. As H is open in G if and only if G/H is discrete, we see that it is so if and only if H_* is compact; consequently, H_* is open in G^* if and only if H is compact. Similarly, H is discrete if and only if G^*/H_* is compact, and G/H is compact if and only if H_* is discrete.

All this may be applied to the additive group of any left vector-space V of finite dimension over a non-discrete locally compact field K (commutative or not). In that case, if V^* is the topological dual of V, and if $v^* \in V^*$, then, for every $a \in K$, the function $v \rightarrow \langle av, v^* \rangle_V$ on V is clearly again a character of V, which we will denote by $v^* a$; one verifies at once, by going back to the definitions, that this makes V^* into a right vector-space over K; by corollary 2 of th. 3, Chap. I-2, its dimension must be finite. In other words, the structure of V^* as a right vector-space over K is defined by the formula

(8) $$\langle av, v^* \rangle_V = \langle v, v^* a \rangle_V.$$

Conversely, if V and V^* are dual groups, and V^* has a structure of right vector-space over K, (8) may be used in order to define V as a left vector-space over K. Thus we may still identify V with the dual of V^* when their structures as vector-spaces over K are taken into account. If L is any closed subgroup of V, the subgroup L_* of V^* associated with L by duality consists of the elements v^* of V^* such that $\langle v, v^* \rangle_V = 1$ for all $v \in L$; in view of (8), this implies that, if L is a left module for some subring of K, L_* is a right module for the same subring, and conversely. In particular, if K is a p-field and R is the maximal compact subring of K, L is a left R-module if and only if L_* is a right R-module. As we have seen that L is compact and open in V if and only if L_* is so in V^*, we see that L is a K-lattice if and only if L_* is one. When that is so, we say that the K-lattices L and L_* are *dual* to each other; then aL and $L_* a^{-1}$ are dual to each other for every $a \in K^\times$. On the other hand, if K is **R**, **C** or **H**, then, clearly, L is an **R**-lattice if and only if L_* is one.

On the other hand, if V is as above, we may consider its *algebraic dual* V', which is the space of K-linear forms on V; as well-known, if we denote by $[v, v']_V$ the value of the linear form v' on V at the point v of V, we can give to V' a "natural" structure of right vector-space over K by means of the formula

$$[av, v'b]_V = a[v, v']_V b,$$

valid for all $v \in V$, $v' \in V'$ and all a, b in K. If χ is any character of the additive group of K, then, for every $v' \in V'$, there is an element v^* of the topological dual V^* such that $\langle v, v^* \rangle_V = \chi([v, v']_V)$ for all $v \in V$. We shall use this operation in order to establish the relation between the algebraic and the topological dual.

THEOREM 3. *Let K be a non-discrete locally compact field, and V a left vector-space of finite dimension n over K; let χ be a non-trivial character of the additive group of K. Then the topological dual V^* of V is a right vector-space of dimension n over K; the formula*

$$\langle v,v^*\rangle_V = \chi([v,v']_V) \text{ for all } v \in V$$

defines a bijective mapping $v' \to v^$ of the algebraic dual V' of V onto V^*; if $\chi(xy)=\chi(yx)$ for all x, y in K, this mapping is an isomorphism for the structures of V', V^* as right vector-spaces over K.*

Let X_K be the topological dual of K. The structure of K as a left vector-space of dimension 1 over itself determines on X_K a structure of right vector-space over K; as such, it has a certain finite dimension d. Similarly, the structure of K as a right vector-space over K determines on X_K a structure of left vector-space of a certain dimension d' over K. Let V be as in theorem 3; by taking a basis of V over K, V can be written as the direct sum of n subspaces of dimension 1; therefore its dual V^*, as a right vector-space, is isomorphic to the direct sum of n spaces isomorphic to X_K, and has therefore the dimension nd. Similarly, the dual of V^* is a left vector-space of dimension ndd'; as it is isomorphic to V, with which we have in fact agreed to identify it, we get $ndd'=n$, hence $d=d'=1$. Now let χ be as in theorem 3; this defines an element $c^* \neq 0$ in the additively written group X_K, so that we have $\chi(t)=\langle t,c^*\rangle_K$ for all $t \in K$. As $d'=1$, every element of X_K can be uniquely written as xc^*, with $x \in K$; as $d=1$, every element of X_K can be uniquely written as c^*y, with $y \in K$. Therefore the relation $xc^*=c^*y$ determines a bijection α of K onto itself, and one verifies at once that this is an automorphism of K. In view of (8), c^*y is the character $t \to \chi(yt)$ of K, and similarly xc^* is $t \to \chi(tx)$. Therefore $\chi(tx)=\chi(\alpha(x)t)$ for all x, t in K, and this determines α uniquely; in particular, α induces the identity on the center of K, and it is the identity if and only if $\chi(tx)=\chi(xt)$ for all x, t in K. Now consider the mapping $v' \to v^*$ of V' into V^* which is defined in theorem 3; take $x \in K$, put $w'=v'x$, and assume that the mapping in question maps w' onto w^*. We have

$$\chi([v,w']_V)=\chi([v,v']_V x)=\chi(\alpha(x)[v,v']_V)=\chi([\alpha(x)v,v']_V).$$

In view of the definition of v^* and w^*, this gives

$$\langle v,w^*\rangle_V = \langle \alpha(x)v,v^*\rangle_V = \langle v,v^*\alpha(x)\rangle_V$$

for all v, hence $w^*=v^*\alpha(x)$. It is customary to express this by saying that the mapping $v' \to v^*$ is α-semilinear. At the same time, it is clearly injective;

for $v^* = 0$ means that $\chi([v,v']_V)$ is 1 for all $v \in V$, hence also that $\chi(x[v,v']_V)$ is 1 for all $x \in K$ and all $v \in V$; as χ is not trivial, this implies $[v,v']_V = 0$ for all v, hence $v' = 0$. As V' and V^* have the same dimension n over K, an α-semilinear mapping of V' into V^* cannot be injective without being bijective; this completes the proof. For purposes of reference, we formulate separately the result about the characters of K:

COROLLARY. *Let K and χ be as in theorem 3; then every character of K can be uniquely written as $t \to \chi(tx)$, with $x \in K$, and also as $t \to \chi(yt)$, with $y \in K$.*

A more "intrinsic" way of formulating theorem 3 would be to say that there is a canonical isomorphism, given by the formula in th. 3, between V^* and the tensor-product $V' \otimes_K X_K$ (and similarly between V^* and $X_K \otimes_K V'$ if V is given as a right vector-space); this will be left to the reader. One may also note that there is always a non-trivial character χ of K for which $\chi(xy) = \chi(yx)$ for all x, y; for instance, one may take $\chi = \chi_0 \circ \tau$, where τ is the "reduced trace" in K over its center K_0 (cf. Chap. IX-2), and χ_0 is a non-trivial character of K_0; the same result could be deduced from the fact that, in view of the Skolem-Noether theorem (which will be proved as prop. 4 of Chap. IX-1), α in the proof of th. 3 must be an inner automorphism of K. Of course the distinction between right and left becomes entirely superfluous if one considers only commutative fields.

It is frequently convenient, having chosen once for all a character χ of K with the properties described in theorem 3, to identify the topological and algebraic duals of every vector-space over K by means of the isomorphism described in that theorem; when doing this, one will refer to χ as "*the basic character*". In particular, K will then be identified with its own topological dual, as shown in the corollary of th. 3. When this is done for a p-field K, the subgroup of K associated by duality with each subgroup of the form P^n must be of the same form, since in general the dual of a K-lattice is a K-lattice. In order to give a more explicit statement, we set up a definition:

DEFINITION 4. *Let K be a p-field, R its maximal compact subring and P the maximal ideal of R. Then the order of a non-trivial character χ of K is the largest integer $v \in \mathbf{Z}$ such that χ is 1 on P^{-v}; it will be denoted by $\mathrm{ord}(\chi)$.*

In other words, P^{-v} is the dual K-lattice to R when K is identified with its dual by means of χ; this shows that v is finite.

PROPOSITION 12. *Let K be a p-field and χ a non-trivial character of K of order v. Then, for any $n \in \mathbf{Z}$, $\chi(xt) = 1$ for all $t \in P^n$ if and only if $x \in P^{-n-v}$.*

This is obvious, and amounts to saying that the dual K-lattice to P^n is P^{-n-v} when K is identified with its dual by means of χ.

As to the explicit construction of characters for local fields, the case of \mathbf{R} is well-known; there one may take as basic character the one given by $\chi_0(x) = e(x) = e^{2\pi i x}$; in \mathbf{C} or \mathbf{H}, one may then take as basic any character $\chi_0 \circ f$, where f is an \mathbf{R}-linear form other than 0 (e.g. the trace over \mathbf{R}). If K is a local field of characteristic p, one may write it as a field of formal power-series $x = \sum a_i T^i$ with coefficients in \mathbf{F}_q, and take as basic the character of order 0 given by $\chi(x) = \psi(a_{-1})$, where ψ is a non-trivial character of the additive group of \mathbf{F}_q. For \mathbf{Q}_p, an explicit construction will be given in Chap. IV-2, as part of the proof of theorem 3 of that Chapter.

Chapter III

Places of A-fields

§ 1. **A-fields and their completions.** By an *algebraic number-field*, it is customary to understand a finite algebraic extension of **Q**. One main object of this book, and of number-theory in general, is to study algebraic number-fields by means of their embeddings into local fields. In the last century, however, it was discovered that the methods by which this can be done may be applied with very little change to certain fields of characteristic $p > 1$; and the simultaneous study of these two types of fields throws much additional light on both of them. With this in mind, we introduce as follows the fields which will be considered from now on:

DEFINITION 1. *A field will be called an A-field if it is either a finite algebraic extension of* **Q** *or a finitely generated extension of a finite prime field* F_p, *of degree of transcendency 1 over* F_p.

Thus, if k is an A-field of characteristic $p > 1$, it must contain a transcendental element t over F_p, and it is then a finite algebraic extension of $F_p(t)$. Therefore, if once for all we denote by T an indeterminate, so that $F_p(T)$ is the field of rational functions in T with coefficients in F_p, an A-field of characteristic p is one which is isomorphic to a finite algebraic extension of $F_p(T)$. One should note that such a field always contains infinitely many fields isomorphic to $F_p(T)$.

We shall study A-fields by means of their embeddings into local fields. In view of theorems 5 and 8 of Chap. I, it is permissible, up to isomorphism, to speak of the set of all local fields. In fact, for a given $p > 1$, the local fields of characteristic p are, up to isomorphism, in a one-to-one correspondence with the finite fields F_q with $q = p^n$ elements, while the local p-fields of characteristic 0 are isomorphic to the subfields of an algebraic closure of Q_p which are of finite degree over Q_p. It will be seen later (as a consequence of lemma 1, Chap. XI-3) that there are only enumerably many fields of the latter type; this will not be needed here. It is now legitimate to speak of the set of places of an A-field according to the following definition:

DEFINITION 2. *Let* λ *be an isomorphic embedding of an A-field* k *into a local field* K; *then* (λ, K) *will be called a completion of* k *if* $\lambda(k)$ *is dense in* K. *Two completions* (λ, K), (λ', K') *of* k *will be called equivalent*

if there is an isomorphism ρ of K onto K' such that $\lambda' = \rho \circ \lambda$. By a place of k, we shall understand an equivalence class of completions of k.

DEFINITION 3. *A place of an A-field k, determined by a completion (λ, K) of k, will be called real if K is isomorphic to \mathbf{R}, imaginary if K is isomorphic to \mathbf{C}, infinite in both of these cases, and finite in all other cases.*

Let v be a place of k, as above; clearly, for all completions (λ, K) of k belonging to v, the function $\mathrm{mod}_K \circ \lambda$ on k is the same; this will be written $x \to |x|_v$. If v is imaginary, $\mathrm{mod}_K(x-y)^{1/2}$ is a distance-function on K; in all other cases, $\mathrm{mod}_K(x-y)$ is such a function. Therefore we can always obtain a completion of k, belonging to v, by taking the completion of k with respect to the distance-function $|x-y|_v^\alpha$ with $\alpha = 1/2$ if v is imaginary and $\alpha = 1$ otherwise. This completion will be denoted by k_v, and will be called *the completion of k at v*; for all $x \in k_v$, we shall write $|x|_v = \mathrm{mod}_{k_v}(x)$. If v is a finite place, we write r_v for the maximal compact subring of k_v, and p_v for the maximal ideal in r_v; these are the subsets of k_v defined respectively by $|x|_v \leqslant 1$ and by $|x|_v < 1$.

As shown by th. 5 of Chap. I-2, \mathbf{Q} has one infinite place, corresponding to the embedding of \mathbf{Q} into $\mathbf{R} = \mathbf{Q}_\infty$; this place will be denoted by ∞. The same theorem shows that the finite places of \mathbf{Q} are in a one-to-one correspondence with the rational primes, with which they will usually be identified, the place p corresponding to the embedding of \mathbf{Q} into \mathbf{Q}_p.

The knowledge of the places of \mathbf{Q} provides us with a starting point for determining the places of algebraic number-fields, considered as finite algebraic extensions of \mathbf{Q}. In order to proceed in the same way for A-fields of characteristic $p > 1$, we have to know the places of $\mathbf{F}_p(T)$. Before determining them, we first give some general results about places of algebraic extensions.

PROPOSITION 1. *Let k be any field, k_0 an infinite subfield of k, and λ an isomorphic embedding of k into a local field K. Then the closure K_0 of $\lambda(k_0)$ in K is a local field, and the closure of $\lambda(k)$ in K is the field generated by $\lambda(k)$ over K_0.*

The first assertion follows at once from corollary 3 of prop. 2, Chap. I-2. Then, by corollary 2 of th. 3, Chap. I-2, K must have a finite degree over K_0, so that, by th. 3, Chap. I-2, every vector-space over K_0 in K is closed in K. The field K_1 generated by $\lambda(k)$ over K_0 is such a vector-space; on the other hand, the closure of $\lambda(k)$ in K is clearly a field, and it contains $\lambda(k_0)$, hence K_0, and $\lambda(k)$; therefore it is K_1.

COROLLARY. *Let k be an A-field, k' a finite algebraic extension of k, and w a place of k'. Let λ be the natural injection of k' into its completion*

k'_w at w. Then k'_w is a finite algebraic extension of the closure of $\lambda(k)$ in k'_w, and the injection of k into that closure, induced on k by λ, determines a place v of k.

In view of our definitions, this is a special case of prop. 1; it enables us to set up the following definition:

DEFINITION 4. *If k, k', w and v are as in the corollary of prop. 1, we say that v is the place of k which lies below w, and that w lies above v; and we write $w|v$.*

When that is so, we shall usually identify k_v with the closure of k in k'_w.

THEOREM 1. *Let k be an A-field, k' a finite algebraic extension of k, and v a place of k. Then there is a place of k' which lies above v, and there are only finitely many such places.*

Let K be an algebraic closure of k_v, and k'' the algebraic closure of k in K; as k'' is algebraically closed, there is at least one isomorphism λ of k' into k'' over k. Call K_λ the field generated by $\lambda(k')$ over k_v; this is a finite algebraic extension of k_v, so that, by corollary 1 of th. 3, Chap. I-2, we can give it its topological structure as a vector-space of finite dimension over k_v; this makes it into a local field. Then, by prop. 1, (λ, K_λ) is a completion of k', and it determines a place of k' which clearly lies above v. Conversely, let w be any place of k' above v. Then, by the corollary of prop. 1, k'_w is algebraic over k_v, so that there is at least one isomorphism φ of k'_w into K over k_v; let λ be the isomorphism of k' into K induced on k' by φ; clearly λ maps k' into k''. By prop. 1, k'_w is generated by k' over k_v, so that $\varphi(k'_w)$ is the same as the field denoted above by K_λ; moreover, again by corollary 1 of th. 3, Chap. I-2, φ is a topological isomorphism of k'_w onto K_λ, so that w is the same as the place of k' determined by the completion (λ, K_λ) of k'. Thus there are at most as many places of k' above v as there are distinct isomorphisms λ of k' into k'' over k. As k' is a finite algebraic extension of k, it is well-known (and easily proved) that there are only finitely many such isomorphisms.

COROLLARY. *An A-field has at most a finite number of infinite places; it has at least one if it is of characteristic 0, and none otherwise.*

The last assertion is obvious; the others are a special case of th. 1, since clearly a place of an A-field of characteristic 0 is infinite if and only if it lies above the place ∞ of \mathbf{Q}.

Now we proceed to the determination of the places of $\mathbf{F}_p(T)$; more generally, we will determine those of $\mathbf{F}_q(T)$, where \mathbf{F}_q is any finite field.

It will be convenient to say that a polynomial π in $\mathbf{F}_q[T]$ is *prime* if it is monic and irreducible in $\mathbf{F}_q[T]$ and if its degree is >0.

THEOREM 2. *The field $k=\mathbf{F}_q(T)$ has one and only one place v for which $|T|_v>1$; for this place, T^{-1} is a prime element of k_v, and the module of k_v is q. For each prime polynomial π in $\mathbf{F}_q[T]$, k has one and only one place v such that $|\pi|_v<1$; for this place, π is a prime element of k_v, and the module of k_v is q^δ if δ is the degree of π. All these places are distinct, and k has no other place.*

Let v be a place of k. Assume first that $|T|_v \leqslant 1$; then $\mathbf{F}_q[T]$ is contained in r_v. Call ρ the canonical homomorphism of r_v onto the finite field r_v/p_v; it induces on $\mathbf{F}_q[T]$ a homomorphism of $\mathbf{F}_q[T]$ onto its image, whose kernel $p_v \cap \mathbf{F}_q[T]$ is clearly a prime ideal in $\mathbf{F}_q[T]$. As $\mathbf{F}_q[T]$ is infinite, and r_v/p_v is finite, this ideal cannot be $\{0\}$; therefore it is the ideal $\pi \cdot \mathbf{F}_q[T]$ generated in $\mathbf{F}_q[T]$ by some prime polynomial π. Then $|\pi|_v<1$, and we have $|\alpha|_v=1$ for every polynomial α prime to π in $\mathbf{F}_q[T]$. Every $\xi\in k^\times$ can be written in the form $\xi=\pi^n\alpha/\alpha'$ with $n\in\mathbf{Z}$ and α,α' in $\mathbf{F}_q[T]$ and prime to π; when ξ is so written, we have $|\xi|_v=|\pi|_v^n$; in particular, ξ is in r_v if and only if $n\geqslant 0$, i.e. if and only if it can be written as $\xi=\beta/\alpha$ with α,β in $\mathbf{F}_q[T]$ and α prime to π. As $\mathbf{F}_q(T)$ is dense in k_v, the range of values taken by $|x|_v$ is the same on $\mathbf{F}_q(T)$ as on k_v; this implies that π is a prime element of k_v. Now let δ be the degree of π. The image of $\mathbf{F}_q[T]$ in r_v/p_v is isomorphic to $\mathbf{F}_q[T]/\pi \cdot \mathbf{F}_q[T]$, which is an extension of \mathbf{F}_q of degree δ, hence a field with q^δ elements; clearly the image of every element of $r_v \cap \mathbf{F}_q(T)$ must then be in that same field, which is therefore no other than r_v/p_v, since $\mathbf{F}_q(T)$ is dense in k_v; this shows that q^δ is the module of k_v, and we have $|\pi|_v=q^{-\delta}$. Consequently, the function $|\xi|_v$ on k is uniquely determined by π, so that, when π is given, there can be at most one place v of k with the properties we have described. Assume now that $|T|_v>1$; then $|T^{-1}|_v<1$, and we may proceed exactly as before, substituting the ring $\mathbf{F}_q[T^{-1}]$ for $\mathbf{F}_q[T]$, and T^{-1} for π; then it is easily seen that, if $\xi=\beta/\alpha$ with polynomials α,β in $\mathbf{F}_q[T]$, other than 0, of respective degrees a,b, we have $|\xi|_v=q^{b-a}$. It is now clear that, if π is any prime polynomial, $|\pi|_v$ cannot be <1 except for the place v described above, if there is such a place, and that the same holds for T^{-1}. In order to show the existence of those places, take first the case $\pi=T$; then the ring $\mathbf{F}_q[T]$ can be embedded in an obvious manner into the ring of formal power-series $\sum\limits_{0}^{\infty} a_i T^i$ with coefficients in \mathbf{F}_q; clearly, if we extend this to the corresponding fields, we get a place of k, corresponding to $\pi=T$. Exchanging T with T^{-1}, we get the same result for T^{-1}. Now take a prime polynomial π of degree δ; then $\mathbf{F}_q(T)$ contains

the field $\mathbf{F}_q(\pi)$ and is algebraic over it; its degree d over $\mathbf{F}_q(\pi)$ is $\leqslant \delta$. As we have just proved, there is a place w of $\mathbf{F}_q(\pi)$ for which $|\pi|_w = q^{-1}$. By th. 1, $\mathbf{F}_q(T)$ has a place v lying above w. By corollary 2 of th. 3, Chap. I-2, we have then $|\pi|_v = |\pi|_w^d = q^{-d}$. This completes our proof, and shows also, incidentally, that $d = \delta$.

COROLLARY. *Notations being as in theorem 2, let v be the place of k corresponding to the prime polynomial π of degree δ. Then the polynomials of degree $< \delta$ in $\mathbf{F}_q[T]$ make up a full set of representatives of the classes in r_v modulo p_v.*

This follows at once from what has been proved above and from the fact that these polynomials make up a full set of representatives of the classes in $\mathbf{F}_q[T]$ modulo π.

From now on, it will be convenient to say that a property, involving a place of an A-field k, holds *for almost all places of k* (or, if no confusion is likely, that it holds *almost everywhere*) if it holds for all except a finite number of such places. This will be of use, for instance, in formulating our next result.

THEOREM 3. *Let k be an A-field and ξ any element of k. Then $|\xi|_v \leqslant 1$ for almost all places v of k.*

This is clear for $k = \mathbf{Q}$, since we can then write $\xi = a/b$ with a, b in \mathbf{Z} and $b \neq 0$, and $|\xi|_p \leqslant 1$ for all the primes p which do not divide b. Now let k be an A-field of characteristic 0, i.e. an algebraic number-field. Then ξ satisfies an equation

$$\xi^n + a_1 \xi^{n-1} + \cdots + a_n = 0$$

with coefficients a_i in \mathbf{Q}. Let P be the finite set consisting of ∞ and of all the primes which occur in the denominators of the a_i. By th. 1, the set P' of the places of k which lie above the places of \mathbf{Q} belonging to P is finite. Take any place v of k, not in P'; then the place p of \mathbf{Q} which lies below v is not in P, so that $|a_i|_p \leqslant 1$ for $1 \leqslant i \leqslant n$; therefore ξ is integral over \mathbf{Z}_p. By prop. 6 of Chap. I-4, this implies that ξ is in r_v, i.e. that $|\xi|_v \leqslant 1$. For an A-field k of characteristic $p > 1$, one could give a similar proof; one may also proceed as follows. If ξ is algebraic over the prime field, we have $|\xi|_v = 1$ or 0 for all v, according as $\xi \neq 0$ or $\xi = 0$. If not, k is algebraic over $\mathbf{F}_p(\xi)$. Let v be a place of k, and let w be the place of $\mathbf{F}_p(\xi)$ lying below it. By corollary 2 of th. 3, Chap. I-2, $|\xi|_v > 1$ if and only if $|\xi|_w > 1$. By th. 2, $\mathbf{F}_p(\xi)$ has only one place w with that property. In view of th. 1, this completes the proof.

COROLLARY 1. *Let E be a finite-dimensional vector-space over an A-field k. Let $\varepsilon, \varepsilon'$ be two finite subsets of E, both containing bases of E over k. For each finite place v of k, put $E_v = E \otimes_k k_v$, and call $\varepsilon_v, \varepsilon'_v$ the r_v-modules respectively generated by ε and by ε' in E_v. Then, for almost all v, $\varepsilon_v = \varepsilon'_v$.*

Here, as on all similar occasions from now on, it is understood that E is regarded as embedded in E_v by means of the injection $e \rightarrow e \otimes 1_{k_v}$. Put $\varepsilon = \{e_1, ..., e_r\}$ and $\varepsilon' = \{e'_1, ..., e'_s\}$. As ε contains a basis for E over k, we may write (perhaps not uniquely) $e'_j = \sum c_{ji} e_i$ for $1 \leqslant j \leqslant s$, with coefficients c_{ji} in k. Then $\varepsilon'_v \subset \varepsilon_v$ whenever all the $|c_{ji}|_v$ are $\leqslant 1$, hence for almost all v. Interchanging ε and ε', we get the assertion in our corollary.

COROLLARY 2. *Let \mathscr{A} be a finite-dimensional algebra over an A-field k. Let α be a finite subset of \mathscr{A}, containing a basis of \mathscr{A} over k. For each finite place v of k, put $\mathscr{A}_v = \mathscr{A} \otimes_k k_v$, and call α_v the r_v-module generated by α in \mathscr{A}_v. Then, for almost all v, α_v is a compact subring of \mathscr{A}_v.*

Put $\alpha = \{a_1, ..., a_r\}$ and $\alpha' = \{1, a_1, ..., a_r\}$. As α contains a basis of \mathscr{A} over k, we may write $a_i a_j = \sum c_{ijh} a_h$ for $1 \leqslant i, j \leqslant r$, with coefficients c_{ijh} in k. Then α'_v is a subring of \mathscr{A}_v whenever all the $|c_{ijh}|_v$ are $\leqslant 1$, hence for almost all v; obviously, it is compact; and $\alpha_v = \alpha'_v$ for almost all v.

§ 2. Tensor-products of commutative fields.

If k is an A-field and k' a finite algebraic extension of k, the proof of theorem 1 gives a construction for the places of k' which lie above a given place of k. This will now be replaced by another one, based on the consideration of the tensor-product $k' \otimes_k k_v$. To simplify matters, we shall deal only with the case where k' is separable over k; this is adequate for our purposes because of the following lemma:

LEMMA 1. *Every A-field of characteristic $p > 1$ is isomorphic to a separably algebraic extension of $F_p(T)$ of finite degree.*

Let k be such a field; write it as $F_p(x_1, ..., x_N)$, where at least one of the x_i, say x_1, has to be transcendental over F_p. We will prove, by induction on N, that there is an x_i such that k is separable over $F_p(x_i)$. This is clear if $N = 1$, and also if $x_2, ..., x_N$ are all algebraic over F_p, since in that case, by th. 2 of Chap. I-1, they are separable over F_p, so that k is separable over $F_p(x_1)$. If that is not so, then, by the induction assumption, $F_p(x_2, ..., x_N)$ is separable over $F_p(x_i)$ for some $i \geqslant 2$, say over $F_p(x_2)$, so that k itself is separable over $F_p(x_1, x_2)$. As k has the degree of transcendency 1 over F_p, there is an irreducible polynomial Φ in $F_p[X_1, X_2]$ such that $\Phi(x_1, x_2) = 0$. Then Φ is not of the form Φ'^p with Φ' in $F_p[X_1, X_2]$; as every element α of F_p satisfies $\alpha^p = \alpha$, this is the same as to say that Φ

contains at least one term $\alpha X_1^a X_2^b$ where $\alpha \neq 0$ and a or b is prime to p. If for instance a is prime to p, x_1 is separable over $\mathbf{F}_p(x_2)$, so that also k is separable over $\mathbf{F}_p(x_2)$.

In the rest of this §, we shall be concerned with the purely algebraic properties of tensor-products of the form $k' \otimes_k K$, where k is any field, k' a separably algebraic extension of k of finite degree, and K is any field containing k; in § 4, this will be applied to the case where k is an A-field and K a completion of k. We dispose first of a side-issue.

LEMMA 2. *If a commutative ring B can be written as a direct sum of fields, it can be so written in only one way; and a homomorphism of B into a field must be 0 on all except one of the summands of B.*

Let B be the direct sum of the fields K_1, \ldots, K_r; put $e_i = 1_{K_i}$. Then $K_i = e_i B$, and B has the unit-element $1_B = \sum e_i$. Clearly the solutions of the equation $X^2 = X$ in B (the "idempotents" of B) are the partial sums of the sum $\sum e_i$; consequently the e_i are uniquely characterized as those among the solutions of $X^2 = X$ in B which cannot be written as $e + e'$, where e, e' are solutions of $X^2 = X$, other than 0. If f is a homomorphism of B into a field K', it must map each e_i onto a solution of $X^2 = X$ in K', hence onto 1 or 0. If $f(e_i) = 1$, then $f(e_j) = 0$ for all $j \neq i$, since $e_i e_j = 0$ for $i \neq j$; this implies that f is 0 on K_j.

PROPOSITION 2. *Let k be a field and $k' = k(\xi)$ a separable extension of k generated by a root ξ of an irreducible monic polynomial F of degree n in $k[X]$. Let K be a field containing k; let F_1, \ldots, F_r be the irreducible monic polynomials in $K[X]$ such that $F = F_1 \ldots F_r$, and, for each i, let ξ_i be a root of F_i in some extension of K. Then the algebra $A = k' \otimes_k K$ over K is isomorphic to the direct sum of the fields $K(\xi_i)$.*

As k' is separable over k, F is without multiple roots in all extensions of k, so that the F_i are all distinct. Call ρ the k-linear homomorphism of the ring $k[X]$ onto k', with the kernel $F \cdot k[X]$, which maps X onto ξ; this can be uniquely extended to a K-linear homomorphism ρ' of $K[X]$ onto A, which has then the kernel $F \cdot K[X]$ and determines an isomorphism of $A' = K[X]/F \cdot K[X]$ onto A. We will now show that A' is isomorphic to the direct sum B of the algebras $B_i = K[X]/F_i \cdot K[X]$ over K; as these are respectively isomorphic to the fields $K(\xi_i)$ in our proposition, our proof will then be complete. Let f be any element of $K[X]$; call \bar{f} its image in A', and \bar{f}_i its image in B_i for every i. Clearly each \bar{f}_i is uniquely determined by \bar{f}, so that $\bar{f} \to (\bar{f}_1, \ldots, \bar{f}_r)$ is a homomorphism φ of A' into B. As the F_i are mutually prime, it is well-known (and easy to prove, by induction on r) that there are polynomials p_1, \ldots, p_r

in $K[X]$ such that $F^{-1}=\sum p_i F_i^{-1}$; this implies, for all i and all $j\neq i$:

(1) $\qquad\qquad p_i F_i^{-1} F\equiv 1 \quad (F_i);\qquad\qquad p_i F_i^{-1} F\equiv 0 \quad (F_j).$

Take r polynomials $f_1,...,f_r$ in $K[X]$; for each i, call \bar{f}_i the image of f_i in B_i; put $f=\sum p_i F_i^{-1} F f_i$, and call \bar{f} the image of f in A'; then \bar{f} is uniquely determined by the \bar{f}_i, so that $(\bar{f}_1,...,\bar{f}_r)\rightarrow\bar{f}$ is a mapping ψ of B into A'. Clearly $\psi\circ\varphi$ is the identity on A', and (1) shows that $\varphi\circ\psi$ is the identity on B. Therefore φ is an isomorphism of A' onto B.

Let k, k' and K be as in proposition 2. Clearly an isomorphism λ of k' into an extension K' of K induces the identity on k if and only if it is k-linear. Such an isomorphism will be called *proper above K* if K' is generated by $\lambda(k')$ over K; then (λ,K') will be called a *proper embedding of k' above K*. Two such embeddings (λ,K'), (λ',K'') will be called *equivalent* if there is a K-linear isomorphism ρ of K' onto K'' such that $\lambda'=\rho\circ\lambda$. One will notice that these are the algebraic concepts underlying definition 2 and proposition 1 of § 1.

PROPOSITION 3. *Let k be a field, k' a separably algebraic extension of k of finite degree n, and K a field containing k; put $A=k'\otimes_k K$. Then, up to equivalence, there are only finitely many proper embeddings (λ_i, K_i) $(1\leqslant i\leqslant r)$ of k' above K; the sum of the degrees of the K_i over K is n. The mapping $(\lambda_1,...,\lambda_r)$ of k' into the direct sum B of the fields K_i is a k-linear isomorphism of k' into B, and its K-linear extension φ to A is an isomorphism of A onto B.*

We may write $k'=k(\xi)$, and then, calling F the irreducible monic polynomial in $k[X]$ with the root ξ, apply prop. 2 to k, k', ξ, F and K; this shows that there is a K-linear isomorphism φ of A onto the direct sum B of certain fields K_i. For each i, call β_i the projection from B to K_i; then $\mu_i=\beta_i\circ\varphi$ is a K-linear isomorphism of A onto K_i, and μ_i induces on k' a k-linear isomorphism λ_i of k' into K_i. Clearly μ_i is the K-linear extension of λ_i to A, so that φ, which is the same as $(\mu_1,...,\mu_r)$, is the K-linear extension of $(\lambda_1,...,\lambda_r)$ to A. If λ_i was not proper above K, there would be a field $K''\neq K_i$, between K and K_i, such that λ_i would map k' into K''; then μ_i would map A into K'', and not onto K_i. Now let λ be any k-linear isomorphism of k' into a field K' containing K, and call μ the K-linear extension of λ to A; μ is then a homomorphism of A into K', so that $\mu\circ\varphi^{-1}$ is a homomorphism of B into K'. By lemma 2, this is 0 on all except one of the summands K_i of B, so that we can write it as $\sigma\circ\beta_i$, where σ is a K-linear homomorphism of K_i into K'; as these are fields, and as σ is not 0, σ must be an isomorphism of K_i onto its image K_i' in K'. This gives $\mu=\sigma\circ\mu_i$, hence $\lambda=\sigma\circ\lambda_i$; if $K_i'\neq K'$, λ, which maps k' into K_i', is not proper; therefore, if λ is proper, σ is an iso-

morphism of K_i onto K', so that (λ, K') is equivalent to (λ_i, K_i). Finally, if at the same time we had $\lambda = \sigma' \circ \lambda_j$ with $j \neq i$, σ' being an isomorphism of K_j into K', this would imply $\mu = \sigma' \circ \mu_j$, hence $\mu \circ \varphi^{-1} = \sigma' \circ \beta_j$, and $\mu \circ \varphi^{-1}$ would not be 0 on K_j. In particular, if λ is proper, (λ, K') is not equivalent to more than one of the embeddings (λ_i, K_i); this shows that the latter are all inequivalent, which completes our proof.

COROLLARY 1. *Notations being as above, let λ be any k-linear isomorphism of k' into a field K' containing K. Then there is a unique i, and a unique isomorphism σ of K_i into K', such that $\lambda = \sigma \circ \lambda_i$.*

This was proved above; it is also an immediate consequence of proposition 3 and of the fact that, if K'' is the subfield of K' generated by $\lambda(k')$ over K, (λ, K'') is a proper embedding of k' above K, so that it must be equivalent to one of the (λ_i, K_i).

COROLLARY 2. *Notations being as above, assume also that k' is a Galois extension of k, with the Galois group G. Let (λ, K') be any proper embedding of k' above K. Then K' is a Galois extension of K; to every automorphism ρ of K' over K, there is a unique $\sigma \in G$ such that $\rho \circ \lambda = \lambda \circ \sigma$, and $\rho \to \sigma$ is an isomorphism of the Galois group of K' over K onto a subgroup H of G. The proper embeddings of k' above K, up to equivalence, are all of the form $(\lambda \circ \sigma, K')$ with $\sigma \in G$; if σ, σ' are in G, $(\lambda \circ \sigma', K')$ is equivalent to $(\lambda \circ \sigma, K')$ if and only if $\sigma' \in H \sigma$.*

Clearly $\lambda(k')$ is a Galois extension of k; as K' is generated by $\lambda(k')$ over K, this implies that K' is a Galois extension of K, and that the restriction to $\lambda(k')$ of the automorphisms of K' over K defines an injective morphism of the Galois group H_1 of K' over K into that of $\lambda(k')$ over k; this is equivalent to the first part of our corollary. For $\sigma \in G$, $(\lambda \circ \sigma, K')$ is obviously a proper embedding of k' above K; if σ, σ' are in G, $(\lambda \circ \sigma', K')$ is equivalent to $(\lambda \circ \sigma, K')$ if and only if there is an automorphism ρ of K' over K such that $\lambda \circ \sigma' = \rho \circ \lambda \circ \sigma$, i.e. $\rho \circ \lambda = \lambda \circ (\sigma' \circ \sigma^{-1})$; this is so if and only if $\sigma' \circ \sigma^{-1}$ is in H. Therefore the number of inequivalent proper embeddings of that form is equal to the index of H in G, i.e. to n/n' if n, n' are the degrees of k' over k, and of K' over K, respectively. By proposition 3, the sum of the degrees of the fields K_i over K, in any set of inequivalent proper embeddings (λ_i, K_i) of k' above K, must be $\leqslant n$; therefore, up to equivalence, there can be none except those of the form $(\lambda \circ \sigma, K')$.

A useful special case of corollary 2 is that in which k' is a subfield of K', generating K' over K; one may then take for λ the identity; the proper embeddings of k' above K can all be written in the form (σ, K'), with $\sigma \in G$, and the morphism $\rho \to \sigma$ of the Galois group of K' over K

into that of k' over k is the restriction to k' of the automorphisms of K' over K.

COROLLARY 3. *Let k and k' be as in proposition 3, and let K be an algebraically closed or separably algebraically closed field containing k. Then there are n, and no more than n, distinct k-linear isomorphisms $\lambda_1,...,\lambda_n$ of k' into K; they are linearly independent over K; if λ, λ' are any two of them, and K is an algebraic closure of k, there is an automorphism α of K such that $\lambda' = \alpha \circ \lambda$.*

A field K is said to be separably algebraically closed if it has no separably algebraic extension, other than itself. The first assertion in our corollary, which is obvious, is inserted here for the sake of reference, and as an illustration of proposition 3, of which it is a special case; in fact, if K is as in our corollary, all the K_i in that proposition must be the same as K. The second assertion (a well-known theorem, due to Dedekind, and easily proved directly) can be deduced as follows from proposition 3. Assume that $\sum c_i \lambda_i = 0$, i.e. that $\sum c_i \lambda_i(\xi) = 0$ for all $\xi \in k'$, with $c_i \in K$ for $1 \leqslant i \leqslant n$. The μ_i and β_i being as in the proof of proposition 3, this implies $\sum c_i \mu_i = 0$, hence $\sum c_i \beta_i = 0$, which is clearly impossible unless all the c_i are 0. The last assertion, also inserted here for the sake of reference, follows at once from the unicity, up to an isomorphism, of the algebraic closure of k, which implies that each λ_i can be extended to an isomorphism of an algebraic closure \bar{k} of k' onto K.

COROLLARY 4. *Assumptions and notations being as in corollary 3, assume also that k' is a Galois extension of k. Then all the λ_i map k' onto the same subfield of K.*

This follows at once from corollary 2.

§ 3. **Traces and norms.** We first recall the concept of "polynomial mapping". Let E, E' be two vector-spaces of finite dimension over a field k with infinitely many elements; let $\varepsilon = \{e_1,...,e_n\}$ and $\varepsilon' = \{e'_1,...,e'_m\}$ be bases for these spaces over k. Then a mapping f of E into E' is called a *polynomial mapping* if there are polynomials P_j in $k[X_1,...,X_n]$ such that

$$f\left(\sum_i x_i e_i\right) = \sum_j P_j(x_1,...,x_n)e'_j$$

for all values of the x_i in k. This is clearly independent of the choice of the bases $\varepsilon, \varepsilon'$; moreover, since k has infinitely many elements, the polynomials P_j are uniquely determined by f, ε and ε'. If $E' = k$, f is called a polynomial function; the degree of the corresponding polynomial P is then independent of ε and is called the degree of f. If K is any field

containing k, put $E_K = E \otimes_k K$ and $E'_K = E' \otimes_k K$; then there is one and only one polynomial mapping of E_K into E'_K which coincides with f on E; this will be called the extension of f to E_K and E'_K (or more briefly to K) and will again be denoted by f; with respect to the bases $\varepsilon, \varepsilon'$ of E_K, E'_K over K, it is given by the same polynomials P_j as before.

If E is as above, we write $\mathrm{End}(E)$ for the ring of endomorphisms of E, considered as an algebra over k. If $a \in \mathrm{End}(E)$, we write $\mathrm{tr}(a)$ and $\det(a)$ for the trace and the determinant of a; the former is a linear form, and the latter is a polynomial function of degree equal to the dimension of E, on $\mathrm{End}(E)$ considered as a vector-space over k.

Now let \mathscr{A} be an algebra of finite dimension over k; as always, it is tacitly assumed to have a unit element 1. For every $a \in \mathscr{A}$, call $\rho(a)$ the endomorphism $x \to ax$ of \mathscr{A} when \mathscr{A} is viewed as a vector-space over k; writing $\mathrm{End}(\mathscr{A})$ for the algebra of all endomorphisms of that vector-space, one may thus consider ρ as a homomorphism of \mathscr{A} into $\mathrm{End}(\mathscr{A})$; it is known as the *regular representation* of \mathscr{A}; as \mathscr{A} has a unit, it is an isomorphism of \mathscr{A} onto a subalgebra of $\mathrm{End}(\mathscr{A})$. The trace and the determinant of ρ are known as *the regular trace* and *the regular norm*, taken in \mathscr{A} over k, and are denoted by $\mathrm{Tr}_{\mathscr{A}/k}$ and $N_{\mathscr{A}/k}$, or (when there can be no confusion) by Tr and by N, respectively; the former is a linear form on \mathscr{A} viewed as a vector-space over k, and the latter is a polynomial function, of degree equal to the dimension of \mathscr{A} over k. If K is a field containing k, and \mathscr{A} is extended to the algebra $\mathscr{A}_K = \mathscr{A} \otimes_k K$ over K, the regular trace and the regular norm in \mathscr{A}_K over K are the extensions of $\mathrm{Tr}_{\mathscr{A}/k}$ and $N_{\mathscr{A}/k}$ to \mathscr{A}_K, and will still be denoted by $\mathrm{Tr}_{\mathscr{A}/k}$ and $N_{\mathscr{A}/k}$. When \mathscr{A} is a field k' of finite degree over k, one drops the word "regular" and calls $\mathrm{Tr}_{k'/k}$, $N_{k'/k}$ the trace and the norm in k' over k. These concepts will now be applied to the situation described in § 2.

PROPOSITION 4. *Let k be a field, k' a separably algebraic extension of k of finite degree n, and K a field containing k. Put $A = k' \otimes_k K$; let $(\lambda_i, K_i)_{1 \le i \le r}$ be a maximal set of inequivalent proper embeddings of k' above K, and let μ_i, for each i, be the K-linear extension of λ_i to A. Then, for all $a \in A$:*

$$\mathrm{Tr}_{k'/k}(a) = \sum_{i=1}^r \mathrm{Tr}_{K_i/K}(\mu_i(a)), \qquad N_{k'/k}(a) = \prod_{i=1}^r N_{K_i/K}(\mu_i(a)).$$

In fact, let notations be the same as in prop. 3 of § 2 and its proof, and put $b = \varphi(a)$. For every i, b has the projection $\beta_i(b) = \mu_i(a)$ on K_i. Then $\mathrm{Tr}_{k'/k}(a)$ and $N_{k'/k}(a)$ are the trace and the determinant of $y \to by$ regarded as an endomorphism of B. Taking for B a basis consisting of the union of bases for the K_i over K, we get the formula in proposition 4.

COROLLARY 1. *If k and k' are as in proposition 4, the k-linear form $Tr_{k'/k}$ on k' is not 0.*

In proposition 4, take for K an algebraically closed field containing k; then $K_i = K$ for all i, and proposition 4 gives $Tr_{k'/k}(a) = \sum \mu_i(a)$. With the same notations as before, put $b = \varphi(a)$, hence $\beta_i(b) = \mu_i(a)$; as the projections $\beta_i(b)$ of b on the summands of B can be chosen arbitrarily, we can choose them so that $Tr_{k'/k}(a)$ is not 0. As $Tr_{k'/k}$ on A is the extension to A of the k-linear form $Tr_{k'/k}$ on k', and the former is not 0, the latter is not 0.

COROLLARY 2. *Notations and assumptions being as in proposition 4, we have, for all $x \in k'$:*

$$Tr_{k'/k}(x) = \sum_i Tr_{K_i/K}(\lambda_i(x)), \quad N_{k'/k}(x) = \prod_i N_{K_i/K}(\lambda_i(x)).$$

COROLLARY 3. *Let k, k' be as in proposition 4; let K be an algebraically closed field containing k, and call $\lambda_1, \ldots, \lambda_n$ the distinct k-linear isomorphisms of k' into K. Then, for all $x \in k'$:*

$$Tr_{k'/k}(x) = \sum_i \lambda_i(x), \quad N_{k'/k}(x) = \prod_i \lambda_i(x).$$

This follows at once from proposition 4 and corollary 3 of prop. 3, § 2.

COROLLARY 4. *Let k and k' be as in proposition 4, and let k'' be a separably algebraic extension of k' of finite degree. Then:*

$$Tr_{k''/k} = Tr_{k'/k} \circ Tr_{k''/k'}, \quad N_{k''/k} = N_{k'/k} \circ N_{k''/k'}.$$

Take for K an algebraic closure of k''; define the λ_i as in corollary 3; similarly, call n' the degree of k'' over k', and call λ'_j, for $1 \leqslant j \leqslant n'$, the distinct k'-linear isomorphisms of k'' into K. Each λ_i can be extended to an automorphism φ_i of K. Put $\lambda''_{ij} = \varphi_i \circ \lambda'_j$ for $1 \leqslant i \leqslant n$, $1 \leqslant j \leqslant n'$; these are k-linear isomorphisms of k'' into K. Clearly $\lambda''_{ij} = \lambda''_{hl}$ implies $i = h$, since λ''_{ij} induces λ_i on k', and $j = l$, since $\varphi_i^{-1} \circ \lambda''_{ij} = \lambda'_j$. Moreover, if λ'' is any k-linear isomorphism of k'' into K, it must induce on k' one of the isomorphisms λ_i, and then $\varphi_i^{-1} \circ \lambda''$ is k'-linear and must be one of the λ'_j, so that $\lambda'' = \lambda''_{ij}$. Now corollary 3 gives, for $x \in k''$:

$$Tr_{k''/k}(x) = \sum_{i,j} \lambda''_{ij}(x) = \sum_i \varphi_i \left(\sum_j \lambda'_j(x) \right)$$
$$= \sum_i \varphi_i(Tr_{k''/k'}(x)) = \sum_i \lambda_i(Tr_{k''/k'}(x)) = Tr_{k'/k}(Tr_{k''/k'}(x)).$$

This proves our first assertion. The formula for the norm can be proved in exactly the same manner.

For the sake of completeness, we will also deal briefly with the trace and the norm for inseparable extensions. Let k' be any algebraic extension of k of finite degree; it is well known that it contains a unique maximal separable extension k'_0 of k, and that it is purely inseparable over it; let $q = p^m$ be the degree of k' over k'_0, p being the characteristic; it is easily seen that $x^q \in k'_0$ for all $x \in k'$. Take a basis $\{\xi_1, \ldots, \xi_q\}$ of k' over k'_0; take $a \in k'$. Then k', as a vector-space over k, is the direct sum of the subspaces $\xi_i k'_0$ for $1 \leqslant i \leqslant q$, and these are invariant under $x \to a^q x$ since $a^q \in k'_0$. Therefore we have

$$N_{k'/k}(a^q) = N_{k'_0/k}(a^q)^q,$$

which obviously implies

$$N_{k'/k}(a) = N_{k'_0/k'}(a^q).$$

Call n_0 the degree of k'_0 over k, so that the degree of k' over k is $n = n_0 q$. If K is an algebraically closed field containing k, each k-linear isomorphism of k'_0 into K can be uniquely extended to one of k' into K; therefore, by corollary 3 of prop. 3, § 2, there are n_0 such isomorphisms λ_i $(1 \leqslant i \leqslant n_0)$, and the above formula for $N_{k'/k}$, together with corollary 3 of prop. 4 applied to k'_0 and k, gives, for all $x \in k'$:

$$N_{k'/k}(x) = \prod_i \lambda_i(x)^{n/n_0}.$$

Now let k'' be any finite extension of k'. Proceeding exactly as in the proof of corollary 4 of prop. 4, we get again

$$N_{k''/k} = N_{k'/k} \circ N_{k''/k'},$$

which is therefore valid, whether k' and k'' are separable over k or not.

As to the trace, the elementary properties of the determinant, and the definition of the trace and the norm, show that, if \mathscr{A} is any algebra over k, $Tr_{\mathscr{A}/k}(x)$, as a linear form on \mathscr{A}, is the sum of the terms of degree 1 in the polynomial function $N_{\mathscr{A}/k}(1 + x)$ when the latter is expressed as a polynomial in the coordinates of $x \in \mathscr{A}$ with respect to some basis of \mathscr{A} over k. This, applied to the present situation, shows that $Tr_{k'/k}(x)$ is the sum of the terms of degree 1 in $N_{k'/k}(1 + x)$. As the latter is equal to $N_{k'_0/k}(1 + x^q)$, it contains only terms whose degree is a multiple of q. This shows that $Tr_{k'/k} = 0$ if $q > 1$, and therefore, in view of corollary 1 of prop. 4, that $Tr_{k'/k} \neq 0$ if and only if k' is separable over k.

PROPOSITION 5. *Let k' be a separably algebraic extension of k of degree n, and let $\{a_1, \ldots, a_n\}$ be a basis of k' over k. Then the determinant of the matrix*

$$(Tr_{k'/k}(a_i a_j))_{1 \leqslant i, j \leqslant n}$$

is not 0.

In view of corollary 1 of prop. 4, this is contained in the following lemma, which will also be useful later:

LEMMA 3. *Let k' be any extension of k of degree n; let E be the vector-space over k underlying k', and let λ be any linear form on E, other than 0. Then $(x, y) \to \lambda(xy)$ is a non-degenerate bilinear form on $E \times E$; one can identify E with its algebraic dual E' by putting $[x, y] = \lambda(xy)$; and, if $a_1, ..., a_n$ is a basis of k' over k, the determinant of the matrix $(\lambda(a_i a_j))$ is not 0.*

As λ is not 0, there is $a \in k'$ such that $\lambda(a) \neq 0$. For each $y \in k'$, define a k-linear form λ_y on k' by $\lambda_y(x) = \lambda(xy)$ for all $x \in k'$. Then $y \to \lambda_y$ is a morphism of E into its dual E'. This has the kernel 0, since $y \neq 0$ implies $\lambda_y(ay^{-1}) \neq 0$, hence $\lambda_y \neq 0$. As E and E' have the same dimension over k, this shows that $y \to \lambda_y$ is an isomorphism of E onto E'; identifying E and E' by means of that isomorphism, we get $[x, y] = \lambda(xy)$. By definition, this is the same as to say that $(x, y) \to \lambda(xy)$ is non-degenerate. Finally, if the matrix $(\lambda(a_i a_j))$ had the determinant 0, one could find $y_1, ..., y_n$ in k, not all 0, so that $\sum_j \lambda(a_i a_j) y_j = 0$, hence, putting $y = \sum_j a_j y_j$, $\lambda_y(a_i) = 0$ for all i, and therefore $\lambda_y = 0$, which contradicts what has been proved above.

§ 4. Tensor-products of A-fields and local fields.

Let k be an A-field and k' a separable extension of k; let v be a place of k, and k_v the completion of k at v. Then, by prop. 1 of § 1 and its corollary, the completions (λ, K') of k' which induce on k its natural injection into k_v are the same as the "proper embeddings" of k' above k_v as defined in § 2. We may therefore use propositions 2 and 3 of § 2 in order to determine the places of k' above v; this will be done now.

THEOREM 4. *Let k be an A-field, k' a separably algebraic extension of k of finite degree n, and α a basis of k' over k. For every place v of k, let k_v be the completion of k at v, and put $A_v = k' \otimes_k k_v$; for every finite place v of k, call r_v the maximal compact subring of k_v, and α_v the r_v-module generated by α in A_v. Let $w_1, ..., w_r$ be the places of k' which lie above v; for each i, call k'_i the completion of k' at w_i, λ_i the natural injection of k' into k'_i and μ_i the k_v-linear extension of λ_i to A_v. Then the mapping $\Phi_v = (\mu_1, ..., \mu_r)$ is an isomorphism of A_v onto the direct sum B_v of the fields k'_i, and, for almost all v, it maps α_v onto the sum of the maximal compact subrings r'_i of the fields k'_i.*

The first assertion is just a special case of prop. 3 of § 2, obtained by taking $K = k_v$ in that proposition; more briefly, but less accurately, it

can be expressed by saying that the completions k_i' of k' at the places of k' which lie above v are the summands of $k' \otimes_k k_v$ when this is written as a direct sum of fields. Now take for v any finite place of k; clearly the sum of the r_i' is the maximal compact subring of B_v; therefore its image ρ_v under Φ_v^{-1} is the maximal compact subring of A_v, and we have to show that this is the same as α_v for almost all v. As each of the r_i' contains r_v, ρ_v is a k_v-lattice in A_v; by th. 1 of Chap. II-2, we can find a basis $\{u_{v,1}, ..., u_{v,n}\}$ of A_v over k_v such that ρ_v is the r_v-module generated by that basis. For almost all v, by corollary 2 of th. 3, § 1, α_v is a compact subring of A_v, hence contained in ρ_v; call P the finite set of places of k for which this is not so. Put $\alpha = \{a_1, ..., a_n\}$; for v not in P, α_v is contained in ρ_v, so that we can write $a_i = \sum c_{v,ij} u_{v,j}$ with $c_{v,ij} \in r_v$ for $1 \leq i, j \leq n$; the matrix $C_v = (c_{v,ij})$ is then in $M_n(r_v)$, and we have $\alpha_v = \rho_v$ if and only if C_v is invertible in $M_n(r_v)$, i.e. if and only if its determinant is invertible in r_v. Now, writing Tr for the trace $Tr_{k'/k}$, call Δ the determinant of the matrix

$$M = (Tr(a_i a_j))_{1 \leq i, j \leq n};$$

Δ is in k, and, by prop. 5 of § 3, it is not 0. Applying th. 3 of § 1 to Δ and to Δ^{-1}, we see that $|\Delta|_v = 1$ for almost all v. On the other hand, if u is any element of A_v, $Tr(u)$ is the trace of $x \to ux$ in A_v; writing $u \cdot u_{v,i} = \sum d_{ij} u_{v,j}$ with $d_{ij} \in k_v$ for $1 \leq i, j \leq n$, we get $Tr(u) = \sum d_{ii}$. As ρ_v is a ring, all the d_{ij} are in r_v if $u \in \rho_v$; this shows that Tr maps ρ_v into r_v. Therefore, if we write N_v for the matrix $(Tr(u_{v,i} u_{v,j}))$, N_v is in $M_n(r_v)$. Substituting now $\sum c_{v,ij} u_{v,j}$ for a_i in the matrix M, we get $M = C_v N_v {}^t C_v$, hence $\Delta = = \det(N_v) \det(C_v)^2$. Here N_v is in $M_n(r_v)$, and so is C_v if v is not in P; and $|\Delta|_v = 1$ for almost all v. Clearly this implies that $|\det(C_v)|_v = 1$ for almost all v, as was to be proved.

In Chap. VIII, it will be shown that theorem 4 remains valid even if k' is not assumed to be separable over k.

COROLLARY 1. *Assumptions and notations being as in theorem 4, the sum of the degrees over k_v of the completions k_i' of k' at the places w_i of k' which lie above v is equal to the degree n of k' over k.*

In fact, this sum is the dimension of B_v over k_v, while that of A_v over k_v is n.

COROLLARY 2. *Let k be an algebraic extension of \mathbf{Q} of degree n; call r_1 the number of the real places of k, and r_2 the number of its imaginary places. Then $r_1 + 2r_2 = n$.*

We get this by replacing k, k', v by \mathbf{Q}, k, ∞ in corollary 1.

COROLLARY 3. *Assumptions and notations being as in theorem 4, the extensions of $Tr_{k'/k}$ and $N_{k'/k}$ to A_v are given by*

$$Tr_{k'/k}(x) = \sum_i Tr_{k_i/k_v}(\mu_i(x)), \quad N_{k'/k}(x) = \prod_i N_{k_i/k_v}(\mu_i(x)).$$

This follows at once from the application of prop. 4 of § 3 to the situation described in theorem 4.

COROLLARY 4. *Assumptions and notations being as in theorem 4, assume also that k' is a Galois extension of k, with the Galois group G. Let w be one of the places w_i of k'. Then the completion k'_w of k' at w is a Galois extension of k_v; the restriction to k' of the Galois group H of k'_w over k_v determines an isomorphism of H onto the subgroup of G, consisting of the automorphisms of k' over k which leave w invariant; the w_i are the images of w under G, and all the k'_i are isomorphic to k'_w.*

Let λ be any isomorphic embedding of k' into a local field K, such that $\lambda(k')$ is dense in K; then, by definition, this determines a place of k', and the image of that place by an automorphism σ of k' is to be understood as the place determined by the embedding $\lambda \circ \sigma$ of k' into K. That being so, we get our corollary by combining theorem 4 with corollary 2 of prop. 3, § 2, the latter being applied to the natural injection of k' into k'_w.

Chapter IV

Adeles

§ 1. Adeles of A-fields. Throughout this Chapter, k will denote an A-field; if v is a place of k, k_v will denote the completion of k at v; if v is a finite place of k, we write r_v for the maximal compact subring of k_v and p_v for the maximal ideal of r_v, these being the subsets of k_v respectively defined by $|x|_v \leqslant 1$ and by $|x|_v < 1$. We write P_∞ for the set of the infinite places of k, and P for any finite set of places of k, containing P_∞. For any such set P, put

$$(1) \qquad k_A(P) = \prod_{v \in P} k_v \times \prod_{v \notin P} r_v,$$

where the second product is taken over all the places of k, not in P. With the usual product topology, this is locally compact, since the k_v are so and the r_v are compact. On $k_A(P)$, we put a ring structure by defining addition and multiplication componentwise; clearly this makes $k_A(P)$ into a topological ring. Set-theoretically, $k_A(P)$ could be defined as the subset of the product $\prod k_v$ consisting of the elements $x = (x_v)$ of that product such that $|x_v|_v \leqslant 1$ for all v not in P. If P' is also a finite set of places of k, and $P' \supset P$, then $k_A(P)$ is contained in $k_A(P')$; moreover, its topology and its ring structure are those induced by those of $k_A(P')$, and $k_A(P)$ is an open subset of $k_A(P')$.

Now we define a locally compact topological ring k_A, the "ring of adeles" of k. Set-theoretically, this is to be the union of all the sets $k_A(P)$; in other words, it consists of the elements $x = (x_v)$ of the product $\prod k_v$ which satisfy $|x_v|_v \leqslant 1$ for almost all v. The topological ring structure of k_A will be defined by prescribing that each $k_A(P)$ is to be an open subring of k_A. This means firstly that, if $x = (x_v)$ and $y = (y_v)$ are in k_A, then $x + y = (x_v + y_v)$ and $xy = (x_v y_v)$; it is clear, in fact, that these are both in k_A. Secondly, we get a fundamental system of neighborhoods of 0 in the additive group of k_A by taking such a system in any one of the $k_A(P)$, for instance in $k_A(P_\infty)$ which is the smallest one of the sets $k_A(P)$; equivalently, we get such a system by taking all the sets of the form $\prod U_v$, where U_v is a neighborhood of 0 in k_v for all v, and $U_v = r_v$ for almost all v.

DEFINITION 1. *By the adele ring k_A of the A-field k, we understand the union of the sets $k_A(P)$ defined by (1), when one takes for P all the finite sets of places of k which contain the set of all infinite places. The topological ring structure of k_A is that for which each $k_A(P)$ is an open subring of k_A.*

The elements of k_A will be called *the adeles of k.*

Take a place v of k; when P contains v, one can write $k_A(P)$ as the product of k_v with an infinite product; denoting the latter by $k'_A(P,v)$, we may proceed with the products $k'_A(P,v)$ just as we have done for the products $k_A(P)$, taking now for P all the finite sets of places of k which contain P_∞ and v. The union of all the $k'_A(P,v)$ is then a locally compact ring $k'_A(v)$, and k_A is obviously isomorphic to the product $k_v \times k'_A(v)$; by means of this isomorphism, the first factor k_v of the latter product is obviously mapped onto the set of the adeles $x = (x_v)$ for which $x_w = 0$ at all places $w \neq v$; this set will be called the *quasifactor* of k_A belonging to v, and will always be identified with k_v. The mapping $(x_v) \to x_v$ of k_A onto k_v, which corresponds to the projection from the product $k_v \times k'_A(v)$ onto its first factor, will be called the *projection* from k_A onto the quasi-factor k_v; it is obviously continuous. Clearly, too, instead of one place v of k, one could start with any finite set P_0 of such places so as to write k_A as the product of the fields k_v for $v \in P_0$ and of one more factor.

Take any character χ of the additive group of k_A; it induces on $k_A(P)$, for every P, a character χ_P of $k_A(P)$, and on the quasifactor k_v, for every v, a character χ_v of k_v. It is well-known that a character of an infinite product of compact groups must induce the trivial character 1 on almost all the factors; this, applied to the character induced by χ_P on the product $\prod r_v$ in (1), shows that χ_v is trivial on r_v for almost all v; then we have, for all $x = (x_v)$ in k_A:

$$(2) \qquad \chi(x) = \prod_v \chi_v(x_v);$$

the product here is taken over all the places v of k; for each $x = (x_v)$ in k_A, almost all the factors are equal to 1.

Let ξ be an element of k. In view of th. 3 of Chap. III-1, we define an adele $x = (x_v)$ by putting $x_v = \xi$ for all v; we write this $\varphi(\xi)$, and call φ the *canonical injection* of k into k_A; we will frequently identify k with its image in k_A by means of φ when there is no danger of confusion.

Let E be a vector-space of finite dimension n over k. For each place v of k, we will write $E_v = E \otimes_k k_v$; as usual, we take E to be "naturally" embedded in E_v by the injection $e \to e \otimes 1_{k_v}$. On the other hand, since k has been embedded in k_A by the canonical injection φ defined above, we may consider the tensor-product $E_A = E \otimes_k k_A$, and regard E as being "naturally" embedded in it by the mapping $e \to e \otimes \varphi(1)$. We define the

topology of E_A as the coarsest one for which the extensions to k_A of the linear forms on E are continuous. Equivalently, take a basis ε of E over k; this determines an isomorphism of k^n onto E, hence an isomorphism of $(k_A)^n$ onto E_A; the topology of E_A is that which is obtained by transferring to E_A the topology of $(k_A)^n$ by means of that isomorphism; it would be easy to verify directly that this does not depend upon ε.

Let E and E' be vector-spaces of finite dimension over k, and let f be a polynomial mapping of E into E'; then f can be extended in an obvious manner to a mapping of E_A into E'_A, viz., the one which is defined by the same polynomial equations if E, E' are identified with spaces k^n, k^m, and consequently E_A, E'_A with $(k_A)^n$, $(k_A)^m$ by the choice of bases for E, E' over k. This extension of f will again be denoted by f; it is clearly continuous, since addition and multiplication are continuous in k_A.

PROPOSITION 1. *Let E be a vector-space of finite dimension n over k. Let ε be a finite subset of E, containing a basis of E over k. For each finite place v of k, call ε_v the r_v-module generated by ε in E_v. For each finite set P of places of k, containing P_∞, write*

$$E_A(P,\varepsilon) = \prod_{v \in P} E_v \times \prod_{v \notin P} \varepsilon_v.$$

Then each $E_A(P,\varepsilon)$ is an open subgroup of E_A, and E_A is the union of these subgroups.

This should be understood in the sense that each product $E_A(P,\varepsilon)$ is endowed with its product-topology, and that the latter coincides with the one induced by that of E_A. Clearly ε_v is a k_v-lattice in E_v, hence open and compact in E_v, for all finite places v. Therefore $E_A(P,\varepsilon)$ is an open subgroup of $E_A(P',\varepsilon)$ whenever $P \subset P'$. Take a basis ε' of E over k, and use it to define an isomorphism of k^n onto E, hence one of $(k_A)^n$ onto E_A; then our definitions show at once that E_A is the union of the sets $E_A(P,\varepsilon')$, and that these are open in E_A. By corollary 1 of th. 3, Chap. III-1, there is a finite set P_0 of places of k, containing P_∞, such that $\varepsilon_v = \varepsilon'_v$ when v is not in P_0. This shows that E_A is the union of the sets $E_A(P,\varepsilon)$, and also, for $P' \supset P \cup P_0$, that $E_A(P,\varepsilon)$ is open in $E_A(P',\varepsilon')$, hence in E_A. Of course one could use proposition 1 to define directly the topology of E_A, just as the topology of k_A has been defined above; corollary 1 of th. 3, Chap. III-1, would then show this to be independent of ε.

COROLLARY 1. *Assumptions and notations being as in proposition 1, let C be a compact subset of E_A. Then there is a finite set P of places of k, such that $C \subset E_A(P,\varepsilon)$.*

As C is contained in the union of the open sets $E_A(P, \varepsilon)$, it must be contained in the union of finitely many such sets $E_A(P_i, \varepsilon)$, hence in $E_A(P, \varepsilon)$ for $P = \bigcup P_i$.

If \mathscr{A} is any algebra of finite dimension over k, we will denote by \mathscr{A}_A the topological ring obtained by extending the multiplication law of \mathscr{A} to the space \mathscr{A}_A in the manner explained above. Clearly this may be regarded as an algebra over k_A, and $k_A \cdot 1_{\mathscr{A}}$ is a closed subspace and a subring of \mathscr{A}_A, isomorphic to k_A.

COROLLARY 2. *Let \mathscr{A} be an algebra of finite dimension over k, and α a finite subset of \mathscr{A}, containing a basis of \mathscr{A} over k. For each finite place v of k, call α_v the r_v-module generated by α in \mathscr{A}_v. For each finite set P of places of k, containing P_∞, write*

$$\mathscr{A}_A(P, \alpha) = \prod_{v \in P} \mathscr{A}_v \times \prod_{v \notin P} \alpha_v.$$

Then there is such a set P_0 with the property that $\mathscr{A}_A(P, \alpha)$ is an open subring of \mathscr{A}_A whenever $P \supset P_0$; and \mathscr{A}_A is the union of these subrings.

This follows at once from corollary 2 of th. 3, Chap. III-1, and from proposition 1.

Take now an algebraic extension k' of k, of finite degree. As k' is an A-field, we may apply to it our general construction, obtaining thus its adele ring k'_A. On the other hand, we may regard k' as an algebra over k and apply to this algebra the construction given above; this gives a ring which we write as $(k'/k)_A$; as we have seen, it is an algebra over k_A, and contains the closed subring $k_A \cdot 1_{k'}$, which we identify with k_A in the obvious manner. It is a central fact in the theory of adeles that the rings k'_A, $(k'/k)_A$ defined in this way are canonically isomorphic; this will be proved now, but only for the case where k' is separable over k. The inseparable case will be treated in Chap. VIII-6.

THEOREM 1. *Let k be an A-field and k' a separably algebraic extension of k of finite degree. Then there is a unique isomorphism Φ of $(k'/k)_A$ onto k'_A with the following properties: (i) Φ induces the identity on k' when k' is naturally embedded both in $(k'/k)_A$ and in k'_A; (ii) on each quasifactor $(k'/k)_v$ of $(k'/k)_A$, Φ induces a k_v-linear isomorphism Φ_v of $(k'/k)_v$ onto the product of the quasifactors k'_w of k'_A corresponding to the places w of k' which lie above v.*

Write \mathscr{A} for the algebra k'/k, i.e. for k' considered as an algebra over k. Then \mathscr{A}_A, in the notation explained above, is the same as $(k'/k)_A$, and \mathscr{A}_v the same as $(k'/k)_v$, i.e. as the algebra $k' \otimes_k k_v$ over k_v which was studied in Chap. III-4. For a finite number of summands, a "direct sum"

is the same as a product; we may therefore interpret th. 4 of Chap. III-4 as defining an isomorphism Φ_v of $(k'/k)_v$ onto the product $\prod k'_w$ of the fields k'_w for the places w lying above v; this is k_v-linear and maps every $\xi \in k'$ onto the element $(\xi, ..., \xi)$ of $\prod k'_w$, and it is uniquely characterized by these properties. Similarly, if we take a basis α of k' over k, the same theorem shows that, for almost all v, Φ_v maps α_v onto the product $\prod r'_w$ of the maximal compact subrings of the fields k'_w; let P_0 be a finite set of places of k, containing P_∞, such that Φ_v has that property for all v not in P_0. For each place w of k', call $f(w)$ the place of k lying below it. Then, for $P \supset P_0$, the mappings Φ_v determine in an obvious manner an isomorphism Φ_P of $\mathscr{A}_A(P,\alpha)$ onto $k'_A(f^{-1}(P))$, where $\mathscr{A}_A(P,\alpha)$ is the open subring of $\mathscr{A}_A = (k'/k)_A$ defined as in corollary 2 of prop. 1. As every set $f^{-1}(P)$ is finite, and every finite set P' of places of k' is contained in $f^{-1}(P)$ for $P = f(P')$, k'_A is the union of the sets $k'_A(f^{-1}(P))$ for $P \supset P_0$. As Φ_{P_1} coincides with Φ_P on the domain of definition of Φ_P whenever $P_1 \supset P$, there is an isomorphism Φ of \mathscr{A}_A onto k'_A which coincides with Φ_P on that domain whenever $P \supset P_0$. It is now clear that Φ has the properties stated in our theorem and that it is uniquely characterized by these properties.

COROLLARY 1. *Assumptions and notations being as in theorem 1, call $f(w)$, for every place w of k', the place of k lying below w. Then, if $x = (x_v)$ is in k_A, $\Phi(x)$ is the element $y = (y_w)$ of k'_A such that $y_w = x_{f(w)}$ for every place w of k'.*

This follows at once from the fact that $\Phi(1) = 1$ and that Φ_v is k_v-linear for every v.

From now on, k_A will usually be identified with its image in k'_A by means of the isomorphism, induced on k_A by Φ, which is described in corollary 1. Clearly k_A is thus a closed subring of k'_A.

COROLLARY 2. *Let k and k' be as in theorem 1; let E/k' be a vector-space of finite dimension over k', and call E/k the underlying vector-space over k. Then the identity mapping of E/k onto E/k' can be uniquely extended to a k_A-linear mapping of $(E/k)_A$ into $(E/k')_A$, and this is an isomorphism of $(E/k)_A$ onto $(E/k')_A$.*

In view of corollary 1, this is merely a restatement of theorem 1 if $E = k'$; the case $E = k'^n$ follows from this immediately, hence also the general case, since E can always be identified with a space k'^n by the choice of a basis.

According to the definitions given above, the k-linear form $Tr_{k'/k}$ and the polynomial function $N_{k'/k}$ on the space k', considered as a vector-space over k, may be extended to mappings $Tr_{k'/k}$, $N_{k'/k}$ of $(k'/k)_A$ into k_A;

then $Tr_{k'/k} \circ \Phi^{-1}$ and $N_{k'/k} \circ \Phi^{-1}$ are mappings of k'_A into k_A. We will simplify the formulation of the next corollary by identifying in it $(k'/k)_A$ with k'_A by means of Φ, so that the latter mappings may be written simply as $Tr_{k'/k}$ and $N_{k'/k}$.

COROLLARY 3. *Let* $x' = (x'_w)$ *be any element of* k'_A; *put* $y = Tr_{k'/k}(x')$ *and* $z = N_{k'/k}(x')$. *Then* y, z *are the elements* (y_v), (z_v) *of* k_A *respectively given by*

$$y_v = \sum_{w|v} Tr_{k'_w/k_v}(x'_w), \qquad z_v = \prod_{w|v} N_{k'_w/k_v}(x'_w).$$

for every place v *of* k, *the sum and the product being taken over all the places* w *of* k' *which lie above* v.

This is an immediate consequence of prop. 4, Chap. III-3, and th. 1.

§ 2. **The main theorems.** In view of lemma 1 of Chap. III-2, every A-field is a separably algebraic extension of one of the fields \mathbf{Q} and $\mathbf{F}_p(T)$. Theorem 1 of § 1 enables us now to prove properties of adele spaces by dealing first with the special cases $k = \mathbf{Q}$ and $k = \mathbf{F}_p(T)$. This method will presently yield some important results; in stating them, we simplify notations by identifying A-fields, and vector-spaces over such fields, with their natural images in the corresponding adele spaces, as explained in § 1; in the proofs, we shall again use φ to denote the canonical injection of an A-field k into k_A.

THEOREM 2. *Let* k *be an A-field and* E *a vector-space of finite dimension over* k. *Then* E *is discrete in* E_A, *and* E_A/E *is compact*.

In view of corollary 2 of th. 1, § 1, and of lemma 1 of Chap. III-2, it is enough to prove this for $k = \mathbf{Q}$ and $k = \mathbf{F}_p(T)$. If n is the dimension of E, E is isomorphic to k^n, so that, if the theorem is proved for $E = k$, it must be true in general. Thus we need only treat the cases $E = k = \mathbf{Q}$ and $E = k = \mathbf{F}_p(T)$. We begin with \mathbf{Q}.

For each prime p, call $\mathbf{Q}^{(p)}$ the set of the elements ξ of \mathbf{Q} such that $|\xi|_{p'} \leqslant 1$ for all the primes p' other than p. Clearly this is a subring of \mathbf{Q}, consisting of the numbers of the form $p^{-n}a$ with $n \in \mathbf{N}$ and $a \in \mathbf{Z}$.

LEMMA 1. *For every prime* p, *we have* $\mathbf{Q}_p = \mathbf{Q}^{(p)} + \mathbf{Z}_p$ *and* $\mathbf{Q}^{(p)} \cap \mathbf{Z}_p = \mathbf{Z}$.

The first assertion follows at once from corollary 2 of th. 6, Chap. I-4, applied to \mathbf{Q}_p, to the prime element p and to the set of representatives $\{0, 1, \ldots, p-1\}$. The second one is obvious.

LEMMA 2. *Put* $A_\infty = \mathbf{R} \times \prod \mathbf{Z}_p$, *and call* φ *the canonical injection of* \mathbf{Q} *into* \mathbf{Q}_A. *Then* $\mathbf{Q}_A = \varphi(\mathbf{Q}) + A_\infty$ *and* $\varphi(\mathbf{Q}) \cap A_\infty = \varphi(\mathbf{Z})$.

With the notation of (1), § 1, A_∞ is the same as $\mathbf{Q_A}(\{\infty\})$; it is there-fore an open subring of $\mathbf{Q_A}$. The second assertion in the lemma is obvious. Now take any $x=(x_v)$ in $\mathbf{Q_A}$; call P the set of the primes p such that x_p is not in \mathbf{Z}_p; it is a finite set. For each $p \in P$, the first part of lemma 1 shows that we may write $x_p = \xi_p + x_p'$ with $\xi_p \in \mathbf{Q}^{(p)}$ and $x_p' \in \mathbf{Z}_p$. For p not in P, put $\xi_p = 0$ and $x_p' = x_p$. Put now $\xi = \sum \xi_p$, the sum being extended to all p, and $y = x - \varphi(\xi)$. If $y = (y_v)$, we have, for every prime p:

$$y_p = x_p - \xi_p - \sum_{p' \neq p} \xi_{p'} = x_p' - \sum_{p' \neq p} \xi_{p'}.$$

By the definition of $\mathbf{Q}^{(p)}$, all the terms in the right-hand side are in \mathbf{Z}_p. This shows that y is in A_∞, hence x in $\varphi(\mathbf{Q}) + A_\infty$.

We can now prove our theorem for $E = k = \mathbf{Q}$. As A_∞ is open in $\mathbf{Q_A}$, the first assertion will be proved if we show that $\varphi(\mathbf{Q}) \cap A_\infty$, i.e. $\varphi(\mathbf{Z})$, is discrete in A_∞; this is clear, since its projection onto the factor \mathbf{R} of the product A_∞ is \mathbf{Z}, which is discrete in \mathbf{R}. Now call I the closed interval $[-1/2, 1/2]$ in \mathbf{R}, and put $C = I \times \prod \mathbf{Z}_p$. Clearly $A_\infty = \varphi(\mathbf{Z}) + C$, hence $\mathbf{Q_A} = \varphi(\mathbf{Q}) + C$. As C is compact, this completes the proof.

For $E = k = \mathbf{F}_p(T)$, the proof is similar but simpler. For each place v of k, call $k^{(v)}$ the set of the elements ξ of k such that $|\xi|_w \leqslant 1$ for all the places w of k, other than v.

LEMMA 3. *For every place v of k, $k_v = k^{(v)} + r_v$ and $k^{(v)} \cap r_v = \mathbf{F}_p$.*

The last assertion is obvious in view of the definition of the func-tions $|\xi|_v$ on k which was given in the proof of th. 2, Chap. III-1. As to the first one, it is enough to consider a place attached to a prime poly-nomial π of $\mathbf{F}_p[T]$, since otherwise we merely interchange T and T^{-1}. Then it follows at once from corollary 2 of th. 6, Chap. I-4, applied to k_v, to the prime element π and to the set of representatives supplied by the corollary of th. 2, Chap. III-1.

LEMMA 4. *Put $A_0 = \prod r_v$. Then $k_A = \varphi(k) + A_0$ and $\varphi(k) \cap A_0 = \varphi(\mathbf{F}_p)$.*

With the notation of (1), § 1, A_0 is the same as $k_A(\emptyset)$; it is a compact open subring of k_A. The last assertion is again obvious. Now take $x = (x_v)$ in k_A. For every v for which $|x_v|_v > 1$, lemma 3 shows that we may write $x_v = \xi_v + x_v'$ with $\xi_v \in k^{(v)}$ and $x_v' \in r_v$. For all other places v, put $\xi_v = 0$ and $x_v' = x_v$. Put $\xi = \sum \xi_v$ and $y = x - \varphi(\xi)$. Just as in the proof of lemma 2, we get $y \in A_0$.

The theorem is now obvious for $E = k = \mathbf{F}_p(T)$, since A_0 is compact and open in k_A and \mathbf{F}_p is finite. This completes the proof.

Now we consider a vector-space E over an A-field k, its algebraic dual E', and the corresponding adele spaces E_A, E_A'. We write $[e, e']$ for

the value at a point e of E of the linear form determined by a point e' of E', and we use the same notation for the extension of this bilinear form to $E_A \times E'_A$. As the additive group of E_A is a locally compact commutative group, we may consider its topological dual, which we denote by E^*_A; and we write $\langle e, e^* \rangle$ for the value at $e \in E_A$ of the character determined by $e^* \in E^*_A$. With these notations:

THEOREM 3. *Let k be an A-field and χ a non-trivial character of k_A, trivial on k. Let E be a vector-space of finite dimension over k; let E' be its algebraic dual, and E^*_A the topological dual of E_A. Then the formula*

$$\langle e, e^* \rangle = \chi([e, e']) \quad \text{for all } e \in E_A \quad (e' \in E'_A, e^* \in E^*_A)$$

determines an isomorphism $e' \to e^$ of E'_A onto E^*_A. Moreover, if e' is such that $\chi([e, e']) = 1$ for all $e \in E$, then $e' \in E'$.*

The last statement amounts to saying that the isomorphism $e' \to e^*$ of E'_A onto E^*_A defined in our theorem maps E' onto the subgroup of E^*_A associated by duality with the discrete subgroup E of E_A.

We begin by treating the case $E = k = Q$. Use again the same notations as in the first part of the proof of th. 2. In view of lemma 2, every character of A_∞, trivial on $\varphi(Z)$, can be uniquely extended to a character of Q_A, trivial on $\varphi(Q)$. We get such a character χ by putting $\chi(x) = \mathbf{e}(-x_\infty)$ for $x = (x_v) \in A_\infty$ (we recall that we write $\mathbf{e}(t) = e^{2\pi i t}$ for $t \in R$). If we extend this to a character χ of Q_A, trivial on $\varphi(Q)$, and call χ_v, for every place v of Q, the character induced by χ on the quasifactor Q_v of Q_A, then χ is obviously characterized by the following facts: it is trivial on $\varphi(Q)$, χ_p is trivial on Z_p for every prime p, and $\chi_\infty(x) = \mathbf{e}(-x)$ for $x \in R$. In order to calculate χ_p, consider again the group $Q^{(p)}$ defined in the proof of th. 2, and take any $\xi \in Q^{(p)}$. Then $\xi \in Z_{p'}$ for all primes $p' \neq p$, so that we have, by (2) of § 1:

$$1 = \chi(\varphi(\xi)) = \chi_\infty(\xi) \chi_p(\xi) = \mathbf{e}(-\xi) \chi_p(\xi)$$

and therefore $\chi_p(\xi) = \mathbf{e}(\xi)$. By lemma 1, χ_p is completely determined by this and by the fact that it is trivial on Z_p, and its kernel is Z_p; it is therefore of order 0 in the sense of def. 4 of Chap. II-5.

Now let χ' be any character of Q_A; for every place v of Q, call χ'_v the character induced by χ' on the quasifactor Q_v of Q_A. By the corollary of th. 3, Chap. II-5, we can write χ'_v uniquely in the form $\chi'_v(x) = \chi_v(a_v x)$ with $a_v \in Q_v$. As we observed when writing formula (2) of § 1, χ'_p must be trivial on Z_p for almost all p if χ is to be continuous on Q_A; this implies $\chi_p(a_p) = 1$, hence $a_p \in Z_p$ for almost all p; therefore $a = (a_v)$ is in Q_A, so that, by (2) of § 1, χ' is the character χ_a of Q_A given by $\chi_a(x) = \chi(ax)$ for all $x \in Q_A$. We have thus shown that the mapping $a \to \chi_a$ of Q_A into the

topological dual $G = \mathbf{Q}_\mathbf{A}^*$ of $\mathbf{Q}_\mathbf{A}$ is surjective. One sees at once that it is continuous and injective, so that it is a bijective morphism of $\mathbf{Q}_\mathbf{A}$ onto its dual G. Call Γ the subgroup of G associated by duality with $\varphi(\mathbf{Q})$, i.e. consisting of the characters of $\mathbf{Q}_\mathbf{A}$, trivial on $\varphi(\mathbf{Q})$; as χ has that property, the same is true of χ_a for all $a \in \varphi(\mathbf{Q})$, so that $a \to \chi_a$ maps $\varphi(\mathbf{Q})$ into Γ. Conversely, let b be such that $\chi_b \in \Gamma$. As in the proof of th. 2 for \mathbf{Q}, put $C = I \times \prod \mathbf{Z}_p$ with $I = [-1/2, 1/2]$. We have shown there that $\mathbf{Q}_\mathbf{A} = \varphi(\mathbf{Q}) + C$; therefore we may write $b = \varphi(\xi) + c$ with $\xi \in \mathbf{Q}$, $c \in C$, and then $\chi_c \in \Gamma$. Writing $c = (c_v)$, we have now, since $c_p \in \mathbf{Z}_p$ for all p:

$$1 = \chi_c(\varphi(1)) = \chi(c) = \chi_\infty(c_\infty) = \mathbf{e}(-c_\infty),$$

hence $c_\infty = 0$ since $c_\infty \in I$. Therefore χ_c is trivial on $A_\infty = \mathbf{R} \times \prod \mathbf{Z}_p$; as it is trivial on $\varphi(\mathbf{Q})$, lemma 2 shows that it is trivial on $\mathbf{Q}_\mathbf{A}$, so that $c = 0$, hence $b \in \varphi(\mathbf{Q})$. Therefore $a \to \chi_a$ maps $\varphi(\mathbf{Q})$ onto Γ. Finally, as $\varphi(\mathbf{Q})$ is discrete in $\mathbf{Q}_\mathbf{A}$, and $\mathbf{Q}_\mathbf{A}/\varphi(\mathbf{Q})$ is compact, the duality theory shows that Γ is discrete in G and that G/Γ is compact. Consequently $a \to \chi_a$ determines a bijective morphism of the compact group $\mathbf{Q}_\mathbf{A}/\varphi(\mathbf{Q})$ onto the compact group G/Γ; it is well-known that this must be an isomorphism. As G is "locally isomorphic" to G/Γ, and $\mathbf{Q}_\mathbf{A}$ to $\mathbf{Q}_\mathbf{A}/\varphi(\mathbf{Q})$, this implies that $a \to \chi_a$ is bicontinuous, so that it is an isomorphism. This completes the proof for $E = k = \mathbf{Q}$.

Now take $E = k = \mathbf{F}_p(T)$. In analogy with \mathbf{Q}, call ∞ the place of k for which T^{-1} is a prime element (although this is of course not an infinite place). Then $|T^{-1}|_\infty = p^{-1}$. We may now apply corollary 2 of th. 6, Chap. I-4, to k_∞, to the prime element T^{-1} and to the set of representatives \mathbf{F}_p, and therefore identify k_∞ with the field of the formal power-series

$$(3) \qquad\qquad\qquad x = \sum_{i=n}^{+\infty} a_i T^{-i}$$

where $n \in \mathbf{Z}$ and $a_i \in \mathbf{F}_p$ for all $i \geq n$. Call ψ the character of the additive group of \mathbf{F}_q given by $\psi(1) = \mathbf{e}(1/p)$; call χ_∞ the character of k_∞ defined by putting $\chi_\infty(x) = \psi(-a_1)$ when x is given by (3); for $x \in \mathbf{F}_p[T]$, we have $a_1 = 0$, hence $\chi_\infty(x) = 1$. Now put $A_\infty = k_\infty \times \prod r_v$, the product being taken over all the places v of k other than ∞; with the notation of (1), § 1, this is $k_\mathbf{A}(\{\infty\})$; it is an open subring of $k_\mathbf{A}$ and contains the set A_0 defined in lemma 4, so that, by that lemma, $k_\mathbf{A} = \varphi(k) + A_\infty$. When $\xi \in k$, $\varphi(\xi)$ is in A_∞ if and only if $|\xi|_v \leq 1$ for all the places v of k attached to prime polynomials in $\mathbf{F}_p[T]$, hence if and only if ξ is in $\mathbf{F}_p[T]$. This means that $\varphi(k) \cap A_\infty = \varphi(\mathbf{F}_p[T])$. Accordingly, every character of A_∞, trivial on $\varphi(\mathbf{F}_p[T])$, can be uniquely extended to one of $k_\mathbf{A}$, trivial on $\varphi(k)$. Applying this to the character χ of A_∞, given by $\chi(x) = \chi_\infty(x_\infty)$ for $x = (x_v) \in A_\infty$,

we get a character χ of k_A which can be characterized by the following facts: χ is trivial on $\varphi(k)$; for every $v \neq \infty$, the character χ_v induced by χ on k_v is trivial on r_v, and χ induces on k_∞ the character χ_∞ defined above. In order to calculate χ_v for a place v attached to a prime polynomial π of degree δ in $\mathbf{F}_p[T]$, call $k_0^{(v)}$ the set of the elements ξ of k such that $|\xi|_w \leqslant 1$ for all the places w of k other than v, and $|\xi|_\infty < 1$; the same proof which was given for lemma 3 shows now that k_v is the direct sum of $k_0^{(v)}$ and r_v; as χ_v is trivial on r_v, it is therefore completely determined by its values on $k_0^{(v)}$. Take $\xi \in k_0^{(v)}$; this can be written as $\xi = \pi^{-n}\alpha$, where $n \in \mathbf{N}$ and α is a polynomial of degree $< n\delta$ in $\mathbf{F}_p[T]$. Call a_1 the coefficient of $T^{n\delta-1}$ in α. As π is monic, it can be written as $T^\delta \omega$, where ω is in $\mathbf{F}_p[T^{-1}]$ and has the constant term 1. This gives

$$\xi = \pi^{-n}\alpha = \omega^{-n}T^{-n\delta}\alpha \equiv a_1 T^{-1} \quad (T^{-2})$$

in the ring r_∞, hence $\chi_\infty(\xi) = \psi(-a_1)$ by the definition of χ_∞. Now we have, by (2) of § 1:

$$1 = \chi(\varphi(\xi)) = \chi_\infty(\xi)\chi_v(\xi) = \psi(-a_1)\chi_v(\xi)$$

and therefore $\chi_v(\xi) = \psi(a_1)$, which completes the determination of χ_v. Furthermore, if ξ is as above and not 0, call d the degree of the polynomial α, and a the coefficient of T^d in α; then $\chi_v(\xi T^{n\delta-1-d})$ has the value $\psi(a)$, which is not 1 since $a \neq 0$. This shows that, if ξ is in $k_0^{(v)}$ and not 0, $\chi_v(\xi t)$ cannot be 1 for all $t \in r_v$. As χ_v is trivial on r_v, and as $k_v = k_0^{(v)} + r_v$, we conclude now from prop. 12 of Chap. II-5 that the character χ_v is of order 0 in the sense of def. 4 of Chap. II-5. In other words, if x is in k_v and such that $\chi_v(xt) = 1$ for all $t \in r_v$, x must be in r_v.

Now we can proceed just as in the case of \mathbf{Q}. Let χ' be any character of k_A. For each place v of k, the character χ'_v induced by χ on k_v can be written as $\chi'_v(x) = \chi_v(a_v x)$ with $a_v \in k_v$; then, from the fact that χ'_v must be trivial on r_v for almost all v, we conclude that $a = (a_v)$ must be in k_A, so that χ' is the character χ_a defined by $\chi_a(x) = \chi(ax)$. As before, we see that $a \rightarrow \chi_a$ is a bijective morphism of k_A onto the topological dual $G = k_A^*$ of k_A, and that it maps $\varphi(k)$ into the subgroup Γ of G associated by duality with $\varphi(k)$. Assume that $\chi_b \in \Gamma$ for some $b \in k_A$; by lemma 4, we may write $b = \varphi(\xi) + c$ with $\xi \in k$, $c \in A_0$; then χ_c is trivial on $\varphi(k)$. Put $c = (c_v)$, so that $c_v \in r_v$ for all v; then there is $\gamma \in \mathbf{F}_p$ such that $c_\infty \equiv \gamma (T^{-1})$; replacing ξ by $\xi + \gamma$ and c by $c - \varphi(\gamma)$, we get $c_\infty \equiv 0 (T^{-1})$. We have now

$$1 = \chi_c(\varphi(1)) = \chi(c) = \chi_\infty(c_\infty),$$

which implies, in view of the definition of χ_∞, that c_∞ is in $T^{-2}r_\infty$, and therefore that $\chi_\infty(c_\infty t) = 1$ for all $t \in r_\infty$. Consequently, χ_c is trivial on A_0, hence on k_A by lemma 4. This gives $c = 0$, hence $b \in \varphi(k)$. The proof can now be completed just as in the case of \mathbf{Q}.

We can now complete the proof of our theorem by a purely formal argument. Denote by $T(E/k,\chi)$ the statement in theorem 3. What we have proved above can be expressed by saying that, for each one of the fields $k=\mathbf{Q}$ and $k=\mathbf{F}_p(T)$, there is a character χ of k_A for which $T(k/k,\chi)$ is true. Obviously this implies that $T(k^n/k,\chi)$ is true for every n, so that $T(E/k,\chi)$ is true for every vector-space E over k. In particular, take a finite algebraic extension k' of k; as in lemma 3 of Chap. III-3, write E for the underlying vector-space over k; choose a k-linear form λ on E, other than 0, and identify E with its algebraic dual E' by putting $[x,y]=\lambda(xy)$. We can then extend λ to a mapping of E_A into k_A, the identification between E and E' to one between E_A and E'_A, and then we have again $[x,y]=\lambda(xy)$ for x, y in $E_A=(k'/k)_A$. If we write $\chi'=\chi\circ\lambda$, this is clearly a non-trivial character of E_A, trivial on E. If now we assume that k' is separable over k, we can identify E_A with k'_A by means of the isomorphism Φ described in th. 1 of § 1. When this is done, χ' becomes a non-trivial character on k'_A, trivial on k', and the statement $T(E/k,\chi)$ becomes exactly $T(k'/k',\chi')$. As we can take for k' any A-field, taking for k either \mathbf{Q} or $\mathbf{F}_p(T)$, we see that, for every A-field k, theorem 3 is true for at least one choice of χ. Now assume $T(k/k,\chi)$ for such a field, and let χ_1 be another character with the properties stated in theorem 3; $T(k/k,\chi)$ implies that χ_1 is of the form $\chi_1(x)=\chi(ax)$ with $a\in k$ and $a\neq 0$. Then the mapping $e'\to e^*$ defined as in theorem 3, but by means of χ_1, is composed of the similar mapping defined by χ and of the mapping $e'\to ae'$ of E'_A into itself. As the latter is clearly an automorphism of E'_A, mapping E' onto itself, we see that $T(E/k,\chi)$ is equivalent with $T(E/k,\chi_1)$. This completes the proof.

COROLLARY 1. *Let χ be as in theorem 3, and call χ_v, for every place v of k, the character induced by χ on the quasifactor k_v of k. Then, for every v, χ_v is non-trivial, and, for almost all finite places v of k, χ_v is of order 0 in the sense of def. 4, Chap. II-5.*

For each $a\in k_A$, call χ_a the character of k_A defined by $\chi_a(x)=\chi(ax)$. If χ was trivial on the quasifactor k_v, that quasifactor would be in the kernel of the morphism $a\to\chi_a$ of k_A into its topological dual; as theorem 3 says that this is an isomorphism, this would be a contradiction. In particular, for every finite place v of k, we may put $\nu(v)=\mathrm{ord}(\chi_v)$ in the sense of def. 4, Chap. II-5. For each mapping $v\to n(v)$ of the set of finite places of k into \mathbf{Z}, call $G(n)$ the group of the elements $x=(x_v)$ of k_A such that $\mathrm{ord}(x_v)\geqslant n(v)$ for all finite places v, and $H(n)$ the subgroup of $G(n)$ consisting of the elements $x=(x_v)$ of $G(n)$ such that $x_w=0$ for all infinite places w of k. In view of the definition of the topology in k_A in § 1, it is obvious that $G(n)$ is open in k_A if and only if $n(v)\leqslant 0$ for almost all v.

It is also clear that $H(n)$ is compact if $n(v) \geqslant 0$ for almost all v; conversely, by corollary 1 of prop. 1, § 1, and with the notation of (1), § 1, every compact subset of $k_\mathbf{A}$ is contained in one of the sets $k_\mathbf{A}(P)$, so that $H(n)$ cannot be compact unless $n(v) \geqslant 0$ for almost all v; therefore this is necessary and sufficient for the compacity of $H(n)$. Now prop. 12 of Chap. II-5, combined with the fact that χ_w is not trivial for any infinite place of k, shows that the set of elements x of $k_\mathbf{A}$ such that $\chi(xy)=1$ for all $y \in G(0)$ is $H(-v)$, and that the set of elements x such that $\chi(xy)=1$ for all $y \in H(0)$ is $G(-v)$. If we identify $k_\mathbf{A}$ with its topological dual by means of the isomorphism described in theorem 3, this means that $H(-v)$ and $G(-v)$ are the subgroups of $k_\mathbf{A}$ respectively associated by duality with $G(0)$ and $H(0)$. As $G(0)$ is open and $H(0)$ is compact, duality theory shows that $H(-v)$ must be compact and $G(-v)$ open. As we have seen, this implies that $-v(v) \geqslant 0$ for almost all v and that $-v(v) \leqslant 0$ for almost all v.

COROLLARY 2. *Let E be a vector-space of finite dimension over k, and let v be any place of k. Then $E + E_v$ is dense in $E_\mathbf{A}$.*

If this is true for $E=k$, it is clearly true for $E=k^n$ and therefore for every E. If $k+k_v$ were not dense in $k_\mathbf{A}$, there would be a non-trivial character of $k_\mathbf{A}$ which would be trivial both on k and on k_v; this contradicts corollary 1.

As in the case of local fields, it is frequently convenient, having chosen once for all a *"basic character"* χ with the properties described in theorem 3, to identify the topological dual of $E_\mathbf{A}$ with the space $E_\mathbf{A}'$ by means of the isomorphism in that theorem, for all vector-spaces E of finite dimension over k. For every quasifactor k_r of $k_\mathbf{A}$, one will then take as "basic character" the character χ_v induced by χ on k_r, and use this to identify the topological and algebraic duals of vector-spaces over k_v as explained in Chap. II-5. This being understood, we have:

COROLLARY 3. *Let assumptions and notations be as in proposition 1 of § 1. Let E be a vector-space over k, and E' its algebraic dual. Let $\varepsilon, \varepsilon'$ be finite subsets of E and of E', respectively, containing bases of these spaces over k. For each place v of k, identify E_v' with the topological dual of E_v as explained above. Then, for almost all finite places v of k, ε_v' is the dual k_v-lattice to ε_v.*

For $E=E'=k$ and $\varepsilon = \varepsilon' = \{1\}$, this is just a restatement of corollary 1; it is an immediate consequence of that corollary if $\varepsilon = \{e_1, \dots, e_n\}$ is a basis of E and $\varepsilon' = \{e_1', \dots, e_n'\}$ is the dual basis to ε, determined by $[e_i, e_j']=1$ if $i=j$ and 0 if $i \neq j$. The general case follows from this at once by corollary 1 of th. 3, Chap. III-1.

§ 3. Ideles. As before (cf. Chap. III-3), if E is a vector-space of finite dimension over any field k, we write $\mathrm{End}(E)$ for the ring of endomorphisms of E, considered as an algebra over k. We will write $\mathrm{Aut}(E)$ for the group of automorphisms of E; this is the same as the group $\mathrm{End}(E)^\times$ of invertible elements of $\mathrm{End}(E)$, and it is the subset of $\mathrm{End}(E)$ determined by $\det(a)\neq 0$; therefore, if k is a topological field, $\mathrm{Aut}(E)$ is an open subset of $\mathrm{End}(E)$; clearly it is a topological group for the topology induced on it by that of $\mathrm{End}(E)$. If K is a field containing k, $\mathrm{End}(E_K)$ is the same as $\mathrm{End}(E)_K = \mathrm{End}(E)\otimes_k K$, and the determinant in $\mathrm{End}(E_K)$ is the extension to that space of the determinant in $\mathrm{End}(E)$.

Let \mathscr{A} be an algebra of finite dimension over k; call ρ its regular representation into $\mathrm{End}(\mathscr{A})$, as defined in Chap. III-3, and write \mathscr{A}^\times, as usual, for the group of invertible elements of \mathscr{A}. Take any $a\in\mathscr{A}$; then $\rho(a)$ is the endomorphism $x\rightarrow ax$ of the vector-space underlying \mathscr{A}; if it is an automorphism, it is surjective, so that there is $b\in\mathscr{A}$ such that $ab=1_{\mathscr{A}}$; then $b=a^{-1}$, and $a\in\mathscr{A}^\times$. As the converse is obvious, this shows that \mathscr{A}^\times is the subset of \mathscr{A} determined by $N_{\mathscr{A}/k}(a)\neq 0$. Therefore, if k is a topological field, \mathscr{A}^\times is open in \mathscr{A}; moreover, ρ is then a topological isomorphism of \mathscr{A} onto a subalgebra of $\mathrm{End}(\mathscr{A})$, which maps \mathscr{A}^\times onto $\rho(\mathscr{A})\cap\mathrm{Aut}(\mathscr{A})$; this implies that \mathscr{A}^\times is then a topological group for the topology induced on it by that of \mathscr{A}.

Now, \mathscr{A} being an algebra of finite dimension over an A-field k, consider the group \mathscr{A}_A^\times of invertible elements of the ring \mathscr{A}_A. The simplest examples, e.g. $\mathscr{A}=k$, show that $x\rightarrow x^{-1}$ is not continuous on that group for the topology induced on it by that of \mathscr{A}_A. We will give it the coarsest topology for which the injection into \mathscr{A}_A and $x\rightarrow x^{-1}$ are both continuous; this is more conveniently stated as follows:

DEFINITION 2. *Let \mathscr{A} be an algebra of finite dimension over the A-field k. Then we denote by \mathscr{A}_A^\times the group of invertible elements of \mathscr{A}_A with the topology for which $x\rightarrow(x,x^{-1})$ is a homeomorphism of \mathscr{A}_A^\times onto its image in $\mathscr{A}_A\times\mathscr{A}_A$.*

It is customary (particularly in the case $\mathscr{A}=k$) to call \mathscr{A}_A^\times, with this topology, *the idele group of \mathscr{A}*, and to call its elements *the ideles of \mathscr{A}*. Obviously $(x,y)\rightarrow xy$ and $x\rightarrow x^{-1}$ are continuous on \mathscr{A}_A^\times, so that our definition does make it into a topological group. At the same time, if we call f the mapping $(x,y)\rightarrow xy$ of $\mathscr{A}\times\mathscr{A}$ into \mathscr{A} and its natural extension to $\mathscr{A}_A\times\mathscr{A}_A$, our definition says that \mathscr{A}_A^\times is homeomorphic to the subset $f^{-1}(\{1\})$ of the latter space; as f is continuous, this is a closed set, so that \mathscr{A}_A^\times is locally compact. It is also clear that \mathscr{A}^\times is canonically embedded in \mathscr{A}_A^\times; as $x\rightarrow(x,x^{-1})$ maps it onto the intersection of $f^{-1}(\{1\})$ with the discrete subset $\mathscr{A}\times\mathscr{A}$ of $\mathscr{A}_A\times\mathscr{A}_A$, it is a discrete subgroup of \mathscr{A}_A^\times.

One can give an alternative definition of the idele group of \mathscr{A}, equivalent to definition 2, by using corollary 2 of th. 3, Chap. III-1, and corollary 2 of prop. 1, § 1. As in these results, take a finite subset α of \mathscr{A}, containing a basis of \mathscr{A} over k, and call α_v, for each finite place v of k, the r_v-module generated by α in \mathscr{A}_v. By corollary 2 of th. 3, Chap. III-1, there is a finite set P_0 of places of k, containing P_∞, such that, for all v not in P_0, α_v is a compact subring of \mathscr{A}_v (containing the unit element). For each v, as we have seen, \mathscr{A}_v^\times is an open subset of \mathscr{A}_v, and $x \to x^{-1}$ is continuous on it; therefore $x \to (x, x^{-1})$ maps it homeomorphically onto its image in $\mathscr{A}_v \times \mathscr{A}_v$. For v not in P_0, α_v^\times is the set of the elements of \mathscr{A}_v^\times which are mapped into $\alpha_v \times \alpha_v$ by $x \to (x, x^{-1})$; therefore it is an open compact subgroup of \mathscr{A}_v^\times and an open compact subset of α_v. We shall now prove the following result, analogous to corollary 2 of prop. 1, § 1:

PROPOSITION 2. *Let \mathscr{A}, α, α_v and P_0 be as explained above. Let P be any finite set of places of k, containing P_0. Then the group*

$$(4) \qquad \mathscr{A}_A(P, \alpha)^\times = \prod_{v \in P} \mathscr{A}_v^\times \times \prod_{v \notin P} \alpha_v^\times$$

is an open subgroup of \mathscr{A}_A^\times; the topologies induced on it by those of \mathscr{A}_A^\times and of \mathscr{A}_A are both the same as the product topology for the right-hand side of (4); and \mathscr{A}_A^\times is the union of these groups.

Let $\mathscr{A}_A(P, \alpha)$ be defined as in corollary 2 of prop. 1, § 1. The topology induced on $\mathscr{A}_A(P, \alpha)^\times$ by that of \mathscr{A}_A is the same as that induced by that of $\mathscr{A}_A(P, \alpha)$, hence the same as the product topology for the right-hand side of (4). For each v, \mathscr{A}_v^\times is open in \mathscr{A}_v, and $x \to x^{-1}$ is continuous on it; therefore $x \to x^{-1}$ is continuous on $\mathscr{A}_A(P, \alpha)^\times$ for that product topology. This implies that $x \to (x, x^{-1})$ is a homeomorphism of $\mathscr{A}_A(P, \alpha)^\times$ onto its image in $\mathscr{A}_A \times \mathscr{A}_A$; therefore the product topology on that set is also that induced by \mathscr{A}_A^\times. Furthermore, $\mathscr{A}_A(P, \alpha)^\times$ is the subset of \mathscr{A}_A^\times which is mapped by $x \to (x, x^{-1})$ into $\mathscr{A}_A(P, \alpha) \times \mathscr{A}_A(P, \alpha)$; as the latter set is open in $\mathscr{A}_A \times \mathscr{A}_A$, and as $\mathscr{A}_A \times \mathscr{A}_A$ is the union of sets of that form, this completes the proof.

COROLLARY. *An element $a = (a_v)$ of k_A is in k_A^\times if and only if $a_v \neq 0$ for all v and $|a_v|_v = 1$ for almost all v. For every finite set P of places of k, containing P_∞, the group*

$$k_A(P)^\times = \prod_{v \in P} k_v^\times \times \prod_{v \notin P} r_v^\times$$

is an open subgroup of k_A^\times, and k_A^\times is the union of these groups.

The first statement is obvious; the rest is a special case of proposition 2.

For every element $a = (a_v)$ of k_A^\times, we will write

$$|a|_{k_A} = \prod_v |a_v|_v,$$

the product being taken over all the places v of k; in view of the corollary of prop. 2, almost all the factors of that product are equal to 1 whenever a is in k_A^\times. Usually, when there is no danger of confusion about the field of reference, we will write $|a|_A$ instead of $|a|_{k_A}$ for this product; it is sometimes called *the module* of a.

PROPOSITION 3. *Let E be a vector-space of finite dimension n over k. Put $\mathscr{A} = \mathrm{End}(E)$, and let $a = (a_v)$ be an element of \mathscr{A}_A. Then the following assertions are equivalent:* (i) *a is in \mathscr{A}_A^\times;* (ii) *$\det(a)$ is in k_A^\times;* (iii) *$e \to ae$ is an automorphism of E_A. When that is so, the module of the latter automorphism is $|\det(a)|_A$. Moreover, the mappings $a \to \det(a)$ and $a \to |\det(a)|_A$ are morphisms of \mathscr{A}_A^\times into k_A^\times and into \mathbf{R}_+^\times, respectively.*

Take a basis ε for E over k; we will use it to identify E with k^n and \mathscr{A} with $M_n(k)$. Then a basis α for A over k is given by the "matrix units" $a_{\lambda\mu}$ for $1 \leqslant \lambda, \mu \leqslant n$, where $a_{\lambda\mu}$ is the matrix (x_{ij}) given by $x_{\lambda\mu} = 1$ and $x_{ij} = 0$ for $(i,j) \neq (\lambda, \mu)$. For every place v of k, an element a_v of $M_n(k_v)$ is invertible in $M_n(k_v)$ if and only if $\det(a_v) \neq 0$; for every finite place v of k, an element a_v of $M_n(r_v)$ is invertible in $M_n(r_v)$ if and only if $\det(a_v)$ is invertible in r_v, i.e. if and only if $|\det(a_v)|_v = 1$. With the notations of prop. 2 and its corollary, this amounts to saying that a is in $\mathscr{A}_A(P, \alpha)^\times$ if and only if $\det(a)$ is in $k_A(P)^\times$; clearly this implies the equivalence of (i) and (ii) in our proposition, and it also shows that the mapping $a \to \det(a)$ of \mathscr{A}_A^\times into k_A^\times is continuous on $\mathscr{A}_A(P, \alpha)^\times$ for every P, hence on \mathscr{A}_A^\times. As it is clear that the mapping $z \to |z|_A$ of k_A^\times into \mathbf{R}_+^\times is continuous on $k_A(P)^\times$ for every P, hence also on k_A^\times, $a \to |\det(a)|_A$ is a continuous morphism of \mathscr{A}_A^\times into \mathbf{R}_+^\times. If a is in \mathscr{A}_A^\times, it has an inverse a^{-1} in \mathscr{A}_A, and then the endomorphism $e \to ae$ of E_A has the inverse $e \to a^{-1}e$, so that it is an automorphism. Conversely, take any $a = (a_v)$ in \mathscr{A}_A; prop. 1 of § 1, applied to \mathscr{A} and α, shows that a_v is in $M_n(k_v)$ for all v and in $M_n(r_v)$ for almost all v. The same proposition, applied to E and ε, shows that a fundamental system of neighborhoods of 0 in E_A is given by the sets $U = \prod U_v$, where U_v is a neighborhood of 0 in $E_v = (k_v)^n$ for all v, and $U_v = (r_v)^n$ for almost all v. If $e \to ae$ is an automorphism of E_A, it must map every neighborhood of 0 onto a neighborhood of 0; this implies that a_v is invertible in $M_n(k_v)$ for all v, and that, for almost all v, the image of $(r_v)^n$ under a_v contains $(r_v)^n$, i.e. that a_v^{-1} is in $M_n(r_v)$ for almost all v. As we have observed above, this is the same as to say that a is in \mathscr{A}_A^\times. Let then P be a finite set of places of k, containing P_∞, such that a_v is in $M_n(r_v)^\times$

for all v not in P. As the set $E_A(P,\varepsilon)$ is open in E_A and invariant under $e \to ae$, the module of $e \to ae$ in E_A is the same as its module in that set; this, in view of the definition of that set in prop. 1 of § 1, is the product of the modules of the automorphisms $e_v \to a_v e_v$ of its factors; these, by corollary 3 of th. 3, Chap. I-2, are respectively equal to $|\det(a_v)|_v$, which completes our proof.

COROLLARY. *Let \mathscr{A} be an algebra of finite dimension over k, and let a be an element of \mathscr{A}_A. Then the following assertions are equivalent:* (i) *a is in \mathscr{A}_A^\times;* (ii) *$N_{\mathscr{A}/k}(a)$ is in k_A^\times;* (iii) *$x \to ax$ is an automorphism of the additive group of \mathscr{A}_A. When that is so, the module of that automorphism is $|N_{\mathscr{A}/k}(a)|_A$. Moreover, $a \to N_{\mathscr{A}/k}(a)$ and $a \to |N_{\mathscr{A}/k}(a)|_A$ are morphisms of \mathscr{A}_A^\times into k_A^\times and into \mathbf{R}_+^\times, respectively.*

As we are always assuming that \mathscr{A} contains a unit, (iii) implies (i). All our other assertions follow at once from proposition 3, applied to the underlying vector-space E of \mathscr{A} over k and to the embedding of \mathscr{A} into $\mathrm{End}(E)$ given by the regular representation ρ.

Of course all that has been said about the endomorphisms $x \to ax$ of an algebra \mathscr{A} applies equally well to the endomorphisms $x \to xa$; the determinant $N'(a)$ of the latter, sometimes called the "coregular norm" on \mathscr{A}, is again a polynomial function, of degree equal to the dimension of \mathscr{A} over k, and the module of the automorphism $x \to xa$ of \mathscr{A}_A, for $a \in \mathscr{A}_A^\times$, is equal to $|N'(a)|_A$. Obviously $N' = N_{\mathscr{A}/k}$ when \mathscr{A} is commutative; the same is known to be true for all semisimple algebras and will be proved in Chap. IX for simple algebras and in particular for division algebras; this will not be needed here.

THEOREM 4. *Let D be a division algebra of finite dimension over k. For every real number $\mu \geqslant 1$, call D_μ the set of the elements d of D_A^\times such that the modules of the automorphisms $x \to dx$ and $x \to xd$ of D_A are respectively $\leqslant \mu$ and $\geqslant \mu^{-1}$. Then D_μ is a closed subset of D_A^\times whose image in D_A^\times/D^\times is compact.*

* Write N for the regular norm $N_{D/k}$, and N' for the "coregular norm" as defined above; by the corollary of prop. 3, $d \to |N(d)|_A$ is continuous on D_A^\times, and the same is true of $d \to |N'(d)|_A$ for similar reasons; in view of that same corollary, this implies that D_μ is closed. By th. 2 of § 2, D is discrete in D_A, and D_A/D is compact; therefore there is a Haar measure α on D_A such that $\alpha(D_A/D) = 1$, this being defined in the manner explained in Chap. II-4. As D_A is not compact, we can choose a compact subset C of D_A such that $\alpha(C) > \mu$. Call C' the image of $C \times C$ under the mapping $(x,y) \to x - y$ of $D_A \times D_A$ into D_A, and C'' the image of $C' \times C'$ under the mapping $(x,y) \to xy$ of $D_A \times D_A$ into D_A; as these mappings are continuous,

C' and C'' are compact. Take any $d \in D_\mu$; as the module of $x \to xd$ is $\geq \mu^{-1}$, it maps C onto a set Cd whose measure is > 1; therefore, by lemma 1 of Chap. II-4, there are two elements x, y of C such that $xd - yd$ is in D and is not 0, i.e. such that it is in D^\times. Write $c_1 = x - y$ and $\delta_1 = c_1 d$; then $c_1 \in C'$ and $\delta_1 \in D^\times$. Similarly, $x \to d^{-1}x$, being the inverse of $x \to dx$, has a module $\geq \mu^{-1}$, so that it maps C onto a set $d^{-1}C$ of measure > 1; as before, we conclude that there is $c_2 \in C'$ such that $\delta_2 = d^{-1}c_2$ is in D^\times. Then $\delta_1 \delta_2 = c_1 c_2$, so that $\delta_1 \delta_2$ is in $D^\times \cap C''$, which is a finite set since D is discrete and C'' compact in D_A. Call $\gamma_1, \ldots, \gamma_N$ all the distinct elements of $D^\times \cap C''$; $c_1 c_2$ is equal to one of these, say γ_i, so that $\gamma_i^{-1} c_1 c_2 = 1$. This shows that c_2 is invertible in D_A and has the inverse $c_2^{-1} = \gamma_i^{-1} c_1$. As $d\delta_2 = c_2$, we see that $d\delta_2$ belongs to the set X of the elements x of D_A^\times whose image under the mapping $x \to (x, x^{-1})$ is in the union of the sets $C' \times (\gamma_i^{-1} C')$ for $1 \leq i \leq N$. In view of def. 2, X is a compact subset of D_A^\times; as $D_\mu \subset X \cdot D^\times$, the image of D_μ in D_A^\times / D^\times is contained in that of X, which proves our theorem.

§ 4. Ideles of A-fields. We will now consider more in detail the case $\mathscr{A} = k$.

THEOREM 5. *Let k be any A-field; then the morphism $z \to |z|_A$ of k_A^\times into \mathbf{R}_+^\times induces the constant 1 on k^\times.*

If $\xi \in k^\times$, $x \to \xi x$ is an automorphism of k_A which maps k onto itself. By th. 2 of § 2, k is discrete in k_A, and k_A/k is compact. Therefore the module of $x \to \xi x$, which is $|\xi|_A$ by prop. 3 of § 3 (if one takes $E = k$ in that proposition), is equal to 1, e.g. by lemma 2 of Chap. I-2.

Theorem 5 is known as "Artin's product formula". From now on, we will write k_A^1 for the kernel of the morphism $z \to |z|_A$, i.e. for the subgroup of k_A^\times given by $|z|_A = 1$; by theorem 5, this contains k^\times.

COROLLARY 1. *If k is of characteristic $p > 1$, k_A^\times is the direct product of k_A^1 and of a discrete subgroup isomorphic to \mathbf{Z}.*

For every place v of k, k_v is of characteristic p, so that $|x|_v$, for every $x \in k_v^\times$, is in the subgroup of \mathbf{R}_+^\times generated by p; therefore the same is true of $|z|_A$ for every $z \in k_A^\times$. This is the same as to say that the image of k_A^\times under the morphism $z \to |z|_A$ is a subgroup of the group in question; as it is clearly not reduced to $\{1\}$, it is generated by some integer $Q = p^N$, where N is an integer ≥ 1. Take $z_1 \in k_A^\times$ such that $|z_1|_A = Q$; then k_A^\times is the direct product of k_A^1 and of the subgroup generated by z_1, which is clearly discrete and isomorphic to \mathbf{Z}.

COROLLARY 2. *Assume that k is of characteristic 0; for each $\lambda \in \mathbf{R}_+^\times$, call $z(\lambda)$ the idele (z_v) such that $z_v = 1$ for every finite place v and $z_w = \lambda$*

for every infinite place w of k. Then $\lambda \rightarrow z(\lambda)$ *is an isomorphism of* \mathbf{R}_+^\times
onto a closed subgroup M of $k_{\mathbf{A}}^\times$, *and* $k_{\mathbf{A}}^\times$ *is the direct product of* $k_{\mathbf{A}}^1$ *and of M.*

With the notation of the corollary of prop. 2, §3, it is clear that
$\lambda \rightarrow z(\lambda)$ is an isomorphism of \mathbf{R}_+^\times onto a subgroup M of $k_{\mathbf{A}}(P_\infty)^\times$. The
definition of $|z|_{\mathbf{A}}$, together with corollary 2 of th. 4, Chap. III-4, shows
that $|z(\lambda)|_{\mathbf{A}} = \lambda^n$, n being the degree of k over \mathbf{Q}. The last assertion is now
obvious.

THEOREM 6. *Let* $k_{\mathbf{A}}^1$ *be the subgroup of* $k_{\mathbf{A}}^\times$ *defined by* $|z|_{\mathbf{A}} = 1$. *Then*
k^\times *is a discrete subgroup of* $k_{\mathbf{A}}^1$; *the factor-group* $k_{\mathbf{A}}^1/k^\times$ *is compact; and*
$k_{\mathbf{A}}^\times/k^\times$ *is the direct product of that compact group and of a group isomorphic*
to \mathbf{R}_+^\times *or to* \mathbf{Z} *according as k is of characteristic* 0 *or not.*

The first assertion is contained in th. 5; the second one is the special
case $D = k$, $\mu = 1$ of th. 4 of §3; the others follow at once from the corollaries
of th. 5.

We will now investigate more closely the structure of various sub-
groups of $k_{\mathbf{A}}^\times$ and of k^\times and of some of their factor-groups. It will be
convenient to write $\Omega(P)$ for the group denoted by $k_{\mathbf{A}}(P)^\times$ in the corollary
of prop. 2, §3. In other words, we will write, from now on:

$$(5) \qquad\qquad \Omega(P) = \prod_{v \in P} k_v^\times \times \prod_{v \notin P} r_v^\times.$$

As always, P is assumed to be a finite set of places of k, containing the
set P_∞ of the infinite places; it may be empty, but only if k is not of
characteristic 0. We recall that $\Omega(P)$ is always an open subgroup of
$k_{\mathbf{A}}^\times$; clearly it is compact if and only if P is empty. We will also write:

$$\Omega_1(P) = \Omega(P) \cap k_{\mathbf{A}}^1;$$

here we may take $P = \emptyset$ if k is of characteristic $p > 1$, and then we have
$\Omega_1(\emptyset) = \Omega(\emptyset)$.

THEOREM 7. *If P is not empty, the group* $k_{\mathbf{A}}^\times/k^\times \Omega(P)$ *is finite. When k*
is of characteristic $p > 1$, $k_{\mathbf{A}}^1/k^\times \Omega(\emptyset)$ *is finite, and* $k_{\mathbf{A}}^\times/k^\times \Omega(\emptyset)$ *is the direct*
product of that group and of a group isomorphic to \mathbf{Z}.

In all cases, $k_{\mathbf{A}}^1/k^\times \Omega_1(P)$ is isomorphic to the quotient of $k_{\mathbf{A}}^1/k^\times$ by
the image of $\Omega_1(P)$ in $k_{\mathbf{A}}^1/k^\times$. As $\Omega_1(P)$ is open in $k_{\mathbf{A}}^1$, that image is open;
as $k_{\mathbf{A}}^1/k^\times$ is compact by th. 6, the quotient in question is finite. If k is of
characteristic 0, $\Omega(P)$ contains the group M defined in corollary 2 of
th. 5; that corollary shows then that $\Omega(P)$ is the direct product of $\Omega_1(P)$
and of M, so that $k_{\mathbf{A}}^\times/k^\times \Omega(P)$ may be identified with $k_{\mathbf{A}}^1/k^\times \Omega_1(P)$.
Assume now that k is of characteristic $p > 1$. As $\Omega(\emptyset) = \Omega_1(\emptyset)$, corollary 1
of th. 5 shows that $k_{\mathbf{A}}^\times/k^\times \Omega(\emptyset)$ is the direct product of the finite group

$g = k_A^1/k^\times \, \Omega(\emptyset)$ and of a group γ isomorphic to **Z**. If $P \neq \emptyset$, $\Omega(P)$ contains $\Omega(\emptyset)$ and is not contained in k_A^1; therfore $k_A^\times/k^\times \, \Omega(P)$ is the quotient of $k_A^\times/k^\times \, \Omega(\emptyset)$, i. e. of $g \times \gamma$, by the image of $k^\times \, \Omega(P)$ in that group, and that image is not contained in the image g of k_A^1; it is then obvious that this quotient is a finite group.

COROLLARY. *Notations being as in theorem 7, one can choose P so that* $k_A^\times = k^\times \, \Omega(P)$.

Take any non-empty P', and take a full set of representatives z_1, \ldots, z_N for the classes in k_A^\times modulo $k^\times \, \Omega(P')$. As k_A^\times is the union of all the groups $\Omega(P)$, one can choose $P \supset P'$ so that all the z_i are in $\Omega(P)$. Then P has the required property.

In the case when k is an algebraic number-field, and $P = P_\infty$, theorem 7, as will be seen in the next Chapter, is in substance the classical theorem of the finiteness of the number of ideal-classes in k.

THEOREM 8. *Let F be the set of the elements ξ of k such that $|\xi|_v \leqslant 1$ for all places v of k, and put $E = F - \{0\}$. Then E is a finite cyclic group consisting of all the roots of 1 in k.*

The set F is the intersection of k and of the set of the elements (x_v) of k_A such that $|x_v|_v \leqslant 1$ for all v; clearly the latter set is compact, and, by th. 2 of § 2, k is discrete in k_A; therefore F is finite. If $\xi \in E$, th. 5 shows that we must have $|\xi|_v = 1$ for all v; therefore E is a subgroup of k^\times of finite order, hence cyclic by lemma 1 of Chap. I-1. Conversely, it is obvious that every root of 1 in k must be in E.

COROLLARY. *If k is of characteristic $p > 1$, the set F defined in theorem 8 is a finite field, the algebraic closure of the prime field in k.*

Here the definition of F can be written as $F = k \cap (\prod r_v)$, where the product is taken over all the places v of k; this shows that F is a ring; as $E = F - \{0\}$ is a group, F is a field. By th. 2 of Chap. I-1, if an element of k, other than 0, is algebraic over the prime field, it is a root of 1, so that, by th. 8, it is in E.

When k is of characteristic $p > 1$, the finite field F defined in the corollary of th. 8 is called *the field of constants of k*.

Now, the set P being as before, we define a subgroup $E(P)$ of k^\times by putting

$$E(P) = k^\times \cap \Omega(P) = k^\times \cap \left(\prod_{v \in P} k_v^\times \times \prod_{v \notin P} r_v^\times \right).$$

This consists of the elements ξ of k^\times such that $|\xi|_v = 1$ for all v not in P. Obviously $E(P)$ contains the group E defined in theorem 8. As k^\times is

discrete in k_A^\times, $E(P)$ is a discrete subgroup of $\Omega(P)$, and also, in view of th. 5, of $\Omega_1(P)$. One may also describe $E(P)$ as the group $k(P)^\times$ of invertible elements (or, as one says traditionally, of "units") of the subring $k(P)$ of k given by

$$k(P) = k \cap \Big(\prod_{v \in P} k_v \times \prod_{v \notin P} r_v \Big),$$

and consisting of the elements ξ of k such that $|\xi|_v \leqslant 1$ for all v not in P. In order to determine the structure of $E(P)$, we need an elementary lemma:

LEMMA 5. *Let G be a group, isomorphic to $\mathbf{R}^r \times \mathbf{Z}^{s+1-r}$, with $s \geqslant r \geqslant 0$. If $r > 0$, let λ be a morphism of G into \mathbf{R}, non-trivial on \mathbf{R}^r; otherwise let λ be a non-trivial morphism of G into \mathbf{Z}. Let G_1 be the kernel of λ, and let Γ be a discrete subgroup of G_1, such that G_1/Γ is compact. Then Γ is isomorphic to \mathbf{Z}^s.*

We may assume that $G = \mathbf{R}^r \times \mathbf{Z}^{s+1-r}$; then every element x of G can be written as (x_0, \ldots, x_s), with $x_i \in \mathbf{R}$ for $0 \leqslant i < r$ and $x_i \in \mathbf{Z}$ for $i \geqslant r$, and λ can be written as

$$x = (x_0, \ldots, x_s) \to \lambda(x) = \sum_{i=0}^{s} a_i x_i$$

with $a_i \in \mathbf{R}$ for all i, if $r > 0$, and $a_i \in \mathbf{Z}$ for all i, if $r = 0$; in both cases, in view of our assumptions about λ, we may assume that $a_0 \neq 0$, and in the former case we may assume that $a_0 = 1$. Consider G as embedded in the obvious manner in the vector-space $V = \mathbf{R}^{s+1}$ over \mathbf{R}; then the above formula defines λ as a linear form on V; let V_1 be the subspace of V defined by $\lambda(x) = 0$, so that $G_1 = G \cap V_1$. For $1 \leqslant j \leqslant s$, call e_j the point (x_i) in V given by $x_0 = -a_j$, $x_j = a_0$, and $x_i = 0$ for $i \neq 0$ and $i \neq j$. As $\{e_1, \ldots, e_s\}$ is a basis for V_1, it generates an \mathbf{R}-lattice H in V_1, so that V_1/H is compact; as $H \subset G_1$, and G_1 is closed in V_1, this implies that V_1/G_1 is compact. Consequently, if Γ is as in the lemma, V_1/Γ is compact, so that Γ is an \mathbf{R}-lattice in V_1, hence isomorphic to \mathbf{Z}^s by prop. 11 of Chap. II-4.

THEOREM 9. *Let P be any finite set of places of k, containing P_∞; let $E(P)$ be the subgroup of k^\times consisting of the elements ξ of k^\times such that $|\xi|_v = 1$ for all v not in P. Then $E(P)$ is the direct product of the group E of all roots of 1 in k, and of a group isomorphic to \mathbf{Z}^s, with $s = 0$ if P is empty, and $s = \mathrm{card}(P) - 1$ otherwise.*

If P is empty, this is contained in th. 8; therefore we may assume $P \neq \emptyset$. Call ν the morphism of $\Omega(P)$ into \mathbf{R}_+^\times induced by $z \to |z|_A$; its kernel is $\Omega_1(P)$ and is open in k_A^1. The canonical morphism of k_A^1 onto k_A^1/k^\times induces on $\Omega_1(P)$ a morphism of $\Omega_1(P)$ onto its image in k_A^1/k^\times, with the

kernel $E(P)$ since $k^\times \cap \Omega_1(P)$ is the same as $k^\times \cap \Omega(P)$. Therefore $\Omega_1(P)/E(P)$ is isomorphic to an open subgroup of k_A^1/k^\times, hence compact by th. 6. On the other hand, for each place v of k, call U_v the compact subgroup of k_v^\times defined by $|x|_v = 1$, this being the same as r_v^\times when v is a finite place; put $U = \prod U_v$, the product being taken over all the places of k; this is a compact subgroup of $\Omega(P)$ and of $\Omega_1(P)$. Put $G = \Omega(P)/U$; clearly this is isomorphic to the product of the groups k_v^\times/U_v for $v \in P$; as k_v^\times/U_v is isomorphic to \mathbf{R}_+^\times, or, what amounts to the same, to \mathbf{R}, when v is an infinite place, and to \mathbf{Z} otherwise, G is isomorphic to $\mathbf{R}^r \times \mathbf{Z}^{s+1-r}$, where r is the number of infinite places of k, and s is as defined in our theorem. As U is contained in the kernel $\Omega_1(P)$ of v in $\Omega(P)$, v determines on G a morphism of G into \mathbf{R}_+^\times, or, what amounts to the same, a morphism λ of G into \mathbf{R}, which is clearly non-trivial on each one of the factors k_v^\times/U_v of G, and in particular on those which are isomorphic to \mathbf{R} if there are such factors, i.e. if $r > 0$. On the other hand, if $r = 0$, we know, by corollary 1 of th. 5, that $|z|_A$ takes its values in a group isomorphic to \mathbf{Z}, so that, up to an isomorphism, λ maps G into \mathbf{Z}. Therefore G and λ satisfy the assumptions in lemma 5; the kernel G_1 of λ is here the image of $\Omega_1(P)$ in G, i.e. $\Omega_1(P)/U$. Call now Γ the image of $E(P)$ in G. If W is any compact neighborhood of 1 in $\Omega(P)$, WU is compact and has therefore a finite intersection with $E(P)$. As the image of that intersection in G is the intersection of Γ with the image of WU in G, and as the latter is a neighborhood of 1 in G, this shows that Γ is discrete in G. The factor-group G_1/Γ is isomorphic to $\Omega_1(P)/E(P)U$, hence to a factor-group of the compact group $\Omega_1(P)/E(P)$, and is therefore compact. We can now apply lemma 5 to G, λ and Γ; it shows that Γ is isomorphic to \mathbf{Z}^s. As $E(P) \cap U = E$, the morphism of $E(P)$ onto Γ, induced by the canonical morphism of $\Omega(P)$ onto G, has the kernel E. Let now e_1, \ldots, e_s be representatives in $E(P)$ of a set of s free generators of Γ; obviously they generate a subgroup of $E(P)$, isomorphic to \mathbf{Z}^s, and $E(P)$ is the direct product of E and of that group. This proves our theorem; we have also proved the following:

COROLLARY. *Assume that P is not empty; let $E(P)$ be as in theorem 9; put $\Omega_1(P) = \Omega(P) \cap k_A^1$ and $G_1 = \Omega_1(P)/U$, where U is the group of the elements (z_v) of k_A^\times such that $|z_v|_v = 1$ for all v. Then the image Γ of $E(P)$ in G_1 is discrete in G_1, and G_1/Γ is compact.*

In the case when k is an algebraic number-field, and $P = P_\infty$, theorem 9, as will be seen in the next Chapter, is Dirichlet's famous "theorem of the units".

Chapter V

Algebraic number-fields

§ 1. Orders in algebras over Q. We shall need some elementary results about vector-spaces over **Q**, involving the following concept:

DEFINITION 1. *Let E be a vector-space of finite dimension over* **Q**. *By a* **Q**-*lattice in E, we understand a finitely generated subgroup of E which contains a basis of E over* **Q**.

PROPOSITION 1. *Let E be a vector-space of finite dimension over* **Q**; *let L, L' be two* **Q**-*lattices in E. Then there is an integer* $m > 0$ *such that* $mL \subset L'$.

Let $\{e_1, ..., e_r\}$ and $\{e'_1, ..., e'_s\}$ be finite sets of generators for L and for L', respectively. As the latter must contain a basis for E over **Q**, we can write (perhaps not uniquely) $e_i = \sum a_{ij} e'_j$ for $1 \leqslant i \leqslant r$, with coefficients $a_{ij} \in$ **Q**. Take for m an integer > 0 such that $m a_{ij} \in$ **Z** for all i, j. Then $mL \subset L'$.

COROLLARY 1. *Let E be as in proposition 1. Then every* **Q**-*lattice L in E has a set of generators which is a basis of E over* **Q**.

Let β be a basis of E over **Q**, contained in L; let L' be the **Q**-lattice generated by β; by proposition 1, there is an integer $m > 0$ such that $mL \subset L'$. Consider E as embedded in $E_{\mathbf{R}} = E \otimes_{\mathbf{Q}} \mathbf{R}$. By prop. 11 of Chap. II-4, L' is an **R**-lattice in $E_{\mathbf{R}}$; as L is contained in $m^{-1} L'$, the same proposition shows, firstly, that L is also an **R**-lattice in $E_{\mathbf{R}}$, and secondly that it is generated by a basis of $E_{\mathbf{R}}$ over **R**; as this basis is contained in E, it is clearly a basis of E over **Q**.

COROLLARY 2. *Let E and L be as in corollary 1. Then every subgroup L' of L which contains a basis of E over* **Q** *is a* **Q**-*lattice in E.*

Let β' be a basis of E over **Q**, contained in L'; let L'' be the **Q**-lattice generated by β'. By proposition 1, there is an integer $m > 0$ such that $mL \subset L''$. Then $m^{-1} L'' \supset L \supset L' \supset L''$. Clearly, if n is the dimension of E over **Q**, L'' has the index m^n in $m^{-1} L''$. Therefore L' is of finite index in L'; as L' is generated by β' and any full set of representatives of the classes modulo L'' in L', this proves our corollary.

DEFINITION 2. *Let \mathscr{A} be an algebra of finite dimension over* **Q**. *A sub-ring of \mathscr{A} will be called an order of \mathscr{A} if it is a* **Q**-*lattice in \mathscr{A} when \mathscr{A} is viewed as a vector-space over* **Q**.

Here, as always, a subring of \mathscr{A} is understood to contain the unit of \mathscr{A}.

PROPOSITION 2. *Every algebra \mathscr{A} of finite dimension over* **Q** *contains at least one order.*

Let $\{a_1,...,a_N\}$ be a finite subset of \mathscr{A}, containing a basis of \mathscr{A} over **Q**; then we can write $a_i a_j = \sum c_{ijh} a_h$ for all i,j, with coefficients c_{ijh} in **Q**. Let m be an integer >0, such that $mc_{ijh}\in\mathbf{Z}$ for all i,j,h. Then the **Q**-lattice generated by $1, ma_1,...,ma_N$ is an order.

Take for instance $\mathscr{A}=\mathbf{Q}$. By corollary 1 of prop. 1, every **Q**-lattice in **Q** is of the form $a\mathbf{Z}$, with $a\in\mathbf{Q}^\times$. If this is an order, we must have $a^2\in a\mathbf{Z}$, hence $a\in\mathbf{Z}$, and $1\in a\mathbf{Z}$, hence $a^{-1}\in\mathbf{Z}$; this gives $a=\pm 1$, which shows that **Z** is the only order in **Q**.

PROPOSITION 3. *Let a be any element of an order in an algebra \mathscr{A} of finite dimension over* **Q**. *Then a is integral over* **Z**, *and $Tr_{\mathscr{A}/k}(a)$ and $N_{\mathscr{A}/k}(a)$ are in* **Z**.

Let R be an order containing a, and let $\{a_1,...,a_N\}$ be a finite set of generators for R. Then we can write $a\cdot a_i=\sum c_{ij}a_j$ for $1\leqslant i\leqslant N$, with coefficients $c_{ij}\in\mathbf{Z}$; this can be written as $\sum(\delta_{ij}a-c_{ij})a_j=0$, where (δ_{ij}) is the unit matrix 1_N. Write $D(T)$ for the determinant of the matrix $(\delta_{ij}T-c_{ij})$, where T is an indeterminate, and $D_{ij}(T)$ for its minors for $1\leqslant i,j\leqslant N$; these are polynomials in $\mathbf{Z}[T]$, and we have

$$\sum_i D_{ih}(T)\cdot(\delta_{ij}T-c_{ij})=\delta_{hj}D(T)$$

for $1\leqslant h,j\leqslant N$. Substitute a for T, multiply to the right with a_j, and sum over j for $1\leqslant j\leqslant N$; we get $D(a)a_h=0$ for all h, hence $D(a)x=0$ for all x; for $x=1$, this gives $D(a)=0$, which proves our first assertion since $D(T)$ is monic. By corollary 1 of prop. 1, we may assume that we have taken for $\{a_1,...,a_N\}$ a basis of \mathscr{A} over **Q**; then $Tr_{\mathscr{A}/k}(a)$ and $N_{\mathscr{A}/k}(a)$ are the trace and the determinant of the matrix (c_{ij}), so that they are integers.

§ 2. Lattices over algebraic number-fields.

From now on, until the end of this Chapter, k will denote an algebraic number-field. We keep the notations explained in Chapter IV. In particular, if v is any place of k, k_v is the completion of k at v; if v is a finite place, r_v is the maximal compact subring of k_v, and p_v the maximal ideal of r_v. We write $k_\mathbf{A}$ for the adele ring of k, and φ for the canonical injection of k into $k_\mathbf{A}$. We will write φ_E for the canonical injection of any finite-dimensional vector-space E

over k into its adele space $E_\mathbf{A}$, this being defined by $e \to e \otimes \varphi(1)$ as explained in Chap. IV-1.

Consider now the algebra $k \otimes_\mathbf{Q} \mathbf{R}$ over \mathbf{R}; this is the same as $(k/\mathbf{Q})_\infty$ in the notation of th. 1, Chap. IV-1, and it has an isomorphism Φ_∞ onto the direct product $\prod k_w$ of the completions of k at its infinite places w, this being fully characterized by the properties stated in th. 4 of Chap. III-4. We will simplify notations by identifying $(k/\mathbf{Q})_\infty$ with that product by means of Φ_∞, and by writing k_∞ for both. Similarly, if E is any finite-dimensional vector-space over k, we will write E_∞ for $E \otimes_\mathbf{Q} \mathbf{R}$, which is the same as $(E/\mathbf{Q})_\infty$ in the notation of corollary 2 of th. 1, Chap. IV-1; as this is also the same as $E \otimes_k k_\infty$, we identify it with the product $\prod E_w$ taken over the infinite places w of k.

With this notation, the open subgroup $k_\mathbf{A}(P_\infty)$ of $k_\mathbf{A}$, given by formula (1) of Chap. IV-1, can be written as $k_\infty \times (\prod r_v)$, where the latter product is taken over all the finite places v of k and is compact. Here, and in similar situations, the following group-theoretic lemma will be found useful:

LEMMA 1. *Let G be a locally compact group with an open subgroup G_1 of the form $G_1 = G' \times G''$, where G' is locally compact and G'' is compact. Let Γ be a discrete subgroup of G such that G/Γ is compact, and call Γ' the projection of $\Gamma \cap G_1$ onto G'. Then Γ' is discrete in G', and G'/Γ' is compact.*

Let W be a compact neighborhood of the neutral element in G' (we need not assume that G, G', G'' are commutative, although only this case will be used). As $W \times G''$ is compact, its intersection with Γ is finite; as the projection of that intersection onto G' is $W \cap \Gamma'$, this shows that Γ' is discrete. As G_1 is open in G, $G_1 \Gamma$ and $G - G_1 \Gamma$ are open, since they are unions of left cosets for G_1; therefore the image of G_1 in G/Γ is open and closed there, hence compact. As it is isomorphic to G_1/Γ_1 with $\Gamma_1 = \Gamma \cap G_1$, this implies that there is a compact subset C of G_1 such that $G_1 = C \cdot \Gamma_1$. Then, if C' is the projection of C onto G', $G' = C' \cdot \Gamma'$, which shows that G'/Γ' is compact.

THEOREM 1. *Let k be an algebraic number-field; put $\mathfrak{r} = \bigcap_v (k \cap r_v)$, where v runs through all the finite places of k. Then \mathfrak{r} is an order of k; it is the unique maximal order of k, and it is the integral closure of \mathbf{Z} in k.*

As explained above, write $k_\mathbf{A}(P_\infty)$ as $k_\infty \times (\prod r_v)$. Clearly an element ξ of k is in \mathfrak{r} if and only if $\varphi(\xi)$ is in that product; when that is so, write $\varphi_\infty(\xi)$ and $\psi(\xi)$ for the projections of $\varphi(\xi)$ onto k_∞ and onto $\prod r_v$, respectively. Obviously \mathfrak{r} is a subring of k. Now apply lemma 1 to $G = k_\mathbf{A}$, $G_1 = k_\mathbf{A}(P_\infty)$, $G' = k_\infty$, $G'' = \prod r_v$, $\Gamma = \varphi(k)$; then, with the notations of

that lemma, Γ' is $\varphi_\infty(\mathfrak{r})$, and the lemma shows that this is an \mathbf{R}-lattice in k_∞. As φ_∞ is also the same as the injection induced on \mathfrak{r} by the natural injection of k into $k_\infty = k \otimes_\mathbf{Q} \mathbf{R}$, this implies that \mathfrak{r} is a \mathbf{Q}-lattice in k, hence an order. Let \mathfrak{r}' be any subring of k whose additive group is finitely generated; clearly the r_v-module generated by \mathfrak{r}' in k_v is a compact subring of k_v; it contains r_v, since \mathfrak{r}' contains 1; therefore it is r_v, so that $\mathfrak{r}' \subset r_v$. As this is true for all v, we get $\mathfrak{r}' \subset \mathfrak{r}$. By prop. 3 of § 1, \mathfrak{r} is contained in the algebraic closure of \mathbf{Z} in k. Conversely, if an element of k is integral over \mathbf{Z}, prop. 6 of Chap. I-4 shows that it is in r_v for all v, hence in \mathfrak{r}.

The mapping ψ of \mathfrak{r} into $\prod r_v$, defined in the proof of theorem 1, will be called *the canonical injection* of \mathfrak{r} into $\prod r_v$; it maps every $\xi \in \mathfrak{r}$ onto the element (x_v) of that product given by $x_v = \xi$ for all v. It is a ring-isomorphism of \mathfrak{r} onto $\psi(\mathfrak{r})$, addition and multiplication in $\prod r_v$ being defined coordinatewise. With this notation, we have:

COROLLARY 1. *Let k, \mathfrak{r} and ψ be as above defined. Then $\psi(\mathfrak{r})$ is dense in $\prod r_v$, and its projection onto every partial product of that product is dense there. In particular, r_v is the closure of \mathfrak{r} in k_v.*

Let G, G_1, G', G'', Γ be as in the proof of theorem 1. By corollary 2 of th. 3, Chap. IV-2, $k_\infty + \varphi(k)$, which is the same as $G'\Gamma$ in that notation, is dense in $G = k_A$, so that its intersection with G_1 must be dense in G_1; as that intersection is $k_\infty + \varphi(\mathfrak{r})$, this implies that its projection onto $G'' = \prod r_v$, which is the same as the projection $\psi(\mathfrak{r})$ of $\varphi(\mathfrak{r})$ onto G'', is dense there. The other statements in our corollary follow trivially from this.

COROLLARY 2. *If k' is a finite algebraic extension of k, the maximal order of k' is the integral closure of \mathfrak{r} in k'.*

This follows again from prop. 6 of Chap. I-4, just as in the proof of theorem 1.

DEFINITION 3. *Let k be an algebraic number-field, \mathfrak{r} its maximal order, and E a vector-space of finite dimension over k. An \mathfrak{r}-module in E will be called a k-lattice in E if it is finitely generated and contains a basis of E over k.*

If k' is a finite algebraic extension of k, \mathfrak{r}' its maximal order, and E a vector-space of finite dimension over k', it is clear that an \mathfrak{r}'-module in E is a k'-lattice if and only if it is a k-lattice when E is viewed as a vector-space over k.

Let E be a vector-space of finite dimension over k. Let L be a k-lattice in E; let ε be a finite subset of E such that L is the \mathfrak{r}-module generated by ε in E. Then, for every finite place v of k, the r_v-module ε_v generated

by ε is the same as the r_v-module L_v generated by L, and prop. 1 of Chap. IV-1 shows that $E_A(P_\infty,\varepsilon)$ is the same as $E_\infty \times \prod L_v$ and is an open subgroup of E_A. For every $e \in L$, we can define an element (e_v) of $\prod L_v$ by putting $e_v = e$ for all v; if we call this element $\psi_L(e)$, ψ_L will be called the canonical injection of L into $\prod L_v$. Then:

PROPOSITION 4. *Let E be a vector-space of finite dimension over k. Let L be a k-lattice in E; for every finite place v of k, let L_v be the r_v- module generated by L in E_v; and let ψ_L be the canonical injection of L into $\prod L_v$. Then $\psi_L(L)$ is dense in $\prod L_v$; its projection onto every partial product of $\prod L_v$ is dense there; in particular, for every v, L_v is the closure of L in E_v.*

Let $\varepsilon = \{e_1,\ldots,e_N\}$ be a finite subset of L such that L is the r-module generated by ε. Take any element (e_v) of $\prod L_v$; then, for every v, we can write $e_v = \sum x_v^{(i)} e_i$ with coefficients $x_v^{(i)}$ in r_v. Put $x_i = (x_v^{(i)})$ for $1 \leqslant i \leqslant N$; the x_i are elements of $\prod r_v$. By corollary 1 of th. 1, we can find elements ξ_i of r such that, for every i, $\psi(\xi_i)$ is arbitrarily close to x_i; clearly, then, $\psi_L(\sum \xi_i e_i)$ can be made to be arbitrarily close to (e_v).

THEOREM 2. *Let k be an algebraic number-field, E a vector-space of finite dimension over k, and L a k-lattice in E. For each finite place v of k, let L_v be the closure of L in E_v and M_v any k_v-lattice in E_v. Then there is a k-lattice M in E whose closure in E_v is M_v for every v if and only if $M_v = L_v$ for almost all v; when that is so, there is only one such k-lattice, and it is given by $M = \bigcap_v (E \cap M_v)$.*

Assume that there is such a k-lattice M; in view of prop. 4, the fact that then $M_v = L_v$ for almost all v is merely a restatement of corollary 1 of th. 3, Chap. III-1. Now assume that $M_v = L_v$ for almost all v; in view of prop. 1 of Chap. IV-1, this implies that $E_\infty \times \prod M_v$ is open in E_A. We can therefore apply lemma 1 to $G = E_A$, $G' = E_\infty$, $G'' = \prod M_v$ and $\Gamma = \varphi_E(E)$, where φ_E is the canonical injection of E into E_A. Clearly, if we put $M = \bigcap (E \cap M_v)$, $\varphi_E(M)$ is the same as $\varphi_E(E) \cap G_1$ with $G_1 = G' \times G''$; lemma 1 shows now that M is an R-lattice in E_∞, hence a Q-lattice in E; as it is obviously an r-module, it is a k-lattice. By corollary 2 of th. 3, Chap. IV-2, $E_\infty + \varphi_E(E)$ is dense in E_A; therefore its intersection $E_\infty + \varphi_E(M)$ with G_1 is dense in G_1. This is the same as to say that the projection of $\varphi_E(M)$ onto $G'' = \prod M_v$ is dense there, and implies that M is dense in M_v for every v. As above, call ψ_M the canonical injection of M into $\prod M_v$. Assume now that there is another k-lattice M' in E with the closure M_v in E_v for every v; clearly M' is contained in M; moreover, by prop. 4, $\psi_M(M')$ is dense in $\prod M_v$, hence also in $\psi_M(M)$.

By prop. 1, there is an integer $m > 0$ such that $M' \supset mM$. Call G_m the image of G_1 under the automorphism $e \to \varphi(m)e$ of E_A; this can be written as $G_m = G' \times G''_m$ with $G''_m = \prod (mM_v)$; clearly $mM_v = M_v$ for almost all v (viz. for all the finite places of k which do not lie above some prime divisor of m in \mathbf{Z}), and G''_m is an open subgroup of G''. Then $\varphi_E(mM)$ is the same as $\varphi_E(E) \cap G_m$, hence also the same as $\varphi_E(M) \cap G_m$, and is contained in $\varphi_E(M')$; this is the same as to say that $\psi_M(M) \cap G''_m$ is contained in $\psi_M(M')$. Now take any $\mu \in M$; as $\psi_M(M')$ is dense in $\psi_M(M)$, there is $\mu' \in M'$ such that $\psi_M(\mu - \mu')$ is in G''_m; then it must be in $\psi_M(M')$, so that $\mu - \mu' \in M'$ and $\mu \in M'$. This shows that $M = M'$, which completes the proof.

COROLLARY. *Let L, L' be two k-lattices in E. Then $L + L'$ and $L \cap L'$ are k-lattices in E, and, for every finite place v of k, their closures in E_v are given in terms of the closures L_v, L'_v of L, L' by*

$$(L + L')_v = L_v + L'_v, \qquad (L \cap L')_v = L_v \cap L'_v.$$

The assertions about $L + L'$ follow at once from prop. 4. As to $L \cap L'$, put $M_v = L_v \cap L'_v$ for every v; for every v, this is a k_v-lattice in E_v, and it is the same as L_v for almost all v. Therefore there is a k-lattice M in E with the closure M_v in E_v for every v, and it is given by $M = \bigcap (E \cap M_v)$; in view of th. 2, this is the same as $L \cap L'$.

§ 3. **Ideals.** In this §, k will denote an algebraic number-field and \mathfrak{r} its maximal order; the results of § 2 will be applied to the case $E = k$. Clearly an \mathfrak{r}-module other than $\{0\}$ in k is a k-lattice if and only if it is finitely generated. By prop. 1 of § 1, if \mathfrak{a} is a k-lattice in k, there is an integer $m > 0$ such that $m\mathfrak{a}$ is contained in \mathfrak{r}; then, clearly, $m\mathfrak{a}$ is an ideal in the ring \mathfrak{r}. Conversely, by corollary 2 of prop. 1, § 1, every ideal in \mathfrak{r}, other than $\{0\}$, is a k-lattice. This shows that a subset of k is a k-lattice if and only if it is of the form $\xi \mathfrak{a}$, where \mathfrak{a} is an ideal in \mathfrak{r}, other than $\{0\}$, and $\xi \in k^\times$.

DEFINITION 4. *Any k-lattice in k will be called a fractional ideal in k; a fractional ideal in k is said to be integral if it is contained in \mathfrak{r}.*

Accordingly, $\{0\}$ is not a fractional ideal.

Let \mathfrak{a} be a fractional ideal in k, and let L be a k-lattice in a vector-space E of finite dimension over k. By $\mathfrak{a}L$, one understands the subgroup of E generated by the elements αe with $\alpha \in \mathfrak{a}$, $e \in L$; this is clearly a k-lattice in E. Let v be any finite place of k; as before, write \mathfrak{a}_v for the closure of \mathfrak{a} in k_v, and L_v, $(\mathfrak{a}L)_v$ for those of L, $\mathfrak{a}L$ in E_v; by prop. 4 of § 2, these are the same as the \mathfrak{r}_v-modules generated respectively by \mathfrak{a}, L

and aL; this makes it clear that $(aL)_v$ is the same as the subgroup $a_v L_v$ of E_v generated by the elements αe with $\alpha \in a_v$, $e \in L_v$.

In particular, if a, b are two fractional ideals in k, ab is the subgroup of the additive group of k generated by the elements $\alpha\beta$ with $\alpha \in a$, $\beta \in b$; it is a fractional ideal, and, for every finite place v of k, we have $(ab)_v = a_v b_v$. If p_v is the maximal ideal in r_v, every k_v-lattice in k_v is of the form p_v^n with $n \in \mathbf{Z}$; in particular, we can write $a_v = p_v^a$, $b_v = p_v^b$ with $a \in \mathbf{Z}$, $b \in \mathbf{Z}$, and then it is obvious that $a_v b_v = p_v^{a+b}$.

THEOREM 3. *Let k be an algebraic number-field and r its maximal order. For every finite place v of k, put $p_v = r \cap p_v$. Then $v \to p_v$ is a bijection of the set of finite places of k onto the set of the prime ideals in r, other than $\{0\}$. For the law $(a,b) \to ab$, the set of the fractional ideals in k is a group with the neutral element r; it is the free abelian group generated by the prime ideals in r; the ideals in r, other than $\{0\}$, make up the monoid generated by these prime ideals.*

For every fractional ideal a in k, we can define a mapping $v \to a(v)$ of the set of finite places of k into \mathbf{Z} by writing $a_v = p_v^{a(v)}$. For $a = r$, all the $a(v)$ are 0. Theorem 2 of § 2 shows now that a given mapping $v \to a(v)$ belongs to a fractional ideal a if and only if $a(v) = 0$ for almost all v, and that it determines a uniquely when that is so, a being then given by $a = \bigcap (k \cap p_v^{a(v)})$. If b corresponds similarly to $v \to b(v)$, we have seen above that ab corresponds to $v \to a(v) + b(v)$; it is also clear that $a \subset b$ if and only if $a(v) \geq b(v)$ for all v; in particular, a is integral if and only if $a(v) \geq 0$ for all v. For any given v, put $a(v) = 1$ and $a(v') = 0$ for all $v' \neq v$; if we call p_v the corresponding ideal, we have $p_v = r \cap p_v$, and it is clear that the fractional ideals make up the free abelian group generated by the p_v. As p_v is prime in r_v, p_v is prime in r. As to the converse, take any a in r, so that $a(v) \geq 0$ for all v; if it is neither r nor any one of the p_v, we can write it as $a'a''$, where a', a'' are ideals in r, other than r. Then a' contains a and is not a, so that $a' - a$ is not empty; the same is true of $a'' - a$. Take $\alpha' \in a' - a$ and $\alpha'' \in a'' - a$. Then $\alpha'\alpha''$ is in a, while neither α' nor α'' is in a, so that a is not prime. This completes the proof.

COROLLARY 1. *Let a, b be two fractional ideals in k; for each v, call $a(v)$ and $b(v)$ the exponents of p_v in a and in b when these are expressed as products of powers of prime ideals of r. Then $a + b$ and $a \cap b$ are fractional ideals in k, and when they are similarly expressed, the exponents of p_v in them are $\min(a(v), b(v))$ and $\max(a(v), b(v))$, respectively.*

This follows at once from th. 3 and the corollary of th. 2, § 2.
As usual, two ideals a, b in r are called *mutually prime* if $a + b = r$.

COROLLARY 2. *Every fractional ideal* \mathfrak{a} *in* k *can be written in one and only one way in the form* $\mathfrak{b}\mathfrak{c}^{-1}$, *where* \mathfrak{b} *and* \mathfrak{c} *are mutually prime ideals in* \mathfrak{r}.

This follows at once from th. 3 and corollary 1. By analogy with the case of \mathbf{Q}, the ideals \mathfrak{b}, \mathfrak{c} in corollary 2 are called the *numerator* and the *denominator* of \mathfrak{a}, respectively.

We will denote by $I(k)$ the group of fractional ideals of k. If $a = (a_v)$ is any element of $k_{\mathbf{A}}^{\times}$, then, by the corollary of prop. 2, Chap. IV-3, we have $|a_v|_v = 1$, hence $a_v r_v = r_v$, for almost all finite places v of k; therefore, by th. 3, there is one and only one fractional ideal \mathfrak{a} of k such that $\mathfrak{a}_v = a_v r_v$ for all finite places v; we will write $\mathfrak{a} = \mathrm{id}(a)$ for this ideal. Clearly the mapping $a \to \mathrm{id}(a)$ of $k_{\mathbf{A}}^{\times}$ into $I(k)$ is surjective; we will write Ω_{∞} for its kernel, which is obviously $k_{\infty}^{\times} \times (\prod r_v^{\times})$, i.e. $k_{\mathbf{A}}(P_{\infty})^{\times}$ in the notation of the corollary of prop. 2, Chap. IV-3, and $\Omega(P_{\infty})$ in the notation of formula (5), Chap. IV-4; as this is an open subgroup of $k_{\mathbf{A}}^{\times}$, $a \to \mathrm{id}(a)$ is a morphism of $k_{\mathbf{A}}^{\times}$ onto $I(k)$ if $I(k)$ is provided with the discrete topology. We may then identify $I(k)$ with $k_{\mathbf{A}}^{\times}/\Omega_{\infty}$.

In particular, for every $\xi \in k^{\times}$, we have $\mathrm{id}(\xi) = \xi \mathfrak{r}$; this is the \mathfrak{r}-module generated by ξ in k, and is frequently denoted by (ξ); its numerator and denominator, as defined above, are called *the numerator* and *the denominator of* ξ. A fractional ideal is called *principal* if it is of the form $\xi \mathfrak{r}$ with $\xi \in k^{\times}$; such ideals make up a subgroup $P(k)$ of $I(k)$, which is the image of k^{\times} under the morphism induced by $a \to \mathrm{id}(a)$. Identifying $I(k)$ with $k_{\mathbf{A}}^{\times}/\Omega_{\infty}$, we see that $P(k)$ is the image of k^{\times} in the latter group; therefore we may identify $I(k)/P(k)$ with $k_{\mathbf{A}}^{\times}/k^{\times}\Omega_{\infty}$, which is a finite group by th. 7 of Chap. IV-4. The elements of $I(k)/P(k)$, or in other words the classes modulo $P(k)$ in $I(k)$, are known as the *ideal-classes* of k. The number of such classes, i.e. the index of $P(k)$ in $I(k)$, will be denoted by h.

THEOREM 4. *Let* k *be an algebraic number-field,* E *a vector-space of finite dimension over* k, *and* L, M *two* k-*lattices in* E *such that* $L \supset M$. *For every finite place* v *of* k, *call* L_v, M_v *the closures of* L, M *in* E_v, *and call* λ_v *the natural homomorphism of* L/M *into* L_v/M_v. *Then* $x \to (\lambda_v(x))$, *where* v *runs through all finite places of* k, *is an isomorphism of* L/M *onto* $\prod_v (L_v/M_v)$ *for their structures as* \mathfrak{r}-*modules,* \mathfrak{r} *being the maximal order of* k.

Call λ that mapping; it is obviously a homomorphism of \mathfrak{r}-modules. Let x be any element of L/M, and e a representative of x in L. If $\lambda(x) = 0$, e must be in M_v for all v; by th. 2 of § 2, this implies that $e \in M$ and $x = 0$. Therefore λ is injective. Now take any element $y = (y_v)$ of $\prod (L_v/M_v)$; for every v, take a representative e_v of y_v in L_v, and put $e = (e_v)$. As $M_v = L_v$ for almost all v, $\prod M_v$ is open in $\prod L_v$; therefore prop. 4 of § 2 shows that there is an element e_0 of L such that $\psi_L(e_0) - e$ is in

$\prod M_v$, ψ_L being the canonical injection of L into $\prod L_v$. This is the same as to say that $\lambda(x_0) = y$ if x_0 is the image of e_0 in L/M, which proves that λ is surjective.

COROLLARY 1. *Assumptions and notations being as in theorem 4, we have* $[L:M] = \prod [L_v:M_v]$.

This is obvious. One should observe that $L_v = M_v$ for almost all v, and that, for all v, M_v is an open subgroup of the compact group L_v, so that $[L_v:M_v]$ is always finite and almost always 1. The fact that $[L:M]$ is finite is implicit in prop. 1 of § 1, or also in lemma 2 of Chap. II-4.

COROLLARY 2. *Let v be a finite place of k; let $\mathfrak{p}_v = \mathfrak{r} \cap \mathfrak{p}_v$ be the prime ideal in the maximal order \mathfrak{r} of k, corresponding to v. Then the natural homomorphism of $\mathfrak{r}/\mathfrak{p}_v$ into r_v/\mathfrak{p}_v is an isomorphism of $\mathfrak{r}/\mathfrak{p}_v$ onto the residual field r_v/\mathfrak{p}_v of r_v.*

COROLLARY 3. *Let \mathfrak{a}, \mathfrak{b} be two fractional ideals in k, such that $\mathfrak{a} \supset \mathfrak{b}$. Let $\mathfrak{a}^{-1}\mathfrak{b} = \prod \mathfrak{p}_v^{n(v)}$ be the expression of $\mathfrak{a}^{-1}\mathfrak{b}$ as a product of prime ideals of \mathfrak{r}. Then $[\mathfrak{a}:\mathfrak{b}] = \prod [\mathfrak{r}:\mathfrak{p}_v]^{n(v)}$.*

By corollary 1, $[\mathfrak{a}:\mathfrak{b}]$ is the product of the indices $[\mathfrak{a}_v:\mathfrak{b}_v]$ for all v. For a given v, we can write $\mathfrak{a}_v = p_v^a$, $\mathfrak{b}_v = p_v^b$, and then we have $b - a = n(v)$. Corollary 2 of th. 6, Chap. I-4, shows that $[p_v^a:p_v^b] = q^{b-a}$ with $q = [r_v:\mathfrak{p}_v]$. Our conclusion follows at once from this and corollary 2.

DEFINITION 5. *Let k be an algebraic number-field and \mathfrak{r} its maximal order. Let $\mathfrak{a} \to \mathfrak{N}(\mathfrak{a})$ be the homomorphism of the group of fractional ideals of k into \mathbf{Q}^\times which is such that $\mathfrak{N}(\mathfrak{p}) = [\mathfrak{r}:\mathfrak{p}]$ for every prime ideal \mathfrak{p} in \mathfrak{r}. Then $\mathfrak{N}(\mathfrak{a})$ is called the norm of the fractional ideal \mathfrak{a} in k.*

Corollary 3 of theorem 5 can now be expressed by saying that, if \mathfrak{a} and \mathfrak{b} are fractional ideals in k, and $\mathfrak{a} \supset \mathfrak{b}$, then $[\mathfrak{a}:\mathfrak{b}]$ is equal to $\mathfrak{N}(\mathfrak{b})/\mathfrak{N}(\mathfrak{a})$. In particular, if \mathfrak{a} is integral, $[\mathfrak{r}:\mathfrak{a}] = \mathfrak{N}(\mathfrak{a})$.

PROPOSITION 5. *Let $a = (a_v)$ be any element of $k_{\mathbf{A}}^\times$. Then $\mathfrak{N}(\mathrm{id}(a))$ is equal to the product $\prod |a_v|_v^{-1}$, taken over all the finite places v of k.*

In view of def. 5, it is enough to verify this for the case when id (a) is a prime ideal of \mathfrak{r}; this is so if and only if, for some finite place v of k, a_v is a prime element of k_v, and $|a_{v'}|_{v'} = 1$ for all finite places $v' \neq v$. Then it is obvious.

COROLLARY 1. *For each $\xi \in k^\times$, we have $N_{k/\mathbf{Q}}(\xi) = (-1)^\rho \mathfrak{N}(\mathrm{id}(\xi))$, ρ being the number of the real places w of k such that the image of ξ in k_w is $\dot{<}0$.*

Combining proposition 5 with th. 5 of Chap. IV-4, we see at once that $\mathfrak{N}(\mathrm{id}(\xi))$ is equal to the product $\prod |\xi_w|_w$, taken over the infinite places w of k. For each real place w of k, and each $x \in k_w^\times$, we have $x = (\mathrm{sgn}\, x) \cdot |x|_w$; for each imaginary place w of k, and each $x \in k_w^\times$, we have $N_{k_w/\mathbf{R}}(x) = x\bar{x} = = |x|_w$. Our conclusion follows now at once from corollary 3 of th. 4, Chap. III-4, applied to k, \mathbf{Q} and the place ∞ of \mathbf{Q}.

COROLLARY 2. *An element ξ of the maximal order \mathfrak{r} of k is invertible in \mathfrak{r} if and only if $N_{k/\mathbf{Q}}(\xi) = \pm 1$.*

Clearly, it is invertible in \mathfrak{r} if and only if $\xi\mathfrak{r} = \mathfrak{r}$; as $\xi\mathfrak{r}$ is the same as $\mathrm{id}(\xi)$, our conclusion follows now at once from corollary 1, combined with the fact that $[\mathfrak{r}:\mathfrak{a}] = \mathfrak{N}(\mathfrak{a})$ for every ideal \mathfrak{a} in the ring \mathfrak{r}.

Traditionally, the elements of \mathfrak{r}^\times, i.e. the invertible elements of \mathfrak{r}, are known as "the units" of k. In the notation of Chap. IV-4, \mathfrak{r}^\times is the same as the group $E(P_\infty)$, as defined in theorem 9 of Chap. IV-4; its structure is given by that theorem; if $r+1$ is the number of the infinite places of k, it is isomorphic to the direct product of the cyclic group E of the roots of 1 in k, and of a group isomorphic to \mathbf{Z}^r. This is Dirichlet's "unit-theorem".

§ 4. Fundamental sets.

Let Γ be a discrete subgroup of a locally compact group G; by a "fundamental set" of G modulo Γ, one understands traditionally a full set X of representatives of the cosets modulo Γ in G, which at the same time is measurable, and which is usually expected to have some additional properties, e.g. to be a Borel set, etc. Then formula (6) of Chap. II-4, applied to G, Γ, to a Haar measure α on G, and to the characteristic function of X, shows that $\alpha(X) = \alpha(G/\Gamma)$; thus the calculation of $\alpha(G/\Gamma)$ may sometimes be effected by constructing a convenient fundamental set. More generally, let us say that a measurable subset X of G is *fundamental of order v* modulo Γ if it has exactly v points in common with every coset modulo Γ; then the same formula gives $\alpha(X) = v\alpha(G/\Gamma)$. This will now be applied to $k_{\mathbf{A}}$ and to $k_{\mathbf{A}}^\times$.

Let k and \mathfrak{r} be as before; call n the degree of k over \mathbf{Q}. As \mathfrak{r} is a \mathbf{Q}-lattice in k when k is viewed as a vector-space over \mathbf{Q}, prop. 11 of Chap. II-4 shows that it has a set of generators $\{\xi_1, \ldots, \xi_n\}$ which is a basis of k over \mathbf{Q}. Then this is also a basis of $k_\infty = k \otimes_{\mathbf{Q}} \mathbf{R}$ over \mathbf{R}; therefore, if we write, for $u = (u_1, \ldots, u_n)$ in \mathbf{R}^n, $\theta(u) = \sum u_i \xi_i$, this defines an isomorphism θ of \mathbf{R}^n onto k_∞.

PROPOSITION 6. *Let k, \mathfrak{r} and θ be as above; call I the interval $0 \leqslant t < 1$ of \mathbf{R}. Then $\theta(I^n) \times \prod_v \mathfrak{r}_v$, where the product is taken over all the finite places v of k, is a fundamental set modulo k in $k_{\mathbf{A}}$.*

Call that set X; it is obviously measurable; we have to show that every element x of k_A can be written in one and only one way as $x_0 + \xi$ with $x_0 \in X$ and $\xi \in k$. By corollary 2 of th. 3, Chap. IV-2, $k_\infty + k$ is dense in k_A; as $k_\infty \times \prod r_v$ is open, this shows that, for a given $x \in k_A$, there is $\eta \in k$ such that $x - \eta$ is in $k_\infty \times \prod r_v$, and the definition of r shows that an element η' of k has the same property if and only if $\eta' - \eta \in r$. Write $y = x - \eta$, and call y_∞ the projection of y onto k_∞ in the product $k_\infty \times \prod r_v$; then we can write $y_\infty = \theta(u)$ with $u = (u_1, ..., u_n)$ in \mathbf{R}^n. For each i, take $a_i \in \mathbf{Z}$ such that $a_i \leqslant u_i < a_i + 1$, i.e. $u_i - a_i \in I$; put $\xi = \eta - \sum_i a_i \xi_i$ and $x_0 = x - \xi$. As $\xi - \eta$ is in r, x_0 is in $k_\infty \times \prod r_v$; moreover, the projection of x_0 onto k_∞ is

$$y_\infty - \sum_i a_i \xi_i = \sum_i (u_i - a_i) \xi_i$$

and is therefore in $\theta(I^n)$. It is also clear that the latter condition could not have been fulfilled by any other choice of the integers a_i. This proves our assertion.

This will now be applied to the calculation of $\alpha(k_A/k)$ for an explicitly given Haar measure α on k_A. Such measures can be constructed as follows. For each place v of k, choose a Haar measure α_v on k_v; if $\alpha_v(r_v) = 1$ for almost all v, the product measure $\prod \alpha_v$ is well defined and is a Haar measure on each one of the open subgroups $k_A(P)$ of k_A given by formula (1) of Chap. IV-1; clearly there is one and only one Haar measure on k_A which coincides with these measures wherever they are defined; this will be denoted by $\prod \alpha_v$. In particular, we will write $\beta = \prod \beta_v$ for the Haar measure obtained by taking $\beta_v(r_v) = 1$ for all finite places v of k, and proceeding as follows at the infinite places. If w is a real place, we have $k_w = \mathbf{R}$ and we take $d\beta_w(x) = dx$, so that β_w is the Lebesgue measure on \mathbf{R}. If w is an imaginary place, we have $k_w = \mathbf{C}$, and we take $d\beta_w(x) = = |dx \wedge d\bar{x}|$; by this we mean that, if we put $x = u + iv$ with u, v in \mathbf{R}, so that $dx \wedge d\bar{x} = -2i(du \wedge dv)$, β_w is the measure corresponding to the differential form $2 du \wedge dv$; in other words, $\beta_w/2$ is the Lebesgue measure in the (u, v)-plane.

In order to calculate $\beta(k_A/k)$, we need another definition. Notations being as above, consider the matrix

(1) $M = (Tr_{k/\mathbf{Q}}(\xi_i \xi_j))_{1 \leqslant i, j \leqslant n}$,

and call D its determinant. By prop. 5 of Chap. III-3, $D \neq 0$; by prop. 3 of § 1, M is in $M_n(\mathbf{Z})$, so that $D \in \mathbf{Z}$. If $k = \mathbf{Q}$, we have $r = \mathbf{Z}$, so that we have to take $\xi_1 = \pm 1$, hence $D = 1$. If $\{\eta_1, ..., \eta_n\}$ is another set of generators for r, and N the matrix obtained by substituting the η_i for the ξ_i in (1), we can write $\eta_i = \sum a_{ij} \xi_j$ with $a_{ij} \in \mathbf{Z}$ for all i, j; then we have $N = A M\, {}^t\!A$, where A is the matrix (a_{ij}). Similarly we can write $\xi_i = \sum b_{ij} \eta_j$ with

$b_{ij} \in \mathbf{Z}$ for all i, j; calling B the matrix (b_{ij}), we have $AB = 1_n$, hence $\det(A)\det(B) = 1$; as $\det(A)$ and $\det(B)$ are in \mathbf{Z}, this gives $\det(A) = \pm 1$, hence $\det(N) = \det(M)$. In other words, the determinant D of M does not depend upon the choice of the basis (ξ_i). This justifies the following:

DEFINITION 6. *Let k and $\{\xi_1, \ldots, \xi_n\}$ be as above. Then the determinant D of the matrix M given by (1) is called the discriminant of k.*

PROPOSITION 7. *Let $\beta = \prod \beta_v$ be the Haar measure on k_A obtained by taking $\beta_v(r_v) = 1$ for all finite places v, $d\beta_w(x) = dx$ for all real places w, and $d\beta_w(x) = |dx \wedge d\bar{x}|$ for all imaginary places w of k. Then $\beta(k_A/k) = |D|^{1/2}$, where D is the discriminant of k.*

Call β_∞ the measure $\prod \beta_w$ on $k_\infty = \prod k_w$, the products being taken over the infinite places of k. By prop. 6, $\beta(k_A/k)$ is the same as $\beta_\infty(\theta(I^n))$; therefore our proposition will be proved if we show that

$$d\beta_\infty(\theta(u)) = |D|^{1/2} du_1 \ldots du_n.$$

Call r_1, r_2 the numbers of real and of imaginary places of k, respectively; put $r = r_1 + r_2 - 1$; let w_0, \ldots, w_r be the infinite places of k, ordered so that w_i is real for $i < r_1$ and imaginary for $i \geq r_1$. For each i, write k_i for the completion of k at w_i, λ_i for the natural injection of k into k_i, and μ_i for the \mathbf{R}-linear extension of λ_i to k_∞; if we identify k_∞ with $\prod k_i$ as above, th. 4 of Chap. III-4 shows that μ_i is the projection from k_∞ onto k_i. By corollary 1 of prop. 3, Chap. III-2, every isomorphic embedding λ' of k into \mathbf{C} is of the form $\sigma \circ \lambda_i$, where σ is an \mathbf{R}-linear isomorphism of k_i into \mathbf{C}; obviously σ is the natural injection of k_i into \mathbf{C} if $k_i = \mathbf{R}$, i.e. if $i < r_1$, and it is one of the two mappings $x \to x$, $x \to \bar{x}$ of \mathbf{C} onto \mathbf{C} if $k_i = \mathbf{C}$, i.e. $i \geq r_1$. Therefore, if we put $\lambda'_i = \lambda_i$ for $0 \leq i \leq r$. and $\lambda'_{r_2+i} = \bar{\lambda}_i$ for $r_1 \leq i \leq r$, the λ'_h for $0 \leq h \leq n-1$ are all the distinct isomorphisms of k into \mathbf{C}. Writing now μ'_h for the \mathbf{R}-linear extension of λ'_h to k_∞, we have, for $u = (u_1, \ldots, u_n) \in \mathbf{R}^n$ and $0 \leq h \leq n-1$:

$$\mu'_h(\theta(u)) = \sum_{i=1}^{n} \lambda'_h(\xi_i) u_i.$$

Call $N = (\lambda'_h(\xi_i))$ the matrix of the coefficients in the right-hand sides. By corollary 3 of prop. 4, Chap. III-3, we have, for all $\xi \in k$, $Tr_{k/\mathbf{Q}}(\xi) = \sum \lambda'_h(\xi)$, and therefore, since the λ'_h are isomorphisms:

$$M = (\sum_h \lambda'_h(\xi_i) \lambda'_h(\xi_j)) = {}^t N \cdot N,$$

hence $D = \det(N)^2$. At the same time, we have, in the exterior algebra of differential forms on \mathbf{R}^n:

(2) $\quad \prod_h d\mu'_h(\theta(u)) = \pm \prod_{0 \leqslant i < r_1} d\mu_i(\theta(u)) \wedge \prod_{r_1 \leqslant j \leqslant r} (d\mu_j(\theta(u)) \wedge d\bar{\mu}_j(\theta(u)))$

$$= \pm \det(N) du_1 \wedge \ldots \wedge du_n.$$

In view of the definition of the measures β_w, this completes the proof. At the same time, one may note that one gets a real differential form on \mathbf{R}^n by multiplying (2) with i^{r_2}; therefore $i^{r_2} \det(N)$ is real, which is the same as to say that $(-1)^{r_2} D > 0$.

COROLLARY 1. *If* $k \neq \mathbf{Q}, |D| > 1$.

Notations being as above, choose $c_i \in \mathbf{R}_+^\times$ for $0 \leqslant i \leqslant r$, and call $Y(c)$ the set of the elements $y = (y_v)$ of $k_\mathbf{A}$ such that $|y_v|_v \leqslant 1$ for all finite places v of k, and $|y_{w_i}|_{w_i} \leqslant c_i/2$ for $0 \leqslant i \leqslant r$. For each infinite place w, and each $c \in \mathbf{R}_+^\times$, the subset of k_w given by $|x|_w \leqslant c/2$ is an interval of length c if w is real, and a circle of β_w-measure πc if w is imaginary. In view of the definition of β, this gives $\beta(Y(c)) = \pi^{r_2} \prod c_i$. If this is $> |D|^{1/2}$, lemma 1 of Chap. II-4, combined with proposition 7, shows that there are y, y' in $Y(c)$ such that $\eta = y' - y$ is in k^\times. Then we have $|\eta|_v \leqslant 1$ for all finite places v of k, $|\eta|_{w_i} \leqslant c_i$ if w_i is a real place, and, as one sees at once, $|\eta|_{w_i} \leqslant 2c_i$ if w_i is imaginary; in view of th. 5 of Chap. IV-4, this implies $2^{r_2} \prod c_i \geqslant 1$. Therefore, if $r_2 > 0$, we get a contradiction if we assume that $|D| = 1$ and choose the c_i so that $\prod c_i$ is $> \pi^{-r_2}$ and $< 2^{-r_2}$. Now assume that $r_2 = 0$, hence $r_1 = n$, and $|D| = 1$. Then, for every choice of the c_i such that $\prod c_i > 1$, there is $\eta \in k^\times$ with the properties stated above. Clearly the set of elements $x = (x_v)$ of $k_\mathbf{A}$ which satisfy $|x_v|_v \leqslant 1$ for all finite places v, and $|x_w|_w \leqslant 2$ for all infinite places w, is compact and therefore contains only finitely many elements η_1, \ldots, η_N of k; therefore we can choose $c' > 1$ such that none of these satisfies $1 < |\eta_v|_{w_0} < c'$. Choose now the c_i so that $\prod c_i > 1$, $1 < c_0 < c'$, $c_0 < 2$, and $c_i < 1$ for $1 \leqslant i \leqslant n-1$; then there is $\eta \in k^\times$ such that $|\eta|_v \leqslant 1$ for all finite places, and $|\eta|_{w_i} \leqslant c_i$ for $0 \leqslant i \leqslant n-1$. In view of the definition of c' and of our assumptions about the c_i, this implies $|\eta|_{w_i} < 1$ for $i > 0$, and $|\eta|_{w_0} \leqslant 1$. This contradicts th. 5 of Chap. IV-4, unless $n = 1$.

COROLLARY 2. *There are only finitely many algebraic number-fields* k *of given degree* n *over* \mathbf{Q} *and given discriminant* D.

As this will not be of any further use to us, we merely sketch the proof. Proceeding just as above, one sees that there is $\eta \in k^\times$ such that $|\eta|_v \leqslant 1$ for all finite places v of k, $|\eta|_w < 1$ for all infinite places w except one such place w_0, and the image $\lambda_0(\eta)$ of η in k_{w_0} is in the interval $|x| \leqslant 2|D|^{1/2}$ if w_0 is real, and in the rectangle given by $x = u + iv$, $|u| \leqslant 1$, $|v| \leqslant |D|^{1/2}$ if w_0 is imaginary. As then we must have $|\eta|_{w_0} > 1$, the latter condition implies that $\lambda_0(\eta)$ is not real if w_0 is imaginary. This implies that $k = \mathbf{Q}(\eta)$; for, if not, call u the place of $\mathbf{Q}(\eta)$ lying below w_0; then $|\eta|_w > 1$ for all the

places w of k above u, if there is more than one such place, and $\lambda_0(\eta)$ must be real if u is real and w_0 imaginary; as this is not so, corollary 1 of th. 4, Chap. III-4, shows that the degree of k over $\mathbf{Q}(\eta)$ cannot be >1. This implies that the $\lambda'_h(\eta)$, for $0 \leqslant h \leqslant n-1$, are all distinct, so that $\prod(X - \lambda'_h(\eta))$ is the irreducible monic polynomial in $\mathbf{Q}[X]$ with the root η; its coefficients are obviously bounded in terms of $|D|$; they are all in \mathbf{Z} since $|\eta|_v \leqslant 1$ for all finite places v of k, which is the same as to say that η is in \mathfrak{r}, i.e. integral over \mathbf{Z}. Therefore the polynomial in question, hence also η, can take only finitely many values when D is given.

Now we will treat the corresponding problems for k_A^\times/k^\times. As above, we write Ω_∞ for the kernel of $a \to \mathrm{id}(a)$ in k_A, this being the group $k_\infty^\times \times \prod r_v^\times$, i.e. the same as $\Omega(P_\infty)$ in the notation of Chap. IV-4. We will write Ω_1, instead of $\Omega_1(P_\infty)$, for $\Omega_\infty \cap k_A^1$. As in Chap. IV-4, we write U for the group of the elements (z_v) of k_A^\times such that $|z|_v = 1$ for all v, finite or not; this is a compact subgroup of Ω_1. As we have observed in § 3, \mathfrak{r}^\times is the same as the group denoted by $E(P_\infty)$ in the notation of Chap. IV-4. We again write E for the cyclic group of the roots of 1 in k. Call again w_0, \ldots, w_r the infinite places of k, in any ordering. For each $z = (z_v)$ in Ω_∞, put

(3) $l(z) = (\log(|z_{w_0}|_{w_0}), \ldots, \log(|z_{w_r}|_{w_r})).$

The mapping l of Ω_∞ into \mathbf{R}^{r+1}, defined by (3), is obviously a morphism of the (multiplicatively written) group Ω_∞ onto the (additively written) group \mathbf{R}^{r+1}, with the kernel U. Let λ be the linear form on \mathbf{R}^{r+1} given by $\lambda(x) = \sum x_i$ for $x = (x_0, \ldots, x_r)$. Then, for $z \in \Omega_\infty$, we have $\log(|z|_A) = \lambda(l(z))$; therefore, if H is the hyperplane defined by $\lambda(x) = 0$ in \mathbf{R}^{r+1}, the set $l^{-1}(H)$, which is the kernel of $\lambda \circ l$, is the same as Ω_1, and l induces on it a morphism of Ω_1 onto H with the kernel U, which we can use to identify the group $G_1 = \Omega_1/U$ with the vector-space H. Put $\Gamma = l(\mathfrak{r}^\times)$; by the corollary of th. 9, Chap. IV-4, this is a discrete subgroup of H, and H/Γ is compact; in other words, it is an \mathbf{R}-lattice in H. It is then obvious (just as in the proof of th. 9, Chap. IV-4) that, if we take r elements $\varepsilon_1, \ldots, \varepsilon_r$ of \mathfrak{r}^\times, these will be free generators of a subgroup of \mathfrak{r}^\times if and only if their images $l(\varepsilon_i)$ in \mathbf{R}^{r+1} make up a basis for H, and that \mathfrak{r}^\times will then be the direct product of E and of that subgroup if and only if these images generate Γ; when that is so, we will say that the ε_i make up *a set of free generators for* \mathfrak{r}^\times *modulo* E. Assume now that they have been so chosen. For $0 \leqslant i \leqslant r$, call δ_i the degree of k_{w_i} over \mathbf{R}; this is 1 or 2, according as w_i is real or imaginary, and, by corollary 2 of th. 4, Chap. III-4, we have $\sum_i \delta_i = n$, i.e. $\lambda(\delta) = n$ if we write δ for the vector $(\delta_0, \ldots, \delta_r)$ in \mathbf{R}^{r+1}. This

implies that δ, together with the vectors $l(\varepsilon_i)$ for $1 \leqslant i \leqslant r$, makes up a basis for \mathbf{R}^{r+1}, so that we can define an automorphism F of \mathbf{R}^{r+1} by:

$$(4) \qquad t = (t_0, ..., t_r) \to F(t) = n^{-1} t_0 \delta + \sum_{i=1}^{r} t_i l(\varepsilon_i).$$

We have then $\lambda(F(t)) = t_0$, and also, for $\eta \in E$ and $(n_1, ..., n_r)$ in \mathbf{Z}^r:

$$(5) \qquad l(\eta \prod_i \varepsilon_i^{n_i}) = F(0, n_1, ..., n_r).$$

PROPOSITION 8. *Put* $\Omega_\infty = k_\infty^\times \times \prod r_v^\times$, *and let* l *be the morphism of* Ω_∞ *onto* \mathbf{R}^{r+1} *given by (3); let* $\{a_1, ..., a_h\}$ *be a full set of representatives for the cosets modulo* $k^\times \Omega_\infty$ *in* $k_\mathbf{A}^\times$. *Let* E *be the group of the roots of* 1 *in* k; *call* e *its order,* $\{\varepsilon_1, ..., \varepsilon_r\}$ *a set of free generators for* \mathfrak{r}^\times *modulo* E, *and* F *the automorphism of* \mathbf{R}^{r+1} *given by (4). Then, if* I *is the interval* $0 \leqslant t < 1$ *in* \mathbf{R}, *the union of the sets* $a_i l^{-1}(F(\mathbf{R} \times I^r))$ *for* $1 \leqslant i \leqslant h$ *is a fundamental set of order* e *modulo* k^\times *in* $k_\mathbf{A}^\times$.

Take any $z = (z_v)$ in $k_\mathbf{A}^\times$; there is one and only one i such that $a_i^{-1} z$ is in $k^\times \Omega_\infty$, and then we can write $z = a_i \xi z'$ with $\xi \in k^\times$, $z' \in \Omega_\infty$; moreover, z' is uniquely determined modulo $k^\times \cap \Omega_\infty$, i.e. modulo \mathfrak{r}^\times. Put $F^{-1}(l(z')) = = (t_0, ..., t_r)$; for $1 \leqslant i \leqslant r$, take $n_i \in \mathbf{Z}$ such that $n_i \leqslant t_i < n_i + 1$; put $\varepsilon = \prod \varepsilon_i^{n_i}$, $z'' = \varepsilon^{-1} z'$ and $\xi' = \xi \varepsilon$. Then we have $z = \xi' a_i z''$, and, in view of (4) and (5), $l(z'') \in F(\mathbf{R} \times I^r)$. Moreover, it is clear that z'' is uniquely determined modulo E by these conditions. This proves our proposition.

As we have seen in § 3, the morphism $z \to \mathrm{id}(z)$ of $k_\mathbf{A}^\times$ onto $I(k)$ determines an isomorphism of $k_\mathbf{A}^\times / k^\times \Omega_\infty$ onto the group $I(k)/P(k)$ of ideal-classes of k; therefore the number h, occurring in proposition 8, is the order of that group, and the ideles a_i in that proposition may also be characterized by saying that the fractional ideals $\mathrm{id}(a_i)$ are representatives of the ideal-classes of k.

Now we define a Haar measure γ on $k_\mathbf{A}^\times$. Just as in the case of $k_\mathbf{A}$, this may be done by choosing, for each v, a Haar measure γ_v on k_v^\times, in such a way that $\gamma_v(r_v^\times) = 1$ for almost all v; then we define γ by prescribing that it should coincide with $\prod \gamma_v$ on every one of the groups $k_\mathbf{A}(P)^\times$, and we write $\gamma = \prod \gamma_v$ for this. As in the case of $k_\mathbf{A}$, we need a definition:

DEFINITION 7. *Notations being as above, call* L *the matrix whose rows are the vectors* $n^{-1}\delta$, $l(\varepsilon_1), ..., l(\varepsilon_r)$. *Then* $R = |\det(L)|$ *is called the regulator of* k.

As L is the matrix of the automorphism F of \mathbf{R}^{r+1} given by (4), F has then the determinant $\pm R$. Our definition would have to be justified by showing that R is independent of the choice of the ε_i; this could be done

easily by applying the same argument which we used for the discriminant. As the same fact will emerge presently as a consequence of proposition 9, we leave it aside for the moment.

PROPOSITION 9. *Let $\gamma=\prod\gamma_v$ be the Haar measure on k_A^\times obtained by taking $\gamma_v(r_v^\times)=1$ for all finite places v of k, $d\gamma_w(x)=|x|^{-1}dx$ for each real place w, and $d\gamma_w(x)=(x\bar{x})^{-1}|dx\wedge d\bar{x}|$ for each imaginary place w. For each $m>1$ in \mathbf{R}, call $C(m)$ the image in k_A^\times/k^\times of the subset of k_A^\times defined by $1\leqslant|z|_A\leqslant m$. Then we have $\gamma(C(m))=c_k\log(m)$, with c_k given by*

$$c_k=2^{r_1}(2\pi)^{r_2}hR/e.$$

Here, as before, r_1 and r_2 are the numbers of real and of imaginary places of k, respectively; h is the number of ideal-classes; R is the regulator, as defined above, and e is the order of the group E of roots of 1 in k, this being always an even integer since ±1 are in k. Clearly $e=2$ if $r_1>0$, since \mathbf{R} contains no root of 1 except ±1.

We begin by modifying the representatives a_i of the cosets modulo $k^\times\Omega_\infty$ in k_A^\times, introduced in prop. 8, by replacing, for each i, a_i by $a_ib_i^{-1}$ with $b_i\in\Omega_\infty$ and $|b_i|_A=|a_i|_A$; once this is done, we have $|a_i|_A=1$ for $1\leqslant i\leqslant h$, and prop. 8 shows that $e\gamma(C(m))=h\gamma(X)$, where X is the intersection of $l^{-1}(F(\mathbf{R}\times I^r))$ with the set $1\leqslant|z|_A\leqslant m$ in k_A^\times. As we have seen above, if $z\in\Omega_\infty$ and $F^{-1}(l(z))=(t_0,...,t_r)$, we have

$$\log(|z|_A)=\lambda(l(z))=\lambda(F(t))=t_0.$$

Therefore the set X can be written as $l^{-1}(F(J\times I^r))$, where J is the interval $0\leqslant t\leqslant\log(m)$ of \mathbf{R}. Now l is a morphism of Ω_∞ onto \mathbf{R}^{r+1} with the compact kernel U; therefore, if Y is any compact subset of \mathbf{R}^{r+1}, $l^{-1}(Y)$ is a compact subset of Ω_∞, and $Y\to\gamma(l^{-1}(Y))$ is a Haar measure on \mathbf{R}^{r+1}, hence a multiple $c\alpha(Y)$ of the Lebesgue measure α on \mathbf{R}^{r+1}, with some constant $c>0$. This gives $\gamma(X)=c\alpha(F(J\times I^r))$. By the definition of the regulator R, it is the module of the automorphism F of \mathbf{R}^{r+1}; therefore we get:

$$\gamma(X)=c\alpha(F(J\times I^r))=cR\alpha(J\times I^r)=cR\log(m).$$

It only remains for us to determine c. Take $Y=J^{r+1}$, so that $\alpha(Y)=(\log m)^{r+1}$. Then $l^{-1}(Y)$ is the set of the elements (z_v) of Ω_∞ such that $1\leqslant|z|_w\leqslant m$ for all infinite places w of k. In view of the definition of γ, we have then $\gamma(l^{-1}(Y))=a^{r_1}b^{r_2}$, with a,b given by

$$a=2\int_1^m x^{-1}dx=2\log(m),\qquad b=\iint_{1\leqslant x\bar{x}\leqslant m}(x\bar{x})^{-1}|dx\wedge d\bar{x}|=2\pi\log(m).$$

This gives $c=2^{r_1}(2\pi)^{r_2}$, which completes the proof. Our conclusion shows that R is independent of the choice of the ε_i, as had been stated above.

Chapter VI

The theorem of Riemann-Roch

The classical theory of algebraic number-fields, as described above in Chapter V, rests upon the fact that such fields have a non-empty set of places, the infinite ones, singled out by intrinsic properties. It would be possible to develop an analogous theory for A-fields of characteristic $p > 1$ by arbitrarily setting apart a finite number of places; this was the point of view adopted by Dedekind and Weber in the early stages of the theory. Whichever method is followed, the study of such fields leads very soon to results which cannot be properly understood without the use of concepts belonging to algebraic geometry; this lies outside the scope of this book. The results to be given here should be regarded chiefly as an illustration for the methods developed above and as an introduction to a more general theory.

From now on, in this Chapter, k will be an A-field of characteristic $p > 1$. In the corollary of th. 8, Chap. IV-4, we have defined a finite field F, which we have called the *field of constants* of k; this is the algebraic closure of the prime field in k, and may consequently also be described as the maximal finite field contained in k; from now on, the number of its elements will be denoted by q, and F will be identified with \mathbf{F}_q. Then, for every place v of k, the completion k_v of k at v contains \mathbf{F}_q; in view of corollary 1 of th. 7, Chap. I-4, and of corollary 2 of th. 2, Chap. I-1, this implies that the module q_v of k_v is of the form q^d, where d is an integer $\geqslant 1$ which is called *the degree of* v and is denoted by $\deg(v)$.

By the *divisors* of k, one understands the elements of the free abelian group $D(k)$ generated by the places of k; this being written additively, it consists of the formal sums $\sum_v a(v) \cdot v$, where $a(v) \in \mathbf{Z}$ for every place v of k, and $a(v) = 0$ for almost all v. If $\mathfrak{a} = \sum a(v) \cdot v$ is such a divisor, we will write $\mathfrak{a} \succ 0$ when $a(v) \geqslant 0$ for all v; if $\mathfrak{a}, \mathfrak{b}$ are two divisors, we write $\mathfrak{a} \succ \mathfrak{b}$ for $\mathfrak{a} - \mathfrak{b} \succ 0$. For every divisor $\mathfrak{a} = \sum a(v) \cdot v$, we write $\deg(\mathfrak{a}) = \sum a(v) \deg(v)$, and call this the *degree* of \mathfrak{a}. Clearly $\mathfrak{a} \rightarrow \deg(\mathfrak{a})$ is a nontrivial morphism of $D(k)$ into \mathbf{Z}; in Chap. VII-5, it will be shown that it is surjective; its kernel, i.e. the group of the divisors of k of degree 0, will be denoted by $D_0(k)$. Obviously $\mathfrak{a} \succ 0$ implies $\deg(\mathfrak{a}) \geqslant 0$, and even $\deg(\mathfrak{a}) > 0$ unless $\mathfrak{a} = 0$, and $\mathfrak{a} \succ \mathfrak{b}$ implies $\deg(\mathfrak{a}) \geqslant \deg(\mathfrak{b})$.

Let $a=(a_v)$ be any element of $k_\mathbf{A}^\times$; for each v, we can write $a_v r_v = p_v^{a(v)}$ with $a(v)=\mathrm{ord}_v(a_v)$; for almost all v, we have $|a_v|_v=1$, hence $a(v)=0$, so that $\sum a(v)\cdot v$ is a divisor of k; this divisor will be denoted by $\mathrm{div}(a)$. Clearly $a\to\mathrm{div}(a)$ is a surjective morphism of $k_\mathbf{A}^\times$ onto $D(k)$, whose kernel is $\prod_v r_v^\times$ and is the same as the group denoted by $\Omega(\emptyset)$ in Chap. IV-4; we may therefore use this morphism to identify $D(k)$ with $k_\mathbf{A}^\times/\Omega(\emptyset)$. The definition of $|a|_\mathbf{A}$ shows at once that, if $a\in k_\mathbf{A}^\times$ and $\mathfrak{a}=\mathrm{div}(a)$, then $|a|_\mathbf{A}=q^{-\deg(\mathfrak{a})}$; therefore $D_0(k)$ is the image of $k_\mathbf{A}^1$ in $D(k)$ under the morphism $a\to\mathrm{div}(a)$; in particular, the image $P(k)$ of k^\times in $D(k)$ under that morphism is contained in $D_0(k)$. The group $P(k)$ is known as the group of the *principal divisors*. Clearly the morphism $a\to\mathrm{div}(a)$ determines isomorphisms of the groups $k_\mathbf{A}^1/\Omega(\emptyset)$, $k_\mathbf{A}^1/k^\times\,\Omega(\emptyset)$ and $k_\mathbf{A}^\times/k^\times\,\Omega(\emptyset)$ onto $D_0(k)$, $D_0(k)/P(k)$ and $D(k)/P(k)$, respectively; $D(k)/P(k)$ is known as the group of the *divisor-classes* of k, and $D_0(k)/P(k)$ as the group of the divisor-classes of degree 0; th. 7 of Chap. IV-4 shows that the latter is finite, and that the former is the direct product of the latter and of a group isomorphic to \mathbf{Z}.

Now we consider vector-spaces over k; we have the following result, a special case of which occurred already in Chapter IV:

PROPOSITION 1. *Let E be a vector-space of finite dimension over k. Let ε be a basis of E over k; for each place v of k, let ε_v be the r_v-module generated by ε in E_v, and let L_v be any k_v-lattice in E_v. Then $\prod L_v$ is an open and compact subgroup of $E_\mathbf{A}$ if and only if $L_v=\varepsilon_v$ for almost all v.*

If P is a finite set of places such that $L_v\subset\varepsilon_v$ for all v not in P, $\prod L_v$ is a compact subgroup of $E_\mathbf{A}(P,\varepsilon)$, hence of $E_\mathbf{A}$; the converse follows at once from corollary 1 of prop. 1, Chap. IV-1. Now assume that this is so. Then $\prod L_v$ is a subgroup of $E_\mathbf{A}$; it is open if and only if it contains a neighborhood of 0; prop. 1 of Chap. IV-1 shows that this is so if and only if $L_v\supset\varepsilon_v$ for almost all v, which completes the proof.

With the notations of proposition 1, put $L=(L_v)$; this will be called a *coherent system* of k_v-lattices, or more briefly a coherent system, belonging to E, if $L_v=\varepsilon_v$ for almost all v. When that is so, we will write $U(L)=\prod L_v$ and $\Lambda(L)=E\cap U(L)$. By prop. 1, $U(L)$ is open and compact; it is also a module over the open and compact subring $\prod r_v$ of $k_\mathbf{A}$. As to $\Lambda(L)$, it is a finite subgroup of E, since E is discrete and $U(L)$ compact in $E_\mathbf{A}$; it is also a module over the ring $k\cap(\prod r_v)$; as this ring, by th. 8 of Chap. IV-4 and its corollary, is the field of constants \mathbf{F}_q of k, this shows that $\Lambda(L)$ is a vector-space over \mathbf{F}_q, whose dimension will be denoted by $\lambda(L)$. Then $\Lambda(L)$ has $q^{\lambda(L)}$ elements.

PROPOSITION 2. *Put $\mathscr{A} = \text{End}(E)$, and let $L=(L_v)$, $M=(M_v)$ be two coherent systems beloning to E. Then there is $a=(a_v)$ in $\mathscr{A}_{\mathbf{A}}^{\times}$ such that $M_v=a_v L_v$ for all v; moreover, the divisor $\text{div}(\det(a))$ is uniquely determined by L and M.*

For each v, by th. 1 of Chap. II-2, there are bases α_v, β_v of E_v over k_v such that L_v, M_v are the r_v-modules respectively generated by α_v and by β_v. Call a_v the automorphism of E_v which maps α_v onto β_v; then $M_v=a_v L_v$. Put $d_v=\det(a_v)$; if μ_v is any Haar measure on E_v, we have $|d_v|_v=\mu_v(M_v)/\mu_v(L_v)$, by corollary 3 of th. 3, Chap. I-2, so that $|d_v|_v$ is independent of the choice of the bases α_v, β_v. Moreover, we have $L_v = M_v$, hence $|d_v|_v = 1$, for almost all v. By prop. 3 of Chap. IV-3, this shows that $a=(a_v)$ is in $\mathscr{A}_{\mathbf{A}}^{\times}$ and $d=(d_v)=\det(a)$ in $k_{\mathbf{A}}^{\times}$. As $|d_v|_v$ depends only upon L_v and M_v, we see that $\text{div}(d)$ depends only upon L and M.

We will write $M=aL$ when L, M and a are as in proposition 2.

COROLLARY 1. *Let ε be a basis of E over k; put $L_0=(\varepsilon_v)$, and let L be any coherent system belonging to E. Then there is $a\in\mathscr{A}_{\mathbf{A}}^{\times}$ such that $L=aL_0$; the divisor $\mathfrak{d}=\text{div}(\det(a))$ depends only upon L and ε, and its class and degree depend only upon L.*

Only the last assertion needs a proof. Replace ε by another basis ε'; put $L_0'=(\varepsilon_v')$, and call α the automorphism of E over k which maps ε' onto ε. Then $L_0=\alpha L_0'$, hence $L=a\alpha L_0'$, so that \mathfrak{d} has to be replaced by $\mathfrak{d}+\text{div}(\det(\alpha))$; the second term in the latter sum is a principal divisor, so that its degree is 0.

COROLLARY 2. *There is a Haar measure μ on $E_{\mathbf{A}}$ such that $\mu(\prod\varepsilon_v)=1$ for every basis ε of E over k; for this measure, if L and \mathfrak{d} are as in corollary 1, we have $\mu(U(L))=q^{-\delta(L)}$ with $U(L)=\prod L_v$ and $\delta(L)=\deg(\mathfrak{d})$.*

Choose one basis ε, and take μ such that $\mu(\prod\varepsilon_v)=1$. If a is as in corollary 1, $U(L)$ is the image of $U(L_0)=\prod\varepsilon_v$ under $e\to ae$. Therefore $\mu(U(L))$ is equal to the module of that automorphism, which is $|\det(a)|_{\mathbf{A}}$ by prop. 3 of Chap. IV-3; in view of our definitions, this is $q^{-\delta(L)}$, as stated in our corollary. By corollary 1, this does not depend upon ε; therefore, replacing ε by another basis ε', we get a measure μ' such that $\mu'(U(L))$ is the same as $\mu(U(L))$; this gives $\mu'=\mu$, so that $\mu(\prod\varepsilon_v')=1$.

As in Chap. IV-2, choose now a non-trivial character χ of $k_{\mathbf{A}}$, trivial on k, and call χ_v the character induced by χ on k_v, which is non-trivial for every v, by corollary 1 of th. 3, Chap. IV-2. Let E be as above, and call E' its algebraic dual. As explained in Chap. IV-2, we use χ to identify $E_{\mathbf{A}}'$ with the topological dual of $E_{\mathbf{A}}$ by means of the isomorphism described in th. 3 of Chap. IV-2, and, for each v, we use χ_v to identify E_v' with the

topological dual of E_v by means of the isomorphism described in th. 3 of Chap. II-5. Let $L = (L_v)$ be a coherent system belonging to E; for every v, call L'_v the dual lattice to L_v. In view of the identifications which have just been made, L'_v is a k_v-lattice in E'_v, and $\prod L'_v$ is the subgroup of E'_A associated by duality with the subgroup $U(L) = \prod L_v$ of E_A. As $U(L)$ is compact, $\prod L'_v$ is open; as $U(L)$ is open, $\prod L'_v$ is compact; by prop. 1, this shows that $L' = (L'_v)$ is a coherent system belonging to E' (a fact which is also implied by corollary 3 of th. 3, Chap. IV-2); we call it the *dual system* to L.

THEOREM 1. *To every A-field k of characteristic $p > 1$, there is an integer $g \geqslant 0$ with the following property. Let E be any vector-space of finite dimension n over k; let L be any coherent system belonging to E, and let L' be the dual system to L. Then:*

$$\lambda(L) = \lambda(L') - \delta(L) - n(g - 1).$$

Put $U = U(L)$, $U' = U(L')$; as we have just seen, U' is the subgroup of E'_A associated by duality with the subgroup U of E_A. By definition, $\lambda(L)$ and $\lambda(L')$ are the dimensions of the vector-spaces $\Lambda = E \cap U$ and $\Lambda' = E' \cap U'$, respectively, over the field of constants \mathbf{F}_q of k. By th. 3 of Chap. IV-2, the subgroup of E'_A associated by duality with the subgroup E of E_A is E'. Therefore the subgroup of E'_A associated by duality with $E + U$ is Λ', so that $E_A/(E + U)$ is the dual group to Λ' and has the same number of elements $q^{\lambda(L')}$ as Λ'. Clearly $E_A/(E + U)$ is isomorphic to $(E_A/E)/(E + U/E)$. Take the Haar measure μ on E_A defined by corollary 2 of prop. 2, and write again μ for its image in E_A/E, as explained in Chap. II-4. As $q^{\lambda(L')}$ is the index of $(E + U)/E$ in E_A/E, we have

$$\mu(E_A/E) = q^{\lambda(L')} \mu(E + U/E).$$

The canonical morphism of E_A onto E_A/E maps U onto $(E + U)/E$, with the finite kernel $\Lambda = E \cap U$; as Λ has $q^{\lambda(L)}$ elements, this gives, e.g. by lemma 2 of Chap. II-4:

$$\mu(U) = q^{\lambda(L)} \mu(E + U/E).$$

Combining these formulas with corollary 2 of prop. 2, which gives $\mu(U) = q^{-\delta(L)}$, we get:

$$\mu(E_A/E) = q^{\lambda(L') - \lambda(L) - \delta(L)}.$$

This shows that $\mu(E_A/E)$ is of the form q^r with $r \in \mathbf{Z}$. In particular, if we apply corollary 2 of prop. 2 to $E = k$ and to the basis $\varepsilon = \{1\}$, we get a Haar measure μ_1 on k_A, such that $\mu_1(\prod r_v) = 1$, and we see that we can write $\mu_1(k_A/k) = q^\gamma$ with $\gamma \in \mathbf{Z}$. Now identify our space E with k^n by means of a basis ε of E over k; it is clear that the measure μ in E_A, defined by corollary 2 of prop. 2, is the product $(\mu_1)^n$ of the measures μ_1 for the n factors of the product $E_A = (k_A)^n$, and then that $q^r = (q^\gamma)^n$, i.e. $r = \gamma n$. This

proves the formula in our theorem, with $g = \gamma + 1$; it only remains for us to show that $g \geq 0$. As to this, apply that formula to the case $E = k$, $L_v = r_v$ for all v. Then $\Lambda = \mathbf{F}_q$, $\lambda(L) = 1$, and clearly $\delta(L) = 0$; this gives $g = \lambda(L')$, which is ≥ 0 by definition.

COROLLARY 1. *Let μ be the Haar measure in E_A defined by corollary 2 of proposition 2; then $\mu(E_A/E) = q^{n(g-1)}$. In particular, if μ_1 is the Haar measure in k_A for which $\mu_1(\prod r_v) = 1$, we have $\mu_1(k_A/k) = q^{g-1}$.*

This was proved above.

COROLLARY 2. *Notations being as in theorem 1, we have $E_A = E + U$ if and only if $\lambda(L') = 0$.*

This is a special case of what has been proved above.

DEFINITION 1. *The integer g defined by theorem 1 is called the genus of k.*

The results obtained above will now be made more explicit in the case $E = k$. Then a coherent system $L = (L_v)$ is given by taking $L_v = p_v^{-a(v)}$ for all v, with $a(v) = 0$ for almost all v; such systems are therefore in a one-to-one correspondence with the divisors of k. Accordingly, if $\mathfrak{a} = \sum a(v) \cdot v$ is such a divisor, we will write $L(\mathfrak{a})$ for the coherent system $(p_v^{-a(v)})$; $L(0)$ being then the coherent system (r_v), we see that $L(\mathfrak{a})$ is the coherent system $a^{-1} L(0)$ when $a \in k_A^\times$ and $\mathfrak{a} = \mathrm{div}(a)$. For $L = L(\mathfrak{a})$, we will also write $U(\mathfrak{a})$, $\Lambda(\mathfrak{a})$, $\lambda(\mathfrak{a})$, $\delta(\mathfrak{a})$ instead of $U(L)$, $\Lambda(L)$, $\lambda(L)$, $\delta(L)$; obviously we have $\delta(\mathfrak{a}) = -\deg(\mathfrak{a})$. The definition of $\Lambda(\mathfrak{a})$ shows that it can be written as $\bigcap_v (k \cap p_v^{-a(v)})$; in other words, it consists of 0 and of the elements ξ of k^\times such that $\mathrm{ord}_v(\xi) \geq -a(v)$ for all v, or, what amounts to the same, such that $\mathrm{div}(\xi) \succ -\mathfrak{a}$. As the degree of $\mathrm{div}(\xi)$ is 0 for all $\xi \in k^\times$, this shows that $\Lambda(\mathfrak{a}) = \{0\}$, hence $\lambda(\mathfrak{a}) = 0$, whenever $\deg(\mathfrak{a}) < 0$.

Now let the "basic" character χ of k_A be chosen as above; for each place v of k, call $\nu(v)$ the order of the character χ_v induced by χ on k_v, this being as defined in def. 4 of Chap. II-5. By corollary 1 of th. 3, Chap. IV-2, we have $\nu(v) = 0$ for almost all v, so that $\mathfrak{c} = \sum \nu(v) \cdot v$ is a divisor of k; we call this *the divisor of χ*, and denote it by $\mathrm{div}(\chi)$. If χ_1 is another such character, then, by th. 3 of Chap. IV-2, it can be written as $x \to \chi(\xi x)$ with $\xi \in k^\times$, and one sees at once that $\mathrm{div}(\chi_1) = \mathrm{div}(\chi) + \mathrm{div}(\xi)$. Thus, when one takes for χ all the non-trivial characters of k_A, trivial on k, the divisors $\mathrm{div}(\chi)$ make up a class of divisors modulo the group $P(k)$ of principal divisors of k. This is known as *the canonical class*, and its elements as *the canonical divisors*.

As before, identify k_A with its topological dual by means of χ, and put $\mathfrak{c} = \mathrm{div}(\chi)$. Using prop. 12 of Chap. II-5, one sees at once that the dual system to $L(\mathfrak{a})$ is $L(\mathfrak{c} - \mathfrak{a})$. Theorem 1 gives now:

THEOREM 2. *Let \mathfrak{c} be a canonical divisor of k. Then, for every divisor \mathfrak{a} of k, we have:*

$$\lambda(\mathfrak{a}) = \lambda(\mathfrak{c} - \mathfrak{a}) + \deg(\mathfrak{a}) - g + 1.$$

COROLLARY 1. *If \mathfrak{c} is as above, $\deg(\mathfrak{c}) = 2g - 2$ and $\lambda(\mathfrak{c}) = g$.*

We get the first relation by replacing \mathfrak{a} by $\mathfrak{c} - \mathfrak{a}$ in theorem 2, and the second one by taking $\mathfrak{a} = 0$.

COROLLARY 2. *If \mathfrak{a} is a divisor of degree $> 2g - 2$, $\lambda(\mathfrak{a}) = \deg(\mathfrak{a}) - g + 1$.*

In fact, we have then $\deg(\mathfrak{c} - \mathfrak{a}) < 0$, and, as we have observed above, this implies $\lambda(\mathfrak{c} - \mathfrak{a}) = 0$, hence our conclusion, by theorem 2.

COROLLARY 3. *Let $\mathfrak{a} = \sum a(v) \cdot v$ be a divisor of degree $> 2g - 2$. Then $k_{\mathbf{A}} = k + \left(\prod_v p_v^{-a(v)} \right)$.*

This is the special case $E = k$, $L = L(\mathfrak{a})$ of corollary 2 of th. 1, since in this case, as shown above, we have $L' = L(\mathfrak{c} - \mathfrak{a})$ and $\lambda(L') = 0$.

Theorem 2 is the "theorem of Riemann-Roch" for a "function-field" k when the field of constants is finite. A proof for the general case can be obtained on quite similar lines; for the concept of compacity, one has to substitute the concept of "linear compacity" for vector-spaces over an arbitrary field K, K itself being discretely topologized; instead of a Haar measure, one has to use a "relative dimension" for compact and open subspaces of locally linearly compact vector-spaces over K. This will not be considered here.

Another point of some importance will merely be mentioned. Instead of identifying the topological dual G of $k_{\mathbf{A}}$ with $k_{\mathbf{A}}$ by means of a "basic" character, consider it as a $k_{\mathbf{A}}$-module by writing, for every $x^* \in G$ and every $a \in k_{\mathbf{A}}$, $\langle x, ax^* \rangle = \langle ax, x^* \rangle$ for all $x \in k_{\mathbf{A}}$. Call Γ the subgroup of G associated by duality with k. Then th. 3 of Chap. IV-2 can be expressed as follows: if γ is any element of Γ, other than 0, $x \to x\gamma$ is an isomorphism of $k_{\mathbf{A}}$ onto G which maps k onto Γ. In particular, Γ has an "intrinsic" structure of vector-space of dimension 1 over k. It is now possible to define "canonically" a differentiation of k into Γ, i.e. a mapping $x \to dx$ of k into Γ such that $d(xy) = x \cdot dy + y \cdot dx$ for all x, y in k, and that Γ may thus be identified with the k-module of all formal sums $\sum y_i dx_i$, where the x_i, y_i are in k. This remains true for every separably algebraic extension of finite degree of any field $K(T)$, where T is an indeterminate over the groundfield K. Even for the case studied here, that of a finite field of constants, this topic can hardly be dealt with properly except by enlarging the groundfield to its algebraic closure, and we will not pursue it any further.

Chapter VII

Zeta-functions of A-fields

§ 1. Convergence of Euler products. From now on, k will be an A-field of any characteristic, either 0 or $p > 1$. Notations will be as before; if v is a place of k, k_v is the completion of k at v; if v is a finite place, r_v is the maximal compact subring of k_v, and p_v the maximal ideal in r_v. Moreover, in the latter case, we will agree once for all to denote by q_v the module of the field k_v and by π_v a prime element of k_v, so that, by th. 6 of Chap. I-4, r_v/p_v is a field with q_v elements, and $|\pi_v|_v = q_v^{-1}$. If k is of characteristic $p > 1$, we will denote by q the number of elements of the field of constants of k and identify that field with \mathbf{F}_q; then, according to the definitions in Chap. VI, we have $q_v = q^{\deg(v)}$ for every place v.

By an *Euler product* belonging to k, we will understand any product of the form

$$\prod_v (1 - \theta_v q_v^{-s})^{-1}$$

where $s \in \mathbf{C}$, $\theta_v \in \mathbf{C}$ and $|\theta_v| \leqslant 1$ for all v, the product being taken over all or almost all the finite places of k. The same name is in use for more general types of products, but these will not occur here. The basic result on the convergence of such products is the following:

PROPOSITION 1. *Let k be any A-field. Then the product*

$$\zeta_k(\sigma) = \prod_v (1 - q_v^{-\sigma})^{-1},$$

where $\sigma \in \mathbf{R}$ and v runs through all the finite places of k, is convergent for $\sigma > 1$, and tends to the limit 1 for σ tending to $+\infty$.

Assume first that k is of characteristic 0, and call n its degree over \mathbf{Q}. By corollary 1 of th. 4, Chap. III-4, there are at most n places v of k above any given place p of \mathbf{Q}; for each of these, k_v is a p-field, so that q_v is of the form p^ν with $\nu \geqslant 1$ and is therefore $\geqslant p$. This gives, for $\sigma > 0$:

$$1 < \zeta_k(\sigma) \leqslant \prod (1 - p^{-\sigma})^{-n}$$

where the product is taken over all rational primes p. Now write:

$$\zeta(\sigma) = \prod (1 - p^{-\sigma})^{-1} = \prod (1 + p^{-\sigma} + p^{-2\sigma} + \cdots).$$

Expanding the last product, we get

$$\zeta(\sigma) = \sum_{v=1}^{+\infty} v^{-\sigma}$$

since every integer $v \geqslant 1$ can be uniquely expressed as a product of powers of rational primes. Furthermore, we have, for $\sigma > 1$:

$$1 < \zeta(\sigma) < 1 + \sum_{v=2}^{+\infty} \int_{v-1}^{v} t^{-\sigma} dt = 1 + \int_{1}^{+\infty} t^{-\sigma} dt = 1 + (\sigma-1)^{-1},$$

which shows that $\zeta(\sigma)$ is convergent for $\sigma > 1$ and tends to 1 for $\sigma \to +\infty$. This proves our proposition when k is a number-field.

Now assume that k is of characteristic $p > 1$; then, by lemma 1 of Chap. III-2, we may write it as a separably algebraic extension of $\mathbf{F}_p(T)$ of finite degree n. By th. 2 of Chap. III-1, $\mathbf{F}_p(T)$ has one place ∞ corresponding to the prime element T^{-1}, while its other places are in a one-to-one correspondence with the prime polynomials π in $\mathbf{F}_p[T]$. It will clearly be enough if we prove the assertion in our proposition, not for the product $\zeta_k(\sigma)$, but for the similar product $\eta(\sigma)$ taken over the places v of k which do not lie above the place ∞ of $\mathbf{F}_p(T)$. Then, just as in the case of characteristic 0, we see that $1 < \eta(\sigma) \leqslant \zeta_p(\sigma)^n$, where $\zeta_p(\sigma)$ denotes the product

$$\zeta_p(\sigma) = \prod (1 - p^{-\deg(\pi)\sigma})^{-1} = \prod (1 + p^{-\deg(\pi)\sigma} + p^{-2\deg(\pi)\sigma} + \cdots)$$

taken over all the prime polynomials π in $\mathbf{F}_p[T]$. As every monic polynomial in $\mathbf{F}_p[T]$ can be uniquely written as a product of powers of prime polynomials, this gives

$$\zeta_p(\sigma) = \sum p^{-\deg(\mu)\sigma}$$

where the sum is taken over all the monic polynomials μ in $\mathbf{F}_p[T]$. As there are p^{δ} monic polynomials of degree δ for every $\delta \geqslant 0$, we get

$$\zeta_p(\sigma) = \sum_{\delta=0}^{+\infty} p^{\delta(1-\sigma)} = (1 - p^{1-\sigma})^{-1},$$

which completes the proof in the present case.

COROLLARY 1. *Let P be a finite set of places of k, containing P_∞; for every v not in P, take $\theta_v \in \mathbf{C}$ such that $|\theta_v| \leqslant 1$. For $s \in \mathbf{C}$, put:*

$$E(s) = \prod_{v \notin P} (1 - \theta_v q_v^{-s})^{-1}.$$

Then $E(s)$ is absolutely convergent, holomorphic in s, and $\neq 0$, for $\operatorname{Re}(s) > 1$, and it tends to 1, uniformly with respect to $\operatorname{Im}(s)$, for $\operatorname{Re}(s)$ tending to $+\infty$.

In fact, for $\sigma = \mathrm{Re}(s)$, the series $\log E(s)$ is majorized by the series $\log \zeta_k(\sigma)$. Our conclusion follows now at once from proposition 1 and the well known elementary theorems on uniformly convergent series of holomorphic functions.

COROLLARY 2. *Let k_0 be an A-field contained in k; let M be a set of finite places of k such that, for almost all $v \in M$, the modular degree $f(v)$ of k_v over the closure of k_0 in k_v is > 1. Then the product*

$$p(M,\sigma) = \prod_{v \in M} (1 - q_v^{-\sigma})^{-1}$$

is absolutely convergent for $\sigma > 1/2$.

If k is of characteristic 0, both k and k_0 are of finite degree over \mathbf{Q}; if it is of characteristic $p > 1$, and T is any element of k_0, not algebraic over the prime field \mathbf{F}_p, k and k_0 are of finite degree over $\mathbf{F}_p(T)$; in both cases, k has a finite degree n over k_0. Let v be a finite place of k, and u the place of k_0 lying below v; then the closure of k_0 in k_v is $(k_0)_u$, and k_v is generated over it by k; therefore the degree of k_v over $(k_0)_u$ is $\leqslant n$, so that $1 \leqslant f(v) \leqslant n$. This shows that M is the union of the sets $M_1, ..., M_n$, consisting respectively of the places $v \in M$ for which $f(v) = f$, with $1 \leqslant f \leqslant n$. Our assumption about M means that M_1 is finite, so that it is enough to prove our assertion for each one of the sets M_f with $f \geqslant 2$. By corollary 1 of th. 4, Chap. III-4, there are at most n/f places $v \in M_f$ over each finite place u of k_0. Therefore the product $p(M_f, \sigma)$ is majorized by $\zeta_{k_0}(f\sigma)^{n/f}$; by proposition 1, this is absolutely convergent for $\sigma > 1/f$.

COROLLARY 3. *Let M be as in corollary 2; for every $v \in M$, take $\theta_v \in \mathbf{C}$ such that $|\theta_v| \leqslant 1$; then the product*

$$\prod_{v \in M} (1 - \theta_v q_v^{-s})^{-1}$$

is absolutely convergent, holomorphic in s, and $\neq 0$, for $\mathrm{Re}(s) > 1/2$.

In view of corollary 2, the proof is similar to that of corollary 1.

§ 2. **Fourier transforms and standard functions.** The theory of zeta-functions depends essentially on the concept of Fourier transforms, applied to the groups k_v, k_A attached to an A-field k. We begin by recalling the results to be used here.

As in Chap. II-5, let G be a commutative locally compact group, G^* its dual, and let $\langle g, g^* \rangle$ be as defined there. Let Φ be a continuous function on G, integrable for a Haar measure α given on G. Then the function Φ^* defined on G^* by

$$\Phi^*(g^*) = \int_G \Phi(g) \langle g, g^* \rangle \, d\alpha(g)$$

is called *the Fourier transform of Φ* with respect to α; one verifies at once that it is continuous on G^*. Clearly, if one replaces α by $c\alpha$, with $c \in \mathbf{R}^\times_+$, this replaces Φ^* by $c\Phi^*$.

LEMMA 1. *Let $g \to \lambda g$ be an automorphism of G, with the module* $\mathrm{mod}_G(\lambda)$. *Let $g^* \to g^*\lambda^*$ be the automorphism of G^* such that $\langle \lambda g, g^* \rangle = \langle g, g^*\lambda^* \rangle$ for all $g \in G$, $g^* \in G^*$. Then, if Φ^* is the Fourier transform of Φ, that of $g \to \Phi(\lambda^{-1}g)$ is $g^* \to \mathrm{mod}_G(\lambda)\Phi^*(g^*\lambda^*)$.*

In the integral which defines the Fourier transform of $\Phi(\lambda^{-1}g)$, substitute λg for g; the conclusion follows at once.

By the theory of Fourier transforms, there is a Haar measure α^* on G^*, such that, whenever the function Φ^* defined as above is integrable on G^*, Φ is given by "Fourier's inversion formula"

$$\Phi(g) = \int_{G^*} \Phi^*(g^*)\langle -g, g^* \rangle \, d\alpha^*(g^*).$$

Then we say that Φ is the inverse Fourier transform of Φ^*. The measure α^* is called *the dual measure to α*. Clearly, for $c \in \mathbf{R}^\times_+$, the dual measure to $c\alpha$ is $c^{-1}\alpha^*$. In particular, assume that G^* has been identified with G by means of some isomorphism of G onto G^*; then $\alpha^* = m\alpha$ with some $m \in \mathbf{R}^\times_+$, and, as the dual of $c\alpha$ is $c^{-1}m\alpha$, there is one and only one Haar measure on G, viz., $m^{1/2}\alpha$, which coincides with its own dual for the given identification of G and G^*; this is then called the *self-dual Haar measure* on G.

If G is compact, G^* is discrete. Then, by taking $\Phi = 1$, one sees at once that the dual of the Haar measure α given by $\alpha(G) = 1$ on G is the one given by $\alpha^*(\{0\}) = 1$ on G^*.

A function Φ on G will be called *admissible* for G if it is continuous, integrable, and if its Fourier transform Φ^* is integrable on G^*. Now let Γ be a discrete subgroup of G, such that G/Γ is compact. Let Γ_* be the subgroup of G^* associated by duality with Γ; as G/Γ is compact, Γ_* is discrete; as Γ is discrete, G^*/Γ_* is compact. Take for α the Haar measure on G determined by $\alpha(G/\Gamma) = 1$. The function Φ on G will be called *admissible for (G, Γ)* if it is admissible for G and if the two series

$$\sum_{\gamma \in \Gamma} \Phi(g+\gamma), \quad \sum_{\gamma^* \in \Gamma_*} \Phi^*(g^*+\gamma^*)$$

are absolutely convergent, uniformly on each compact subset with respect to the parameters g, g^*. The first one of these series defines then a continuous function F on G, constant on cosets modulo Γ; this may be regarded in an obvious manner as a function on G/Γ. As Γ_* is the dual group to G/Γ, F has then the Fourier transform

$$\gamma^* \to \int_{G/\Gamma} \left(\sum_{\gamma\in\Gamma} \Phi(g+\gamma) \right) \langle g, \gamma^* \rangle \, d\alpha(\dot{g}),$$

where, as usual, \dot{g} is the image of g in G/Γ under the canonical homomorphism of G onto G/Γ, and the integrand, which is written as a function of g but is constant on cosets modulo Γ, is regarded as a function of \dot{g}. According to formula (6) of Chap. II-4, this integral has then the value $\Phi^*(\gamma^*)$, so that the Fourier transform of F, when F is regarded as a function on G/Γ, is the function induced by Φ^* on Γ_*. Since Φ has been assumed to be admissible for (G,Γ), this is integrable on Γ_*, so that we get, by Fourier's inversion formula for G/Γ and Γ_*:

$$F(g) = \sum_{\gamma\in\Gamma} \Phi(g+\gamma) = \sum_{\gamma^*\in\Gamma_*} \Phi^*(\gamma^*)\langle -g, \gamma^* \rangle.$$

For $g=0$, this gives:

(1) $$\sum_{\gamma\in\Gamma} \Phi(\gamma) = \sum_{\gamma^*\in\Gamma_*} \Phi^*(\gamma^*).$$

This is known as *Poisson's summation formula*, which is thus shown to be valid whenever Φ is admissible for (G,Γ), and α is such that $\alpha(G/\Gamma)=1$.

Assume that there are admissible functions Φ for (G,Γ) for which both sides of (1) are not 0; this assumption (an easy consequence of the general theory of Fourier transforms) will be verified by an explicit construction in the only case in which we are interested, viz., the case $G=E_A$, $\Gamma=E$ when E is a vector-space of finite dimension over an A-field. Call then α^* the dual measure to α; put $\alpha^*(G^*/\Gamma_*)=c$, and interchange the roles of G, G^* in the above calculation, starting with Φ^* and taking its inverse Fourier transform by means of the Haar measure $c^{-1}\alpha^*$ on G^*; as this is $c^{-1}\Phi$, we find as end-result the same formula as (1), except that Φ has been replaced by $c^{-1}\Phi$. A comparison with (1) gives now $c=1$. This shows that the Haar measures α, α^* given on G, G^* by $\alpha(G/\Gamma)=1$, $\alpha^*(G^*/\Gamma_*)=1$ are dual to each other. In particular, if there is an isomorphism of G onto G^* which maps Γ onto Γ_*, and this is used to identify G and G^*, the self-dual measure on G is the one given by $\alpha(G/\Gamma)=1$.

Now we construct special types of admissible functions for the groups in which we are interested; these will be called "standard functions". On any space, a function is called *locally constant* if every point has a neighborhood where the function is constant. If f is such, $f^{-1}(\{a\})$ is open for every a; it is also closed, since its complement is the union of the open sets $f^{-1}(\{b\})$ for $b\neq a$. In a connected space, e.g. any vector-space over \mathbf{R}, only the constant functions are locally constant.

DEFINITION 1. *Let E be a vector-space of finite dimension over a p-field K. By a standard function on E, we understand a complex-valued locally constant function with compact support on E.*

It will be enough to consider the case when K is commutative. Let E^* be the "topological dual" of E, i.e. the dual of E when E is regarded as a locally compact group. On E^*, we put a structure of vector-space over K in the manner described in Chap. II-5; as we proved there, E^* has the same dimension as E over K. With these notations, we have:

PROPOSITION 2. *A function Φ on E is standard if and only if there are K-lattices L, M in E such that $L \supset M$ and that Φ is 0 outside L and constant on cosets modulo M in L. Then, if L_* and M_* are the dual K-lattices to L and M, we have $M_* \supset L_*$, and the Fourier transform Φ^* of Φ is 0 outside M_* and constant on cosets modulo L_* in M_*.*

If Φ, L, M have the properties stated in our proposition, it is clear that Φ is standard. Conversely, assume that it is such. Take a K-norm N on E, and call μ an upper bound for N on the support of Φ; then, as we have seen in Chap. II-2, the set L defined by $N(e) \leqslant \mu$ is a K-lattice, and it contains the support of Φ. As the sets $\Phi^{-1}(\{a\})$, for $a \in C$, are open, and L is compact, L is contained in the union of finitely many such sets; in other words, Φ takes only finitely many distinct values a_1, \ldots, a_n on L. Take $\varepsilon > 0$ such that $|a_i - a_j| > \varepsilon$ whenever $i \neq j$. As Φ is uniformly continuous on L, there is $\delta > 0$ such that $N(e - e') \leqslant \delta$, for e and e' in L, implies $|\Phi(e) - \Phi(e')| \leqslant \varepsilon$. Then the set M defined by $N(e) \leqslant \delta$ is a K-lattice, contained in L if we have taken $\delta \leqslant \mu$, and Φ is constant on cosets modulo M in L. Now consider the Fourier transform

$$\Phi^*(e^*) = \int_E \Phi(e) \langle e, e^* \rangle \, d\alpha(e),$$

where α is any Haar measure on E. As Φ is 0 outside L, this integral is not changed by taking it over L. Replace e^* by $e^* + e_1^*$ with $e_1^* \in L_*$; by definition, the latter assumption means that $\langle e, e_1^* \rangle = 1$ for all $e \in L$, so that the integral is not changed; therefore Φ^* is constant on cosets modulo L_* in E^*. On the other hand, as M is an open subgroup of the compact group L, L is the union of finitely many cosets $e_i + M$. As Φ is constant on each one of these, we have

$$(2) \qquad \Phi^*(e^*) = \sum_i \Phi(e_i) \int_M \langle e_i + e, e^* \rangle \, d\alpha(e) = \sum_i \Phi(e_i) \langle e_i, e^* \rangle \int_M \langle e, e^* \rangle \, d\alpha(e).$$

As the last integral is clearly 0 unless the character $e \to \langle e, e^* \rangle$ is trivial on M, i.e. unless $e^* \in M_*$, we see that Φ^* is 0 outside M_*.

COROLLARY 1. *If Φ is the characteristic function of the K-lattice L in E, $\alpha(L)^{-1} \Phi^*$ is the characteristic function of the K-lattice L_* dual to L in E^*, and $\alpha^*(L_*) = \alpha(L)^{-1}$ if α^* is the dual measure to α.*

The first assertion follows at once from (2) for $L=M$, $\Phi(0)=1$. It implies that the inverse Fourier transform of Φ^* is $\alpha^*(L_*)\alpha(L)\Phi$; as this must be Φ, we get the last assertion.

COROLLARY 2. *Every standard function on E is admissible for E.*

This is an obvious consequence of proposition 2 and the definitions.

In the next corollary, we identify K with its topological dual by means of a character χ of K in the manner explained in Chap. II-5, i.e. by writing $\langle x,y \rangle = \chi(xy)$ for x, y in K; for this identification, we may then speak of a self-dual measure on K.

COROLLARY 3. *Let R be the maximal compact subring of K, and φ the characteristic function of R. Let χ be a non-trivial character of K, of order v, and let α be the self-dual Haar measure on K for the identification of K with its dual, based on χ. Let $a \in K^\times$ be such that $\mathrm{ord}_K(a)=v$. Then $\alpha(R)=\mathrm{mod}_K(a)^{1/2}$, and the Fourier transform of φ is $y \to \mathrm{mod}_K(a)^{1/2}\varphi(ay)$.*

Apply corollary 1 to $E=K$, $L=R$; then, by prop. 12 of Chap. II-5, $L_*=P^{-v}$, i.e. $L_*=a^{-1}R$ if a is as defined above; then the characteristic function of L_* is $\varphi(ay)$, and we have $\alpha(L_*)=\mathrm{mod}_K(a)^{-1}\alpha(R)$. Our assertions follow now at once from corollary 1.

DEFINITION 2. *Let E be a vector-space of finite dimension over \mathbf{R}. By a standard function on E, we understand any function of the form $e \to p(e)\exp(-q(e))$, where p is a complex-valued polynomial function on E and q a real-valued positive-definite quadratic form on E.*

PROPOSITION 3. *Let E be as in definition 2; then every standard function on E has a Fourier transform which is a standard function, and is admissible for (E,L) if L is any \mathbf{R}-lattice in E.*

Choose a basis for E over \mathbf{R}, such that, when E is identified with \mathbf{R}^n by means of that basis, the quadratic form q is given by $q(x)=\pi\sum x_v^2$. It is clearly enough to prove our first assertion for a function $M(x)\exp(-q(x))$, where $M(x)$ is a monomial in the x_v. By th. 3 of Chap. II-5, we may identify \mathbf{R}^n with its dual by putting $\langle x,y \rangle = \mathbf{e}(\sum x_v y_v)$; then we see that it is enough to deal with the case $n=1$, i.e. to show that the Fourier transform of $x^m\exp(-\pi x^2)$ is standard on \mathbf{R} for every integer $m \geqslant 0$. The Fourier transform of $\exp(-\pi x^2)$ is $\exp(-\pi y^2)$, as shown by the well-known formula

$$\exp(-\pi y^2)=\int \exp(-\pi x^2+2\pi ixy)dx.$$

Differentiating both sides m times with respect to y, one sees at once, by induction on m, that the left-hand side is of the form $p_m(y)\exp(-\pi y^2)$,

where p_m is a polynomial of degree m, and that the differentiation may be carried out inside the integral in the right-hand side. This gives

$$p_m(y)\exp(-\pi y^2)=\int(2\pi i x)^m\exp(-\pi x^2+2\pi i x y)dx,$$

which proves our first assertion. Now let L be an \mathbf{R}-lattice in E. By prop. 11 of Chap. II-4, there is a basis of E over \mathbf{R} which generates the group L; in other words, identifying E with \mathbf{R}^n by means of that basis, we may assume that $E=\mathbf{R}^n$ and $L=\mathbf{Z}^n$. In order to prove that standard functions in \mathbf{R}^n are admissible for $(\mathbf{R}^n,\mathbf{Z}^n)$, it is now enough to show that, if Φ is such a function, $\sum|\Phi(x+v)|$, taken over all $v\in\mathbf{Z}^n$, is uniformly convergent on every compact subset C of \mathbf{R}^n. Put $\Phi(x)=p(x)\exp(-q(x))$ and $r(x)=\sum x_v^2$; take $\delta>0$ such that the quadratic form $q-\delta r$ is positive-definite; this will be so provided $\delta<\mu$, if we call μ the lower bound of q on the sphere $r=1$. Then the function $y\to\Phi(y)\exp(\delta r(y-x))$ tends to 0, uniformly in x for $x\in C$, when $r(y)$ tends to $+\infty$. This implies that this function is bounded for $x\in C$ and all $y\in\mathbf{R}^n$, and therefore, replacing y by $x+u$, that, for a suitable $A>0$, we have

$$|\Phi(x+u)|\leqslant A\exp(-\delta r(u))$$

for all $x\in C$. This gives

$$\sum_v|\Phi(x+v)|\leqslant A\sum_v\exp(-\delta\sum_i v_i^2)=A\left(\sum_{v=-\infty}^{+\infty}\exp(-\delta v^2)\right)^n,$$

which completes our proof.

We will also need a more explicit statement for some special cases of prop. 3, corresponding to $E=\mathbf{R}$ or \mathbf{C}; in each case we choose a "basic" character χ, and identify \mathbf{R} (resp. \mathbf{C}) with its topological dual by means of that character, just as we have done above for p-fields, according to Chap. II-5. The self-dual measures to be considered now are taken with reference to that identification.

PROPOSITION 4. On \mathbf{R}, the self-dual Haar measure, with reference to the basic character $\chi(x)=\mathbf{e}(-ax)$ with $a\in\mathbf{R}^\times$, is $d\alpha(x)=|a|^{1/2}dx$. If $\varphi_A(x)=x^A\exp(-\pi x^2)$ with $A=0$ or 1, the Fourier transform of φ_A is $\varphi_A'(y)=i^{-A}|a|^{1/2}\varphi_A(ay)$.

Put $d\alpha(x)=c\cdot dx$ with $c\in\mathbf{R}_+^\times$; then φ_0' is given by

$$\varphi_0'(y)=c\int_{\mathbf{R}}\exp(-\pi x^2-2\pi i a x y)dx.$$

As recalled above, this is equal to $c\varphi_0(ay)$. Applying now Fourier's inversion formula and lemma 1, we get $c=|a|^{1/2}$. Differentiating both sides of the above formula for $\varphi_0'(y)$ with respect to y, we get the Fourier transform of φ_1.

PROPOSITION 5. *On* **C**, *the self-dual Haar measure, with reference to the basic character* $\chi(x) = \mathbf{e}(-ax - \bar{a}\bar{x})$, *with* $a \in \mathbf{C}^{\times}$, *is* $d\alpha(x) = (a\bar{a})^{1/2}|dx \wedge d\bar{x}|$. *If* $\varphi_A(x) = x^A \exp(-2\pi x\bar{x})$, *A being any integer* $\geqslant 0$, *the Fourier transform of* φ_A *is* $i^{-A}(a\bar{a})^{1/2} \overline{\varphi_A(ay)}$, *and that of* $\overline{\varphi_A}$ *is* $i^{-A}(a\bar{a})^{1/2} \varphi_A(ay)$.

The proof of the assertions about α and about the Fourier transform of φ_0 is quite similar to that in prop. 4. Differentiating A times, with respect to y, the formula for the Fourier transform of φ_0, we get that of φ_A; that of $\overline{\varphi_A}$ follows from this at once.

DEFINITION 3. *Let* E *be a vector-space of finite dimension over an A-field* k. *Let* ε *be a basis of* E *over* k; *for each finite place* v *of* k, *let* ε_v *be the* r_v-*module generated by* ε *in* E_v. *By a standard function on* $E_\mathbf{A}$, *we understand any function of the form*

$$e = (e_v) \to \Phi(e) = \prod_v \Phi_v(e_v)$$

where Φ_v *is, for every place* v *of* k, *a standard function on* E_v, *and, for almost every* v, *the characteristic function of* ε_v.

Corollary 1 of th. 3, Chap. III-1, shows that the latter condition is independent of the choice of ε. The formula which defines Φ, for which we will write more briefly $\Phi = \prod \Phi_v$, is justified by prop. 1 of Chap. IV-1, which shows that almost all the factors in the right-hand side are equal to 1 whenever e is in $E_\mathbf{A}$; the same proposition shows also that Φ is 0 outside $E_\mathbf{A}(P, \varepsilon)$ for a suitable P, and that it is continuous.

Just as in the case of $k_\mathbf{A}$ in Chap. V-4, a Haar measure on $E_\mathbf{A}$ can be defined by choosing a Haar measure α_v on E_v for each v, so that $\alpha_v(\varepsilon_v) = 1$ for almost all v; when the α_v satisfy the latter condition, we will say that they are *coherent*. Then there is a unique measure α on $E_\mathbf{A}$ which coincides with the product measure $\prod \alpha_v$ on every one of the open subgroups $E_\mathbf{A}(P, \varepsilon)$ of $E_\mathbf{A}$; this will be written as $\alpha = \prod \alpha_v$. Clearly, if a Haar measure α is given on $E_\mathbf{A}$, one can find coherent measures α_v such that $\alpha = \prod \alpha_v$ by choosing any set of coherent measures on the spaces E_v and suitably modifying one of them.

From now on, we also choose, once for all, a "basic" character χ of $k_\mathbf{A}$, i.e. a non-trivial character of $k_\mathbf{A}$, trivial on k; we denote by χ_v the character induced by χ on k_v, which is non-trivial by corollary 1 of th. 3, Chap. IV-2. If E is any vector-space of finite dimension over k, we call E' its algebraic dual, and we use χ and χ_v for identifying the topological dual of $E_\mathbf{A}$ with $E'_\mathbf{A}$, and that of E_v with E'_v for each v, in the manner described in Chap. IV-2, i.e. by applying th. 3 of Chap. IV-2 to the former space and th. 3 of Chap. II-5 to the latter.

THEOREM 1. *Let E be a vector-space of finite dimension over the A-field k. Let the α_v be coherent Haar measures on the spaces E_v, and let $\Phi = \prod \Phi_v$ be a standard function on E_A. Then the Fourier transform of Φ, with respect to the measure $\alpha = \prod \alpha_v$ on E_A, is a standard function on E'_A, given by $\Phi' = \prod \Phi'_v$, where Φ'_v, for every v, is the Fourier transform of Φ_v with respect to α_v. Moreover, Φ is admissible for (E_A, E).*

Let $\varepsilon, \varepsilon'$ be bases for E and for E' over k; for each finite place v, let ε_v be as before, and let ε'_v be similarly defined for E'_v. By corollary 3 of th. 3, Chap. IV-2, there is a finite set P of places of k, containing P_∞, such that ε'_v is the dual k_v-lattice to ε_v when v is not in P; in view of our assumption on the α_v, we may also assume that P has been so chosen that $\alpha_v(\varepsilon_v) = 1$ for v not in P. Then, by corollary 1 of prop. 2, the Fourier transform of the characteristic function of ε_v is the characteristic function of ε'_v, and the dual measure α'_v to α_v is given by $\alpha'_v(\varepsilon'_v) = 1$, for all v not in P. Now let $\Phi = \prod \Phi_v$ be a standard function on E_A; for each v, call Φ'_v the Fourier transform of Φ_v with respect to α_v. From what has just been said, and from propositions 2 and 3, it follows that $\Phi' = \prod \Phi'_v$ is a standard function on E'_A; we will show that it is the Fourier transform of Φ. Replacing P if necessary by some larger set, we may assume that Φ_v is the characteristic function of ε_v for v not in P; in particular, the support of Φ is contained in $E_A(P, \varepsilon)$, so that the Fourier transform Φ'' of Φ is given by the integral

$$\Phi''(e') = \int \Phi(e) \chi([e, e']) \, d\alpha(e)$$

taken on $E_A(P, \varepsilon)$. In view of our definitions, the integrand here, for $e = (e_v)$, $e' = (e'_v)$, is given by

$$\Phi(e) \chi([e, e']) = \prod_v (\Phi_v(e_v) \chi_v([e_v, e'_v]));$$

moreover, when e' is given, the factor in the right-hand side corresponding to v has, for almost all v, the constant value 1 on ε_v. In view of the definition of $E_A(P, \varepsilon)$ in prop. 1 of Chap. IV-1, this implies that $\Phi''(e')$ is the same as $\Phi'(e')$.

Now, in order to prove that Φ is admissible for (E_A, E), it is enough to show that, for each compact subset C of E_A, the series

$$(3) \qquad \sum_{\eta \in E} |\Phi(e + \eta)| = \sum_{\eta \in E} |\prod_v \Phi_v(e_v + \eta)|$$

is uniformly convergent for $e \in C$. By corollary 1 of prop. 1, Chap. IV-1, C is contained in some set $E_A(P, \varepsilon)$; take P such that this is so and that $E_A(P, \varepsilon)$ also contains the support of Φ. For each place v in P, call C_v the projection of C onto E_v; for each finite place $v \in P$, call C'_v the support

of Φ_v; for v not in P, put $C_v = C'_v = \varepsilon_v$. As $\prod C_v$ is compact and contains C, it will be enough if we prove our assertion for $C = \prod C_v$. Assume first that k is of characteristic $p > 1$. Then Φ is 0 outside the compact set $C' = \prod C'_v$, so that all the terms of (3) are 0 for $e \in C$ except those corresponding to $\eta \in E \cap C''$, where C'' is the image of $C \times C'$ under the mapping $(e, e') \to e' - e$; as C'' is compact, $E \cap C''$ is finite, and the assertion becomes obvious. Now let k be of characteristic 0. For each finite $v \in P$, take a k_v-norm N_v on E_v, and call L_v the k_v-lattice given by $N_v(e_v) \leqslant \mu$, where μ is an upper bound for the values of N_v on the compact set $C_v \cup C'_v$. For v not in P, put $L_v = \varepsilon_v$. Put $L = \bigcap (E \cap L_v)$, where v runs through all the finite places of k; by th. 2 of Chap. V-2, this is the k-lattice in E with the closure L_v in E_v for all finite v. Clearly, for $e = (e_v)$ in C, $\Phi_v(e_v + \eta)$ is 0 unless η is in $E \cap L_v$, so that $\Phi(e + \eta)$ is 0 unless η is in L. Furthermore, if A_v is the upper bound of $|\Phi_v|$ for each finite place v of k, we have $A_v = 1$ for almost all v; putting $A = \prod A_v$, we see now that (3), for $e \in C$, is majorized by the series

$$ A \sum_{\eta \in L} |\prod_w \Phi_w(e_w + \eta)| $$

where the product is taken over the infinite places w of k. As explained in Chap. V-2, put $E_\infty = E \otimes_{\mathbf{Q}} \mathbf{R}$, and identify this with the product $\prod E_w$ taken over the infinite places of k. It is then obvious that the function Φ_∞ on E_∞, defined for $e_\infty = (e_w)$ by

$$ \Phi_\infty(e_\infty) = \prod_w \Phi_w(e_w), $$

is standard. As L is a k-lattice in E, it is a \mathbf{Q}-lattice in E regarded as a vector-space over \mathbf{Q}, hence an \mathbf{R}-lattice in E_∞. Our assertion is now contained in prop. 3.

COROLLARY 1. *If the α_v are coherent measures on the spaces E_v, their duals α'_v are coherent; the dual of $\alpha = \prod \alpha_v$ is $\alpha' = \prod \alpha'_v$; if $\alpha(E_A/E) = = 1$, then $\alpha'(E'_A/E') = 1$.*

The first assertion has been proved above; the second one follows at once from theorem 1 and the definitions. As to the last one, we know, by th. 3 of Chap. IV-2, that E' is the subgroup of E'_A associated by duality with the subgroup E of E_A; therefore, as we have seen, our assertion follows from Poisson's formula provided we can exhibit a function Φ, admissible for (E_A, E), for which the left-hand side of (1) is not 0; by theorem 1, any standard function $\Phi \geqslant 0$ such that $\Phi(0) > 0$ has these properties.

An important special case is that in which $E = E' = k$, $[x, y] = xy$; then we identify k_A and k_v with their topological duals by means of χ, χ_v, as explained before, and we have:

COROLLARY 2. *Let* α, α_v *be the self-dual measures on* k_A, k_v. *Then the* α_v *are coherent,* $\alpha = \prod \alpha_v$, *and* $\alpha(k_A/k) = 1$.

Take any coherent measures β_v on the groups k_v; by corollary 1, their duals β'_v are coherent, which implies that $\beta_v = \beta'_v$ for almost all v; in other words, β_v coincides with the self-dual measure α_v for almost all v; this implies that the α_v are coherent. Our other assertions follow now at once from corollary 1.

Notations being as in corollary 1, the measure α on E_A for which $\alpha(E_A/E) = 1$ is known as *the Tamagawa measure* on E_A; corollary 1 shows that its dual is the Tamagawa measure on E'_A. In particular, on k_A, the Tamagawa measure and the self-dual measure are the same.

Now, for each finite place v of k, call $v(v)$ the order of χ_v, which is 0 for almost all v by corollary 1 of th. 3, Chap. IV-2, and choose $a_v \in k_v^\times$ such that $\mathrm{ord}_v(a_v) = v(v)$. On the other hand, for each real place v of k, apply to $x \to e(-x)$ the corollary of th. 3, Chap. II-5; it shows that there is one and only one $a_v \in k_v^\times$ such that $\chi_v(x) = e(-a_v x)$ for all $x \in k_v$. Similarly, for each imaginary place v, there is one and only one $a_v \in k_v^\times$ such that $\chi_v(x) = e(-a_v x - \bar{a}_v \bar{x})$ for all $x \in k_v$. As $v(v) = 0$ for almost all v, (a_v) is in k_A^\times.

DEFINITION 4. *Let* χ *be a non-trivial character of* k_A, *trivial on* k, *inducing* χ_v *on* k_v *for every* v. *An idele* $a = (a_v)$ *of* k *will be called a differential idele attached to* χ *if* $\mathrm{ord}_v(a_v)$ *is equal to the order* $v(v)$ *of* χ_v *for every finite place* v *of* k, $\chi_v(x) = e(-a_v x)$ *for every real place* v, *and* $\chi_v(x) = e(-a_v x - \bar{a}_v \bar{x})$ *for every imaginary place* v *of* k.

Clearly, when χ is given, the differential idele a is uniquely determined modulo $\prod r_v^\times$, the latter product being taken over all the finite places v of k. If χ_1 is another character such as χ, then, by th. 3 of Chap. IV-2, it can be written as $\chi_1(x) = \chi(\xi x)$ with $\xi \in k^\times$; if a is as above, ξa is then a differential idele attached to χ_1. Consequently, the set of all differential ideles is a coset modulo $k^\times \prod r_v^\times$ in k_A^\times. If k is of characteristic $p > 1$, a is a differential idele attached to χ if and only if $\mathrm{div}(a) = \mathrm{div}(\chi)$, in the sense explained in Chap. VI; this implies that $\mathrm{div}(a)$ belongs to the canonical class.

PROPOSITION 6. *Let* a *be a differential idele. Then, if* k *is of characteristic 0,* $|a|_A = |D|^{-1}$, *where* D *is the discriminant of* k; *if* k *is of characteristic* $p > 1$, *and if* F_q *is its field of constants and* g *its genus,* $|a|_A = q^{2-2g}$.

The latter statement is equivalent to $\deg(\mathrm{div}(a)) = 2g - 2$; as $\mathrm{div}(a)$ is a canonical divisor, this is corollary 1 of th. 2, Chap. VI. In the case of characteristic 0, let α, α_v be the self-dual measures in k_A, k_v, so that

$\alpha = \prod \alpha_v$ by corollary 2 of th. 1; let $\beta = \prod \beta_v$ be as in prop. 7 of Chap. V-4. Applying corollary 3 of prop. 2, and propositions 4 and 5, we get $\alpha = |a|_A^{1/2} \beta$. As $\alpha(k_A/k) = 1$ by corollary 2 of th. 1, and $\beta(k_A/k) = |D|^{1/2}$ by prop. 7 of Chap. V-4, we get $|a|_A = |D|^{-1}$.

§ 3. Quasicharacters.

We first insert here some auxiliary results. As before, if $z \in C$, we denote by $\mathrm{Re}(z)$, $\mathrm{Im}(z)$ its real and imaginary parts, and we put $|z| = (z\bar{z})^{1/2}$, $|z|_\infty = \mathrm{mod}_C(z) = z\bar{z}$.

LEMMA 2. *A character ω of a group G is trivial if $\mathrm{Re}(\omega(g)) > 0$ for all $g \in G$.*

If $z \in C$, $|z| = 1$, $z \neq 1$ and $\mathrm{Re}(z) > 0$, we can write $z = e(t)$ with $t \in R$, $0 < |t| < 1/4$. Call n the smallest integer such that $n|t| > 1/4$; then $(n-1)|t| \leqslant 1/4$, hence $1/4 < n|t| < 1/2$ and $\mathrm{Re}(z^n) < 0$. Therefore the subset of C determined by $|z| = 1$, $\mathrm{Re}(z) > 0$ contains no subgroup of C^\times except $\{1\}$.

LEMMA 3. *Every homomorphism ω of a compact group G into C^\times is a character of G.*

In fact, $g \to |\omega(g)|$ must map G onto a compact subgroup of R_+^\times, and there is none except $\{1\}$.

A group G is called *totally disconnected* if there is a fundamental system of neighborhoods of the neutral element in G, consisting of subgroups of G. For instance, if K is a p-field, with the maximal compact subring R, and the maximal ideal P in R, the groups K and K^\times are totally disconnected, since the subgroups P^n in K, and the subgroups $1 + P^n$ in K^\times, for $n \geqslant 1$, make up such fundamental systems.

LEMMA 4. *Let the group G be locally compact and totally disconnected; then every representation of G into C^\times is locally constant. If G is compact, every such representation is a character of G of finite order. Conversely, if G is a compact commutative group, and if every character of G is of finite order, G is totally disconnected.*

If G is locally compact and totally disconnected, lemmas 2 and 3 show that every representation of G into C^\times is trivial on some open subgroup of G, hence locally constant. If G is compact, any open subgroup of G is of finite index, hence the second assertion. If G is commutative and compact, its dual G^* is discrete. As G may be identified with the dual of G^*, there is then a fundamental system of neighborhoods of 0 in G, consisting of sets defined by conditions of the form $|\omega_i(g) - 1| \leqslant \varepsilon$ $(1 \leqslant i \leqslant N)$, where the ω_i are characters of G. If all the ω_i are of finite order, we can take ε such that these inequalities imply $\omega_i(g) = 1$ for $1 \leqslant i \leqslant N$; then the neighborhood which is so defined is a subgroup of G.

From now on, we shall be chiefly concerned with representations into \mathbf{C}^\times of groups of the form K^\times, where K is a local field, and k_A^\times/k^\times, where k is an A-field. All these groups have the property stated in the following definition:

DEFINITION 5. *A group G will be called quasicompact if it is the direct product of a compact commutative group G_1 and of a group isomorphic to \mathbf{R} or to \mathbf{Z}; a representation of G into \mathbf{C}^\times will then be called a quasi-character of G.*

It would be easy to show that a group G is quasicompact if and only if it is commutative and locally compact, and its dual G^* is locally isomorphic to \mathbf{R}, i.e. if it has an open subgroup isomorphic to \mathbf{R} or to \mathbf{R}/\mathbf{Z}; the latter condition may even be replaced by the weaker requirement that G^* should have a neighborhood of 0, homeomorphic to \mathbf{R}. From this, one concludes easily that G is quasicompact if and only if it has a compact subgroup G_1 such that G/G_1 is isomorphic to \mathbf{R} or \mathbf{Z}. These facts will not be needed in the sequel. It is clear that, if G has the property described in definition 5, G_1 is its unique maximal compact subgroup.

DEFINITION 6. *If G is a quasicompact group, a quasicharacter of G will be called principal if it is trivial on the maximal compact subgroup G_1 of G.*

The quasicharacters of a quasicompact group G make up a group in an obvious manner; this will be denoted by $\Omega(G)$ and written multiplicatively. In other words, if ω, ω' are in $\Omega(G)$, we write $\omega\omega'$ for the quasicharacter $g \to \omega(g)\omega'(g)$ of G. Clearly the principal quasicharacters of G make up a subgroup Ω_1 of $\Omega(G)$.

PROPOSITION 7. *Let G be a quasicompact group and G_1 its maximal compact subgroup. Then G has non-trivial representations into \mathbf{R}_+^\times; if ω_1 is such a representation, its kernel is G_1, and every representation of G into \mathbf{R}_+^\times can be written in one and only one way in the form $g \to \omega_1(g)^\sigma$ with $\sigma \in \mathbf{R}$.*

Put $G = G_1 \times N$, with N isomorphic to \mathbf{R} or \mathbf{Z}. By lemma 3, every representation ω of G into \mathbf{R}_+^\times must be trivial on G_1; writing elements of G as (g_1, n) with $g_1 \in G_1$, $n \in N$, we see that ω must then be of the form $(g_1, n) \to \varphi(n)$, where φ is a representation of N into \mathbf{R}_+^\times. Identify N with \mathbf{R} or with \mathbf{Z}, as the case may be. In the former case, the condition for φ amounts to saying that $n \to \log\varphi(n)$ is an endomorphism of \mathbf{R}, hence of the form $n \to an$ with $a \in \mathbf{R}$, so that $\varphi(n) = \exp(an)$. For $N = \mathbf{Z}$, φ is obviously of the form $\varphi(n) = b^n$ with $b \in \mathbf{R}_+^\times$ and may still be written as $\varphi(n) = \exp(an)$ with $a = \log b$. In both cases, φ is non-trivial if $a \neq 0$.

Therefore, if ω_1 is as in our proposition, it can be written as $\omega_1(g_1,n) =$ $= \exp(a_1 n)$ with $a_1 \neq 0$. This has obviously the kernel G_1; moreover, if ω, φ and a are as above, we have $\omega = (\omega_1)^\sigma$ with $\sigma = a/a_1$, and σ is uniquely determined by this.

COROLLARY 1. *Let* G, G_1 *and* ω_1 *be as in proposition 7. Then the group* Ω_1 *of the principal quasicharacters of* G *is isomorphic to* **C** *or to* \mathbf{C}^\times *according as* G/G_1 *is isomorphic to* **R** *or to* **Z**; *every such quasicharacter is of the form*

$$g \rightarrow \omega_s(g) = \omega_1(g)^s$$

with $s \in \mathbf{C}$; *and* $s \rightarrow \omega_s$ *is a morphism of* **C** *onto* Ω_1, *whose kernel is* $\{0\}$ *or of the form* $i\, a\mathbf{Z}$ *with* $a \in \mathbf{R}_+^\times$, *according as* G/G_1 *is isomorphic to* **R** *or to* **Z**.

Let ω be any quasicharacter of G; with the above notation, proposition 7, applied to $g \rightarrow |\omega(g)|$, shows that $|\omega| = \omega_\sigma$ with $\sigma \in \mathbf{R}$; then $\omega' = \omega_\sigma^{-1}\omega$ is a character of G. If ω is trivial on G_1, so is ω'; with the same notations as in the proof of prop. 7, we may then write $\omega'(g_1,n) = \psi(n)$, where ψ is a character of N. As in that proof, identify N with **R** or with **Z**, as the case may be, ω_1 being given by $\omega_1(g_1,n) = \exp(a_1 n)$ in both cases. Every character of N can be written as $\psi(n) = \mathbf{e}(\tau n)$ with $\tau \in \mathbf{R}$; this is obvious for $N = \mathbf{Z}$ and is well known (and a special case of th. 3, Chap. II-5) for $N = \mathbf{R}$. That being so, we get $\omega = \omega_s$, with $s = \sigma + 2\pi i\, \tau/a_1$. Moreover, σ and ψ are uniquely determined by ω; τ is uniquely determined by ψ if $N = \mathbf{R}$, and uniquely determined modulo **Z** if $N = \mathbf{Z}$. This shows that $s \rightarrow \omega_s$ is an isomorphism of **C** onto Ω_1 if $N = \mathbf{R}$; if $N = \mathbf{Z}$, we have $\omega(g_1,n) = u^n$ with $u = \exp(a_1 s)$, and $u \rightarrow \omega$ is an isomorphism of \mathbf{C}^\times onto Ω_1. This completes the proof.

COROLLARY 2. *Let* G *be a quasicompact group, the direct product of the compact group* G_1 *and of a group* N, *isomorphic to* **R** *or* **Z**. *Then the group* $\Omega(G)$ *of quasicharacters of* G *is the direct product of the group* Ω_1 *considered in corollary 1, and of the group of the characters of* G, *trivial on* N; *the latter is isomorphic to the dual of* G_1.

We have already noted above that every quasicharacter ω of G can be uniquely written as $\omega_\sigma \psi$, where ψ is a character of G, and $\sigma \in \mathbf{R}$. Clearly ψ can be uniquely written as $\psi_1 \psi_2$, with ψ_1 trivial on G_1 and ψ_2 trivial on N; then $\omega = (\omega_\sigma \psi_1)\psi_2$, and $\omega_\sigma \psi_1$ is in Ω_1. The last assertion in our corollary is obvious.

So far we have refrained from mentioning any topology on $\Omega(G)$. We will put on Ω_1, not only the topology, but also the complex structure determined by the morphism $s \rightarrow \omega_s$ of **C** onto Ω_1 defined in corollary 1

of proposition 7; we define the topology on $\Omega(G)$ by prescribing that Ω_1 shall be an open subgroup of $\Omega(G)$, and we define a complex structure on $\Omega(G)$ by putting, on every coset modulo Ω_1 in $\Omega(G)$, the complex structure deduced from that of Ω_1 by translation. Then $\Omega(G)/\Omega_1$ is discrete, hence isomorphic to the dual of G_1 also as a topological group since that dual is discrete. The connected components of $\Omega(G)$ are the cosets modulo Ω_1; they are all isomorphic to \mathbf{C} or to \mathbf{C}^\times, as the case may be.

Clearly the above concepts and results can be applied to $G = K^\times$ if K is any local field, with $\omega_1(x) = \mathrm{mod}_K(x)$; we can take for N the subgroup \mathbf{R}_+^\times of K^\times if K is \mathbf{R} or \mathbf{C}, and the group generated by any prime element π of K if K is a p-field. In the latter case, this gives:

PROPOSITION 8. *Let K be a p-field and π a prime element of K. Then the principal quasicharacters of K^\times are those of the form $x \rightarrow \mathrm{mod}_K(x)^s$ with $s \in \mathbf{C}$; the group $\Omega(K^\times)$ of quasicharacters of K^\times is the direct product of the group of principal quasicharacters and of the group of the characters ψ of K^\times such that $\psi(\pi) = 1$.*

By lemma 4, every quasicharacter of K^\times is locally constant. If R and P have their usual meaning, the groups R^\times and $1 + P^n$ for $n \geqslant 1$ are open in K^\times and make up a fundamental system of neighborhoods of 1. This justifies the following definition:

DEFINITION 7. *Let K be a p-field, R its maximal compact subring and P the maximal ideal of R. Let ω be a quasicharacter of K^\times; let f be the smallest integer $\geqslant 0$ such that $\omega(x) = 1$ for $x \in R^\times$, $x - 1 \in P^f$. Then P^f is called the conductor of ω.*

Obviously ω is principal if and only if $f = 0$, i.e. if and only if its conductor is R; when that is so, we will also say that ω is *unramified*.

For $K = \mathbf{R}$ or \mathbf{C}, we have the following result:

PROPOSITION 9. *Every quasicharacter of \mathbf{R}^\times can be written in one and only one way as $x \rightarrow x^{-A}|x|^s$ with $A = 0$ or 1, and $s \in \mathbf{C}$. Every quasicharacter of \mathbf{C}^\times can be written in one and only one way as $x \rightarrow x^{-A} \bar{x}^{-B}(x\bar{x})^s$, where A and B are integers, $\inf(A, B) = 0$, and $s \in \mathbf{C}$.*

For \mathbf{R}^\times, this is an immediate consequence of prop. 7 and its corollaries, since here $G_1 = \{\pm 1\}$. For $G = \mathbf{C}^\times$, G_1 is the group determined by $x\bar{x} = 1$; as this is the dual of \mathbf{Z}, its characters are the functions $x \rightarrow x^n$ with $n \in \mathbf{Z}$; this can be written as $x \rightarrow (x/|x|)^{-A}$ with $A = -n \geqslant 0$ if $n \leqslant 0$, and as $x \rightarrow (\bar{x}/|x|)^{-B}$ with $B = n \geqslant 0$ if $n \geqslant 0$. Our assertions follow at once from this and prop. 7.

§ 4. Quasicharacters of A-fields. By th. 6 of Chap. IV-4, if k is an A-field, k_A^\times/k^\times is quasicompact. From now on, we will write $G_k = k_A^\times/k^\times$; this group is known as the "idele-class group" of k. We write $\Omega(G_k)$ for the group of quasicharacters of G_k, provided with its topology and its complex structure according to our definitions in § 3. The quasicharacters of G_k will be identified in an obvious manner with the representations of k_A^\times into \mathbf{C}^\times, trivial on k^\times.

As $z \to |z|_A$ is a non-trivial representation of k_A^\times into \mathbf{R}_+^\times, trivial on k^\times, it determines a non-trivial representation of G_k into \mathbf{R}_+^\times, which will be denoted by ω_1, and to which we can apply prop. 7 of § 3 and its corollaries, writing again $\omega_s = (\omega_1)^s$ for $s \in \mathbf{C}$. In particular, the kernel $G_k^1 = k_A^1/k^\times$ of ω_1 is the maximal compact subgroup of C_k; $s \to \omega_s$ is a morphism of \mathbf{C} onto the group Ω_1 of principal quasicharacters of G_k; if ω is any quasicharacter of G_k, there is one and only one $\sigma \in \mathbf{R}$ such that $|\omega| = \omega_\sigma$.

If k is of characteristic 0, corollary 2 of th. 5, Chap. IV-4, shows that G_k is the direct product of G_k^1 and of the image N in G_k of the group M defined in that corollary. On the other hand, if k is of characteristic $p > 1$, we choose an element z_1 of k_A^\times among those for which $|z|_A$ has its smallest value $Q > 1$; as we have seen in Chap. VI that the values of $|z|_A$ are all of the form q^n with $n \in \mathbf{Z}$ if \mathbf{F}_q is the field of constants of k, we have $Q = q^\nu$ with $\nu \geqslant 1$; it will be seen later that $\nu = 1$, $Q = q$ (this is corollary 6 of th. 2, § 5). Then we call M the subgroup of k_A^\times generated by z_1, and N its image in G_k. In all cases, we will identify N with its image in \mathbf{R}_+^\times under ω_1, so that ω_1 may be regarded as the projection from the product $G_k = G_k^1 \times N$ onto the factor N. Thus $N = \mathbf{R}_+^\times$ if k is of characteristic 0; otherwise it is the subgroup of \mathbf{R}_+^\times generated by Q; this implies that in the latter case the morphism $s \to \omega_s$ of \mathbf{C} onto Ω_1 has the same kernel as the morphism $s \to Q^s$ of \mathbf{C} onto \mathbf{C}^\times, i.e. $2\pi i(\log Q)^{-1}\mathbf{Z}$.

Let ω be any quasicharacter of G_k; regarding it as a representation of k_A^\times into \mathbf{C}^\times, trivial on k^\times, we will, for every place v of k, denote by ω_v the quasicharacter of k_v^\times induced on k_v^\times by ω. As the groups $k_A(P)^\times$ defined in the corollary of prop. 2, Chap. IV-3, are open in k_A^\times, every neighborhood of 1 in k_A^\times contains a subgroup of the form $\prod_{v\in P} r_v^\times$; therefore, by lemma 2 of § 3, ω must be trivial on some such group, which is the same as to say that ω_v is unramified for almost all v. Consequently, for all $z = (z_v)$ in k_A^\times we have $\omega(z) = \prod \omega_v(z_v)$, the product being taken over all the places v of k; for each z, almost all the factors in that product have the value 1. We will write this more briefly as $\omega = \prod \omega_v$.

The chief purpose of this Chapter can now be stated; it is to investigate the integrals of the form

$$(4) \qquad Z(\omega, \Phi) = \int_{k_A^\times} \Phi(j(z))\omega(z)\,d\mu(z),$$

where the notations have the following meaning. For μ, we take a Haar measure on k_A^\times; for ω, we take a quasicharacter of $G_k = k_A^\times/k^\times$, regarded as above as a function on k_A^\times. For Φ, we take a standard function on k_A. By j, we denote the natural bijection of k_A^\times onto the set of invertible elements of k_A, which is a continuous mapping of k_A^\times into k_A, by prop. 2 of Chap. IV-3. By abuse of notation, we will usually write $\Phi(z)$ instead of $\Phi(j(z))$ in the future.

As to μ, it has already been observed in Chap. V-4, in the case of characteristic 0, that such a measure can be defined by choosing, for every v, a Haar measure μ_v on k_v^\times, in such a way that $\mu_v(r_v^\times) = 1$ for almost all v. Then we write $\mu = \prod \mu_v$ for the measure on k_A^\times which coincides with the product measure $\prod \mu_v$ on every one of the subgroups $k_A(P)^\times$. The construction of the measures μ_v is contained in the following:

LEMMA 5. *Let K be a local field and α a Haar measure on K. Then the formula $d\mu(x) = \mathrm{mod}_K(x)^{-1} d\alpha(x)$ defines a Haar measure μ on K^\times; moreover, if K is a p-field, q its module, and R its maximal compact subring, then $\mu(R^\times) = (1 - q^{-1})\alpha(R)$.*

By the definition of mod_K, $x \to ax$ leaves μ invariant for $a \in K^\times$; this proves the first assertion. The second one follows at once from th. 6 of Chap. I-4.

PROPOSITION 10. *Let $\Phi = \prod \Phi_v$ be a standard function on k_A, $\omega = \prod \omega_v$ a quasicharacter of $G_k = k_A^\times/k^\times$, and $\mu = \prod \mu_v$ a Haar measure on k_A^\times. Assume that $|\omega| = \omega_\sigma$ with $\sigma > 1$. Then the integral $Z(\omega, \Phi)$ in (4) is absolutely convergent, and its value is also given by the absolutely convergent product*

$$(5) \qquad Z(\omega, \Phi) = \prod_v \left(\int_{k_v^\times} \Phi_v(x)\omega_v(x)\,d\mu_v(x) \right).$$

For each finite place v of k, put $\Psi_v = |\Phi_v|$; for each infinite place w of k, choose a standard function Ψ_w on k_w such that $\Psi_w \geqslant |\Phi_w|$; then, clearly, $\Psi = \prod \Psi_v$ is a standard function on k_A, majorizing $|\Phi|$, and $Z(\omega, \Phi)$ is majorized by $Z(\omega_\sigma, \Psi)$. Call $I(P)$, $J(P)$ the integrals of $\Phi\omega\,d\mu$ and of $\Psi\omega_\sigma\,d\mu$, respectively, on $k_A(P)^\times$. Call I_v, J_v the integrals of $\Phi_v\omega_v\,d\mu_v$ and of $\Psi_v|\omega_v|\,d\mu_v$, respectively, on k_v^\times, and, for every finite v, call I_v', J_v' the same integrals taken on r_v^\times instead of k_v^\times; I_v is the factor corresponding to v in the right-hand side of (5). For almost all finite places v of k, Φ_v is the characteristic function of r_v, ω_v is unramified, and $\mu_v(r_v^\times) = 1$; let P_0 be a finite set of places, containing P_∞, such that this is so for v not in P_0. Then, for v not in P_0, $I_v' = J_v' = 1$. This implies that we have, for all $P \supset P_0$:

$$I(P) = \prod_{v \in P} I_v, \qquad J(P) = \prod_{v \in P} J_v.$$

Therefore $Z(\omega_\sigma, \Psi)$ is $< +\infty$ provided all the integrals J_v, and the infinite product $\prod J_v$, are convergent; moreover, if we show that this is so, it will imply that $Z(\omega, \Phi)$, the integrals I_v and the product $\prod I_v$ are all absolutely convergent, and that $Z(\omega, \Phi)$ is equal to that product, which is what we have to prove. For any v, take a Haar measure α_v on k_v; then, by lemma 5, $d\mu_v(x) = m_v|x|_v^{-1} d\alpha_v(x)$ with some $m_v \in \mathbf{R}_+^\times$. This gives

$$J_v = m_v \int_{k_v^\times} \Psi_v(x)|x|_v^{\sigma-1} d\alpha_v(x).$$

In view of our definition of a standard function, one sees at once that this is convergent for $\sigma \geqslant 1$; it would still be so even for $\sigma > 0$, but this is not needed here. On the other hand, for v not in P_0, we have, since $k_v^\times \cap r_v$ is the disjoint union of the sets $u_v = \pi_v^\nu r_v^\times$ for $\nu \geqslant 0$:

$$J_v = \sum_{\nu=0}^{+\infty} \int_{u_v} |x|_v^\sigma d\mu_v(x) = \sum_{\nu=0}^{+\infty} q_v^{-\nu\sigma} = (1 - q_v^{-\sigma})^{-1}.$$

Prop. 1 of § 1 shows now that $\prod J_v$ is convergent, which completes the proof.

The method of calculation which we have just given for J_v can be applied to I_v; we formulate this as follows:

PROPOSITION 11. *Let K be a p-field, q its module, R its maximal compact subring, and μ the Haar measure on K^\times such that $\mu(R^\times)=1$. Call φ the characteristic function of R. Then, for $\mathrm{Re}(s)>0$:*

$$\int_{K^\times} \varphi(x) \mathrm{mod}_K(x)^s d\mu(x) = (1 - q^{-s})^{-1}.$$

In fact, we can write $K^\times \cap R$ as the disjoint union of the sets $U_\nu = \pi^\nu R^\times = P^\nu - P^{\nu+1}$ for $\nu \geqslant 0$. Then our integral can be written as

$$\sum_{\nu=0}^{+\infty} \int_{U_\nu} \mathrm{mod}_K(x)^s d\mu(x) = \sum_{\nu=0}^{+\infty} q^{-\nu s},$$

which is absolutely convergent for $\mathrm{Re}(s)>0$ and has the value stated above.

§ 5. The functional equation. We will first choose a Haar measure on $k_\mathbf{A}^\times$. On the compact group G_k^1, take the Haar measure μ_1 given by $\mu_1(G_k^1)=1$. On the group N, take the measure ν given by $d\nu(n)=n^{-1} dn$ if $N=\mathbf{R}_+^\times$ and by $\nu(\{1\})=1$ otherwise. On $G_k = G_k^1 \times N$, we take the measure $\mu = \mu_1 \times \nu$. Finally, on $k_\mathbf{A}^\times$, as explained in Chap. II-4, we choose as μ the measure whose image in $G_k = k_\mathbf{A}^\times/k^\times$ is the one we have just defined.

LEMMA 6. *Let F_1 be a measurable function on N such that $0 \leqslant F_1 \leqslant 1$; assume also that there is a compact interval $[t_0, t_1]$ in \mathbf{R}_+^\times such that $F_1(n) = 1$ for $n \in N$, $n < t_0$, and $F_1(n) = 0$ for $n \in N$, $n > t_1$. Then the integral*

$$\lambda(s) = \int_N n^s F_1(n) \, d\nu(n)$$

is absolutely convergent for $\mathrm{Re}(s) > 0$. *The function $\lambda(s)$ can be continued analytically in the whole s-plane as a meromorphic function. If we put $\lambda_0(s) = s^{-1}$ if $N = \mathbf{R}_+^\times$, and $\lambda_0(s) = \frac{1}{2}(1 + Q^{-s})(1 - Q^{-s})^{-1}$ if $N = \{Q^\nu\}_{\nu \in \mathbf{Z}}$, then $\lambda - \lambda_0$ is an entire function of s. Finally, if $F_1(n) + F_1(n^{-1}) = 1$ for all $n \in N$, then $\lambda(s) + \lambda(-s) = 0$.*

Take first for F_1 the function f_1 given by $f_1(n) = 1$ for $n < 1$, $f_1(1) = 1/2$, $f_1(n) = 0$ for $n > 1$. Then λ becomes, for $N = \mathbf{R}_+^\times$, the integral $\int_0^1 n^{s-1} dn$, and, for $N = \{Q^\nu\}$, the series $\frac{1}{2} + \sum_1^{+\infty} Q^{-\nu s}$; in both cases it is absolutely convergent for $\mathrm{Re}(s) > 0$, and equal to $\lambda_0(s)$. This gives, for any F_1:

$$\lambda(s) - \lambda_0(s) = \int_N n^s (F_1(n) - f_1(n)) \, d\nu(n).$$

As $F_1 - f_1$ is a bounded measurable function with compact support on N, the last integral is absolutely convergent for all s, uniformly on every compact subset of the s-plane; this implies that it is an entire function of s. Assume now that $F_1(n) + F_1(n^{-1}) = 1$; as f_1 has the same property, the function $F_2 = F_1 - f_1$ satisfies $F_2(n^{-1}) = -F_2(n)$. Replacing n by n^{-1} in the last integral, and observing that $\lambda_0(-s) = -\lambda_0(s)$, we get $\lambda(-s) = -\lambda(s)$.

Lemma 6 implies that λ has at $s = 0$ a residue equal to 1 if $N = \mathbf{R}_+^\times$ and to $(\log Q)^{-1}$ if $N = \{Q^\nu\}$. Here, and also in the next results, it is understood that residues are taken with respect to the variable s; in other words, if a function $f(s)$ of s has a simple pole at $s = s_0$, its residue there is the limit of $(s - s_0) f(s)$ for $s \to s_0$.

THEOREM 2. *Let Φ be a standard function on k_A. Then the function $\omega \to Z(\omega, \Phi)$ defined by formula (4) of § 4 when the integral in (4) is absolutely convergent can be continued analytically as a meromorphic function on the whole of the complex manifold $\Omega(G_k)$. It satisfies the equation*

$$Z(\omega, \Phi) = Z(\omega_1 \omega^{-1}, \Phi'),$$

where Φ' is the Fourier transform of Φ with respect to the Tamagawa measure on k_A. Moreover, $Z(\omega, \Phi)$ is holomorphic everywhere on $\Omega(G_k)$ except for simple poles at ω_0 and ω_1, with the residues $-\rho \Phi(0)$ at ω_0 and $\rho \Phi'(0)$ at ω_1, where $\rho = 1$ if $N = \mathbf{R}_+^\times$ and $\rho = (\log Q)^{-1}$ if $N = \{Q^\nu\}$.

* On \mathbf{R}_+^\times, choose two continuous functions F_0, F_1 with the following properties: (i) $F_0 \geqslant 0$, $F_1 \geqslant 0$, $F_0 + F_1 = 1$; (ii) there is a compact interval $[t_0, t_1]$ in \mathbf{R}_+^\times such that $F_0(t) = 0$ for $0 < t < t_0$, and $F_1(t) = 0$ for $t > t_1$. Take any $B > 1$. Then, for $\sigma \in \mathbf{R}$, $\sigma \leqslant B$, $t \in \mathbf{R}_+^\times$, we have $t^\sigma F_0(t) \leqslant t_0^{\sigma - B} t^B$. Write now, for $i = 0, 1$:

$$Z_i = Z_i(\omega, \Phi) = \int_{k_\mathbf{A}^\times} \Phi(z)\omega(z)F_i(|z|_\mathbf{A})d\mu(z).$$

As before, put $|\omega| = \omega_\sigma$ with $\sigma \in \mathbf{R}$; by prop. 10 of §4, Z_0 and Z_1 are absolutely convergent for $\sigma > 1$. On the other hand, if $\sigma \leqslant B$, Z_0 is majorized by the integral

$$\int_{k_\mathbf{A}^\times} |\Phi(z)| \cdot |z|_\mathbf{A}^\sigma F_0(|z|_\mathbf{A})d\mu(z) \leqslant t_0^{\sigma-B} \int_{k_\mathbf{A}^\times} |\Phi(z)| \cdot |z|_\mathbf{A}^B d\mu(z),$$

which is convergent by prop. 10 of §4. In particular, $Z_0(\omega_s \omega, \Phi)$ is absolutely convergent for all $s \in \mathbf{C}$, and one verifies easily that this is so uniformly with respect to s on every compact subset of \mathbf{C}. As the quasicharacters $\omega_s \omega$, for $s \in \mathbf{C}$, make up the connected component of ω in $\Omega(G_k)$, with the complex structure determined by the variable s, this shows that $\omega \to Z_0(\omega, \Phi)$ is holomorphic on the whole of $\Omega(G_k)$.

Now apply formula (6) of Chap. II-4 to the group $k_\mathbf{A}^\times$, the discrete subgroup k^\times and the integrals Z_0, Z_1. This gives:

$$Z_i = \int_{G_k} \Big(\sum_{\xi \in k^\times} \Phi(z\xi) \Big) \omega(z)F_i(|z|_\mathbf{A})d\mu(\dot{z}),$$

where \dot{z} is the image of z in $G_k = k_\mathbf{A}^\times / k^\times$, and the integrand is to be understood as a function of \dot{z}. Here the integrals for Z_0, Z_1 are absolutely convergent whenever the original integrals for Z_0, Z_1 are so, i.e. for $\sigma > 1$ in the case of Z_1 and for all σ in the case of Z_0.

For each $z \in k_\mathbf{A}^\times$, we may apply lemma 1 of §2 to the automorphism $x \to z^{-1}x$ of $k_\mathbf{A}$; applying then Poisson's formula, i.e. (1) of §2, to the function $x \to \Phi(zx)$, we get:

$$\Phi(0) + \sum_{\xi \in k^\times} \Phi(z\xi) = |z|_\mathbf{A}^{-1} \Big(\Phi'(0) + \sum_{\xi \in k^\times} \Phi'(\xi z^{-1}) \Big),$$

and therefore:

$$Z_1 = \int_{G_k} \Big(\sum_{\xi \in k^\times} \Phi'(\xi z^{-1}) + \Phi'(0) - |z|_\mathbf{A} \Phi(0) \Big) |z|_\mathbf{A}^{-1} \omega(z)F_1(|z|_\mathbf{A})d\mu(\dot{z}).$$

On the other hand, what we have proved above for Z_0 remains valid if we replace ω by $\omega_1 \omega^{-1}$, Φ by Φ' and F_0 by the function $t \to F_1(t^{-1})$.

Calling Z'_0 the result of this substitution, we get:

$$Z'_0 = \int_{k_A^\times} \Phi'(z)|z|_A \omega(z)^{-1} F_1(|z|_A^{-1}) d\mu(z);$$

therefore this is always absolutely convergent, and holomorphic on the whole of $\Omega(G_k)$. In this integral, replace z by z^{-1}; this changes the Haar measure μ into a Haar measure $c\mu$, where $c^2 = 1$ since it is a homeomorphism of order 2 of k_A^\times onto itself, hence $c = 1$. After this change of variable, apply again formula (6) of Chap. II-4 to k_A^\times and k^\times. This gives:

$$Z'_0 = \int_{G_k} \left(\sum_{\xi \in k^\times} \Phi'(\xi^{-1} z^{-1}) \right) |z|_A^{-1} \omega(z) F_1(|z|_A) d\mu(\dot{z}),$$

this again being always absolutely convergent. As $\xi \to \xi^{-1}$ is a bijection of k^\times onto itself, we get now:

$$Z_1 - Z'_0 = \int_{G_k} (\Phi'(0) - |z|_A \Phi(0)) |z|_A^{-1} \omega(z) F_1(|z|_A) d\mu(\dot{z}),$$

this being absolutely convergent for $\sigma > 1$, since Z_1 and Z'_0 are so. By corollary 2 of prop. 7, § 3, we can write $\omega = \omega_s \psi$, where ψ is a character of G_k, trivial on N. In view of our definition of μ as the measure $\mu_1 \times \nu$ on $G_k = G_k^1 \times N$, our last formula can now be written:

$$Z_1 - Z'_0 = \left(\int_{G_k^1} \psi d\mu_1 \right) \cdot \left(\int_N (\Phi'(0) - n\Phi(0)) n^{s-1} F_1(n) d\nu(n) \right).$$

The first factor in the right-hand side is 1 or 0 according as ψ is trivial or not, i.e. according as ω is principal or not; write δ_ω for this factor. The second one can be evaluated at once by lemma 6. If $\lambda(s)$ is as defined in that lemma, this gives:

$$Z_1 - Z'_0 = \delta_\omega(\Phi'(0)\lambda(s-1) - \Phi(0)\lambda(s)).$$

As $Z(\omega, \Phi) = Z_0 + Z_1$, this proves that $Z(\omega, \Phi)$ can be continued everywhere on $\Omega(G_k)$ outside the connected component Ω_1 of $\omega_0 = 1$ as a holomorphic function, and on that component as a meromorphic function having at most the same poles as $\lambda(s-1)$ and $\lambda(s)$; as to the latter poles and their residues, they are given by lemma 6 and are as stated in our theorem. Finally, assume that we have chosen F_0, F_1 so that $F_0(t) = F_1(t^{-1})$ for all t; this can be done by taking for F_1 a continuous function for $t \geqslant 1$, such that $0 \leqslant F_1(t) \leqslant 1$ for all $t \geqslant 1$, $F_1(1) = 1/2$, and $F_1(t) = 0$ for $t \geqslant t_1$, and then putting $F_1(t) = 1 - F_1(t^{-1})$ for $0 < t < 1$, and $F_0 = 1 - F_1$. That being so, we have $Z'_0 = Z_0(\omega_1 \omega^{-1}, \Phi')$, and therefore

$$Z(\omega, \Phi) = Z_0(\omega, \Phi) + Z_0(\omega_1 \omega^{-1}, \Phi') + \delta_\omega(\Phi'(0)\lambda(s-1) - \Phi(0)\lambda(s)).$$

In this formula, replace ω by $\omega_1 \omega^{-1}$ and Φ by Φ'. In view of Fourier's inversion formula, this replaces Φ' by the function Φ'' given by

$\Phi''(x) = \Phi(-x)$. As ω is trivial on k^\times, we have $\omega(-1) = 1$, hence $\omega(-z) = \omega(z)$ for all z, so that $Z_0(\omega, \Phi'')$ is the same as $Z_0(\omega, \Phi)$; therefore this substitution merely interchanges the first two terms in the right-hand side of our formula; as it changes s into $1-s$, lemma 6 shows that it does not change the last term. This completes the proof of the "functional equation" in theorem 2.

COROLLARY 1. *Let P be a finite set of places of k, containing P_∞; then the product*

$$p(k,P,s) = \prod_{v \notin P} (1 - q_v^{-s})^{-1}$$

is absolutely convergent for $\mathrm{Re}(s) > 1$, and $(s-1)p(k,P,s)$ tends to a finite limit > 0 when s tends to 1.

The first assertion is contained in corollary 1 of prop. 1, § 1. Now take Haar measures α_v on k_v, μ_v on k_v^\times, as explained above; by lemma 5 of § 4, we have, for every v, $d\mu_v(x) = m_v |x|_v^{-1} d\alpha_v(x)$, with some $m_v > 0$. Take the standard function Φ so that Φ_v is the characteristic function of r_v for all v not in P, and that $\Phi_v \geqslant 0$ and $\Phi_v(0) > 0$ for all v. Apply prop. 10 of § 4 to $Z(\omega_s, \Phi)$ for $\mathrm{Re}(s) > 1$; the factor I_v corresponding to v, in the right-hand side of the formula in that proposition, can now be written as

$$I_v = m_v \int_{k_v^\times} \Phi_v(x) |x|_v^{s-1} d\alpha_v(x).$$

For v not in P, by prop. 11 of § 4, this differs from $(1 - q_v^{-s})^{-1}$ only by the scalar factor $\mu_v(r_v^\times)$, which is always > 0, and which is 1 for almost all v. For $v \in P$, one can verify at once that I_v is continuous for $\mathrm{Re}(s) \geqslant 1$ (one could easily show, in fact, that it is holomorphic for $\mathrm{Re}(s) > 0$, and, in the next §, one will obtain a much more precise result for a specific choice of Φ, but this is not needed now); for s tending to 1, it tends to $m_v \int \Phi_v d\alpha_v$, which is > 0. This shows that $Z(\omega_s, \Phi)$ differs from the product $p(k,P,s)$ in our corollary by a factor which tends to a finite limit > 0 when s tends to 1. On the other hand, theorem 2 shows that $Z(\omega_s, \Phi)$ has a simple pole at $s = 1$, with the residue $\rho \Phi'(0)$, and $\rho > 0$; as $\Phi'(0) = \int \Phi d\alpha$, and as this is obviously > 0, this completes the proof.

COROLLARY 2. *Let P be as above; let ω be a non-trivial character of k_A^\times, trivial on k^\times, such that ω_v is unramified for all v not in P; for v not in P, put $\lambda(v) = \omega_v(\pi_v)$, where π_v is a prime element of k_v. Then the product*

$$p(k,P,\omega,s) = \prod_{v \notin P} (1 - \lambda(v) q_v^{-s})^{-1}$$

is absolutely convergent for $\mathrm{Re}(s) > 1$ and tends to a finite limit when s tends to 1; if ω^2 is not trivial, this limit is not 0.

As ω is a character, we have $|\lambda(v)|=1$ for all v not in P, so that the first assertion is again contained in corollary 1 of prop. 1, § 1. Take α_v, μ_v as before, and take Φ so that, for v not in P, Φ_v is the characteristic function of r_v. Apply prop. 10 of § 4 to $Z(\omega_s\omega, \Phi)$ for $\mathrm{Re}(s)>1$; the factor I_v is now

$$I_v = m_v \int_{k_v^\times} \Phi_v(x)\omega_v(x)|x|_v^{s-1}\, d\alpha_v(x).$$

For v not in P, ω_v is unramified and may be written as $\omega_v(x)=|x|_v^{s_v}$, where s_v can be determined by $\lambda(v)=q_v^{-s_v}$; then prop. 11 of § 4 shows that I_v differs from $(1-\lambda(v)q_v^{-s})^{-1}$ only by the scalar factor $\mu_v(r_v^\times)$, which is 1 for almost all v. For $v \in P$, we observe, as before, that I_v is continuous for $\mathrm{Re}(s) \geqslant 1$; taking prop. 9 into account when v is an infinite place, one sees easily that, for each $v \in P$, Φ_v may be so chosen that I_v is not 0 for $s=1$, and we will assume that it has been so chosen (for specific choices of Φ_v, I_v will be computed explicitly in § 7). We see now that $Z(\omega_s\omega, \Phi)$ differs from the product $p(k,P,\omega,s)$ in our corollary by a factor which tends to a finite limit, other than 0, when s tends to 1. In view of theorem 2, this proves the second assertion in our corollary. As to the last one, we need a lemma:

LEMMA 7. *For* $t \in \mathbf{C}$, $\lambda \in \mathbf{C}$, *put* $\varphi(\lambda,t)=(1-t)^3(1-\lambda t)^4(1-\lambda^2 t)$. *Then* $|\varphi(\lambda,t)|<1$ *for* $t \in \mathbf{R}$, $0<t<1$, $\lambda\bar\lambda=1$.

In fact, we have then

$$\log|\varphi(\lambda,t)|^2 = \log(\varphi(\lambda,t)\,\varphi(\bar\lambda,t))$$

$$= -\sum_{n=1}^{\infty} \frac{t^n}{n}(6+4\lambda^n+4\bar\lambda^n+\lambda^{2n}+\bar\lambda^{2n})$$

$$= -\sum_{n=1}^{\infty} \frac{t^n}{n}(2+\lambda^n+\bar\lambda^n)^2 < 0.$$

If now $\varphi(\lambda,t)$ is defined as in the lemma, we have

$$p(k,P,s)^3\, p(k,P,\omega,s)^4\, p(k,P,\omega^2,s) = \prod_{v \notin P} \varphi(\lambda(v), q_v^{-s})^{-1}.$$

By the lemma, this has an absolute value >1 for $s \in \mathbf{R}$, $s>1$, so that it cannot tend to 0 for s tending to 1. For s tending to 1, as shown above, $p(k,P,\omega^2,s)$ tends to a finite limit if ω^2 is not trivial, and $p(k,P,\omega,s)$ is the product of a factor, tending to a finite limit other than 0, and of $Z(\omega_s\omega, \Phi)$, which is holomorphic in a neighborhood of $s=1$; therefore, if $p(k,\omega,P,s)$ tends to 0, it must be of the form $F(s)(s-1)$, with F bounded. In view of corollary 1, this implies that the left-hand side of the last formula tends to 0 for s tending to 1. This completes our proof. It is an important fact that the conclusion of our corollary remains true

* even if $\omega^2 = 1$; the proof for this, which requires quite different methods, will be given in Chap. XIII-12.

COROLLARY 3. *Let k_0 be an A-field contained in k; let V be a set of finite places of k, such that, for almost all the finite places v of k, not in V, the modular degree of k_v over the closure of k_0 in k_v is >1. Then the product*

$$q(k,V,s) = \prod_{v \in V} (1 - q_v^{-s})^{-1}$$

is absolutely convergent for $\mathrm{Re}(s) > 1$, and $(s-1)q(k,V,s)$ tends to a finite limit >0 when s tends to 1.

In fact, with the notation of corollary 1, $p(k,P_\infty,s)$ is the product of $q(k,V,s)$ and of the similar product, taken over the set M of all the finite places of k, not in V; applying corollary 3 of prop. 1, § 1, to the latter product, and corollary 1 to $p(k,P_\infty,s)$, we get our conclusion at once. Of course our corollary implies that V cannot be a finite set, or in other words that there are infinitely many places v of k for which the modular degree in question is 1.

COROLLARY 4. *Let k_0 and V be as in corollary 3; let k' be a separably algebraic extension of k of finite degree n, and assume that there are n distinct places of k' above every place $v \in V$. Then $k' = k$.*

Call V' the set of the places of k' lying above those of V. By corollary 1 of th. 4, Chap. III-4, if $v \in V$, and w lies above v, we have $k'_w = k_v$, hence $q'_w = q_v$. For any place v of k, and any place w of k' above v, the modular degree of k'_w over the closure of k_0 in k_v is at least equal to that of k_v over that closure; therefore, for almost all v, not in V, or, what amounts to the same, for almost all w, not in V', that degree is >1. We can now apply corollary 3 to the products $q(k,V,s)$ and $q(k',V',s)$; as the latter is equal to $q(k,V,s)^n$, this gives $n=1$.

COROLLARY 5. *Let k be an A-field of characteristic $p>1$, and let P be a finite set of places of k. Then there is a divisor $\mathfrak{m} = \sum m(v) \cdot v$ of k of degree 1 such that $m(v) = 0$ for all $v \in P$.*

Call ν the g.c.d. of the degrees of all the places v, not in P; we have to show that $\nu = 1$. Let $F = \mathbf{F}_q$ be the field of constants of k; by th. 2 of Chap. I-1, there is, in an algebraic closure of k, a field F' with q^ν elements, and it is separable over F. Call k' the compositum of k and F', and n its degree over k; k' is separable over k. Let v be any place of k, not in P; let w be a place of k' above v; by prop. 1 of Chap. III-1, k'_w is generated over k_v by k', hence by F'. By the definition of ν, the module of k_v is of the form $q^{\nu r}$, where r is an integer; therefore, by corollary 1

of th. 7, Chap. I-4, combined with corollary 2 of th. 2, Chap. I-1, k_v contains a subfield with q^v elements. By th. 2 of Chap. I-1, k'_w cannot contain more than one field with q^v elements; therefore $F' \subset k_v$, hence $k'_w = k_v$. Corollary 1 of th. 4, Chap. III-4, shows now that there are n distinct places of k' above each place v of k, not in P. Taking $k_0 = k$ in corollary 4, and taking for V the complement of P, we get $k' = k$, hence $F' \subset F$, i.e. $v = 1$.

COROLLARY 6. *Let k be as in corollary 5, and let \mathbf{F}_q be its field of constants. Then the value-group N of $|z|_{\mathbf{A}}$ on $k_{\mathbf{A}}^{\times}$ is generated by q.*

As we have seen in § 4, N is generated by the value-groups of $|x|_v$ on k_v^{\times} for all v, hence by the modules $q_v = q^{\deg(v)}$, so that it has the generator $Q = q^v$, where v is the g.c.d. of all the degrees $\deg(v)$. By corollary 5, $v = 1$.

Taking corollary 6 into account, we can reformulate the last assertion of theorem 2, in the case of characteristic $p > 1$, as follows:

COROLLARY 7. *Let k and \mathbf{F}_q be as in corollary 6; let notations be as in theorem 2. Then $Z(\omega_s, \Phi) + \Phi(0)(1 - q^{-s})^{-1}$ is holomorphic at $s = 0$.*

This follows at once from the results we have just mentioned and from the fact that $(1 - q^{-s})^{-1}$ has the residue $(\log q)^{-1}$ at $s = 0$.

§ 6. The Dedekind zeta-function.

Special choices of Φ in $Z(\omega, \Phi)$ lead to the definition of important functions on the connected components of $\Omega(G_k)$; these will now be investigated more in detail. We begin with the consideration of the connected component Ω_1 of $\omega_0 = 1$ in $\Omega(G_k)$, i.e. of the group of the principal quasicharacters of G_k, choosing Φ as follows. Whenever v is a finite place of k, we take for Φ_v the characteristic function of r_v. When v is real, i.e. $k_v = \mathbf{R}$, we take $\Phi_v(x) = \exp(-\pi x^2)$. When v is imaginary, i.e. $k_v = \mathbf{C}$, we take $\Phi_v(x) = \exp(-2\pi x \bar{x})$. We have now to calculate the factors in the product (5) for $Z(\omega, \Phi)$, for this choice of Φ and for $\omega = \omega_s$; when v is a finite place, these are given by prop. 11 of § 4, up to a scalar factor depending on μ. For the infinite places, they are as follows:

LEMMA 8. *Let G_1, G_2 be defined, for all $s \in \mathbf{C}$, by the formulas*

$$G_1(s) = \pi^{-s/2} \Gamma(s/2), \qquad G_2(s) = (2\pi)^{1-s} \Gamma(s).$$

Then we have, for $\mathrm{Re}(s) > 0$:

$$\int_{\mathbf{R}^{\times}} \exp(-\pi x^2) |x|^{s-1} dx = G_1(s),$$

$$\int_{\mathbf{C}^{\times}} \exp(-2\pi x \bar{x})(x \bar{x})^{s-1} |dx \wedge d\bar{x}| = G_2(s).$$

This can be verified at once by obvious changes of variables, viz., $|x| = t^{1/2}$ in the first integral and $x = t^{1/2} \mathbf{e}(u)$ in the second one, with $t \in \mathbf{R}_+^\times$, $u \in \mathbf{R}$, $0 \leqslant u < 1$; in the latter case, $|dx \wedge d\bar{x}| = 2\pi \, dt \, du$.

Now consider the measure $\gamma = \prod \gamma_v$ on $k_\mathbf{A}^\times$, given by taking $\gamma_v(r_v^\times) = 1$ for every finite place v, $d\gamma_v(x) = |x|^{-1} dx$ when v is real, and $d\gamma_v(x) = (x \bar{x})^{-1} |dx \wedge d\bar{x}|$ when v is imaginary; when k is of characteristic 0, this is the measure occurring in prop. 9 of Chap. V-4. The relation between γ and the measure μ introduced at the beginning of § 5 is as follows:

PROPOSITION 12. *Let μ be as in § 5, and γ as above. If k is of characteristic 0, we have $\gamma = c_k \mu$, where c_k is as defined in proposition 9 of Chapter V-4. If k is of characteristic $p > 1$, with the field of constants \mathbf{F}_q, and if h is the number of divisor-classes of degree 0 of k, then $\gamma = c_k \mu$ with $c_k = h/(q-1)$.*

In view of our definition of μ, the first assertion is merely a restatement of prop. 9, Chap. V-4. Now let k be of characteristic $p > 1$, and put $U = \prod r_v^\times$; this is the same as $\Omega(\emptyset)$ in the notation of Chap. IV-4, and it is an open subgroup of $k_\mathbf{A}^\times$; by definition, we have $\gamma(U) = 1$. As explained in Chap. II-4, we will also write γ for the image of the measure γ in $G_k = k_\mathbf{A}^\times / k^\times$; G_k^1 being, as before, the image of $k_\mathbf{A}^1$ in G_k, μ is defined by $\mu(G_k^1) = 1$, so that we have $\gamma = c_k \mu$ with $c_k = \gamma(G_k^1)$. Call U' the image of U in G_k; by th. 8 of Chap. IV-4 and its corollary, the kernel of the morphism of U onto U', induced by the canonical morphism of $k_\mathbf{A}^\times$ onto G_k, is \mathbf{F}_q^\times, so that we can compute $\gamma(U')$ by taking $G = U$, $\Gamma_1 = \mathbf{F}_q^\times$, $\Gamma = \{1\}$ in lemma 2 of Chap. II-4; this gives $\gamma(U') = (q-1)^{-1}$. Clearly the index of U' in G_k^1 is equal to that of $k^\times U$ in $k_\mathbf{A}^1$; as we have seen in Chap. VI that $k_\mathbf{A}^1 / k^\times U$ may be identified with the group $D_0(k)/P(k)$ of the divisor-classes of degree 0 of k, that index is h. Therefore $\gamma(G_k^1) = h/(q-1)$.

Now, for each infinite place w of k, put $G_w = G_1$ or $G_w = G_2$ according as w is real or imaginary. Combining prop. 10 of § 4, prop. 11 of § 4, lemma 8, and prop. 12, we get for $\mathrm{Re}(s) > 1$, Φ being chosen as explained above:

(6) $$Z(\omega_s, \Phi) = c_k^{-1} \prod_{w \in P_\infty} G_w(s) \prod_{v \notin P_\infty} (1 - q_v^{-s})^{-1},$$

with c_k as in prop. 12. By th. 2 of § 5, the left-hand side can be continued analytically as a meromorphic function over the whole s-plane; as the same is true of the factors G_w, it is also true of the last product in the right-hand side. This justifies the following definition:

DEFINITION 8. *The meromorphic function ζ_k in the s-plane, given for* $\mathrm{Re}(s) > 1$ *by the product*

$$\zeta_k(s) = \prod_v (1 - q_v^{-s})^{-1}$$

taken over all the finite places v of k, is called the Dedekind zeta-function of k.

When Φ is as above, its Fourier transform Φ' is immediately given by th. 1 of § 2 and its corollary 2, combined with corollary 3 of prop. 2, § 2, and propositions 4 and 5 of § 2. This gives

$$\Phi'(y) = |a|_A^{1/2} \Phi(ay)$$

where a is a differential idele attached to the basic character χ. In view of the definition of $Z(\omega, \Phi)$ by formula (4) of § 4, we have now:

$$Z(\omega, \Phi') = |a|_A^{1/2} \omega(a)^{-1} Z(\omega, \Phi),$$

hence in particular, for $\omega = \omega_s$, i.e. $\omega(x) = |x|_A^s$:

(7) $$Z(\omega_s, \Phi') = |a|_A^{1/2 - s} Z(\omega_s, \Phi);$$

moreover, the value of $|a|_A$ is that given in prop. 6 of § 2.

We are now ready to formulate our final results on the zeta-function.

THEOREM 3. *Let k be an algebraic number-field with r_1 real places and r_2 imaginary places. Call ζ_k its zeta-function, and write*

$$Z_k(s) = G_1(s)^{r_1} G_2(s)^{r_2} \zeta_k(s).$$

Then Z_k is a meromorphic function in the s-plane, holomorphic except for simple poles at $s = 0$ and $s = 1$, and satisfies the functional equation

$$Z_k(s) = |D|^{\frac{1}{2} - s} Z_k(1 - s)$$

where D is the discriminant of k. Its residues at $s = 0$ and $s = 1$ are respectively $-c_k$ and $|D|^{-1/2} c_k$, with c_k given by

$$c_k = 2^{r_1} (2\pi)^{r_2} h R / e,$$

where h is the number of ideal-classes of k, R its regulator, and e the number of roots of 1 in k.

This follows immediately from (6), (7), prop. 12, prop. 6 of § 2, and from th. 2 of § 5.

COROLLARY. *The Dedekind zeta-function $\zeta_k(s)$ has the residue $|D|^{-1/2} c_k$ at $s = 1$.*

This follows from th. 3 and the well-known fact that $G_1(1) = G_2(1) = 1$.

THEOREM 4. *Let k be an A-field of characteristic $p > 1$; let \mathbf{F}_q be its field of constants and g its genus. Then its zeta-function can be written in the form*

$$\zeta_k(s) = \frac{P(q^{-s})}{(1 - q^{-s})(1 - q^{1-s})}$$

where P is a polynomial of degree 2g with coefficients in \mathbf{Z}, such that

(8) $P(u) = q^g u^{2g} P(1/qu).$

Moreover, $P(0) = 1$, and $P(1)$ is equal to the number h of divisor-classes of degree 0 of k.

In fact, corollary 6 of th. 2, § 5, shows at once that $s \to \omega_s$ has the same kernel as $s \to q^{-s}$, so that $\zeta_k(s)$ may be written as $R(q^{-s})$, where R is a meromorphic function in \mathbf{C}^\times, with simple poles at 1 and at q^{-1}. Moreover, corollary 1 of prop. 1, § 1, shows that $R(u)$ tends to 1 for u tending to 0, so that R is holomorphic there, and that $R(0) = 1$. We may therefore write $R(u) = P(u)/(1-u)(1-qu)$, where P is an entire function in the u-plane, with $P(0) = 1$. Now (7), combined with (6) and with prop. 6 of § 2, gives formula (8) of our theorem; clearly this implies that P is a polynomial of degree $2g$. Finally, corollary 7 of th. 2, § 5, combined with prop. 12, gives $P(1) = h$.

§ 7. L-functions.

We will now extend the above results to arbitrary quasicharacters of G_k; in order to do this, we adopt the following notations. Let ω be any quasicharacter of G_k; as we have seen in §§ 3–4, we may write $|\omega| = \omega_\sigma$, with $\sigma \in \mathbf{R}$. For every v, we write ω_v for the quasicharacter of k_v^\times induced on k_v^\times by ω. For every finite place v, we write $p_v^{f(v)}$ for the conductor of ω_v; $f(v)$ is 0 if and only if ω_v is unramified, hence, as we have seen in § 4, at almost all finite places of k; when that is so, we write $\omega_v(x) = |x|_v^{s_v}$ with $s_v \in \mathbf{C}$; clearly we have then $\mathrm{Re}(s_v) = \sigma$. At the infinite places of k, we can apply prop. 9 of § 3; this shows that ω_v may be written as $\omega_v(x) = x^{-A} |x|_v^{s_v}$ if v is real, with $A = 0$ or 1 and $s_v \in \mathbf{C}$, and as $\omega_v(x) = x^{-A} \bar{x}^{-B} (x\bar{x})^{s_v}$ if v is imaginary, with $\inf(A, B) = 0$ and $s_v \in \mathbf{C}$; in the former case we put $N_v = A$, and we have $\mathrm{Re}(s_v) = N_v + \sigma$, and in the latter case we put $N_v = \sup(A, B)$, and we have $\mathrm{Re}(s_v) = (N_v/2) + \sigma$. As the connected component of ω, in the group $\Omega(G_k)$ of the quasicharacters of G_k, consists of the quasicharacters $\omega_s \omega$ for $s \in \mathbf{C}$, the integers $f(v)$, N_v have the same values for all the quasicharacters in that component. They are all 0 if ω is principal, or, more generally, if ω is trivial on the group U of the ideles (z_v) such that $|z_v|_v = 1$ for all places v of k; the structure of the group of the quasicharacters with that property can easily be determined by the method used in the proof of th. 9, Chap. IV-4.

Furthermore, with the same notations as above, we attach to ω a standard function $\Phi_\omega = \prod \Phi_v$ on $k_\mathbf{A}$, as follows. For each finite place v where $f(v) = 0$, i.e. where ω_v is unramified, we take for Φ_v, as before, the characteristic function of r_v. For each finite place v where $f(v) \geqslant 1$, we take Φ_v equal to ω_v^{-1} on r_v^\times and to 0 outside r_v^\times. At each infinite place v, we take $\Phi_v(x) = x^A \exp(-\pi x^2)$ if v is real, and $\Phi_v(x) = x^A \bar{x}^B \exp(-2\pi x\bar{x})$ if v is imaginary, the integers A, B being as explained above. Then Φ_ω will be called *the standard function attached to* ω; it is clear that it does not change if ω is replaced by $\omega_s\omega$, with any $s \in \mathbf{C}$, and also that the function attached in this manner to $\bar{\omega}$, or to $\omega^{-1} = \omega_{-2\sigma}\bar{\omega}$, or to $\omega' = \omega_1 \omega^{-1}$, is $\bar{\Phi}_\omega$.

We need to know the Fourier transform of Φ_ω, or, what amounts to the same in view of th. 1 of § 2, those of the functions Φ_v defined above. The latter are given by our earlier results except when v is a finite place where ω_v is ramified. For that case, we have:

PROPOSITION 13. *Let K be a p-field; let R be its maximal compact subring, P the maximal ideal of R, and ω a quasicharacter of K^\times with the conductor P^f, where $f \geqslant 1$. Let χ be a character of K of order v, α the self-dual measure on K with reference to χ, and let $b \in K^\times$ be such that $\mathrm{ord}_K(b) = v + f$. Let φ be the function on K, equal to ω^{-1} on R^\times and to 0 outside R^\times. Then the Fourier transform of φ is*

$$\varphi'(y) = \kappa \, \mathrm{mod}_K(b)^{1/2} \, \overline{\varphi(b\,y)},$$

where κ is such that $\kappa\bar{\kappa} = 1$ and is given by

$$\kappa = \mathrm{mod}_K(b)^{-1/2} \int_{R^\times} \omega(x)^{-1} \chi(b^{-1}x)\,d\alpha(x).$$

By prop. 12 of Chap. II-5, the dual of the K-lattice P^f in K is P^{-f-v}; as φ is constant on classes modulo P^f in K, prop. 2 of § 2 shows that φ' is 0 outside $P^{-f-v} = b^{-1}R$. The definition of φ gives

$$(9) \qquad \varphi'(y) = \int_{R^\times} \omega(x)^{-1} \chi(x\,y)\,d\alpha(x).$$

Obviously the measure induced by α on R^\times is a Haar measure on R^\times (this may also be regarded as a consequence of lemma 5, § 4). Take y such that $\mathrm{ord}_K(y) \geqslant -f - v + 1$; then, by prop. 12 of Chap. II-5, $x \to \chi(x\,y)$ is constant on classes modulo P^{f-1}. Assume first that $f = 1$; then $\chi(x\,y) = 1$ on R, so that (9) is the integral of $\omega^{-1}\,d\alpha$ on R^\times, which is 0 since ω is a non-trivial character of the compact group R^\times. Assume now $f > 1$; then (9) is the sum of the similar integrals taken over the classes modulo P^{f-1} contained in R^\times, which are the same as the cosets of the subgroup $1 + P^{f-1}$ in R^\times; since the definition of the conductor implies that ω is non-trivial on $1 + P^{f-1}$, the same argument as before gives again

$\varphi'(y)=0$ in this case. Now take $y=b^{-1}u$ in (9), with $u\in R^{\times}$; substituting $u^{-1}x$ for x, we get $\varphi'(b^{-1}u)=\omega(u)\varphi'(b^{-1})$. This proves that φ' is of the form $c\,\overline{\varphi(by)}$ with $c\in\mathbf{C}^{\times}$. Applying to this Fourier's inversion formula and lemma 1 of §2, we get $c\bar{c}=\mathrm{mod}_K(b)$. As $c=\varphi'(b^{-1})$, we get for κ the formula in our proposition. It would be easy to verify directly that $\kappa\bar{\kappa}=1$ when κ is defined by that formula; moreover, as the integrand there is constant on classes modulo P^f in R, we can rewrite the integral as a sum over R/P^f; sums of that type are known as "Gaussian sums".

PROPOSITION 14. *Let ω be a quasicharacter of G_k, and Φ_ω the standard function attached to ω. Then the Fourier transform of Φ_ω, with reference to the basic character χ of $k_{\mathbf{A}}$, is given by*

$$\Phi'(y)=\kappa|b|_{\mathbf{A}}^{1/2}\overline{\Phi_\omega(by)}=\kappa|b|_{\mathbf{A}}^{1/2}\Phi_{\bar{\omega}}(by)$$

where $\kappa=\prod\kappa_v$, $\kappa_v\in\mathbf{C}$ and $\kappa_v\bar{\kappa}_v=1$ for all v, $b=(b_v)\in k_{\mathbf{A}}^{\times}$, and κ_v, b_v are as follows. Let $a=(a_v)$ be a differential idele attached to χ; then $b_v=a_v$ at each infinite place v, and, for each finite place v of k, $\mathrm{ord}_v(b_v a_v^{-1})=f(v)$. At every infinite place v of k, $\kappa_v=i^{-N_v}$; at every finite place v where $f(v)=0$, $\kappa_v=1$; at all other places:

$$\kappa_v=|b_v|_v^{-1/2}\int_{r_v^{\times}}\omega_v(x)^{-1}\chi_v(b_v^{-1}x)d\alpha_v(x),$$

where α_v is the self-dual Haar measure on k_v with reference to χ_v.

This follows at once from prop. 13, propositions 4 and 5 of §2, and corollary 3 of prop. 2, §2.

COROLLARY. *Let ω be as in proposition 14, and put $\omega'=\omega_1\omega^{-1}$. Then $Z(\omega,\Phi_\omega)=\kappa|b|_{\mathbf{A}}^{-1/2}\omega(b)Z(\omega',\Phi_{\omega'})$.*

For all ω, by th. 2 of §5, $Z(\omega,\Phi_\omega)$ is equal to $Z(\omega',\Phi')$, where Φ' is as in prop. 14. Express $Z(\omega',\Phi')$ by the integral in (4), §4, under the assumption that it is convergent, which, as one sees at once, amounts to $\sigma<0$. Expressing Φ' by proposition 14, and making the change of variable $z\to b^{-1}z$ in that integral, one gets the right-hand side of the formula in our corollary. By th. 2 of §5, both sides can then be continued analytically over the whole of the connected component of ω in $\Omega(G_k)$, so that the result is always true.

Now apply prop. 10 of §4 to $Z(\omega,\Phi_\omega)$; for $\sigma>1$, this gives an infinite product whose factors are all known to us except those corresponding to the finite places v of k where $f(v)>0$; as to these, our choice of Φ_v makes it obvious that they are respectively equal to $\mu_v(r_v^{\times})$. As in §6, put $G_w=G_1$ when w is a real place, and $G_w=G_2$ when it is an imaginary place. Taking into account prop. 11 of §4, lemma 8 of §6, and prop. 12 of §6, we get, for $\sigma>1$:

$$(10) \qquad Z(\omega, \Phi_\omega) = c_k^{-1} \prod_{w \in P_\infty} G_w(s_w) \prod_{v \notin P} (1 - q_v^{-s_v})^{-1},$$

where P is the set consisting of the infinite places and of the finite places where $f(v) > 0$.

For every place v of k, not in the set P which we have just defined, put $\lambda(v) = q_v^{-s_v}$; these are the finite places where ω_v is unramified, and the definition of s_v for such places shows that we can also write this as $\lambda(v) = \omega_v(\pi_v)$, where π_v is a prime element of k_v, or even as $\lambda(v) = \omega(\pi_v)$ if k_v^\times is considered as embedded as a quasifactor in k_A^\times. Clearly we have $|\lambda(v)| = q_v^{-\sigma}$.

In (10), replace now ω by $\omega_s \omega$, with $s \in \mathbf{C}$; as observed above, this does not change Φ_ω; it replaces the right-hand side of (10) by a product which is absolutely convergent for $\mathrm{Re}(s) > 1 - \sigma$. As th. 2 of § 5 shows that this can be continued analytically over the whole s-plane (as a holomorphic function if ω is not principal), and as the same is true of the factors G_w when they occur, we may now introduce a meromorphic function $L(s, \omega)$, given, for $\mathrm{Re}(s) > 1 - \sigma$, by the product

$$(11) \qquad L(s, \omega) = \prod_v (1 - \lambda(v) q_v^{-s})^{-1}$$

taken over all the finite places v where ω_v is unramified.

In order to formulate our final result in the case of characteristic 0, we introduce the ideal in \mathfrak{r} given by $\mathfrak{f} = \prod \mathfrak{p}_v^{f(v)}$, which is called *the conductor of* ω.

THEOREM 5. *Let k be an algebraic number-field, and ω a non-principal quasicharacter of $G_k = k_A^\times / k^\times$, with the conductor \mathfrak{f}. Then*

$$\Lambda(s, \omega) = \prod_{w \in P_\infty} G_w(s + s_w) \cdot L(s, \omega)$$

is an entire function of s, and satisfies the functional equation

$$\Lambda(s, \omega) = \kappa \, \omega(b) \left(|D| \, \mathfrak{N}(\mathfrak{f}) \right)^{\frac{1}{2} - s} \Lambda(1 - s, \omega^{-1}),$$

where κ and b are as in proposition 14.

This is an immediate consequence of the corollary of prop. 14, when one replaces ω in it by $\omega_s \omega$, taking into account the definitions of a, b and \mathfrak{f} and the fact that $|a|_A = |D|^{-1}$. As it is well-known that $\Gamma(s)^{-1}$ is an entire function, the same is true of the functions $G_w(s + s_w)^{-1}$; therefore theorem 5 implies that $L(s, \omega)$ is an entire function of s.

According to their definition, the above functions do not depend essentially upon the choice of ω in a given connected component of $\Omega(G_k)$; more precisely, they are independent of that choice, up to a translation in the s-plane, since, for every $t \in \mathbf{C}$, $L(s, \omega_t \omega)$ is the same as

$L(s+t,\omega)$, this being also true for $\Lambda(s,\omega)$. In view of corollary 2 of prop. 7, § 3, one may therefore always assume, after replacing ω by $\omega_{-t}\omega$ with a suitable $t\in\mathbf{C}$ if necessary, that ω is a character of $k_\mathbf{A}^\times$, trivial on k^\times and also on the group M defined in corollary 2 of th. 5, Chap. IV-4. The latter assumption can be written as $\sum(\delta_v s_v - N_v) = 0$, where the sum is taken over the infinite places of k, s_v and N_v are as above, and $\delta_v = 1$ or 2 according as k_v is \mathbf{R} or \mathbf{C}. Since this implies that ω is a character, we have then $\sigma = 0$.

On the other hand, if k is of characteristic $p > 1$, we introduce the divisor $\mathfrak{f} = \sum f(v)\cdot v$, and call this *the conductor of* ω. Then:

THEOREM 6. *Let k be an A-field of characteristic $p > 1$; let \mathbf{F}_q be its field of constants, g its genus, and ω a non-principal quasicharacter of $G_k = k_\mathbf{A}^\times/k^\times$ with the conductor \mathfrak{f}. Then one can write $L(s,\omega) = P(q^{-s},\omega)$, where $P(u,\omega)$ is a polynomial of degree $2g - 2 + \deg(\mathfrak{f})$ in u; and we have*

$$P(u,\omega) = \kappa\,\omega(b)\cdot(q^{1/2}u)^{2g-2+\deg(\mathfrak{f})}\cdot P(1/qu,\omega^{-1})$$

where κ and b are as in proposition 14.

The fact that we can write $L(s,\omega) = P(q^{-s},\omega)$, where $P(u,\omega)$ is holomorphic in the whole u-plane, is proved just as the corresponding fact in theorem 4. The last formula in our theorem is then an immediate consequence of the corollary of prop. 14 when one replaces ω by $\omega_s\omega$ there, provided one takes into account the definitions of a, b and \mathfrak{f} and the fact that $|a|_\mathbf{A} = q^{2-2g}$. Then that formula shows that $P(u,\omega)$ is a polynomial whose degree is as stated.

Here again one will observe that, for $t\in\mathbf{C}$, $P(u,\omega_t\omega)$ is the same as $P(q^{-t}u,\omega)$. In this case, we have written $k_\mathbf{A}^\times = k_\mathbf{A}^1 \times M$, where (if one takes corollary 6 of th. 2, § 5, into account) M is the subgroup of $k_\mathbf{A}^\times$ generated by an element z_1 such that $|z_1|_\mathbf{A} = q$, i.e. such that $\mathrm{div}(z_1)$ has the degree -1. Then corollary 2 of prop. 7, § 3, shows that, after replacing ω by $\omega_{-t}\omega$ with a suitable $t\in\mathbf{C}$, if necessary, one may assume that $\omega(z_1) = 1$; the corollary in question shows also that ω is then a character of $k_\mathbf{A}^\times$, i. e. that $\sigma = 0$; furthermore, if one combines it with lemma 4 of § 3, and with the obvious fact that in the present case the group $k_\mathbf{A}^\times$, hence also the groups $k_\mathbf{A}^1$, G_k, G_k^1 are totally disconnected, it shows that ω is then a character of finite order of $k_\mathbf{A}^\times$.

§ 8. **The coefficients of the L-series.** When an Euler product such as the right-hand side of (11) is given, the question arises whether it can be derived from a quasicharacter ω of $k_\mathbf{A}^\times/k^\times$. The answer to this, and to a somewhat more general problem which will be stated presently, depends on the following result:

PROPOSITION 15. *Let P be a finite set of places of k, containing P_∞; let G_P be the subgroup of k_A^\times, consisting of the ideles (z_v) such that $z_v = 1$ for all $v \in P$. Then $k^\times G_P$ is dense in k_A^\times.*

Put $k_P = \prod k_v$, the product being taken over the places $v \in P$; write A_P for the subgroup of k_A consisting of the adeles (x_v) such that $x_v = 0$ for all $v \in P$; then $k_A = k_P \times A_P$ and $k_A^\times = k_P^\times \times G_P$, and our assertion amounts to saying that the projection from k_A^\times onto k_P^\times maps k^\times onto a dense subgroup of k_P^\times. In fact, k_P^\times is an open subset of k_P, and its topology is the one induced by that of k_P; our assertion follows now at once from corollary 2 of th. 3, Chap. IV-2, which shows that the projection from k_A onto k_P maps k onto a dense subset of k_P.

From prop. 15, it follows at once that a continuous representation ω of k_A^\times into any group Γ, trivial on k^\times, is uniquely determined when its values on the groups k_v^\times are known for almost all v. In particular, if $\Gamma = \mathbf{C}^\times$, or more generally if Γ is such that every morphism of k_A^\times into Γ is trivial on r_v^\times for almost all v, ω is uniquely determined when the $\omega(\pi_v)$ are given for almost all v. Clearly every finite group Γ has that property, since the kernel of every morphism of k_A^\times into a finite group is open in k_A^\times and therefore contains $\prod_{v \notin P} r_v^\times$ for some P; the same is true of every group Γ without arbitrarily small subgroups, for the same reason for which it is true for $\Gamma = \mathbf{C}^\times$. Another case of interest is given by the following:

PROPOSITION 16. *Let K be a p-field, and assume that k is not of characteristic p. Then every morphism ω of k_A^\times into K^\times is trivial on r_v^\times for almost all v, and is locally constant on k_v^\times whenever k_v is not a p-field.*

As k is not of characteristic p, we have $|p|_v = 1$ for almost all v, and then k_v is not a p-field. As every morphism of a connected group into a totally disconnected one must obviously be trivial, ω is trivial on k_v^\times when $k_v = \mathbf{C}$, and on \mathbf{R}_+^\times when $k_v = \mathbf{R}$. Call R the maximal compact subring of K, and P its maximal ideal. Let v be any finite place of k such that k_v is not a p-field; let $m \geqslant 1$ be such that ω maps $1 + p_v^m$ into $1 + P$. For every $n \geqslant 0$, by prop. 8 of Chap. II-3, every $z \in 1 + p_v^m$ can be written as z'^{p^n} with $z' \in 1 + p_v^m$; therefore $\omega(z)$ is in $(1 + P)^{p^n}$, hence in $1 + P^{n+1}$ by lemma 5 of Chap. I-4; as n is arbitrary, this shows that ω is trivial on $1 + p_v^m$, hence locally constant on k_v^\times. By th. 7 of Chap. I-4 if K is of characteristic p, and by that theorem and prop. 9 of Chap. II-3 if it is of characteristic 0, there are only finitely many roots of 1 in K, and we can choose $\nu > 0$ so that there is no root of 1, other than 1, in $1 + P^\nu$. Take a neighborhood of 1 in k_A^\times which is mapped into $1 + P^\nu$ by ω; as this contains r_v^\times for almost all v, we see now that, for almost all v, ω is trivial on $1 + p_v$ and also on the group of all roots of 1 in k_v, and therefore on r_v^\times.

For every finite set P of places of k, containing P_∞, we will write $G'_P = \prod_{v \notin P} r_v^\times$; this is an open subgroup of the group G_P defined in prop. 15; it consists of the ideles (z_v) such that $z_v = 1$ for $v \in P$, and $z_v \in r_v^\times$, i. e. $|z_v|_v = 1$, for v not in P; if Γ is any group with the property described above, and ω is any morphism of k_A^\times into Γ, there is a set P such that ω is trivial on G'_P and therefore determines a morphism φ of G_P/G'_P into Γ; if at the same time ω is trivial on k^\times, prop. 15 shows that ω is uniquely determined by φ. We will discuss now the conditions on φ for such a morphism ω to exist.

If k is an algebraic number-field, and P is as above, we will say that a fractional ideal of k is *prime to P* if no prime ideal \mathfrak{p}_v, corresponding to a place $v \in P$, occurs in it with an exponent $\neq 0$. Similarly, if k is of characteristic $p > 1$, we say that a divisor is *prime to P* if no place $v \in P$ occurs in it with a coefficient $\neq 0$. We will write $I(P)$ (resp. $D(P)$) for the group of the fractional ideals of k (resp. of the divisors of k) prime to P. Clearly the morphism $z \to \mathrm{id}(z)$ of k_A^\times onto $I(k)$.(resp. the morphism $z \to \mathrm{div}(z)$ of k_A^\times onto $D(k)$) determines an isomorphism of G_P/G'_P onto $I(P)$ (resp. $D(P)$), which may be used to identify these groups with each other, or, what amounts to the same, with the free abelian group generated by the places of k, not in P. In particular, every mapping $v \to \lambda(v)$ of the set of these places into a commutative group Γ can be uniquely extended to a morphism φ of $I(P)$ (resp. $D(P)$) into Γ; then $\varphi \circ (\mathrm{id})$ (resp. $\varphi \circ (\mathrm{div})$) is a morphism of G_P into Γ, trivial on G'_P.

PROPOSITION 17. *Let φ be a morphism of $I(P)$ (resp. $D(P)$) into a commutative group Γ; for each $v \in P$, let g_v be an open subgroup of k_v^\times, contained in r_v^\times whenever v is finite. Then the morphism $\varphi \circ (\mathrm{id})$ (resp. $\varphi \circ (\mathrm{div})$) of G_P into Γ can be extended to a morphism ω of k_A^\times into Γ, trivial on k^\times, if and only if one can find, for every $v \in P$, a morphism ψ_v of g_v into Γ, so that $\varphi(\mathrm{id}(\xi))$ (resp. $\varphi(\mathrm{div}(\xi))$) is equal to $\prod \psi_v(\xi)$ for all $\xi \in \bigcap (k^\times \cap g_v)$. When that is so, ω is unique and induces ψ_v^{-1} on g_v for every $v \in P$.*

Put $g = \prod_{v \in P} g_v$; this being considered as a subgroup of k_A^\times in the obvious manner, $g \cdot G_P$ is an open subgroup of k_A^\times and is the direct product of g and G_P. Then $k^\times g \cdot G_P$ is an open subgroup of k_A^\times, so that, in view of prop. 15, it is k_A^\times. It is now obvious that a morphism of $g \cdot G_P$ into Γ can be extended to one of $k_A^\times = k^\times g \cdot G_P$, trivial on k^\times, if and only if it is trivial on the group $\gamma = k^\times \cap (g \cdot G_P)$, and that the extension is then unique. Clearly γ is the same as the group $\bigcap (k^\times \cap g_v)$ in our proposition. As $z \to \mathrm{id}(z)$ (resp. $z \to \mathrm{div}(z)$) is trivial on g, it maps $g \cdot G_P = g \times G_P$ onto

$I(P)$ (resp. $D(P)$); therefore if we write φ_1 for the morphism $\varphi \circ (\mathrm{id})$ (resp. $\varphi \circ (\mathrm{div})$) of G_P into Γ, its extensions to $g \cdot G_P$ are the morphisms of the form $\psi^{-1}\varphi_1$, where ψ is any morphism of g into Γ. Writing ψ_v for the morphism induced by ψ on g_v, we get our conclusion.

Obviously, if the condition in proposition 17 is satisfied for some choice of the groups g_v and of the morphisms ψ_v, it remains so when one substitutes, for each g_v, any open subgroup g'_v of g_v, and then for ψ_v the morphism induced by ψ_v on g'_v. For instance, one may always take $g_v = \mathbf{R}^{\times}_+$ when $k_v = \mathbf{R}$, and take for g_v one of the groups $1 + p^m_v$ with $m \geqslant 1$ when v is a finite place. The same idea gives the following:

COROLLARY. *In proposition 17, assume that Γ is (a) discrete, or (b) the group \mathbf{C}^{\times}, or (c) the group K^{\times}, where K is a local p-field. Then the extension ω exists if and only if groups g_v and morphisms ψ_v can be found with the properties stated in proposition 17 and the following additional one: in case (a), $\psi_v = 1$ for all $v \in P$; in case (b), $\psi_v = 1$ for all the finite places $v \in P$; in case (c), $\psi_v = 1$ for all the places $v \in P$ for which k_v is not a p-field.*

In fact, assume that the conditions in proposition 17 are fulfilled for some choice of the groups g_v and of the morphisms ψ_v. Then, in case (a), we can replace g_v by the kernel g'_v of ψ_v for each $v \in P$, since this is an open subgroup of g_v, and then ψ_v by 1. In case (b), we can do this for every finite place $v \in P$, by lemma 4 of § 3; this can also be done, for similar reasons, whenever Γ is a group without arbitrarily small subgroups. Case (c) can be treated similarly, with the help of prop. 16.

Instead of verifying the condition in proposition 17 for all ξ in the group $\gamma = \bigcap (k^{\times} \cap g_v)$, it is clearly enough to verify it for a set of generators of γ; in this connection, the following result is occasionally useful:

PROPOSITION 18. *Notations being as in proposition 17, assume that k is an algebraic number-field, and call \mathfrak{r} its maximal order. Then the group $\gamma = \bigcap (k^{\times} \cap g_v)$ is generated by $\gamma \cap \mathfrak{r}$.*

Take any $\xi \in \gamma$, and write $\xi \mathfrak{r} = \mathfrak{b} \mathfrak{a}^{-1}$, where $\mathfrak{a}, \mathfrak{b}$ are two ideals in \mathfrak{r}, prime to each other. For every finite place $v \in P$, ξ is in r^{\times}_v, so that \mathfrak{p}_v is not a prime factor of \mathfrak{a} or of \mathfrak{b}. Apply corollary 1 of th. 1, Chap. V-2, to the projection of k onto the product $\prod r_v$ taken over the finite places v of k which either belong to P or correspond to the prime ideals dividing \mathfrak{a}; it shows that there is $\alpha \in \mathfrak{r}$ such that $\alpha \in g_v$ for every finite $v \in P$, $\alpha \in \mathfrak{a}$, and $\alpha \neq 0$; then α^2 satisfies the same conditions and is in g_v for every infinite place v, so that it is in γ, hence in $\gamma \cap \mathfrak{r}$. That being so, also $\xi \alpha^2$ is in $\gamma \cap \mathfrak{r}$; this proves our proposition.

In particular, assume that $g_v = 1 + p_v^{m(v)}$, with $m(v) \geqslant 1$, for every finite place $v \in P$; put $\mathfrak{m} = \prod p_v^{m(v)}$; let v_1, \dots, v_ρ be all the real places of k for which $g_v = \mathbf{R}_+^\times$. Then one sees at once that the set $\gamma \cap \mathfrak{r}$, in prop. 17, consists of the elements of \mathfrak{r} which are $\equiv 1$ (\mathfrak{m}) and whose image in k_{v_i} is > 0 for $1 \leqslant i \leqslant \rho$.

Chapter VIII

Traces and norms

§ 1. Traces and norms in local fields. In §§ 1–3, we will consider exclusively local fields (assumed to be commutative). We denote by K a local field and by K' an algebraic extension of K of finite degree n over K. If K is an **R**-field and $K' \neq K$, we must have $K = \mathbf{R}$, $K' = \mathbf{C}$, $n = 2$; then, by corollary 3 of prop. 4, Chap. III-3, $Tr_{\mathbf{C}/\mathbf{R}}(x) = x + \bar{x}$ and $N_{\mathbf{C}/\mathbf{R}}(x) = x\bar{x}$; $Tr_{\mathbf{C}/\mathbf{R}}$ maps **C** onto **R**, and $N_{\mathbf{C}/\mathbf{R}}$ maps \mathbf{C}^{\times} onto \mathbf{R}_{+}^{\times}, which is a subgroup of \mathbf{R}^{\times} of index 2.

From now on, until the end of § 3, we assume K to be a p-field and adopt our usual notations for such fields, denoting by q the module of K, by R its maximal compact subring, by P the maximal ideal in R, and by π a prime element of K. The field K' being as stated above, we adopt similar notations, viz., q', R', P', π', for K'. We write f for the modular degree of K' over K and e for the order of ramification of K' over K, as defined in def. 4 of Chap. I-4; then $q' = q^{f}$ and $n = ef$, by corollary 6 of th. 6, Chap. I-4. As $e = \mathrm{ord}_{K'}(\pi)$, the R'-module generated in K' by $P^{v} = \pi^{v} R$, for any $v \in \mathbf{Z}$, is P'^{ev}; for this, we will write $\iota(P^{v})$.

By corollary 1 of prop. 4, Chap. III-3, and the remarks following that proposition, $Tr_{K'/K}$ is $\neq 0$ if and only if K' is separable over K; then, being K-linear, it maps K' onto K. By the definition of the norm, and by corollary 3 of th. 3, Chap. I-2, we have, for all $x' \in K'$:

$$(1) \qquad \mathrm{mod}_{K'}(x') = \mathrm{mod}_{K}(N_{K'/K}(x')).$$

In view of th. 6 of Chap. I-4, this implies that $x' \in R'$ if and only if $N_{K'/K}(x') \in R$, and $x' \in R'^{\times}$ if and only if $N_{K'/K}(x') \in R^{\times}$. As $\mathrm{mod}_{K}(\pi) = q^{-1}$ and $\mathrm{mod}_{K'}(\pi') = q^{-f}$, (1) may also be written as follows, for $x' \neq 0$:

$$(2) \qquad \mathrm{ord}_{K}(N_{K'/K}(x')) = f \cdot \mathrm{ord}_{K'}(x').$$

From now on, we will write Tr, N instead of $Tr_{K'/K}$, $N_{K'/K}$, except when there are more fields to be considered than K and K'. For every $v \in \mathbf{Z}$, we will write $\mathfrak{N}(P'^{v}) = P^{fv}$; by (2), this is the R-module generated in K by the image of P'^{v} under N.

PROPOSITION 1. *Let K' be separable over K. Then, if $x' \in R'$, $Tr(x') \in R$; if $x' \in P'$, $Tr(x') \in P$ and $N(1+x') = 1 + Tr(x') + y$ with $y \in R \cap x'^{2} R'$.*

Let \bar{K} be an algebraic closure of K'; call $\lambda_1, \ldots, \lambda_n$ the distinct K-linear isomorphisms of K' into \bar{K}; then, by corollary 3 of prop. 4, Chap. III-3, we have

$$(3) \qquad Tr(x') = \sum_i \lambda_i(x'), \quad N(1+x') = \prod_i (1 + \lambda_i(x')).$$

Call K'' the compositum of the fields $\lambda_i(K')$, which is the smallest Galois extension of K in \bar{K}, containing K'; define R'', P'' for K'' as R, P are defined for K. By corollary 5 of th. 6, Chap. I-4, we have $\lambda_i(R') \subset R''$ and $\lambda_i(P') \subset P''$ for all i, so that $Tr(x')$ is in R'' if $x' \in R'$, and in P'' if $x' \in P'$; as the same corollary shows that $R = K \cap R''$ and $P = K \cap P''$, this proves our assertions concerning Tr. Now assume $x' \in R'$, $x' \neq 0$, and put

$$y = N(1+x') - 1 - Tr(x');$$

by (3), this is a sum of monomials of degree ≥ 2 in the $\lambda_i(x')$. As one of the λ_i is the identity, and as the λ_i, by corollary 2 of prop. 3, Chap. III-2, differ from one another only by automorphisms of K'' over K, all the $\lambda_i(x')$ have the same order as x' in K'', so that yx'^{-2} is in R'' if x' is in R'. As $R' = K' \cap R''$, this proves our last assertion. In view of the fact that $Tr = 0$ if K' is inseparable over K, and of the remarks about that case in Chap. III-3, our proposition is still valid (but uninteresting) in the inseparable case.

COROLLARY. *If* $x' \in P'^{-e+1}$, $Tr(x') \in R$.

By definition, $e = \text{ord}_{K'}(\pi)$; therefore our assumption amounts to $\pi x' \in P'$, which implies $Tr(\pi x') \in P$ by prop. 1, hence $Tr(x') \in R$ since Tr is K-linear.

DEFINITION 1. *Let K' be separable over K; let d be the largest integer such that $Tr(x') \in R$ for all $x' \in P'^{-d}$. Then P'^d is called the different of K' over K, and d its differental exponent.*

For the different, we will write $D(K'/K)$, or simply D. If K' is inseparable over K, Tr is 0, so that it maps $P'^{-\nu}$ into R for all ν; in that case we put $d = +\infty$, $D(K'/K) = 0$.

By the corollary of prop. 1, we have $d \geq e - 1$. In particular, if $d = 0$, $e = 1$, so that K' is unramified over K. The converse is also true; this will be a consequence of the following results:

PROPOSITION 2. *Let K' be unramified over K; call ρ, ρ' the canonical homomorphisms of R onto $k = R/P$, and of R' onto $k' = R'/P'$, respectively. Then, for $x' \in R'$, we have*

$$\rho(Tr(x')) = Tr_{k'/k}(\rho'(x')), \quad \rho(N(x')) = N_{k'/k}(\rho'(x')).$$

As in th. 7 of Chap. I-4 and its corollaries, call M'^{\times} the group of roots of 1 of order prime to p in K'; by corollary 2 of that theorem, K' is cyclic of degree f over K, and its Galois group is generated by the Frobenius automorphism, which induces on M'^{\times} the permutation $\mu \rightarrow \mu^{q}$. In view of corollary 2 of th. 2, Chap. I-1, this amounts to saying that the automorphisms of K' over K determine on $k' = R'/P'$ the automorphisms which make up its Galois group over k. Our conclusion follows at once from this, the formulas $Tr(x') = \sum \lambda_i(x')$, $N(x') = \prod \lambda_i(x')$ and the similar ones for k and k', i.e. from corollary 3 of prop. 4, Chap. III-3, applied first to K and K', and then to k and k'.

PROPOSITION 3. *Let K' be unramified over K. Then Tr maps P'^{v} surjectively onto P^{v} for every $v \in \mathbf{Z}$, and N maps R'^{\times} surjectively onto R^{\times}.*

Let k, k' be as in prop. 2. As k' is separable over k, $Tr_{k'/k}$ is not 0; the first formula in prop. 2 shows then that the image $Tr(R')$ of R' under Tr is not contained in P; as it is contained in R by prop. 1, and as it is an R-module since R' is an R-module and Tr is K-linear, it is R. As K' is unramified, a prime element π of K is also a prime element of K'; therefore, for $v \in \mathbf{Z}$, $P'^{v} = \pi^{v} R'$. As Tr is K-linear, we get

$$Tr(P'^{v}) = \pi^{v} Tr(R') = \pi^{v} R = P^{v}.$$

As to the norm, put $G_0 = R^{\times}$, $G_0' = R'^{\times}$, $G_v = 1 + P^v$ and $G_v' = 1 + P'^v$ for all $v \geqslant 1$. The last assertion in prop. 1 shows that, for every $v \geqslant 1$, N maps G_v' into G_v, and also, in view of what we have just proved about the trace, that it determines on G_v'/G_{v+1}' a surjective morphism of that group onto G_v/G_{v+1}. On the other hand, call φ the Frobenius automorphism of K' over K, and μ a generator of the group M'^{\times} of the roots of 1 of order prime to p in K'; then μ is of order $q' - 1$, i.e. $q^f - 1$, and its norm is given by

$$N(\mu) = \prod_{i=0}^{f-1} \mu^{\varphi^i} = \prod_{i=0}^{f-1} \mu^{q^i} = \mu^{1 + q + \cdots + q^{f-1}} = \mu^{(q^f - 1)/(q-1)};$$

clearly this is a root of 1 of order $q - 1$, hence a generator of the group M^{\times} of roots of 1 of order prime to p in K. As M^{\times} is a full set of representatives of cosets modulo $G_1 = 1 + P$ in $G_0 = R^{\times}$, this shows that N determines on G_0'/G_1' a surjective morphism of that group onto G_0/G_1. Now, for every $x_0 \in R^{\times}$, we can determine inductively two sequences (x_v), (x_v') such that, for all $v \geqslant 0$, $x_v \in G_v$, $x_v' \in G_v'$, $N(x_v') \in x_v G_{v+1}$ and $x_{v+1} = N(x_v')^{-1} x_v$. Then, for $y_v' = x_0' x_1' \ldots x_{v-1}'$, we have $N(y_v') = x_0 x_v^{-1}$. Clearly the sequence (y_v') tends to a limit $y' \in R'^{\times}$, and $N(y') = x_0$.

COROLLARY. *Let K' be any extension of K of finite degree. Then the different of K' over K is R', i.e. $d = 0$, if and only if K' is unramified over K.*

Proposition 3 shows that $d=0$ if K' is unramified over K. Conversely, if $d=0$, K' is separable over K, and then, as we have already observed above, corollary 1 of prop. 1 gives $e=1$.

PROPOSITION 4. *Let K' be separable over K, and let P'^d be its different over K. Then, for every $v\in\mathbf{Z}$, the image of P'^v under Tr is P^μ, where μ is such that $e\mu\leqslant v+d<e(\mu+1)$.*

As Tr is K-linear and not 0, it maps every K-lattice in K', and in particular every set P'^v, onto a K-lattice in K, i.e. onto a set of the form P^μ. If μ is as stated in our proposition, then, since $\mathrm{ord}_{K'}(\pi)=e$, P'^v is contained in $\pi^\mu P'^{-d}$ and contains $\pi^{\mu+1}P'^{-d-1}$. In view of the definition of d and of the K-linearity of Tr, this implies that $Tr(P'^v)$ is contained in $\pi^\mu R=P^\mu$ and not in $\pi^{\mu+1}R=P^{\mu+1}$. This completes the proof.

COROLLARY 1. *For every $x'\in K'^\times$, we have:*

$$\mathrm{ord}_{K'}(Tr(x'))=e\cdot\mathrm{ord}_K(Tr(x'))\geqslant\mathrm{ord}_{K'}(x')+d-e+1.$$

In fact, if we put $v=\mathrm{ord}_{K'}(x')$, and if we define μ as in proposition 4, the left-hand side of the inequality in our corollary is $\geqslant e\mu$ by that proposition, and the definition of μ shows that this is $>v+d-e$.

COROLLARY 2. *$Tr(R')=R$ if and only if $d=e-1$.*

In fact, by proposition 4, $\mu=v=0$ implies $d<e$. As $d\geqslant e-1$ by the corollary of prop. 1, we get $d=e-1$.

If $d=e-1$, one says that K' is *tamely ramified* over K.

COROLLARY 3. *Let χ be a character of K of order μ; then $\chi\circ Tr$ is a character of K' of order $d+e\mu$.*

Our assumption means that χ is trivial on $P^{-\mu}$ and not on $P^{-\mu-1}$. Put $v=d+e\mu$; proposition 4 shows that $Tr(P'^{-v})=P^{-\mu}$ and that $Tr(P'^{-v-1})=$ $=P^{-\mu-1}$. Therefore $\chi\circ Tr$ is trivial on P'^{-v} and not on P'^{-v-1}, which is what we had to prove.

In the next corollary, we introduce an algebraic extension K'' of K' of finite degree; R'', P'' will have the same meaning for K'' as R, P have for K. For every $v\in\mathbf{Z}$, we will write $\iota'(P'^v)$ for the R''-module generated in K'' by P'^v, which is $P''^{e'v}$ if $e'=\mathrm{ord}_{K''}(\pi')$ is the order of ramification of K'' over K'. With these notations, we have:

COROLLARY 4. *Let K, K', K'' be as above; let $D=P'^d, D'=P''^{d'}, D''=P''^{d''}$ be the differents of K' over K, of K'' over K' and of K'' over K, respectively. Then $D''=\iota'(D)\cdot D'$ and $d''=e'd+d'$, where e' is the order of ramification of K'' over K'.*

This is trivially so if K'' is inseparable over K, since then $D''=0$ and either D or D' must be 0; we may therefore assume that K'' is separable over K, and, putting $\delta=e'd+d'$, we have to prove that $d''=\delta$. In fact, by proposition 4, $Tr_{K''/K'}$ maps $P''^{-\delta}$ onto P'^{-d} and $P''^{-\delta-1}$ onto P'^{-d-1}, and $Tr_{K'/K}$ maps P'^{-d} onto R and P'^{-d-1} onto P^{-1}. Our assertion follows at once from this and from the "transitivity of traces", i.e. corollary 4 of prop. 4, Chap. III-3.

COROLLARY 5. *Let K and K' be as above, and let K_1 be the maximal unramified extension of K, contained in K'. Then K' has the same different over K as over K_1.*

For the definition of K_1, cf. corollary 4 of th. 7, Chap. I-4. Our assertion follows then at once from corollary 4, combined with the corollary of prop. 3.

PROPOSITION 5. *Let K, K' be as in proposition 4; then the norm N determines an open morphism of K'^{\times} onto an open subgroup of K^{\times}.*

As before, call P'^d the different of K' over K, and put $G_\nu=1+P^\nu$, $G'_\nu=1+P'^\nu$ for $\nu\geqslant 1$. Take any $\mu>2d$, and put $\nu=e\mu-d$. By prop. 4, $Tr(P'^\nu)=P^\mu$; moreover, we have $e(\mu-1)\geqslant 2d$, hence $2\nu\geqslant e(\mu+1)$, hence $P'^{2\nu}\subset\pi^{\mu+1}R'$, and therefore $K\cap P'^{2\nu}\subset P^{\mu+1}$. That being so, the last part of prop. 1 shows, firstly, that N maps G'_ν into G_μ, and secondly that it determines a surjective morphism of G'_ν onto $G_\mu/G_{\mu+1}$. Take now any $x_0\in G_\mu$; we can choose inductively two sequences (x_i), (x'_i), so that, for all $i\geqslant 0$, $x_i\in G_{\mu+i}$, $x'_i\in G'_{\nu+ei}$, $N(x'_i)\in x_i G_{\mu+i+1}$ and $x_{i+1}=N(x'_i)^{-1}x_i$. Then, putting $y'_i=x'_0 x'_1\ldots x'_i$, we have $N(y'_i)=x_0 x_{i+1}^{-1}$. Clearly the sequence (y'_i) converges to a limit $y'\in G'_\nu$, and $N(y')=x'_0$. This shows that N maps G'_ν onto G_μ, which proves our proposition, since the groups G_μ, G'_ν, for $\mu>2d$, $\nu=e\mu-d$, make up fundamental systems of neighborhoods of 1 in K^{\times} and in K'^{\times}, respectively. By using corollary 2 of prop. 4, Chap. I-4, and the results of Chap. III-3, it would be easy to show that the conclusion of our proposition remains valid for any extension K' of K of finite degree, separable or not. Obviously it is also valid for **R**-fields.

§ 2. Calculation of the different.

Let assumptions and notations be as in § 1. When K' is regarded as a vector-space of dimension n over K, R' is a K-lattice, to which we can apply th. 1 of Chap. II-2. This shows that there is a basis $\{\alpha_1,\ldots,\alpha_n\}$ of K' over K, such that $R'=\sum R\alpha_i$.

Now assume that K' is separable over K, so that Tr is not 0; then, by lemma 3 of Chap. III-3, we may identify K', as a vector-space over K, with its algebraic dual, by putting $[x',y']=Tr(x'y')$; the dual basis $\{\beta_1,\ldots,\beta_n\}$ to $\{\alpha_1,\ldots,\alpha_n\}$ is then the one given by $Tr(\alpha_i\beta_j)=\delta_{ij}$ for $1\leqslant i,j\leqslant n$.

PROPOSITION 6. *Let K' be separable over K; call $D = P'^d$ its different. Let $\{\alpha_1,\dots,\alpha_n\}$ be a basis of K' over K such that $R' = \sum R\alpha_i$, and let $\{\beta_1,\dots,\beta_n\}$ be the basis of K' over K given by $Tr(\alpha_i\beta_j) = \delta_{ij}$ for $1 \leqslant i,j \leqslant n$. Then $D^{-1} = P'^{-d} = \sum R\beta_i$.*

In fact, take any $x' \in R'$, any $y' \in K'$, and write $x' = \sum x_i \alpha_i$ and $y' = \sum y_i \beta_i$ with $x_i \in R$ and $y_i \in K$ for $1 \leqslant i \leqslant n$. Then $Tr(x'\,y') = \sum x_i y_i$; this shows that $Tr(x'\,y') \in R$ for all $x' \in R'$, i.e. that Tr maps $R'\,y'$ into R, if and only if $y_i \in R$ for all i. By the definition of the different, this means that y' is in P'^{-d} if and only if it is in $\sum R\beta_i$, as was to be proved.

COROLLARY. *Let assumptions be as in proposition 6, and call Δ the determinant of the matrix*

$$M = (Tr(\alpha_i\alpha_j))_{1 \leqslant i,j \leqslant n}.$$

Then $\operatorname{ord}_K(\Delta) = fd$, and $\Delta R = \mathfrak{N}(D)$.

Write $\alpha_i = \sum a_{ij}\beta_j$, with $a_{ij} \in K$ for $1 \leqslant i,j \leqslant n$. Multiplying both sides with α_j and taking the trace, we get $Tr(\alpha_i\alpha_j) = a_{ij}$, hence $M = (a_{ij})$. Therefore the automorphism of the vector-space K' over K which maps $\{\beta_1,\dots,\beta_n\}$ onto $\{\alpha_1,\dots,\alpha_n\}$, hence the K-lattice D^{-1} onto R', is represented by the matrix (a_{ij}) with respect to the first one of these bases, and its module, by corollary 3 of th. 3, Chap. I-2, has the value $\operatorname{mod}_K(\Delta)$. As the mapping $x' \to \pi'^d x'$ also maps $D^{-1} = P'^{-d}$ onto R', its module $\operatorname{mod}_{K'}(\pi'^d)$ must be the same as $\operatorname{mod}_K(\Delta)$. This gives $fd = \operatorname{ord}_K(\Delta)$, hence $\mathfrak{N}(D) = \Delta R$. One will note that our corollary remains valid in the inseparable case, since then $Tr = 0$ and $D = 0$. Clearly our result implies that $\operatorname{ord}_K(\Delta)$ is independent of the choice of α_1,\dots,α_n; this could easily be verified directly, and justifies the following definition:

DEFINITION 2. *Let Δ be as in the corollary of proposition 6; then the ideal ΔR in R is called the discriminant of K' over K.*

Still assuming K' to be separable of degree n over K, call \bar{K} an algebraic closure of K'. As in § 1, let $\lambda_1,\dots,\lambda_n$ be the n distinct K-linear isomorphisms of K' into \bar{K}; as the identity is one of them, we may assume that it is λ_1. Take any $\xi \in K'$, and put $\xi_i = \lambda_i(\xi)$ for $1 \leqslant i \leqslant n$, hence in particular $\xi_1 = \xi$. If ν is the degree of K' over $K(\xi)$, there are ν distinct $K(\xi)$-linear isomorphisms of K' into \bar{K}, hence ν and no more than ν distinct ones among the λ_i which map ξ onto itself. This shows that $K(\xi) = K'$ if and only if $\xi_i \neq \xi$ for all $i \neq 1$.

Take now an indeterminate X over K. We can, in the manner described in Chap. III-3, extend the K-linear mapping Tr of K' into K, and the

polynomial mapping N of K' into K, to mappings of $K'[X] = K' \otimes_K K[X]$ into $K[X]$, which we again denote by Tr and N. Put then:

$$(4) \qquad F(X) = N(X - \xi) = \prod_{i=1}^{n} (X - \xi_i) = X^n + \sum_{i=1}^{n} a_i X^{n-i}.$$

This is a monic polynomial in $K[X]$; calling F' its formal derivative, we have

$$F'(\xi) = \prod_{i=2}^{n} (\xi - \xi_i).$$

In particular, in view of what has been proved above, we have $K(\xi) = K'$ if and only if $F'(\xi) \neq 0$. It is well known, and easily verified, that $F(X)^{-1}$ has in $\bar{K}(X)$ the "partial fraction decomposition" given by

$$\frac{1}{F(X)} = \sum_{i=1}^{n} \frac{1}{F'(\xi_i)(X - \xi_i)}.$$

Considering the field $\bar{K}(X)$ as embedded in the obvious manner in the field of formal power-series in X^{-1} with coefficients in \bar{K}, we get from this:

$$X^{-n}(1 + \sum_{i=1}^{n} a_i X^{-i})^{-1} = \sum_{i=1}^{n} F'(\xi_i)^{-1} \sum_{\nu=0}^{+\infty} \xi_i^\nu X^{-\nu-1}$$

which may also be written as

$$X^{-n} \sum_{\nu=0}^{+\infty} \left(\sum_{i=1}^{n} a_i X^{-i} \right)^\nu = \sum_{\nu=0}^{+\infty} Tr(F'(\xi)^{-1} \xi^\nu) X^{-\nu-1}.$$

Equating coefficients on both sides, we get

$$(5) \qquad P_\nu(a) = Tr(F'(\xi)^{-1} \xi^\nu)$$

for $\nu \geq 0$, where $P_\nu(a)$ is, for all ν, a polynomial in $\mathbf{Z}[a_1, ..., a_n]$, with $P_\nu = 0$ for $0 \leq \nu < n-1$, and $P_{n-1} = 1$.

PROPOSITION 7. *Let K' be separable of degree n over K, and call D its different. For any $\xi \in K'$, let F be the polynomial defined by (4). Then all the coefficients a_i of F are in R if $\xi \in R'$, and in P if $\xi \in P'$; moreover, if $\xi \in R'$, $F'(\xi)D^{-1}$ is contained in $R[\xi]$, and it is the largest R'-module contained in $R[\xi]$.*

The assertions about the a_i are proved exactly as the assertions about the trace in prop. 1. In fact, if the $\xi_i = \lambda_i(\xi)$ are defined as above, the assumption $\xi \in R'$ (resp. $\xi \in P'$) implies that, for every i, ξ_i is in $\lambda_i(R')$ (resp. in $\lambda_i(P')$), hence in the maximal compact subring R'' of the compositum K'' of the fields $\lambda_i(K')$ (resp. in the maximal ideal P'' of R''); (4) shows then that all the a_i are in R'', hence in $R = K \cap R''$ (resp. in P'', hence in $P = K \cap P''$).

As to the assertions about $F'(\xi)$, assume first that $F'(\xi)=0$; as we have seen, this is so if and only if $K(\xi)\neq K'$; then $K(\xi)$, hence also $R[\xi]$, cannot contain any R'-module other than $\{0\}$, which proves our assertion in this case. Assume now that $F'(\xi)\neq 0$; then $K'=K(\xi)$, so that $\{1,\xi,\ldots,\xi^{n-1}\}$ is a basis of K' over K. As F is monic and in $R[X]$, and $F(\xi)=0$, a well known elementary argument shows that $R[\xi]$ is the R-module $\sum_{i=0}^{n-1} R\xi^i$.

Take now any $x'\in K'$; write $F'(\xi)x'=\sum_{i=0}^{n-1} x_i\xi^i$ with $x_i\in K$ for $0\leqslant i\leqslant n-1$.

Multiplying this with $F'(\xi)^{-1}\xi^\nu$ and taking the traces of both sides, we get, in view of (5):

$$(6) \qquad Tr(x'\xi^\nu)=\sum_{i=0}^{n-1} x_i P_{\nu+i}(a)$$

for all $\nu\geqslant 0$, hence in particular, for $0\leqslant\nu\leqslant n-1$:

$$(7) \qquad x_{n-\nu-1}=Tr(x'\xi^\nu) - \sum_{i=n-\nu}^{n-1} x_i P_{\nu+i}(a).$$

Assume first that $x'\in D^{-1}$; then (7) shows, by induction on ν for $0\leqslant\nu\leqslant n-1$, that all x_i are in R, i.e. that $F'(\xi)x'$ is in $R[\xi]$, so that $F'(\xi)D^{-1}\subset R[\xi]$. On the other hand, assume that $x_i\in R$ for $0\leqslant i\leqslant n-1$, i.e. that $F'(\xi)x'\in R[\xi]$; then (6), for $\nu=0$, shows that $Tr(x')\in R$. Replacing x' by $x'y'$ with $y'\in R'$, we see that, if x' is such that $F'(\xi)x'R'\subset R[\xi]$, then $x'\in D^{-1}$. This proves our last assertion.

COROLLARY 1. *Assumptions and notations being as in proposition 7, we have* $D=F'(\xi)R'$ *if and only if* $R'=R[\xi]$.

This follows at once from the second part of proposition 7.

COROLLARY 2. *Let assumptions and notations be as in proposition 7; assume also that* K' *is fully ramified over* K; *put*

$$F(X)=N(X-\pi')=X^n+\sum_{i=1}^{n} a_i X^{n-i},$$

where π' *is any prime element of* K'. *Then* $\operatorname{ord}_K(a_i)\geqslant 1$ *for* $1\leqslant i\leqslant n$, $\operatorname{ord}_K(a_n)=1$, *and* $D=F'(\pi')R'$.

Taking $\xi=\pi'$ in proposition 7, we get the first assertion; the second one is obvious in view of formula (2) of § 1, since $a_n=N(-\pi')$; the last one follows at once from corollary 1, combined with prop. 4 of Chap. I-4.

COROLLARY 3. *Let assumptions and notations be as in corollary 2; then K' is tamely ramified if and only if n is prime to p.*

As K' is fully ramified, we have, in our usual notation, $f=1$ and $n=e$. By corollary 2, we have $d=\operatorname{ord}_{K'}(F'(\pi'))$, and all the terms in $F'(\pi')$ except the first one $n\pi'^{n-1}$ are of order $\geqslant e=\operatorname{ord}_{K'}(\pi)$ in K'. Therefore $d=e-1$, i.e. K' is tamely ramified, if and only if $\operatorname{ord}_{K'}(n)=0$, i.e. if and only if n is prime to p.

A polynomial F satisfying the conditions in corollary 2, i.e. a monic polynomial $X^n + \sum\limits_{i=1}^{n} a_i X^{n-i}$ in $K[X]$ such that $\operatorname{ord}_K(a_i)\geqslant 1$ for all i and $\operatorname{ord}_K(a_n)=1$, is called an *Eisenstein polynomial* over K.

PROPOSITION 8. *Let F be an Eisenstein polynomial over K. Then F is irreducible in $K[X]$, and, if π' is a root of F in any extension of K, $K(\pi')$ is a fully ramified extension of K, having π' as a prime element.*

Assume that $F=GH$, with G and H in $K[X]$. Let a, b be the smallest integers such that $G_1=\pi^a G$ and $H_1=\pi^b H$ are in $R[X]$, and put $F_1 = \pi^{a+b}F$, so that $F_1=G_1 H_1$. Put $k=R/P$, and call F_0, G_0, H_0 the polynomials in $k[X]$ obtained by replacing each coefficient in F_1, G_1, H_1, respectively, by its image in R/P under the canonical homomorphism of R onto R/P. By the definition of a and b, G_0 and H_0 are not 0, so that $F_0\neq 0$; this implies that $a+b=0$, $F_1=F$, and $F_0=X^n$; consequently there is v such that $G_0=X^v$, $H_0=X^{n-v}$. Then the degrees of G_1, H_1 are at least v, $n-v$; as $F_1=G_1 H_1$, they are v, $n-v$. If $v>0$, $n-v>0$, call g, h the constant terms in G_1, H_1; as $G_0=X^v$ and $H_0=X^{n-v}$, g and h are both in P; as the constant term of F is now gh, it is in P^2, which contradicts the definition of an Eisenstein polynomial. Now let π' be a root of F in an extension of K, which we may assume to be algebraically closed; as F is irreducible, the distinct K-linear isomorphisms of $K'=K(\pi')$ into that extension map π' onto all the distinct roots of F, so that $F(X)=N(X-\pi')$, hence, by the definition of an Eisenstein polynomial, $\operatorname{ord}_K(N(\pi'))=1$. By formula (2), § 1, this implies that $f=1$ and that π' is a prime element of K'.

§ 3. Ramification theory. In this §, it will be convenient to write isomorphisms and automorphisms of fields exponentially, i.e. as $x\to x^\lambda$, etc. Furthermore, K being as before, it is convenient to extend ord_K to all algebraic extensions of K as follows. Let x' be any element of such an extension; let K' be any extension of K of finite degree, containing x'; π being as before a prime element of K, put

$$\operatorname{ord}_K(x')=\operatorname{ord}_{K'}(x')/\operatorname{ord}_{K'}(\pi);$$

here, if we replace K' by any similar field K'' containing K', $\mathrm{ord}_{K'}(x')$ and $\mathrm{ord}_{K'}(\pi)$ are both multiplied with the order of ramification of K'' over K', so that our definition of $\mathrm{ord}_K(x')$ is independent of the choice of K'; of course one could take $K' = K(x')$ in that definition. That being so, ord_K coincides on K^\times with the mapping ord_K of K^\times into \mathbf{Z}, as previously defined, and determines a mapping of every algebraic extension of K into $\mathbf{Q} \cup \{+\infty\}$, with $\mathrm{ord}_K(x') = +\infty$ if and only if $x' = 0$.

As before, let K' be an extension of K of degree n, which we assume to be separable; let notations be as in §§ 1–2; in particular, let $D = P'^d$ be the different of K' over K. Call K_1 the maximal unramified extension of K contained in K', this being uniquely defined by corollary 4 of th. 7, Chap. I-4. Then K' has the degree e over K_1, and, by corollary 5 of prop. 4, § 1, it has the different D over K_1. Put

$$F(X) = N_{K'/K_1}(X - \pi');$$

by corollary 2 of prop. 7, § 2, this is an Eisenstein polynomial over K_1, and $D = F'(\pi')R'$.

Let L be any Galois extension of K of finite degree, containing K'; for instance, one may take for L the compositum of the images of K' under all the distinct K-linear isomorphisms of K' into some algebraic closure of K'. For every K-linear isomorphism $x' \to x'^\lambda$ of K' into L, put

$$v(\lambda) = \min_{x' \in R'} \mathrm{ord}_{K'}(x' - x'^\lambda) = \min_{x' \in R'} \mathrm{ord}_L(x' - x'^\lambda)/\mathrm{ord}_L(\pi').$$

Since $\mathrm{ord}_L(x' - x'^\lambda)$ is an integer $\geqslant 0$ or $+\infty$, this is well defined; it is $+\infty$ if and only if λ is the identity; the identity, i.e. the natural injection of K' into L, will be denoted by ε. By th. 7 of Chap. I-4 and its corollaries 3 and 4, K_1 is generated over K by the roots of 1 of order prime to p in K', and these, together with 0, make up a full set of representatives for R'/P' in R'; therefore, if λ does not induce the identity on K_1, there is such a root ζ for which $\zeta^\lambda \neq \zeta$, and then $\zeta - \zeta^\lambda$ is in K' and not in P', so that, taking $x' = \zeta$, we get $v(\lambda) = 0$. Now assume that λ induces the identity on K_1. As K' is fully ramified over K_1, prop. 4 of Chap. I-4 shows that $R' = R_1[\pi']$, R_1 being the maximal compact subring of K_1, so that every $x' \in R'$ can be written as $G(\pi')$ with $G \in R_1[X]$. This gives:

$$x' - x'^\lambda = G(\pi') - G(\pi'^\lambda) = (\pi' - \pi'^\lambda) H(\pi', \pi'^\lambda)$$

with $H \in R_1[X, Y]$. As we have already observed in the proof of prop. 1, π'^λ has the same order as π' in L; this implies that $\mathrm{ord}_{K'}(\pi'^\lambda) = \mathrm{ord}_{K'}(\pi') = 1$, so that we have

$$\mathrm{ord}_{K'}(x' - x'^\lambda) \geqslant \mathrm{ord}_{K'}(\pi' - \pi'^\lambda) \geqslant 1$$

and therefore, whenever λ induces the identity on K_1:

(8) $$v(\lambda) = \operatorname{ord}_{K'}(\pi' - \pi'^\lambda) \geqslant 1,$$

which implies that $\operatorname{ord}_{K'}(\pi' - \pi'^\lambda)$ does not depend upon the choice of π'. Now, F being as defined above, we have, by formula (4) of § 2:

$$F(X) = \prod_\lambda (X - \pi'^\lambda),$$

where the product is taken over all the distinct K_1-linear isomorphisms λ of K' into L, and therefore

$$F'(\pi') = \prod_{\lambda \neq \varepsilon} (\pi' - \pi'^\lambda)$$

where the product is now taken over the same isomorphisms except the identity. This gives

$$d = \operatorname{ord}_{K'}(F'(\pi')) = \sum_{\lambda \neq \varepsilon} v(\lambda),$$

where the sum is taken over those same isomorphisms, and also, since the number of such isomorphisms is $e - 1$:

$$d - e + 1 = \sum_{\lambda \neq \varepsilon} (v(\lambda) - 1).$$

As $v(\lambda) = 0$ when λ does not induce the identity on K_1, it amounts to the same to write:

(9) $$d = \sum_{\lambda \neq \varepsilon} v(\lambda), \quad d - e + 1 = \sum_{\lambda \neq \varepsilon} (v(\lambda) - 1)^+,$$

where the sums are now taken over all the distinct K-linear isomorphisms of K' into L, other than the identity; moreover, the number of terms > 0 in the latter sum is $\leqslant e - 1$.

If K' is itself a Galois extension of K, we may take $L = K'$, and the isomorphisms λ are the automorphisms of K' over K; they make up the Galois group \mathfrak{g} of K' over K. The definition of $v(\lambda)$ shows that it is now an integer or $+ \infty$; if $\lambda \neq \varepsilon$, $v(\lambda)$ is the largest of the integers v such that λ determines the identity on the ring R/P^v. For every $v \geqslant 0$, the automorphisms λ of K' over K for which $v(\lambda) \geqslant v$ make up a subgroup \mathfrak{g}_v of \mathfrak{g}; we have $\mathfrak{g}_0 = \mathfrak{g}$, and the groups \mathfrak{g}_v for $v \geqslant 1$ are known as "the higher ramification groups" of K' over K. As we have seen above, \mathfrak{g}_1, which is traditionally known as "the group of inertia" of K', consists of the automorphisms of K' which induce the identity on K_1; in other words, it is the subgroup of $\mathfrak{g}_0 = \mathfrak{g}$ attached to K_1 in the sense of Galois theory; it is of order e, and $\mathfrak{g}_0/\mathfrak{g}_1$ may be identified with the Galois group of K_1 over K, which, as we know, is cyclic of order f and generated by the Frobenius automorphism of K_1 over K.

Still assuming K' to be a Galois extension of K, call g_v the order of the group \mathfrak{g}_v for each $v \geqslant 0$. Then $g_v - 1$ is the number of the elements λ of \mathfrak{g}, other than ε, for which $v(\lambda) \geqslant v$. We can therefore rewrite (9) as follows:

(10) $$d = \sum_{v=1}^{+\infty} (g_v - 1), \quad d - e + 1 = \sum_{v=2}^{+\infty} (g_v - 1).$$

PROPOSITION 9. *Let K' be a Galois extension of K with the Galois group $\mathfrak{g} = \mathfrak{g}_0$; let the \mathfrak{g}_v, for $v \geqslant 1$, be its higher ramification groups. Put $G_0' = R'^\times$ and $G_v' = 1 + P'^v$ for $v \geqslant 1$. Then, for each $v \geqslant 1$, \mathfrak{g}_v consists of the elements λ of \mathfrak{g}_1 such that $\pi'^\lambda \pi'^{-1}$ is in G_{v-1}'; when that is so, the image $\gamma(\lambda)$ of $\pi'^\lambda \pi'^{-1}$ in the group $\Gamma_v = G_{v-1}'/G_v'$ is independent of the choice of the prime element π' of K', and $\lambda \to \gamma(\lambda)$ is a morphism of \mathfrak{g}_v into Γ_v, with the kernel \mathfrak{g}_{v+1}.*

The first assertion follows at once from (8) and the definitions. Replace π' by another prime element of K'; this can be written as $\pi' u$ with $u \in R'^\times$; for $\lambda \in \mathfrak{g}_v$, this modifies $\pi'^\lambda \pi'^{-1}$ by the factor $u^\lambda u^{-1}$, which, by the definition of \mathfrak{g}_v, is in $1 + P'^v$, i.e. in G_v'; this shows that $\gamma(\lambda)$ is independent of the choice of π'. If λ, μ are in \mathfrak{g}_v, put $u = \pi'^\lambda \pi'^{-1}$, $v = \pi'^\mu \pi'^{-1}$. Then $\pi'^{\lambda\mu} \pi'^{-1} = (u^\mu u^{-1}) u v$; as $u \in R'^\times$, $u^\mu u^{-1}$ is in G_v'; this shows that $\lambda \to \gamma(\lambda)$ is a morphism. It is then obvious that its kernel is \mathfrak{g}_{v+1}.

COROLLARY 1. *For every $v \geqslant 0$, $\mathfrak{g}_v/\mathfrak{g}_{v+1}$ is commutative; for $v = 0$, it is cyclic of order f; for $v = 1$, it is cyclic, and its order e_0 divides $q' - 1$, q' being the module of K'; for $v \geqslant 2$, it is isomorphic to a subgroup of the additive group of R'/P', and its order divides q'.*

For $v = 0$, this was proved above. Now put $k' = R'/P'$; this is a field with q' elements. The canonical morphism of R' onto k' induces on G_0' a morphism of G_0' onto k'^\times with the kernel G_1', so that Γ_1 is cyclic of order $q' - 1$. Similarly, for $v \geqslant 2$, the mapping $x' \to 1 + \pi'^{v-1} x'$ of R' onto G_{v-1}' determines an isomorphism of R'/P' onto Γ_v. Our assertions for $v \geqslant 1$ are immediate consequences of these facts and of proposition 9.

COROLLARY 2. *Assumptions and notations being as in corollary 1, we have $e = e_0 p^N$ with $N \geqslant 0$ and e_0 prime to p.*

This is obvious in view of corollary 1, since \mathfrak{g}_1 is of order e.

COROLLARY 3. *If $v(\lambda)$ has the same value v for all $\lambda \neq \varepsilon$ in \mathfrak{g}, \mathfrak{g} is commutative, with an order dividing $q - 1$ if $v = 1$ and q if $v \geqslant 2$.*

In fact, we have then $\mathfrak{g}_v = \mathfrak{g}$, $\mathfrak{g}_{v+1} = \{\varepsilon\}$; moreover, if $v \geqslant 1$, we have $e = n$, hence $f = 1$ and $q = q'$.

Finally, the numbers $v(\lambda)$ have important "transitivity properties". As above, let K' be a separable extension of K of finite degree n, but not necessarily a Galois extension; let K'' be a separable extension of K' of finite degree; take for L a Galois extension of K of finite degree, containing K''. Notations for K and K' being as before, let K_2 be the maximal unramified extension of K, contained in K''; call K'_2 the compositum of K' and K_2. Call e' the order of ramification of K'' over K', and f' its modular degree over K'. As K' has the same module q' as K_1, and K'' and K'_2 have the same module as K_2, K_2 is the unramified extension of K_1 of degree f', and K'_2 is the maximal unramified extension of K' contained in K'' and is of degree f' over K'. As K' is of degree e over K_1, this implies that K'_2 is of degree ef' over K_1, hence of degree e over K_2. Each K_2-linear isomorphism σ of K'_2 into L induces on K' a K_1-linear isomorphism λ of K' into L; as K'_2 is the compositum of K' and K_2, two such isomorphisms σ, σ' cannot coincide on K' unless $\sigma = \sigma'$; as there are e such isomorphisms, and the same number of K_1-linear isomorphisms of K' into L, $\sigma \to \lambda$ is a bijection of the former onto the latter; in particular, each isomorphism λ of K' into L, inducing the identity on K_1, can be uniquely extended to an isomorphism σ of K'_2 into L, inducing the identity on K_2.

Now, calling π'' a prime element of K'', put:

$$G(X) = N_{K''/K'_2}(X - \pi'') = X^{e'} + \sum_{i=1}^{e'} \alpha_i X^{e'-i}.$$

By corollary 2 of prop. 7, § 2, this is an Eisenstein polynomial over K'_2; in particular, $\alpha_{e'}$ is a prime element of K'_2; so is π', since K'_2 is unramified over K'. Let λ be any isomorphism of K' into L, other than the identity, inducing the identity on K_1; as we have seen above, this can be uniquely extended to an isomorphism σ of K'_2 into L, inducing the identity on K_2. Write G^σ for the polynomial obtained by applying σ to each coefficient of G; we have

$$G(X) - G^\sigma(X) = \alpha_{e'} - \alpha_{e'}^\sigma + \sum_{i=1}^{e'-1} (\alpha_i - \alpha_i^\sigma) X^{e'-i}.$$

As $\alpha_{e'}$ and π' are prime elements of K'_2, and K'_2 is unramified over K', we have, by what we have proved above:

$$\operatorname{ord}_{K'_2}(\alpha_{e'} - \alpha_{e'}^\sigma) = \operatorname{ord}_{K'_2}(\pi' - \pi'^\sigma) = \operatorname{ord}_{K'}(\pi' - \pi'^\lambda) = v(\lambda),$$

$$\operatorname{ord}_{K'_2}(\alpha_i - \alpha_i^\sigma) \geqslant \operatorname{ord}_{K'_2}(\pi' - \pi'^\sigma) = v(\lambda) \quad (1 \leqslant i \leqslant e'),$$

and therefore:

$$\operatorname{ord}_{K'}(G(\pi'') - G^\sigma(\pi'')) = v(\lambda).$$

We have $G(\pi'')=0$. On the other hand, G^σ is the monic polynomial whose roots are the images π''^τ of π'' under the distinct isomorphisms τ of K'' into L which coincide with σ on K'_2. In other words, we have

$$G^\sigma(\pi'')=\prod_\tau(\pi''-\pi''^\tau),$$

where the product is taken over all the distinct isomorphisms τ of K'' into L which induce λ on K' and the identity on K_2. Let now $v'(\tau)$ be defined for K, K'' and τ, just as $v(\lambda)$ has been defined for K, K' and λ; in other words, we put $v'(\tau)=0$ if τ does not induce the identity on K_2, and if it does, we put

$$v'(\tau)=\mathrm{ord}_{K''}(\pi''-\pi''^\tau).$$

Since $\mathrm{ord}_{K''}=e'\cdot\mathrm{ord}_{K'}$, we get now, by comparing the above formulas:

$$(11) \qquad e'v(\lambda)=\sum_\tau v'(\tau),$$

where the sum may be taken over all the isomorphisms τ of K'' into L which coincide with λ on K', since those which do not induce the identity on K_2 make no contribution to the right-hand side; for a similar reason, (11) remains valid when λ is an isomorphism of K' into L which does not induce the identity on K_1. Combining formulas (9) and (11), one gets another proof for corollary 4 of prop. 4, §1.

Let now L be a Galois extension of K, not necessarily of finite degree. Call \mathfrak{G} its Galois group, topologized in the usual manner, i.e. by taking, as a fundamental system of neighborhoods of the identity, all the subgroups of \mathfrak{G} attached to extensions of K of finite degree, contained in L. Then \mathfrak{G} is compact, and (11) and (9), together with corollary 4 of prop. 4, §1, may be interpreted by saying that there is a finitely additive function H, on the family of all open and closed subsets of \mathfrak{G}, with the following property. Let K' be any extension of K of finite degree, contained in L; let e be its order of ramification over K, and d its differential exponent over K; call \mathfrak{H} the open and closed subgroup of \mathfrak{G}, consisting of the elements of \mathfrak{G} which induce the identity on K'. Then $H(\mathfrak{H})=d/e$, and, for every coset $\mathfrak{H}\lambda$ of \mathfrak{H} in \mathfrak{G}, other than \mathfrak{H}, we have $H(\mathfrak{H}\lambda)=-v(\lambda)/e$, where $v(\lambda)$ is as defined above. From this, we derive a linear form $f\to H(f)$, i.e. a "distribution", on the space of all locally constant functions f on \mathfrak{G}, by putting $H(f)=H(\mathfrak{H}\lambda)$ whenever f is the characteristic function of $\mathfrak{H}\lambda$, where λ is any element of \mathfrak{G}, and \mathfrak{H} is as above; as all locally constant functions on \mathfrak{G} can be written as finite linear combinations of such characteristic functions, this determines H uniquely. We will call H *the Herbrand distribution on \mathfrak{G}.* In view of the foregoing results, it is clear that its knowledge implies the full knowledge of the ramification pro-

perties of K'' over K' whenever K', K'' are of finite degree over K, and $K \subset K' \subset K'' \subset L$.

§ 4. Traces and norms in A-fields.

In this §, we consider an A-field k and a separably algebraic extension k' of k, of finite degree n over k. Notations will be as explained in Chap. IV.

THEOREM 1. *Let k be an A-field and k' a separable extension of k of finite degree. Then, for almost all finite places w of k', k'_w is unramified over the closure k_v of k in k'_w.*

Let χ be a "basic character" for k, i.e. a non-trivial character of k_A, trivial on k. Put $\chi' = \chi \circ Tr_{k'/k}$; this is a character of k'_A, trivial on k'. As $Tr_{k'/k}$ is not 0, and as it is k-linear on k', there is $\xi \in k'$ such that $Tr_{k'/k}(\xi) = 1$. As the extension of $Tr_{k'/k}$ to k'_A is k_A-linear, this implies that it maps k'_A surjectively onto k_A, so that χ' is not trivial on k'_A. Let w be a finite place of k', and v the place of k lying below w; call χ_v, χ'_w the characters respectively induced by χ on k_v and by χ' on k'_w. By corollary 3 of th. 1, Chap. IV-1, we have $\chi'_w = \chi_v \circ Tr_{k'_w/k_v}$. By corollary 1 of th. 3, Chap. IV-2, χ_v is of order 0 for almost all v, and χ'_w is of order 0 for almost all w; our conclusion follows now immediately from this and from corollary 3 of prop. 4, § 1.

COROLLARY. *Let assumptions be as in theorem 1; then $N_{k'/k}$ is an open morphism of k'^\times_A onto an open subgroup of k^\times_A.*

By corollary 3 of th. 1, Chap. IV-1, $N_{k'/k}$ induces $N_{k'_w/k_v}$ on k'^\times_w for all places w of k'. By prop. 5 of § 1, this is, for all w, including the infinite places, an open morphism of k'^\times_w onto an open subgroup of k^\times_v; by theorem 1, combined with prop. 3 of § 1, it maps r'^\times_w onto r^\times_v for almost all w. In view of the corollary of prop. 2, Chap. IV-3, our assertion follows immediately from these facts.

If k_v and k'_w are as above, k'_w, being generated over k_v by k', is separable over k_v, so that, if v and hence w are finite places, its different over k_v is not 0 and may be written as $p'^{d(w)}_w$, with $d(w) \geq 0$. This justifies the following definition:

DEFINITION 3. *Let k, k' be as in theorem 1; for every finite place w of k', let $p'^{d(w)}_w$ be the different of k'_w over the closure k_v of k in k'_w. Then, by the different of k' over k, we understand the ideal $\prod p'^{d(w)}_w$ of k' if k, k' are of characteristic 0, and the divisor $\sum d(w) \cdot w$ of k' if they are of characteristic $p > 1$; it will be denoted by $\mathfrak{d}_{k'/k}$, or by \mathfrak{d} if no confusion can arise.*

We will now consider separately the cases of characteristic 0 and of characteristic $p > 1$.

PROPOSITION 10. *Let k be an algebraic number-field, k' a finite alge-braic extension of k, \mathfrak{r} and \mathfrak{r}' their maximal orders, and \mathfrak{d} the different of k' over k. Then \mathfrak{d}^{-1} is the set of the elements $\eta \in k'$ such that $\mathrm{Tr}(\xi\eta)$ is in \mathfrak{r} for all $\xi \in \mathfrak{r}'$.*

Take first any $\xi \in \mathfrak{r}'$ and any $\eta \in \mathfrak{d}^{-1}$; then $\xi\eta \in \mathfrak{d}^{-1}$, which means, by definition, that $\xi\eta \in k'$ and $\xi\eta \in p_w'^{-d(w)}$ for all finite places w of k'. This implies that $\mathrm{Tr}_{k_w'/k_v}(\xi\eta) \in r_v$ for all such places, and therefore, by corollary 3 of th. 1, Chap. IV-1, that $\mathrm{Tr}_{k'/k}(\xi\eta)$ is in $k \cap r_v$ for all v, hence in \mathfrak{r}. Conversely, assume η to be such that this is so for all $\xi \in \mathfrak{r}'$; take $x' = (x_w') \in k_A'$, and put $z = \mathrm{Tr}_{k'/k}(x'\eta)$. Then, by corollary 3 of th. 1, Chap. IV-1, $z = (z_v)$ is given by

$$z_v = \sum_{w|v} \mathrm{Tr}_{k_w'/k_v}(x_w'\eta).$$

Take a finite place v of k; by corollary 1 of th. 1, Chap. V-2, the projection of \mathfrak{r}' on the product $\prod r_w'$, taken over the places w lying above v, is dense there. As z_v is in r_v, by our assumption, whenever x' is in \mathfrak{r}', and depends continuously upon x', it is therefore in r_v whenever $x_w' \in r_w'$ for all w above v. This implies that Tr_{k_w'/k_v} maps $\eta r_w'$ into r_v, hence, by the definition of the different, that η is in $p_w'^{-d(w)}$; as this is so for all w, η must be in \mathfrak{d}^{-1}.

COROLLARY. *If \mathfrak{a}' is any fractional ideal of k', the set of the elements η of k' such that $\mathrm{Tr}_{k'/k}(\xi\eta)$ is in \mathfrak{r} for all $\xi \in \mathfrak{a}'$ is the fractional ideal $\mathfrak{a}'^{-1}\mathfrak{d}^{-1}$.*

In fact, in view of proposition 10, this set consists of all the η such that $\eta\mathfrak{a}' \subset \mathfrak{d}^{-1}$.

Now we introduce two morphisms ι, \mathfrak{N} of the groups $I(k)$, $I(k')$ of fractional ideals of k and of k' into each other, as follows. Consider again the morphism $a \to \mathrm{id}(a)$ of k_A^\times onto $I(k)$, with the kernel $\Omega_\infty = k_A(P_\infty)^\times$, which was defined in Chap. V-3; as pointed out there, we may use it to identify $I(k)$ with k_A^\times/Ω_∞; we recall that Ω_∞ is the group $k_\infty^\times \times \prod r_v^\times$ consisting of the ideles (z_v) such that $|z_v|_v = 1$ for all finite places v of k. If the group Ω_∞' is similarly defined for k', we may also identify $I(k')$ with $k_A'^\times/\Omega_\infty'$. Write now ι for the natural embedding of k_A^\times into $k_A'^\times$; by corollary 1 of th. 1, Chap. IV-1, this maps every $z = (z_v)$ in k_A^\times onto the element $\iota(z) = (z_w')$ of $k_A'^\times$ such that $z_w' = z_v$ whenever w lies above v; then $|z_v|_v = 1$ implies $|z_w'|_w = 1$, so that $\iota(z)$ is in Ω_∞' if and only if $z \in \Omega_\infty$. This shows that ι determines an injective morphism of $I(k)$ into $I(k')$, which we will call the *natural embedding* of $I(k)$ into $I(k')$, and which we will also denote by ι; with this notation, we have $(\mathrm{id}) \circ \iota = \iota \circ (\mathrm{id})$; this may be regarded as defining the injection ι of $I(k)$ into $I(k')$. Clearly, if k'' is an extension of k' of finite degree, and if the morphisms ι' of $k_A'^\times$ into $k_A''^\times$ and ι'' of k_A^\times into $k_A''^\times$

are defined just as ι was defined for k_A^\times and $k_A'^\times$, we have $\iota'' = \iota' \circ \iota$; therefore the corresponding relation holds true for the natural embeddings of $I(k)$ into $I(k'')$, of $I(k')$ into $I(k'')$ and of $I(k)$ into $I(k')$. On the other hand, corollary 3 of th. 1, Chap. IV-1, combined with formula (1) of § 1, shows that $N_{k'/k}$ maps Ω_∞' into Ω_∞; therefore it determines a morphism of $I(k')$ into $I(k)$, also known as the *norm*, which we will denote by $\mathfrak{N}_{k'/k}$; we have $(\mathrm{id})\circ N_{k'/k} = \mathfrak{N}_{k'/k}\circ(\mathrm{id})$, and this may be regarded as defining $\mathfrak{N}_{k'/k}$. If k'' is as above, we have $\mathfrak{N}_{k''/k} = \mathfrak{N}_{k'/k}\circ\mathfrak{N}_{k''/k'}$, as an immediate consequence of the corresponding relation for ordinary norms. Furthermore, if n is the degree of k' over k, we have $N_{k'/k}(x) = x^n$ for all $x \in k$, as an immediate consequence of the definition of $N_{k'/k}$; this implies at once the corresponding relation for the extension of $N_{k'/k}$ to k_A. For $z \in k_A^\times$, we can write it as $N_{k'/k}(\iota(z)) = z^n$, which implies that we have $\mathfrak{N}_{k'/k}(\iota(\mathfrak{a})) = \mathfrak{a}^n$ for all $\mathfrak{a} \in I(k)$.

By th. 3 of Chap. V-3, $I(k)$ and $I(k')$ are the free groups respectively generated by the prime ideals \mathfrak{p}_v, \mathfrak{p}_w' of \mathfrak{r}, \mathfrak{r}'. We will now describe the morphisms ι, $\mathfrak{N}_{k'/k}$ in terms of these generators.

PROPOSITION 11. *For each finite place v of k, and each place w of k' lying above v, call $e(w)$ the order of ramification and $f(w)$ the modular degree of k_w' over k_v. Then we have:*

$$\iota(\mathfrak{p}_v) = \prod_{w|v} \mathfrak{p}_w'^{\,e(w)}, \qquad \mathfrak{N}_{k'/k}(\mathfrak{p}_w') = \mathfrak{p}_v^{f(w)}, \qquad \sum_{w|v} e(w)f(w) = n,$$

where the product in the first formula, and the sum in the last one, are taken over all the places w of k' lying above v.

The first formula follows at once from the definitions, and the second one from the definitions, corollary 3 of th. 1, Chap. IV-1, and formula (1) of § 1. As to the last formula, since the degree of k_w' over k_v is $e(w)f(w)$, it is nothing else than corollary 1 of th. 4, Chap. III-4; it is also an immediate consequence of the first two formulas and of $\mathfrak{N}_{k'/k}(\iota(\mathfrak{p}_v)) = \mathfrak{p}_v^n$.

COROLLARY. *Let k be an algebraic number-field, and \mathfrak{a} a fractional ideal of k. Then $\mathfrak{N}_{k/Q}(\mathfrak{a})$ is the fractional ideal $\mathfrak{N}(\mathfrak{a})\,\mathbf{Z}$ of \mathbf{Q}, where \mathfrak{N} is the norm as defined in definition 5 of Chap. V-3.*

This follows at once from the latter definition and from the second formula in proposition 11, applied to the fields k and \mathbf{Q}.

As every ideal in the ring \mathbf{Z} is of the form $m\mathbf{Z}$ with $m \in \mathbf{N}$, every fractional ideal of \mathbf{Q} can be written in one and only one way as $r\mathbf{Z}$ with $r \in \mathbf{Q}, r > 0$; one may therefore identify the group $I(\mathbf{Q})$ of fractional ideals of \mathbf{Q} with $\mathbf{Q}_+^\times = \mathbf{Q}^\times \cap \mathbf{R}_+^\times$, by means of the isomorphism $r \to r\mathbf{Z}$ of the latter onto the former. Then the norm \mathfrak{N} of definition 5, Chap. V-3, becomes the same as $\mathfrak{N}_{k/Q}$ as defined above.

PROPOSITION 12. *Let χ be the character of $\mathbf{Q_A}$, trivial on \mathbf{Q}, such that $\chi_\infty(x) = \mathbf{e}(-x)$; let k be an algebraic number-field, and put $\chi' = \chi \circ Tr_{k/\mathbf{Q}}$; let $a = (a_v)$ be a differental idele for k, attached to χ'. Then $a_v = 1$ for every infinite place v of k, and $\mathrm{id}(a)$ is the different $\mathfrak{d}_{k/\mathbf{Q}}$ of k over \mathbf{Q}.*

The character χ is the same which has been introduced in the first part of the proof of th. 3, Chap. IV-2; it was shown there that it is uniquely determined by the condition stated above, and that χ_p is of order 0 for every place p of \mathbf{Q}. Our first assertion is now an immediate consequence of the definition of differential ideles in Chap. VII-2, combined with corollary 3 of th. 1, Chap. IV-1. Our last assertion is an immediate consequence of the same results, combined with corollary 3 of prop. 4, § 1.

COROLLARY. *Let k be as in proposition 12, and let D be its discriminant. Then $|D| = \mathfrak{N}(\mathfrak{d}_{k/\mathbf{Q}})$.*

If a is as in proposition 12, we have $|a|_A = |D|^{-1}$, by prop. 6 of Chap. VII-2. On the other hand, since $a_v = 1$ for all infinite places of k, the definition of \mathfrak{N} shows at once that $|a|_A = \mathfrak{N}(\mathrm{i}\,\mathrm{d}(a))^{-1}$; in view of proposition 12, this proves our assertion.

Now we generalize the definition of the discriminant, i.e. definition 6 of Chap. V-4, as follows:

DEFINITION 4. *Let k be an algebraic number-field, k' a finite extension of k, and \mathfrak{d} the different of k' over k. Then the ideal $\mathfrak{D} = \mathfrak{N}_{k'/k}(\mathfrak{d})$, in the maximal order \mathfrak{r} of k, is called the discriminant of k' over k.*

One should note that, according to this, the discriminant of k over \mathbf{Q} is not D, but the ideal $D\mathbf{Z} = |D|\mathbf{Z}$ in \mathbf{Z}. When the latter is given, D is determined by $D = (-1)^{r_2}|D|$, as follows from the remark at the end of the proof of prop. 7, Chap. V-4.

PROPOSITION 13. *Let k, k', k'' be algebraic number-fields such that $k \subset k' \subset k''$; let \mathfrak{d} and \mathfrak{D}, \mathfrak{d}' and \mathfrak{D}', \mathfrak{d}'' and \mathfrak{D}'' be the differents and the discriminants of k' over k, of k'' over k', and of k'' over k, respectively. Then:*

$$\mathfrak{d}'' = \iota'(\mathfrak{d})\mathfrak{d}', \qquad \mathfrak{D}'' = \mathfrak{D}''^{n'}\mathfrak{N}_{k'/k}(\mathfrak{D}'),$$

where ι' is the natural embedding of $I(k')$ into $I(k'')$, and n' is the degree of k'' over k'.

The first formula follows at once from the corresponding local result, i.e. corollary 4 of prop. 4, § 1. The second one follows from this and definition 4, combined with the transitivity property of norms.

Now let k be an A-field of characteristic $p > 1$, and k' a separable extension of k of finite degree n. As $a \to \mathrm{div}(a)$ is a morphism of k_A^\times onto the

group $D(k)$ of divisors of k, with the kernel $\prod r_v^\times$, we see, just as in the case of number-fields, that the natural embedding of k_A^\times into $k_A'^\times$ determines an injective morphism ι of $D(k)$ into $D(k')$, which we call the natural embedding of $D(k)$ into $D(k')$. Similarly, the norm mapping $N_{k'/k}$ of $k_A'^\times$ into k_A^\times determines a morphism of $D(k')$ into $D(k)$, which we denote by $\mathfrak{S}_{k'/k}$ (the notation \mathfrak{N} would be undesirable here, since the groups of divisors are written additively). The properties of ι and \mathfrak{S} are quite similar to those of ι and \mathfrak{N} in the case of number-fields. In particular, we have $\mathfrak{S}_{k'/k}(\iota(\mathfrak{a}))=$ $=n\mathfrak{a}$ for every divisor \mathfrak{a} of k, and, with the same notations as in proposition 11:

$$\iota(v)=\sum_{w|v} e(w)\cdot w, \quad \mathfrak{S}_{k'/k}(w)=f(w)\cdot v, \quad \sum_{w|v} e(w)f(w)=n,$$

the proof being the same as there. Let \mathbf{F}_q, $\mathbf{F}_{q'}$ be the fields of constants of k and of k', and let f_0 be the degree of the latter over the former. Then the definition of $f(w)$, and that of the degree of a place, give $f_0\deg(w)=f(w)\deg(v)$, and consequently, at first for places, and then for arbitrary divisors:

(12) $\deg(\mathfrak{S}_{k'/k}(\mathfrak{a}'))=f_0\deg(\mathfrak{a}'), \quad \deg(\iota(\mathfrak{a}))=(n/f_0)\deg(\mathfrak{a})$

where \mathfrak{a}' is any divisor of k', and \mathfrak{a} any divisor of k.

If \mathfrak{d} is the different of k' over k, we define the *discriminant* of k' over k as being the divisor $\mathfrak{S}_{k'/k}(\mathfrak{d})$ of k. With notations similar to those in prop. 13, we have:

$$\mathfrak{d}''=\iota'(\mathfrak{d})+\mathfrak{d}', \quad \mathfrak{D}''=n'\mathfrak{D}+\mathfrak{S}_{k'/k}(\mathfrak{D}').$$

PROPOSITION 14. *Let k and k' be as above; let \mathfrak{d} be the different of k' over k, and let \mathfrak{c} be a canonical divisor of k. Then the divisor $\iota(\mathfrak{c})+\mathfrak{d}$ is a canonical divisor of k'.*

By the definition of a canonical divisor, there is a "basic character" χ for k, such that $\mathfrak{c}=\operatorname{div}(\chi)$. Then corollary 3 of prop. 4, § 1, combined with corollary 3 of th. 1, Chap. IV-1, and with the definitions, shows at once that the divisor of $\chi\circ \mathit{Tr}_{k'/k}$ is $\iota(\mathfrak{c})+\mathfrak{d}$.

COROLLARY. *Let k, k' and \mathfrak{d} be as in proposition 14: let g be the genus of k; let n be the degree of k' over k, and f_0 the degree of the field of constants of k' over that of k. Then the genus g' of k' is given by*

$$2g'-2=(n/f_0)(2g-2)+\deg(\mathfrak{d}).$$

This follows at once from proposition 14, corollary 1 of th. 2, Chap. VI, and the second formula (12). It implies that the degree of the different is always an even integer; a more precise result will be proved in Chap. XIII-12.

§ 5. Splitting places in separable extensions. Assumptions and nota-
tions being as in theorem 1 of § 4, one can express that theorem by
saying that, for almost all places w of k', the degree of k'_w over k_v is equal
to its modular degree over k_v. Therefore corollaries 2 and 3 of prop. 1,
Chap. VII-1, and corollaries 3 and 4 of th. 2, Chap. VII-5, are valid if
"degree" is substituted for "modular degree", provided one adds there the
assumption that k is separable over k_0. We will now consider some
consequences of these results.

As before, let k be an A-field, k' a separable extension of k of finite
degree n, and v a place of k. We can write $k' = k(\xi)$, where ξ is a root of an
irreducible monic polynomial F of degree n in $k[X]$. Combining th. 4 of
Chap. III-4 with prop. 2 of Chap. III-2, we see that the places w of k'
which lie above v are in a one-to-one correspondence with the irredu-
cible monic polynomials dividing F in $k_v[X]$; if, for each such place w,
we call F_w the corresponding polynomial, the degree of k'_w over k_v is
equal to the degree of F_w; by th. 1 of § 4, that degree, for almost all v, is
equal to the modular degree of k'_w over k_v. We also see that the places w,
lying above v, for which $k'_w = k_v$ are in a one-to-one correspondence with
the roots of F in k_v. By corollary 1 of th. 4, Chap. III-4, there are n distinct
places of k' lying above v if and only if $k'_w = k_v$ for every such place w;
when that is so, one says that v *splits fully in* k'; it does so if and only if F
has n distinct roots in k_v. If L is a Galois extension of k, then, by corollary
4 of th. 4, Chap. III-4, the completions of L at the places of L lying above
v are all isomorphic; therefore, if $L_u = k_v$ for one such place u, v splits fully
in L. Let $k' = k(\xi)$ be a field between k and L; then, if F is defined as above,
it splits into linear factors in $L[X]$, and the smallest Galois extension L'
of k, contained in L and containing k', is the subfield of L generated over k
by the roots of F in L. If now t is a place of L' lying above v, L'_t is generated
over k_v by the roots of F, so that $L'_t = k_v$ if and only if v splits fully in k'; in
that case, as we have seen, it also splits fully in L'.

PROPOSITION 15. *Let k', k'' be two extensions of k, both contained in a
separable extension L of k of finite degree. Let X be the set of the places v
of k such that $k'_w = k_v$ for at least one place w of k' lying above v. If almost
all the places $v \in X$ split fully in k'', k'' is contained in k'.*

We may assume that L is the compositum of k' and k''. Call W the
set of the places w of k' such that the place v of k which lies below w
splits fully in k'' and that $k_v = k'_w$. Let u be a place of L above w, and t
the place of k'' below u; L_u is generated over k_v by L, hence by k'_w and
k''_t; therefore, if $w \in W$, $L_u = k_v$; this shows that all the places in W split
fully in L. Now take a place w of k', not in W; call v the place of k below
w. If then $k_v = k'_w$, v is in X, so that it must be in the finite subset of X,

consisting of the places in X which do not split fully in k''. If $k_v \neq k'_w$, the degree of k'_w over k_v is >1; by th. 1 of §4, this is the same as the modular degree, except for finitely many places. We have thus shown that the modular degree of k'_w over k_v is >1 for almost all the places w of k', not in W. Applying now corollary 4 of th. 2, Chap. VII-5, to k, k' and L (instead of k_0, k and k' of that corollary), we get $k' = L$, i.e. $k'' \subset k'$.

COROLLARY. *Let k', k'' be two Galois extensions of k, contained in some extension of k of finite degree. Let S', S'' be the sets of the places of k which split fully in k' and in k'', respectively. Then k' contains k'' if and only if S'' contains almost all the places $v \in S'$.*

If $k' \supset k''$, it is obvious that a place of k which splits fully in k' does the same in k''. Conversely, as k' is a Galois extension, S' is the same as the set X in proposition 15; our conclusion is now a special case of that proposition. In particular, we see that k' must be the same as k'' if S', S'' differ by no more than finitely many elements.

§ 6. **An application to inseparable extensions.** It will now be shown that one of our main results, the isomorphism between k'_A and $(k'/k)_A$, which was proved for separable extensions as theorem 1 of Chap. IV-1, is still valid without the assumption of separability. For this, we need a lemma:

LEMMA 1. *Let k be an A-field of characteristic $p > 1$; then k is purely inseparable of degree p over its image k^p under the endomorphism $x \to x^p$.*

By lemma 1 of Chap. III-2, we may write k as $k = F_p(x_0, \ldots, x_N)$, where x_0 is transcendental over F_p, and x_i is separably algebraic over $F_p(x_0)$ for $1 \leqslant i \leqslant N$. Then $k^p = F_p(x_0^p, \ldots, x_N^p)$. Put $k' = k^p(x_0) = F_p(x_0, x_1^p, \ldots, x_N^p)$. As each x_i is purely inseparable over $F_p(x_i^p)$ and separable over $F_p(x_0)$, k is at the same time purely inseparable and separable over k', so that $k = k'$; this implies that k is purely inseparable of degree 1 or p over k^p. If k was the same as k^p, it would contain an element y such that $y^p = x_0$. Clearly y cannot be in $F_p(x_0)$, so that it is purely inseparable over $F_p(x_0)$; this contradicts the assumption that k is separable over $F_p(x_0)$.

Now, in order to extend theorem 1 of Chap. IV-1 to the case of an inseparable extension k' of k, it is clearly enough to show the validity of th. 4, Chap. III-4, in that case, since the latter alone is involved in the proof of the former. We will first do this for a purely inseparable extension of k of degree p. Let k' be such an extension; for any $x' \in k'$, there must then be an integer $n \geqslant 0$ such that $x'^{p^n} \in k$, and, if n is the

smallest such integer, the degree of x' over k is p^n; as this must be $\leqslant p$, n is 0 or 1. This shows that $k'^p \subset k \subset k'$, hence, in view of lemma 1, that $k = k'^p$. For that case, we prove the following:

PROPOSITION 16. *Let k' be an A-field of characteristic $p > 1$; put $k = k'^p$. Then, above each place v of k, there lies one and only one place w of k'; it is the image of v under the isomorphism $x \to x^{1/p}$ of k onto k'; we have $k_v = (k'_w)^p$, and the k_v-linear extension Φ_v of the natural injection of k' into k'_w to $A_v = k' \otimes_k k_v$ is an isomorphism of A_v onto k'_w. Moreover, if α is a basis of k' over k, and α_v, for each v, is the r_v-module generated by α in A_v, then, for almost all v, Φ_v maps α_v onto the maximal compact subring r'_w of k'_w.*

Let v be a place of k, and w a place of k' lying above v. By the corollary of prop. 1, Chap. III-1, k'_w is generated over k_v by k', hence purely inseparable of degree 1 or p over k_v. In the former case, every element of k must be a p-th power in k_v; this is impossible, since k is dense in k_v and therefore contains at least one prime element of k_v. Therefore, by corollary 2 of prop. 4, Chap. I-4, k'_w is uniquely determined, up to an isomorphism, and $y \to y^p$ is an isomorphism of k'_w onto k_v. Let λ be the natural injection of k' into k'_w; this must induce on k the natural injection λ_0 of k into k_v; therefore, for every $\xi \in k'$, we have $\lambda_0(\xi^p) = \lambda(\xi)^p$; as this determines $\lambda(\xi)$ uniquely, we see that w is uniquely determined by v, and also that it is the image of v under $x \to x^{1/p}$. If now Φ_v is as in our proposition, it is clearly a surjective homomorphism of A_v onto k'_w; as both of these spaces have the dimension p over k_v, it is an isomorphism. Finally, let α be a basis of k' over k. In view of corollary 1 of th. 3, Chap. III-1, and of lemma 1 of Chap. III-2, we may assume that α contains an element a such that k' is separably algebraic over $F_p(a)$. Let then v and w be as above, and let u be the place of the field $k_0 = F_p(a)$ which lies below w. By th. 1 of § 4, for almost all w, k'_w is unramified over $(k_0)_u$. Take w such that this is so; since th. 2 of Chap. III-1 shows that k_0 has just one place u for which $|a|_u > 1$, we may also assume that w does not lie above that place. Then, by that theorem, there is a polynomial $\pi \in F_p[T]$ such that $\pi(a)$ is a prime element of $(k_0)_u$, hence also of k'_w since k'_w is unramified over $(k_0)_u$. Now, by corollary 2 of th. 3, Chap. III-1, α_v is a compact subring of A_v for almost all v; this implies that it contains 1. hence $r_v \cdot 1$. As it contains a, it contains $\pi(a)$, hence the ring $r_v[\pi(a)]$; by prop. 4 of Chap. I-4 and its corollary 1, this is the same as r'_w.

Clearly proposition 16 implies the validity of th. 4, Chap. III-4, when $k = k'^p$. Now take for k' an arbitrary extension of k, of finite degree. Call k'_0 the maximal separably algebraic extension of k, contained in

k'; let p^m be the degree of k' over k'_0. If x' is any element of k', there is $n \geqslant 0$ such that $x'^{p^n} \in k'_0$, and, if n is the smallest such integer, x' is of degree p^n over k'_0, so that $n \leqslant m$; this shows that $k' \supset k'_0 \supset k'^{p^m}$. Applying lemma 1 to the sequence of fields $k', k'^p, \ldots, k'^{p^m}$, we see that each is of degree p over the next one, so that k' is of degree p^m over k'^{p^m}, which is therefore the same as k'_0. Proceeding now by induction on m, we may assume that theorem 4 of Chap. III-4 is valid for the extension k'^p of k, and we have to show that it is also valid for the extension k' of k. Put $k'' = k'^p$; let v be a place of k; call w'_1, \ldots, w'_r the places of k'' lying above v, and, for each i, call k''_i the completion of k'' at w'_i. By prop. 16, there is, for each i, one and only one place w_i of k', lying above w'_i, and the completion k'_i of k' at w_i may be identified with $k' \otimes_{k''} k''_i$. By the induction assumption, we have an isomorphism \varPhi'_v of $A'_v = k'' \otimes_k k_v$ onto the direct sum of the fields k''_i, with the properties stated in our theorem. By the properties of tensor-products, the tensor-product $A_v = k' \otimes_k k_v$ is canonically isomorphic, in an obvious manner, to $k' \otimes_{k''} A'_v$, hence to the direct sum of the products $k' \otimes_{k''} k''_i$ and therefore to the direct sum of the fields k'_i; it is then easily seen that the isomorphism \varPhi_v of A_v onto the latter sum which has been so defined has the properties required by our theorem. As to the last part, it can be deduced in the same manner from the induction assumption and prop. 16, by taking a basis α' of k'' over k, a basis β of k' over k'', and taking for k' over k the basis α consisting of all the products $a' b$ of an element a' of α' and an element b of β.

Chapter IX

Simple algebras

§ 1. Structure of simple algebras. This Chapter will be purely algebraic in nature; this means that we will operate over a groundfield, subject to no restriction except commutativity, and carrying no additional structure. All fields are understood to be commutative. All algebras are understood to have a unit, to be of finite dimension over their ground-field, and to be central over that field (an algebra A over K is called central if K is its center). If A, B are algebras over K with these properties, so is $A \otimes_K B$; if A is an algebra over K with these properties, and L is a field containing K, then $A_L = A \otimes_K L$ is an algebra over L with the same properties. Tensor-products will be understood to be taken over the groundfield; thus we write $A \otimes B$ instead of $A \otimes_K B$ when A, B are algebras over K, and $A \otimes L$ or A_L, instead of $A \otimes_K L$, when A is an algebra over K and L a field containing K, A_L being always considered as an algebra over L.

Let A be an algebra over K, with the unit 1_A; all modules over A will be understood to be unitary (this means, e.g. for a left module M, that $1_A \cdot m = m$ for all $m \in M$) and of finite dimension over K, when regarded as vector-spaces over K by putting, e.g. for a left module M, $\xi m = (\xi \cdot 1_A) m$ for all $\xi \in K$ and $m \in M$. If M' is a subset of a left A-module M, the *annihilator* of M' in A is the set of all $x \in A$ such that $x m = 0$ for all $m \in M'$; this is a left ideal in A. The annihilator of M in A is a two-sided ideal in A; if it is $\{0\}$, M is called *faithful*.

DEFINITION 1. *Let A be an algebra over K. An A-module is called simple if it is not $\{0\}$ and has no submodule except itself and $\{0\}$. The algebra A is called simple if it has no two-sided ideal except itself and $\{0\}$.*

For a given A, there are always simple left A-modules; for instance, any left ideal of A, other than $\{0\}$, with the smallest dimension over K, will be such a module.

PROPOSITION 1. *Let A be an algebra over K, with a faithful simple left A-module M. Then every left A-module is a direct sum of modules, all isomorphic to M.*

We first prove our assertion for A itself, considered as a left A-module. In M, there are finite subsets with the annihilator $\{0\}$ in A (e.g. any basis

of M over K); take any minimal set $\{m_1, \ldots, m_n\}$ with that property. For $0 \leqslant i \leqslant n$, call A_i the annihilator of $\{m_{i+1}, \ldots, m_n\}$ in A; for $i \geqslant 1$, put $M_i = A_i m_i$. Clearly $A_0 = \{0\}$, $A_n = A$; for $i \geqslant 1$, $A_i \supset A_{i-1}$, and $A_i \neq A_{i-1}$, since otherwise $x m_j = 0$ for $j > i$ would imply $x m_i = 0$, and m_i could be omitted from $\{m_1, \ldots, m_n\}$. For $i \geqslant 1$, A_i is a left ideal, M_i is a submodule of M, and $x \rightarrow x m_i$ induces on A_i a morphism of A_i onto M_i with the kernel A_{i-1}, so that it determines an isomorphism of A_i/A_{i-1} onto M_i for their structures as left A-modules. As $A_i \neq A_{i-1}$, M_i is not $\{0\}$; therefore it is M. By induction on i for $0 \leqslant i \leqslant n$, one sees now at once that $x \rightarrow (x m_1, \ldots, x m_i)$ induces on A_i a bijective mapping of A_i onto the product $M^i = M \times \ldots \times M$ of i modules, all equal to M; this is obviously an isomorphism for the structure of left A-module. For $i = n$, this proves our assertion for A. Now take any left A-module M', and a finite set $\{m'_1, \ldots, m'_r\}$ generating M' (e.g. any basis of M' over K). Then the mapping of A^r into M', given by $(x_i)_{1 \leqslant i \leqslant r} \rightarrow \sum x_i m'_i$, is a surjective morphism of left A-modules; as we have just proved that A, as such, is isomorphic to M^n for some n, this shows that there is a surjective morphism of M^{nr} onto M', or, what amounts to the same, a surjective morphism F, onto M', of a direct sum of $s = nr$ modules M_i, all isomorphic to M. Call N the kernel of F, and take a maximal subset $\{M_{i_1}, \ldots, M_{i_h}\}$ of $\{M_1, \ldots, M_s\}$ such that the sum $N' = N + \sum M_{i_\lambda}$ is direct; after renumbering the M_i if necessary, we may assume that this subset is $\{M_1, \ldots, M_h\}$. Then, for $j > h$, the sum $N' + M_j$ is not direct, so that $N' \cap M_j$ is not $\{0\}$; as it is a submodule of M_j, which is isomorphic to M, it is M_j. This shows that $M_j \subset N'$ for all $j > h$. Therefore F maps N' onto M'; as its kernel is N, it determines an isomorphism of $\sum_{i=1}^{h} M_i$ onto N'.

PROPOSITION 2. *Let A and M be as in proposition 1, and let D be the ring of endomorphisms of M. Then D is a division algebra over K, and A is isomorphic to $M_n(D)$ for some $n \geqslant 1$.*

We recall that here, as explained on p. XV, D should be understood as a ring of right operators on M, the multiplication in it being defined accordingly. As D is a subspace of the ring of endomorphisms of the underlying vector-space of M over K, it is a vector-space of finite dimension over K. Every element of D maps M onto a submodule of M, hence onto M or $\{0\}$; therefore, if it is not 0, it is an automorphism, hence invertible. This shows that D is a division algebra over a center which is of finite dimension over K. By prop. 1, there is, for some $n \geqslant 1$, an isomorphism of A, regarded as a left A-module, onto M^n; this must determine an isomorphism between the rings of endomorphisms of these two

left A-modules. Clearly that of M^n consists of the mappings

$$(m_j)_{1 \leqslant j \leqslant n} \rightarrow (\sum m_i d_{ij})_{1 \leqslant j \leqslant n}$$

with $d_{ij} \in D$ for $1 \leqslant i,j \leqslant n$, and may therefore be identified with the ring $M_n(D)$ of the matrices (d_{ij}) over D. On the other hand, an endomorphism of A regarded as a left A-module is a mapping f such that $f(xy) = x f(y)$ for all x, y in A; for $y = 1_A$, this shows that f can be written as $x \rightarrow xa$ with $a = f(1_A)$; the ring of such endomorphisms may now be identified with A, which is therefore isomorphic to $M_n(D)$. As the center of $M_n(D)$ is clearly isomorphic to that of D, this implies that the latter is K, which completes the proof.

THEOREM 1. *An algebra A over K is simple if and only if it is isomorphic to an algebra $M_n(D)$, where D is a division algebra over K; when A is given, n is uniquely determined, and so is D up to an isomorphism.*

Let A be simple; take any simple left A-module M; as the annihilator of M in A is a two-sided ideal in A and is not A, it is $\{0\}$; therefore M is faithful, and we can apply prop. 2 to A and M; it shows that A is isomorphic to an algebra $M_n(D)$. Conversely, take $A = M_n(D)$. For $1 \leqslant h, k \leqslant n$, call e_{hk} the matrix (x_{ij}) given by $x_{hk} = 1$, $x_{ij} = 0$ for $(i,j) \neq (h,k)$. If $a = (a_{ij})$ is any matrix in $M_n(D)$, we have $e_{ij} a e_{hk} = a_{jh} e_{ik}$ for all i,j,h,k; this shows that, if $a \neq 0$, the two-sided ideal generated by a in A contains all the e_{ik}; therefore it is A, so that A is simple. Let now M be the left ideal generated by e_{11} in A; it consists of the matrices (a_{ij}) such that $a_{ij} = 0$ for $j \geqslant 2$; if a is such a matrix, we have $e_{ij} a = a_{j1} e_{i1}$, which shows that, if $a \neq 0$, the left ideal generated by a is M, which is therefore a minimal left ideal and a simple left A-module. Let now f be an endomorphism of M regarded as a left A-module, and put $f(e_{11}) = a$ with $a = (a_{ij})$, $a_{ij} = 0$ for $j \geqslant 2$. Writing that $f(e_{ij} e_{11}) = e_{ij} a$, we get, for $j \geqslant 2$, $a_{j1} = 0$; then, for $x = (x_{ij})$ with $x_{ij} = 0$ for $j \geqslant 2$, we get $f(x) = f(x e_{11}) = xa = (x_{ij} a_{11})$. This shows that the ring of endomorphisms of M is isomorphic to D. As prop. 1 shows that all simple left A-modules are isomorphic to M, this shows that D is uniquely determined by A up to an isomorphism. As the dimension of A over K is n^2 times that of D, n also is uniquely determined.

We recall now that the *inverse* of an algebra A over K is the algebra A^0 with the same underlying vector-space over K as A, but with the multiplication law changed from $(x,y) \rightarrow xy$ to $(x,y) \rightarrow yx$.

PROPOSITION 3. *Let A be an algebra over K; call A^0 its inverse, and put $C = A \otimes A^0$. For all a, b in A, call $f(a,b)$ the endomorphism $x \rightarrow axb$ of the underlying vector-space of A; let F be the K-linear mapping of C*

into $\text{End}_K(A)$ such that $F(a\otimes b)=f(a,b)$ for all a, b. Then A is simple if and only if F maps C surjectively onto $\text{End}_K(A)$; when that is so, F is an isomorphism of C onto $\text{End}_K(A)$.

One verifies at once that F is a homomorphism of C into $\text{End}_K(A)$. If N is the dimension of A over K, both C and $\text{End}_K(A)$ have the dimension N^2 over K; therefore F is an isomorphism of C onto $\text{End}_K(A)$ if and only if it is surjective, and if and only if it is injective. Assume that A is not simple, i.e. that it has a two-sided ideal I other than $\{0\}$ and A. Then, for all $a,b,f(a,b)$ maps I into I; therefore the same is true of $F(c)$ for all $c\in C$, so that the image of C under F is not the whole of $\text{End}_K(A)$. Assume now that A is simple, and call M the underlying vector-space of A over K, regarded as a left C-module for the law $(c,x)\to F(c)x$. Any submodule M' of M is then mapped into itself by $x\to axb$ for all a,b, so that it is a two-sided ideal in A; as A is simple, this shows that M is simple. An endomorphism φ of M is a mapping φ such that $\varphi(axb)= =a\varphi(x)b$ for all a,x,b in A; for $x=b=1_A$, this gives $\varphi(a)=a\varphi(1_A)$, hence $axb\varphi(1_A)=ax\varphi(1_A)b$, so that $\varphi(1_A)$ must be in the center K of A; in other words, φ is of the form $x\to\xi x$ with $\xi\in K$. Call C' the annihilator of M in C, which is the same as the kernel of F. We can now apply prop. 2 to the algebra C/C', to its center Z, and to the module M; as D *
is then K, it shows that C/C' is isomorphic to some $M_n(K)$, hence Z to K; but then, as has been seen in the proof of th. 1, M must have the dimension n over K, so that $n=N$. As C/C' has then the same dimension N^2 over K as C, we get $C'=\{0\}$, which completes the proof.

COROLLARY 1. *Let L be a field containing K. Then the algebra $A_L=A\otimes L$ over L is simple if and only if A is so.*

In fact, let C_L, F_L be defined for A_L just as C, F are defined for A in proposition 3; one sees at once that $C_L=C\otimes L$, and that F_L is the L-linear extension of F to C_L. Our assertion follows now from proposition 3.

COROLLARY 2. *Let L be an algebraically closed field containing K. Then A is simple if and only if A_L is isomorphic to some $M_n(L)$.*

If D is a division algebra over a field K, the extension of K generated in D by any $\xi\in D-K$ is an algebraic extension of K, other than K. In particular, if L is algebraically closed, there is no division algebra over L, other than L. Therefore, by th. 1, an algebra over L is simple if and only if it is isomorphic to some $M_n(L)$. Our assertion follows now from corollary 1.

COROLLARY 3. *The dimension of a simple algebra A over K is of the form n^2.*

In fact, by corollary 2, A_L is isomorphic to some $M_n(L)$ if L is an algebraic closure of K; its dimension over L is then n^2, and it is the same as that of A over K.

COROLLARY 4. *Let A, B be two simple algebras over K; then $A \otimes B$ is simple over K.*

Take an algebraic closure L of K; $(A \otimes B)_L$ is the same as $A_L \otimes B_L$. Since clearly $M_n(K) \otimes M_m(K)$ is isomorphic to $M_{nm}(K)$ for all m, n, and all fields K, our conclusion follows from corollary 2.

COROLLARY 5. *Let A be a simple algebra of dimension n^2 over K. Let L be a field containing K, and let F be a K-linear homomorphism of A into $M_n(L)$. Then the L-linear extension F_L of F to A_L is an isomorphism of A_L onto $M_n(L)$.*

Clearly F_L is a homomorphism of A_L into $M_n(L)$, so that its kernel is a two-sided ideal in A_L. As A_L is simple by corollary 1, and as F_L is not 0, this kernel is $\{0\}$, i.e. F_L is injective. As A_L and $M_n(L)$ have the same dimension n^2 over L, this implies that it is bijective, so that it is an isomorphism of A_L onto $M_n(L)$.

COROLLARY 6. *Let L be an extension of K of degree n; let A be a simple algebra of dimension n^2 over K, containing a subfield isomorphic to L. Then A_L is isomorphic to $M_n(L)$.*

We may assume that A contains L. Then $(x, \xi) \to x\xi$, for $x \in A$, $\xi \in L$, defines on A a structure of vector-space over L; call V that vector-space, which is clearly of dimension n over L. For every $a \in A$, the mapping $x \to ax$ may be regarded as an endomorphism of V, which, if we choose a basis for V over L, is given by a matrix $F(a)$ in $M_n(L)$. Our assertion follows now from corollary 5.

PROPOSITION 4. *Let A be a simple algebra over K. Then every automorphism α of A over K is of the form $x \to a^{-1}xa$ with $a \in A^\times$.*

Take a basis $\{a_1, \dots, a_N\}$ of A over K. Then every element of $A \otimes A^0$ can be written in one and only one way as $\sum a_i \otimes b_i$, with $b_i \in A^0$ for $1 \leqslant i \leqslant N$. By prop. 3, α can therefore be written as $x \to \sum a_i x b_i$. Writing that $\alpha(xy) = \alpha(x)\alpha(y)$ for all x, y, we get

$$0 = \sum a_i x y b_i - \sum a_i x b_i \alpha(y) = \sum a_i x (y b_i - b_i \alpha(y)).$$

For each $y \in A$, this is so for all x; by prop. 3, we must therefore have $y b_i = b_i \alpha(y)$. In particular, since this gives $y(b_i z) = b_i \alpha(y)z$ for all y and z in A, $b_i A$ is a two-sided ideal in A, hence A or $\{0\}$, for all i, so that b_i is either 0 or invertible in A. As α is an automorphism, the b_i cannot all be 0; taking $a = b_i \neq 0$, we get the announced result.

COROLLARY. *Let α and a be as in proposition 4, and let $a' \in A$ be such that $a' \alpha(x) = x a'$ for all $x \in A$. Then $a' = \xi a$ with $\xi \in K$.*

In fact, the assumption can be written as $a' a^{-1} x = x a' a^{-1}$ for all x; this means that $a' a^{-1}$ is in the center K of A.

Proposition 4 is generally known as "the theorem of Skolem-Noether" (although that name is sometimes reserved for a more complete statement involving a simple subalgebra of A). One can prove, quite similarly, that every derivation of A is of the form $x \to xa - ax$, with $a \in A$.

We will also need a stronger result than corollary 2 of prop. 3; this will appear as a corollary of the following:

PROPOSITION 5. *Let D be a division algebra over K, other than K. Then D contains a separably algebraic extension of K, other than K.*

We reproduce Artin's proof. In D, considered as a vector-space over K, take a supplementary subspace E to $K = K \cdot 1_D$, and call φ the projection from $D = E \oplus K \cdot 1_D$ onto E. Then, for every integer $m \geqslant 1$, $x \to \varphi(x^m)$ is a polynomial mapping of D into E, whose extension to D_L and E_L, if L is any field containing K, is again given by $x \to \varphi(x^m)$, where φ denotes again the L-linear extension of φ to D_L and E_L. Now call N the dimension of D over K. Clearly every $\xi \in D$, not in K, generates over K an extension $K(\xi)$ of degree > 1 and $\leqslant N$; moreover, if this is not purely inseparable over K, it contains a separable extension of K, other than K. Assume now that our proposition is not true for D. Then K has inseparable extensions, which implies that it is of characteristic $p > 1$ and that it is not a finite field; moreover, every $\xi \in D$ must be purely inseparable over K, hence must satisfy an equation $\xi^{p^n} = x \in K$, where p^n is its degree over K. As this degree is $\leqslant N$, it divides the highest power q of p which is $\leqslant N$, so that $\xi^q \in K$. Then, if E and φ are as above defined, the polynomial mapping $x \to \varphi(x^q)$ maps D onto 0. As K is an infinite field, this implies that the same holds true for the extension of that mapping to D_L and E_L, when L is any field containing K. In other words, for all L, $x \to x^q$ maps D_L into its center $L \cdot 1_D$. This is palpably false when L is algebraically closed, for then D_L is isomorphic to an algebra $M_n(L)$, and taking e.g. $x = e_{11}$ in the notation of the proof of th. 1, we have $x^q = e_{11}$, and this is not in the center of $M_n(L)$.

COROLLARY. *Let A be a simple algebra over K, and L a separably algebraically closed field containing K. Then A_L is isomorphic to an algebra $M_n(L)$.*

The assumption means that L has no separably algebraic extension other than itself. Then proposition 5 shows that there is no division

algebra over L, other than L. Our conclusion follows now at once from th. 1, combined with corollary 1 of prop. 3.

§ 2. **The representations of a simple algebra.** Let A be a simple algebra over K; by corollary 3 of prop. 3, § 1, its dimension N over K may be written as $N = n^2$. For any field L containing K, call \mathfrak{M}_L the space of the K-linear mappings of A into $M_n(L)$; every such mapping F can be uniquely extended to an L-linear mapping F_L of A_L into $M_n(L)$. If one takes a basis $\alpha = \{a_1, \ldots, a_N\}$ of A over K, F is uniquely determined by the N matrices $X_i = F(a_i)$, so that, by the choice of this basis, \mathfrak{M}_L is identified with the space of the sets $(X_i)_{1 \leq i \leq N}$ of N matrices in $M_n(L)$, which is obviously of dimension N^2 over L.

By corollary 5 of prop. 3, § 1, a mapping $F \in \mathfrak{M}_L$ is an isomorphism of A into $M_n(L)$, and its extension F_L to A_L is an isomorphism of A_L onto $M_n(L)$, if and only if F is a homomorphism, i.e. if and only if $F(1_A) = 1_n$ and $F(ab) = F(a)F(b)$ for all a, b in A, or, what amounts to the same, for all a, b in the basis α. When that is so, we say that F is an *L-representation* of A; if we write $K(F)$ for the field generated over K by the coefficients of the matrices $F(a)$ for all $a \in A$, or, what amounts to the same, for all $a \in \alpha$, then F is also a $K(F)$-representation of A.

If L is suitably chosen (for instance, by corollary 2·of prop. 3, § 1, if it is algebraically closed, or even, by the corollary of prop. 5, § 1, if it is separably algebraically closed), the set of L-representations of A is not empty. Moreover, if F and F' are in that set, then $F'_L \circ F_L^{-1}$ is an automorphism of $M_n(L)$, hence, by prop. 4 of § 1, of the form $X \to Y^{-1} X Y$ with $Y \in M_n(L)^\times$; this can be written as $F'_L(F_L^{-1}(X)) = Y^{-1} X Y$; for $a \in A$, $X = F(a)$, it implies $F'(a) = Y^{-1} F(a) Y$; we express this by writing $F' = Y^{-1} F Y$. Moreover, when F and F' are given, the corollary of prop. 4, § 1, shows that Y is uniquely determined up to a factor in the center L^\times of $M_n(L)^\times$.

PROPOSITION 6. *Let A be a simple algebra of dimension n^2 over K. Then there is a K-linear form $\tau \neq 0$ and a K-valued function ν on A, such that, if L is any field containing K, and F any L-representation of A, $\tau(a) = \mathrm{tr}(F(a))$ and $\nu(a) = \det(F(a))$ for all $a \in A$; if K is an infinite field, ν is a polynomial function of degree n on A.*

Put $N = n^2$, and take a basis $\{a_1, \ldots, a_N\}$ of A over K. Take first for L a "separable algebraic closure" of K, i.e. the union of all separably algebraic extensions of K in some algebraically closed field containing K; this is always an infinite field. By the corollary of prop. 5, § 1, there is an L-representation F of A, and then, as we have seen above, all such representations can be written as $F' = Y^{-1} F Y$ with $Y \in M_n(L)^\times$. Clearly $a \to \mathrm{tr}(F_L(a))$ is an L-linear form τ on A_L, and $a \to \det(F_L(a))$ is a poly-

nomial function v of degree n on A_L; as F_L is an isomorphism of A_L onto $M_n(L)$, τ is not 0; neither τ nor v is changed if F is replaced by $F' = Y^{-1} F Y$. Writing $a = \sum x_i a_i$ with $x_i \in L$ for $1 \leqslant i \leqslant N$, we can write τ and v as a linear form and as a homogeneous polynomial of degree n, respectively, in the x_i, with coefficients in L. If σ is any automorphism of L over K, we will write τ^σ, v^σ for the polynomials in the x_i, respectively derived from τ, v by substituting for each coefficient its image under σ. Similarly, we write F^σ for the L-representation of A such that, for each a in the basis $\{a_1, \ldots, a_N\}$, $F^\sigma(a)$ is the image $F(a)^\sigma$ of $F(a)$ under σ, i.e. the matrix whose coefficients are respectively the images of those of $F(a)$. Then, clearly, for all $a \in A_L$, $\tau^\sigma(a)$ and $v^\sigma(a)$ are respectively the trace and the determinant of $F^\sigma(a)$; as we have seen above, they must therefore be equal to $\tau(a)$, $v(a)$ for all $a \in A_L$. This implies that all the coefficients in τ and v, when these are written as polynomials in the x_i, are invariant under all automorphisms of L over K, hence that they are in K. This proves our assertion, so far as only L-representations are concerned, with L chosen as above. Obviously it remains true for L'-representations if L' is any field containing L. As every field containing K is isomorphic over K to a subfield of such a field L', this completes the proof.

The functions τ, v defined in proposition 6 are called *the reduced trace* and *the reduced norm* in A. Clearly $\tau(xy) = \tau(yx)$ and $v(xy) = v(x)v(y)$ for all x, y in A; in particular, v determines a morphism of A^\times into K^\times.

COROLLARY 1. *Let A and v be as in proposition 6. Then, for every $a \in A$, the endomorphisms $x \to ax$, $x \to xa$ of the underlying vector-space of A over K have both the determinant $N_{A/K}(a) = v(a)^n$.*

It is clearly enough to verify this for A_L with a suitable L; taking L such that A_L is isomorphic to $M_n(L)$, we see that it is enough to verify it for an algebra $M_n(L)$ over L; but then it is obvious. This is the result announced in the remarks preceding th. 4 of Chap. IV-3.

COROLLARY 2. *Let D be a division algebra over K; let τ_0, v_0 be the reduced trace and the reduced norm in D. For any $m \geqslant 1$, put $A = M_m(D)$, and call τ, v the reduced trace and the reduced norm in A. Then, for every $x = (x_{ij})$ in A, $\tau(x) = \sum_i \tau_0(x_{ii})$; if the matrix $x = (x_{ij})$ in A is triangular, i.e. if $x_{ij} = 0$ for $1 \leqslant j < i \leqslant m$, $v(x) = \prod_i v_0(x_{ii})$.*

Take L such that D has an L-representation F. Then the mapping which, to every matrix $x = (x_{ij})$ in $M_m(D)$, assigns the matrix obtained by substituting the matrix $F(x_{ij})$ for each coefficient x_{ij} in x is an L-representation of A. Using this for defining τ and v, we get at once the conclusion of our corollary.

COROLLARY 3. *Let assumptions and notations be as in corollary* 2. *Then* $v(A^\times) = v_0(D^\times)$.

We may regard A as the ring of endomorphisms of the space $V = D^m$ considered as a left vector-space over D, and consequently A^\times as the group of automorphisms of that space. By an elementary result (already used in the proof of corollary 3 of th. 3, Chap. I-2, but only for a vector-space over a commutative field), every automorphism of V can be written as a product of automorphisms, each of which is either a permutation of the coordinates or of the form

$$(x_1,\ldots,x_m) \to (\sum_i x_i a_i, x_2, \ldots, x_m)$$

with $a_1 \in D^\times$ and $a_i \in D$ for $2 \leqslant i \leqslant m$. By corollary 2, the latter automorphism has the reduced norm $v_0(a_1)$. As to a permutation of coordinates, the same L-representation of A which was used in the proof of corollary 2 shows at once that it has the reduced norm 1 if the dimension d^2 of D over K is even, and ± 1 if it is odd. As $v_0(-1_D) = (-1)^d$, we have thus shown that $v(A^\times)$ contains $v_0(D^\times)$ and is contained in it.

§ 3. Factor-sets and the Brauer group.

Up to an isomorphism, the algebras over a given field K may be regarded as making up a set, since the algebra structures that one can put on a given vector-space over K clearly make up a set, and every such space is isomorphic to K^n for some n.

From now on, we will consider only *simple* algebras over K; it is still understood that they are of finite dimension and central over K. Consider two such algebras A, A'; by th. 1 of § 1, they are isomorphic to algebras $M_n(D)$, $M_{n'}(D')$, where D, D' are division algebras over K which are uniquely determined, up to an isomorphism, by A, A'. One says then that A and A' are *similar*, and that they belong to the same *class*, if D and D' are isomorphic over K. Clearly, in each class of simple algebras, there is, up to an isomorphism, one and only one division algebra, and there is at most one algebra of given dimension over K. An algebra will be called *trivial* over K if it is similar to K, i.e. isomorphic to $M_n(K)$ for some n. We will write $\mathrm{Cl}(A)$ for the class of simple algebras similar to a given one A.

Let A, A' be two simple algebras, respectively isomorphic to $M_n(D)$ and to $M_{n'}(D')$, where D, D' are division algebras over K. By corollary 4 of prop. 3, § 1, $D \otimes D'$ is simple, hence isomorphic to an algebra $M_m(D'')$, where D'' is a division algebra over K which is uniquely determined, up to an isomorphism, by D and D', hence also by A and A'. By the

associativity of tensor-products, $A \otimes A'$ is isomorphic to $M_{nn'm}(D'')$. This shows that the class of $A \otimes A'$ is uniquely determined by those of A and A'. Write now:

$$\mathrm{Cl}(A \otimes A') = \mathrm{Cl}(A) \cdot \mathrm{Cl}(A'),$$

and consider this as a law of composition in the set of classes of simple algebras over K. It is clearly associative and commutative; it has a neutral element, viz., the class $\mathrm{Cl}(K)$ of trivial algebras over K. Moreover, if A^0 is the inverse algebra to A, prop. 3 of § 1 shows that $A \otimes A^0$ is trivial, so that $\mathrm{Cl}(A^0)$ is the inverse of $\mathrm{Cl}(A)$ for our law of composition. Therefore, for this law, the classes of simple algebras over K make up a group; this is known as *the Brauer group of* K; we will denote it by $B(K)$. If K' is any field containing K, and A a simple algebra over K, it is obvious that the class of $A_{K'}$ is determined uniquely by that of A, and that the mapping $\mathrm{Cl}(A) \rightarrow \mathrm{Cl}(A_{K'})$ is a morphism of $B(K)$ into $B(K')$, which will be called *the natural morphism* of $B(K)$ into $B(K')$.

It will now be shown that the Brauer group can be defined in another way, by means of "factor-sets"; this will require some preliminary definitions. We choose once for all an algebraic closure \bar{K} for K; we will denote by K_{sep} the maximal separable extension of K in \bar{K}, i.e. the union of all separable extensions of K of finite degree, contained in \bar{K}. We will denote by \mathfrak{G} the Galois group of K_{sep} over K, topologized as usual by taking, as a fundamental system of neighborhoods of the identity ε, all the subgroups of \mathfrak{G} attached to separable extensions of K of finite degree. Clearly this makes \mathfrak{G} into a totally disconnected compact group. As \bar{K} is purely inseparable over K_{sep}, each automorphism of K_{sep} can be uniquely extended to one of \bar{K}, so that \mathfrak{G} may be identified with the group of all automorphisms of \bar{K} over K.

DEFINITION 2. *Let* $\mathfrak{G}^{(m)}$ *be the product* $\mathfrak{G} \times \cdots \times \mathfrak{G}$ *of m factors equal to* \mathfrak{G}; *let* \mathfrak{H} *be an open subgroup of* \mathfrak{G}. *Then a mapping f of* $\mathfrak{G}^{(m)}$ *into any set S will be called* \mathfrak{H}-*regular if it is constant on left cosets in* $\mathfrak{G}^{(m)}$ *with respect to* $\mathfrak{H}^{(m)}$.

This amounts to saying that $f(\sigma_1, \ldots, \sigma_m)$ depends only upon the left cosets $\mathfrak{H}\sigma_1, \ldots, \mathfrak{H}\sigma_m$ determined by the σ_i in \mathfrak{G}. When that is so, f is locally constant, or, what amounts to the same, it is continuous when S is provided with the discrete topology. Conversely, let f be a mapping of $\mathfrak{G}^{(m)}$ into S; if it is locally constant, it is continuous if S is topologized discretely, hence uniformly continuous since \mathfrak{G} is compact; this implies that there is an open subgroup \mathfrak{H} of \mathfrak{G} such that f is \mathfrak{H}-regular.

DEFINITION 3. *Let* $\mathfrak{G}^{(m)}$ *be as in definition* 2. *Then a mapping* f *of* $\mathfrak{G}^{(m)}$ *into* $M_n(K_{\mathrm{sep}})$, *for any* $n \geqslant 1$, *will be called covariant if it is locally constant and satisfies the condition*

$$f(\sigma_1 \lambda, \ldots, \sigma_m \lambda) = f(\sigma_1, \ldots, \sigma_m)^{\lambda}$$

for all $\sigma_1, \ldots, \sigma_m, \lambda$ *in* \mathfrak{G}.

LEMMA 1. *Let* \mathfrak{H} *be an open subgroup of* \mathfrak{G}; *let* L *be the subfield of* K_{sep}, *consisting of the elements invariant under* \mathfrak{H}. *Then an* \mathfrak{H}-*regular mapping of* \mathfrak{G} *into* K_{sep} *is covariant if and only if it is of the form* $\sigma \to \xi^{\sigma}$, *with* $\xi \in L$.

Let x, i.e. $\sigma \to x(\sigma)$, be a mapping of \mathfrak{G} into K_{sep}, and put $\xi = x(\varepsilon)$. If x is covariant, we have $x(\sigma) = \xi^{\sigma}$ for all σ; if this is \mathfrak{H}-regular, ξ must be in L. The converse is obvious.

LEMMA 2. *Let* \mathfrak{H} *be an open subgroup of* \mathfrak{G}. *Call* X_m *the space of* \mathfrak{H}-*regular covariant mappings of* $\mathfrak{G}^{(m)}$ *into* K_{sep}, *regarded as a vector-space over* K; *call* X'_m *the space of all* \mathfrak{H}-*regular mappings of* $\mathfrak{G}^{(m)}$ *into* K_{sep}, *regarded as a vector-space over* K_{sep}. *Then* $X'_m = X_m \otimes_K K_{\mathrm{sep}}$, *and the dimension of* X_m *over* K, *and of* X'_m *over* K_{sep}, *is* n^m, *if* n *is the index of* \mathfrak{H} *in* \mathfrak{G}.

Let L be as in lemma 1; it has the degree n over K. Take a full set $\mathfrak{a} = \{\alpha_1, \ldots, \alpha_n\}$ of representatives of the cosets $\mathfrak{H}\alpha$ of \mathfrak{H} in \mathfrak{G}; then the isomorphisms $\lambda_1, \ldots, \lambda_n$, respectively induced on L by the α_i, are the n distinct K-linear isomorphisms of L into K_{sep}. Any mapping $x \in X'_m$ is uniquely determined by its values on $\mathfrak{a} \times \cdots \times \mathfrak{a}$, and these can be chosen arbitrarily; therefore X'_m has the dimension n^m over K_{sep}, and every linear form L on X'_m can be written as

$$L(x) = \sum_{(i)} a_{i_1 \cdots i_m} x(\alpha_{i_1}, \ldots, \alpha_{i_m})$$

with coefficients $a_{(i)}$ in K_{sep}. Now we proceed by induction on m. For $m = 1$, lemma 1 shows that X_1, as a vector-space over K, is isomorphic to L, hence of dimension n, so that we need only show that X_1 generates X'_1 as a vector-space over K_{sep}. If not, there would be a linear form L on X'_1, other than 0, which would be 0 on X_1; writing L as above, and making use of lemma 1, we get $0 = \sum a_i \xi^{\lambda_i}$ for all $\xi \in L$; this contradicts the linear independence of the λ_i over K_{sep}, i.e. corollary 3 of prop. 3, Chap. III-2. Now, for any m, consider the tensor-product $Y_m = X_1 \otimes \cdots \otimes X_1$, taken over K, of m factors equal to X_1, and the similar product $Y'_m = X'_1 \otimes \cdots \otimes X'_1$ taken over K_{sep}; as we have just shown that X'_1 is the same as $X_1 \otimes_K K_{\mathrm{sep}}$, we may, in an obvious manner, identify Y'_m with $Y_m \otimes_K K_{\mathrm{sep}}$. Call φ the K_{sep}-linear mapping of Y'_m into X'_m which, to every element $x_1 \otimes \cdots \otimes x_m$ of Y'_m, assigns the mapping

$$(\sigma_1,\ldots,\sigma_m)\to x_1(\sigma_1)\ldots x_m(\sigma_m)$$

of $\mathfrak{G}^{(m)}$ into K_{sep}. This is surjective; for, if a linear form L on X'_m is 0 on $\varphi(Y'_m)$, we must have, for all x_1,\ldots,x_m in X'_1:

$$0=\sum_{(i)}a_{i_1\cdots i_m}x_1(\alpha_{i_1})\ldots x_m(\alpha_{i_m}),$$

which clearly implies that all the $a_{(i)}$ are 0. As Y'_m has the same dimension n^m as X'_m, this shows that φ is an isomorphism of Y'_m onto X'_m. Now take a basis $\{f_1,\ldots,f_n\}$ of X_1 over K. Then the n^m elements $f_{i_1}\otimes\cdots\otimes f_{i_m}$ make up a basis of Y_m over K, hence also of Y'_m over K_{sep}, so that their images under φ make up a basis of X'_m over K_{sep}. This amounts to saying that every element of X'_m can be written uniquely in the form

$$(\sigma_1,\ldots,\sigma_m)\to\sum_{(i)}x_{i_1\cdots i_m}f_{i_1}(\sigma_1)\ldots f_{i_m}(\sigma_m)$$

with coefficients $x_{(i)}$ in K_{sep}. Writing now that this is in X_m, i.e. that it is covariant, we see that this is so if and only if all the $x_{(i)}$ are invariant under \mathfrak{G}, i.e. if and only if they are all in K. Therefore φ maps Y_m onto X_m. This completes the proof.

Let now K' be any field containing K, and let \bar{K}', K'_{sep}, \mathfrak{G}' be defined for K' as \bar{K}, K_{sep}, \mathfrak{G} have been defined for K. As \bar{K} is determined only up to an isomorphism, we will always assume, in such a situation, that we have taken for \bar{K} the algebraic closure of K in \bar{K}'. It is obvious that K_{sep} is then contained in K'_{sep}. Every automorphism σ' of \bar{K}' over K' induces on \bar{K} an automorphism σ of \bar{K} over K (more precisely, over $\bar{K}\cap K'$); clearly the mapping $\sigma'\to\sigma$ is a continuous morphism ρ of \mathfrak{G}' into \mathfrak{G}; this will be called the *restriction morphism*; it is injective if K' is algebraic over K, since then $\bar{K}'=\bar{K}$; in that case one will usually identify \mathfrak{G}' with its image in \mathfrak{G}, which is always a closed subgroup of \mathfrak{G}, and is open in \mathfrak{G} when K' is of finite degree over K. If \mathfrak{H} is any open subgroup of \mathfrak{G}, and L is the corresponding subfield of K_{sep}, i.e. the one consisting of the elements invariant under \mathfrak{H}, the subgroup $\mathfrak{H}'=\rho^{-1}(\mathfrak{H})$ of \mathfrak{G}' is open, and the corresponding subfield of K'_{sep} is the one generated by L over K'.

Let notations be as above, and let f be as in definition 2, i.e. a mapping of $\mathfrak{G}^{(m)}$ into some set S. We will write $f\circ\rho$ for the mapping

$$(\sigma'_1,\ldots,\sigma'_m)\to f(\rho(\sigma'_1),\ldots,\rho(\sigma'_m))$$

of $\mathfrak{G}'^{(m)}$ into S. This is obviously continuous, i.e. locally constant, if f is so; if f is \mathfrak{H}-regular, it is \mathfrak{H}'-regular, with $\mathfrak{H}'=\rho^{-1}(\mathfrak{H})$; if $S=M_n(K_{\mathrm{sep}})$, and f is covariant, $f\circ\rho$ is covariant. If K' is algebraic over K, \mathfrak{G}' is a subgroup of \mathfrak{G}, and ρ is its natural injection into \mathfrak{G}; then $f\circ\rho$ is the restriction of f to $\mathfrak{G}'^{(m)}$.

After these preliminaries, we can now go back to our main topic.

THEOREM 2. *Let A be a simple algebra of dimension n^2 over K. Let \mathfrak{H} be an open subgroup of \mathfrak{G}, L the corresponding subfield of K_{sep}, and F an L-representation of A. Then there is an \mathfrak{H}-regular covariant mapping Y of $\mathfrak{G} \times \mathfrak{G}$ into $M_n(K_{sep})^\times$, such that $F^\sigma = Y(\rho,\sigma)^{-1} F^\rho Y(\rho,\sigma)$ for all ρ, σ in \mathfrak{G}; if Y is such, there is an \mathfrak{H}-regular covariant mapping f of $\mathfrak{G} \times \mathfrak{G} \times \mathfrak{G}$ into K_{sep}^\times such that, for all ρ, σ, τ in \mathfrak{G}:*

(1) $$f(\rho,\sigma,\tau) Y(\rho,\tau) = Y(\rho,\sigma) Y(\sigma,\tau),$$

and this satisfies the condition

(2) $$f(\rho,\sigma,\tau) f(\nu,\rho,\tau) = f(\nu,\sigma,\tau) f(\nu,\rho,\sigma)$$

for all ν,ρ,σ,τ in \mathfrak{G}.

For every $\lambda \in \mathfrak{G}$, F^λ is a K_{sep}-representation of A, hence of the form $Z(\lambda)^{-1} F Z(\lambda)$, with $Z(\lambda) \in M_n(K_{sep})^\times$. As F^λ depends only upon the left coset $\mathfrak{H}\lambda$, we may, to begin with, assume that $\lambda \to Z(\lambda)$ is \mathfrak{H}-regular; it would then be easy to verify that $Y(\rho,\sigma) = Z(\sigma\rho^{-1})^\rho$ satisfies all the conditions of the first part of our theorem, except possibly that of the \mathfrak{H}-regularity. To obtain this, we refine our construction as follows. Take a full set Λ of representatives of the double cosets $\mathfrak{H}\lambda\mathfrak{H}$ in \mathfrak{G} with respect to \mathfrak{H}. For each $\lambda \in \Lambda$, F and F^λ are both L'-representations, if L' is the compositum $L \cdot L^\lambda$ of L and its image L^λ under λ; choose then $Z(\lambda)$ in $M_n(L')^\times$, so that $F^\lambda = Z(\lambda)^{-1} F Z(\lambda)$. Each $\rho \in \mathfrak{G}$ can be written as $\rho = \alpha\lambda\beta$, with a uniquely determined $\lambda \in \Lambda$ and with α, β in \mathfrak{H}. If at the same time we have $\rho = \alpha'\lambda\beta'$ with α', β' in \mathfrak{H}, then:

$$\beta'\beta^{-1} = \lambda^{-1}(\alpha'^{-1}\alpha)\lambda,$$

so that, if we put $\gamma = \beta'\beta^{-1}$, γ is both in \mathfrak{H} and in $\lambda^{-1}\mathfrak{H}\lambda$, which implies that it leaves fixed all the elements of L and of L^λ, hence $L \cdot L^\lambda$ and $Z(\lambda)$. Therefore, if we put $Z(\rho) = Z(\lambda)^\beta$, this depends only upon ρ, not upon the choice of α, β, subject to the conditions stated above. It is then easy to check that $Y(\rho,\sigma) = Z(\sigma\rho^{-1})^\rho$ satisfies all the conditions stated in our theorem. Now, for all ρ,σ,τ:

$$F^\tau = Y(\sigma,\tau)^{-1} F^\sigma Y(\sigma,\tau) = Y(\sigma,\tau)^{-1} Y(\rho,\sigma)^{-1} F^\rho Y(\rho,\sigma) Y(\sigma,\tau).$$

At the same time, we have $F^\tau = Y(\rho,\tau)^{-1} F^\rho Y(\rho,\tau)$. As we have observed above, this implies, by the corollary of prop. 4, § 1, that $Y(\rho,\sigma) Y(\sigma,\tau)$ differs from $Y(\rho,\tau)$ only by a scalar factor $f(\rho,\sigma,\tau)$, which proves (1). One can then verify (2) by a straightforward calculation, and the remaining assertions are obvious.

COROLLARY. *Let assumptions and notations be as in theorem* 2; *let K' be a field containing K*; \mathfrak{G}' *being as before, let* ρ *be the restriction morphism of* \mathfrak{G}' *into* \mathfrak{G}. *Let* $F_{K'}$ *be the K'-linear extension of F to* $A_{K'}$. *Then* $Y \circ \rho$ *and* $f \circ \rho$ *are related to* $F_{K'}$ *in the same manner as Y and f to F.*

This is obvious. When Y and f are related to an L-representation F of A in the manner described in theorem 2, we will say that they *belong to A.*

DEFINITION 4. *A covariant mapping f of* $\mathfrak{G} \times \mathfrak{G} \times \mathfrak{G}$ *into* K_{sep}^{\times} *is called a factor-set of K if it satisfies* (2) *for all* v, ρ, σ, τ *in* \mathfrak{G}.

Clearly the factor-sets of K make up a group $\zeta(K)$ under multiplication. If K', \mathfrak{G}' and ρ are as above, $f \to f \circ \rho$ is obviously a morphism of $\zeta(K)$ into $\zeta(K')$.

Let z be any covariant mapping of $\mathfrak{G} \times \mathfrak{G}$ into K_{sep}^{\times}. Obviously the mapping

$$(3) \qquad (\rho, \sigma, \tau) \to z(\rho, \sigma) z(\sigma, \tau) z(\rho, \tau)^{-1}$$

is covariant, and one verifies immediately that it is a factor-set.

DEFINITION 5. *The factor-set defined by* (3) *will be called the coboundary of z; a factor-set of K will be called trivial if it is the coboundary of a covariant mapping of* $\mathfrak{G} \times \mathfrak{G}$ *into* K_{sep}^{\times}.

The trivial factor-sets make up a subgroup $\beta(K)$ of the group $\zeta(K)$ of all factor-sets of K. The quotient $\zeta(K)/\beta(K)$ will be denoted by $H(K)$, and its elements, i.e. the classes modulo $\beta(K)$ in $\zeta(K)$, will be called the *factor-classes* of K. If K' and ρ are again as before, it is obvious that $f \to f \circ \rho$ maps coboundaries into coboundaries, so that ρ determines a morphism of $H(K)$ into $H(K')$, which we again denote by ρ.

PROPOSITION 7. *The factor-sets belonging to a simple algebra A over K make up a factor-class of K.*

Let \mathfrak{H}, L, F, Y and f be as in th. 2; let z be any covariant mapping of $\mathfrak{G} \times \mathfrak{G}$ into K_{sep}^{\times}; let \mathfrak{H}' be an open subgroup of \mathfrak{H} such that z is \mathfrak{H}'-regular; let L' be the subfield of K_{sep} corresponding to \mathfrak{H}'. Then F is also an L'-representation; $Y' = zY$ is related to F in the same manner as Y, and it determines the factor-set $f' = f_0 f$, where f_0 is the coboundary of z. This shows that all the factor-sets in the class determined by f belong to A. On the other hand, let \mathfrak{H}', L', F', Y', f' be related to A in the same manner as \mathfrak{H}, L, F, Y and f. Put $\mathfrak{H}'' = \mathfrak{H} \cap \mathfrak{H}'$, and call L'' the corresponding subfield of K_{sep}, which is the compositum of L and L'.

Then there is $Z \in M_n(L')^{\times}$ such that $F' = Z^{-1} F Z$. A trivial calculation gives now $F'^{\sigma} = W^{-1} F'^{\rho} W$ with $W = (Z^{\rho})^{-1} Y(\rho, \sigma) Z^{\sigma}$, so that $Y'(\rho, \sigma)$ can differ from W only by a scalar factor. If we write $z(\rho, \sigma)$ for this factor, we have now

$$Y'(\rho, \sigma) = z(\rho, \sigma)(Z^{\rho})^{-1} Y(\rho, \sigma) Z^{\sigma},$$

which implies that z is \mathfrak{H}''-regular and covariant. Then $f' f^{-1}$ is the coboundary of z, which completes the proof.

COROLLARY. *Let K' be a field containing K; then the factor-class of K' determined by $A_{K'}$ is the image of the factor-class of K determined by A under the restriction morphism ρ of \mathfrak{G}' into \mathfrak{G}.*

This is obvious in view of the corollary of th. 2.

If A is a simple algebra over K, the factor-class of K, consisting of the factor-sets belonging to A, will be said to *belong to A* or to be *attached to A*.

THEOREM 3. *The mapping which, to every simple algebra A over K, assigns the factor-class of K attached to A, is constant on classes of simple algebras over K and determines an isomorphism of the group $B(K)$ of such classes onto the group $H(K)$ of factor-classes of K.*

Take first two simple algebras A, A' over K; call n^2, n'^2 their dimensions over K. Let L, F, Y and f be defined for A as in th. 2, and let L', F', Y', f' be similarly related to A'; call L'' the compositum of L and L'. We may identify $M_n(L'') \otimes M_{n'}(L'')$ with $M_{nn'}(L'')$. Then, if we put $A'' = A \otimes A'$, and if we write $F'' = F \otimes F'$ for the K-linear mapping of A'' into $M_{nn'}(L'')$ given by $F''(a \otimes a') = F(a) \otimes F'(a')$ for all $a \in A$ and $a' \in A'$, F'' is an L''-representation of A'', and one sees at once that $Y'' = Y \otimes Y'$ and $f'' = f f'$ are related to A'' and F'' as in th. 2. This shows that the factor-class attached to A'' is the product of those attached to A and to A'. If $A = M_n(K)$, one can take for F the identity mapping of A onto $M_n(K)$ and then take $Y = 1$, hence $f = 1$; therefore the factor-class attached to a trivial algebra is the trivial one, and the factor-classes attached to A' and to $M_n(A')$ are the same. This proves the first assertion in our theorem and shows that the mapping μ of $B(K)$ into $H(K)$ which is thus defined is a morphism. It will now be shown, firstly that μ is injective, and then that it is surjective; this will be done in several steps, which we formulate as lemmas.

LEMMA 3. *Let \mathfrak{H} be an open subgroup of \mathfrak{G}, and L the subfield of K_{sep} corresponding to \mathfrak{H}. Let Y be an \mathfrak{H}-regular covariant mapping of $\mathfrak{G} \times \mathfrak{G}$ into $M_n(K_{\text{sep}})^{\times}$, such that $Y(\rho, \tau) = Y(\rho, \sigma) Y(\sigma, \tau)$ for all ρ, σ, τ in \mathfrak{G}. Then there is $Z \in M_n(L)^{\times}$ such that $Y(\rho, \sigma) = (Z^{\rho})^{-1} Z^{\sigma}$ for all ρ, σ in \mathfrak{G}.*

Take a full set \mathfrak{a} of representatives of the cosets $\mathfrak{H}\alpha$ of \mathfrak{H} in \mathfrak{G}; as we have observed in the proof of lemma 2, they induce on L all the distinct K-linear isomorphisms of L into K_{sep}, and these are linearly independent over K_{sep}, as has been shown in corollary 3 of prop. 3, Chap. III-2. Let $M_n(K_{\text{sep}})$ operate on the right, by matrix multiplication, on the space $M_{1,n}(K_{\text{sep}})$ of row vectors over K_{sep}, and similarly on the left on column vectors. For each $u \in M_{1,n}(L)$, put

$$z = \sum_{\alpha \in \mathfrak{a}} u^\alpha Y(\alpha, \varepsilon).$$

For any $\rho \in \mathfrak{G}$, $\mathfrak{a}\rho$ is again a full set of representatives of the cosets of \mathfrak{H} in \mathfrak{G}; as Y is covariant, we have then

$$z^\rho = \sum_\alpha u^{\alpha\rho} Y(\alpha\rho, \rho) = \sum_\alpha u^\alpha Y(\alpha, \rho);$$

for $\rho \in \mathfrak{H}$, this shows that z is invariant under \mathfrak{H}, i.e. that it is in $M_{1,n}(L)$. Therefore, if we write φ for the mapping $u \to z$ defined above, φ maps $M_{1,n}(L)$ into itself. Now we show that there are n vectors u_1, \ldots, u_n in $M_{1,n}(L)$, such that the vectors $\varphi(u_i)$ are linearly independent over L. In fact, if this were not so, there would be a column vector v in $M_{n,1}(L)$, other than 0, such that $\varphi(u)v = 0$ for all $u \in M_{1,n}(L)$. This can be written as $\sum u^\alpha (Y(\alpha, \varepsilon)v) = 0$, which, in view of the linear independence of the α on L, implies $Y(\alpha, \varepsilon)v = 0$ for all α; hence $v = 0$. Choose now n vectors u_i such that the $\varphi(u_i)$ are linearly independent over L; call U the matrix in $M_n(L)$ whose rows are the u_i, and put $Z = \sum U^\alpha Y(\alpha, \varepsilon)$. As the rows of Z are the $\varphi(u_i)$, Z is invertible in $M_n(L)$. Just as above, we have, for all ρ, σ:

$$Z^\rho = \sum_\alpha U^\alpha Y(\alpha, \rho), \qquad Z^\sigma = \sum_\alpha U^\alpha Y(\alpha, \sigma),$$

and therefore $Z^\sigma = Z^\rho Y(\rho, \sigma)$ in view of the assumption on Y. This shows that Z has the property stated in our lemma.

It is now easy to show that the morphism μ of $B(K)$ into $H(K)$ which has been defined above is injective. In fact, assume a simple algebra A over K to have a trivial factor-set; in view of prop. 7, this implies that we can choose \mathfrak{H}, L, F and Y as in th. 2, so that (1) holds with $f = 1$. Let now Z be as in lemma 3, and put $F' = ZFZ^{-1}$; then one sees at once that $F'^\sigma = F'^\rho$ for all ρ, σ in \mathfrak{G}. This means that F' is a K-representation of A, i.e. an isomorphism of A onto $M_n(K)$, so that A is trivial.

Finally, the surjectivity of μ is contained in the following more precise result:

LEMMA 4. *Let \mathfrak{H} and L be as in lemma 3; call n the degree of L over K. Let f be an \mathfrak{H}-regular factor-set of K. Then one can choose A, F and Y with the properties described in theorem 2, so that the factor-set defined by (1) is the given one and that A contains a subfield isomorphic to L.*

* This will be proved by an explicit construction, due to R. Brauer. We first observe that, if we take $v = \rho = \sigma$ in the formula (2) of th. 2 which defines factor-sets, we get $f(\rho,\rho,\tau) = f(\rho,\rho,\rho)$; as f is covariant, this gives $f(\rho,\rho,\tau) = a^\rho$ with $a = f(\varepsilon,\varepsilon,\varepsilon)$. Now apply lemma 2 to the case $m = 2$; this gives two spaces X_2, X_2' of dimension n^2, over K and over K_{sep} respectively, and $X_2' = X_2 \otimes_K K_{sep}$. Take a full set \mathfrak{a} of representatives of the cosets $\mathfrak{H}\alpha$ of \mathfrak{H} in \mathfrak{G}. For any x, y in X_2', and any ρ, σ in \mathfrak{G}, put

$$z(\rho,\sigma) = \sum_{\alpha \in \mathfrak{a}} f(\sigma,\alpha,\rho) x(\rho,\alpha) y(\alpha,\sigma).$$

Clearly z, i.e. the mapping $(\rho,\sigma) \to z(\rho,\sigma)$, is in X_2', and it is in X_2 if x, y are in X_2; more precisely, $(x,y) \to z$ is a bilinear mapping of $X_2' \times X_2'$ into X_2' which induces on $X_2 \times X_2$ a bilinear mapping of $X_2 \times X_2$ into X_2. It will now be shown that, if we write this as $(x,y) \to xy$, it makes X_2 into an algebra A with the required properties. In fact, for each $\rho \in \mathfrak{G}$, and each $x \in X_2'$, put

$$\Phi_\rho(x) = (f(\beta,\alpha,\rho) x(\alpha,\beta))_{\alpha,\beta \in \mathfrak{a}}.$$

After choosing an ordering on the set \mathfrak{a}, we may identify mappings of $\mathfrak{a} \times \mathfrak{a}$ into K_{sep} with the matrices in $M_n(K_{sep})$; then each Φ_ρ may be regarded as a mapping of X_2' into $M_n(K_{sep})$; as such, it is obviously K_{sep}-linear and bijective. Using formula (2), one verifies at once that $\Phi_\rho(xy) = \Phi_\rho(x)\Phi_\rho(y)$ for all x, y in X_2'. Call e the element of X_2' given by $e(\rho,\sigma) = (a^\rho)^{-1}$, with $a = f(\varepsilon,\varepsilon,\varepsilon)$, whenever σ is in the same coset $\mathfrak{H}\rho$ as ρ, and $e(\rho,\sigma) = 0$ otherwise. Clearly e is in X_2; since $f(\alpha,\alpha,\rho) = a^\alpha$ for all α,ρ, we have $\Phi_\rho(e) = 1_n$. It is now obvious that, for each ρ, Φ_ρ maps X_2', with the multiplication $(x,y) \to xy$, isomorphically onto the algebra $M_n(K_{sep})$, the unit of the former being e. As $X_2' = X_2 \otimes_K K_{sep}$, this implies, by corollary 1 of prop. 3, § 1, that this multiplication makes X_2 into a simple algebra A over K, with $1_A = e$. For any $\xi \in L$, and any $x \in A$, write ξx for the element of X_2 given by $(\rho,\sigma) \to \xi^\rho x(\rho,\sigma)$; this defines on X_2 a structure of left vector-space over L; moreover, it is clear that $(\xi x)y = \xi(xy)$ for all $\xi \in L$ and all x, y in A; therefore $\xi \to \xi e$ is an isomorphism of L into A, and $\xi x = (\xi e)x$ for all $\xi \in L$ and all $x \in A$.

We will now construct F and Y with the properties stated in our lemma. For all ρ, σ in \mathfrak{G}, call $D(\rho,\sigma)$ the diagonal matrix given by

$$D(\rho,\sigma) = (\delta_{\alpha\beta} f(\alpha,\rho,\sigma))_{\alpha,\beta \in \mathfrak{a}}$$

with $\delta_{\alpha\beta} = 1$ or 0 according as $\alpha = \beta$ or not. Using (2), one verifies at once that one has, for all ρ,σ, and all $x \in X_2'$:

(4) $D(\rho,\sigma)\Phi_\sigma(x) = \Phi_\rho(x)D(\rho,\sigma).$

Now choose a basis $\{\xi_1,...,\xi_n\}$ of L over K, and call $\{\eta_1,...,\eta_n\}$ the dual basis to this when L is identified with its own dual by putting $[\xi,\eta] = = Tr_{L/K}(\xi\eta)$. As the elements α of \mathfrak{a} induce on L the n distinct K-linear isomorphisms of L into K_{sep}, we have, for every $\xi\in L$, $Tr_{L/K}(\xi) = \sum_\alpha \xi^\alpha$, so that the definition of the η_i may be written as

$$\delta_{ij} = Tr_{L/K}(\xi_i\eta_j) = \sum_{\alpha\in\mathfrak{a}} \xi_i^\alpha \eta_j^\alpha.$$

Therefore we may put:

$$X = (\xi_i^\alpha)_{1 \leq i \leq n; \alpha\in\mathfrak{a}}, \quad X^{-1} = (\eta_i^\alpha)_{\alpha\in\mathfrak{a}; 1 \leq i \leq n}.$$

Write now, for each ρ, $F_\rho = X \Phi_\rho X^{-1}$; this gives

$$F_\rho(x) = \Big(\sum_{\alpha,\beta\in\mathfrak{a}} \xi_i^\alpha f(\beta,\alpha,\rho) x(\alpha,\beta) \eta_j^\beta \Big)_{1 \leq i,j \leq n}.$$

Assume that $x\in A$, i.e. that it is covariant; as f is covariant, and as $\mathfrak{a}\lambda$, for every λ, is a full set of representatives of the cosets of \mathfrak{H} in \mathfrak{G}, we have

$$F_\rho(x)^\lambda = \Big(\sum_{\alpha,\beta\in\mathfrak{a}} \xi_i^\alpha f(\beta,\alpha,\rho\lambda) x(\alpha,\beta) \eta_j^\beta \Big)_{1 \leq i,j \leq n} = F_{\rho\lambda}(x).$$

In particular, if we put $F = F_\varepsilon$, we have, for every ρ, $F_\rho = F^\rho$. By the definition of F_ρ, this gives $F^\rho = F$ for $\rho\in\mathfrak{H}$, i.e. $F(x)^\rho = F(x)$ for every $x\in A$ and every $\rho\in\mathfrak{H}$; in other words, F maps A into $M_n(L)$, so that it is an L-representation of A. For all $\rho\in\mathfrak{G}$, we have $F^\rho = X \Phi_\rho X^{-1}$. In view of (4), this gives $F^\sigma = Y(\rho,\sigma)^{-1} F^\rho Y(\rho,\sigma)$, where we have put

$$Y(\rho,\sigma) = X D(\rho,\sigma) X^{-1} = \Big(\sum_{\alpha\in\mathfrak{a}} \xi_i^\alpha f(\alpha,\rho,\sigma) \eta_j^\alpha \Big)_{1 \leq i,j \leq n}.$$

One can now verify at once that this, together with A and F, is as required by our lemma. We also note for future use that the reduced trace τ and the reduced norm ν in A can be calculated by means of any one of the K_{sep}-representations Φ_ρ of A, e.g. from Φ_ε; this gives, for all $x\in A$, $\tau(x) = \mathrm{tr}(\Phi_\varepsilon(x))$, $\nu(x) = \det(\Phi_\varepsilon(x))$, and in particular:

$$\tau(\xi\cdot 1_A) = Tr_{L/K}(\xi), \quad \nu(\xi\cdot 1_A) = N_{L/K}(\xi)$$

for all $\xi\in L$.

With lemma 4, the proof of theorem 3 is now complete.

COROLLARY 1. *Let K' be a field containing K, and ρ the restriction morphism of \mathfrak{G}' into \mathfrak{G}. Let A be a simple algebra over K, and f a factor-set belonging to A. Then $A_{K'}$ is trivial if and only if $f\circ\rho$ is so.*

This follows at once from theorem 3 and the corollary of th. 2.

COROLLARY 2. *Let \mathfrak{H} be an open subgroup of \mathfrak{G}, and f an \mathfrak{H}-regular factor-set of K. Then f is trivial if and only if it is the coboundary of an \mathfrak{H}-regular covariant mapping of $\mathfrak{G} \times \mathfrak{G}$ into $K_{\mathrm{sep}}^{\times}$.*

Assume that f is trivial; construct A, F and Y as in lemma 4. By theorem 3, A is trivial, so that there is an isomorphism F' of A onto $M_n(K)$. Then $F = Z^{-1}F'Z$, with some $Z \in M_n(L)^{\times}$, hence $F^{\sigma} = Y'(\rho,\sigma)^{-1}F^{\rho}Y'(\rho,\sigma)$ with $Y'(\rho,\sigma) = (Z^{\rho})^{-1}Z^{\sigma}$. This implies $Y(\rho,\sigma) = z(\rho,\sigma)Y'(\rho,\sigma)$, where z is \mathfrak{H}-regular and covariant. Then f is the coboundary of z.

COROLLARY 3. *Let L be a separable extension of K of degree n. Let A be a simple algebra over K. Then A_L is trivial if and only if there is an algebra A' of dimension n^2 over K, similar to A, containing a subfield isomorphic to L; when A' exists, it is unique, up to an isomorphism.*

The last assertion is obvious. By corollary 6 of prop. 3, § 1, the existence of A' implies the triviality of A'_L, hence that of A_L. Conversely, assume that there is an isomorphism of A_L onto a matrix algebra $M_n(L)$; this induces on A an L-representation F. By th. 2, we can construct an \mathfrak{H}-regular factor-set f belonging to A. Then, by lemma 4, we can construct an algebra A' such as required our corollary.

It is frequently convenient to identify the groups $B(K)$ and $H(K)$ by means of the isomorphism μ described in theorem 3. If this is done, and if K' is any field containing K, the corollaries of th. 2 and of prop. 7 show that the natural morphism of $B(K)$ into $B(K')$, which maps the class of every simple algebra A over K onto the class of $A_{K'}$, coincides with the restriction morphism ρ of $H(K)$ into $H(K')$.

§ 4. Cyclic factor-sets.

We will now discuss in greater detail a type of factor-sets of particular importance, attached to the cyclic extensions of the groundfield K. Here, as always, we understand "cyclic" as meaning a Galois extension (hence, by definition, a separable one) with a finite cyclic Galois group. With the same notations as in § 3, the cyclic extensions of K are the subfields L of K corresponding to the open subgroups \mathfrak{H} of \mathfrak{G} with cyclic factor-group. If L and \mathfrak{H} are such, and if n is the degree of L over K, $\mathfrak{G}/\mathfrak{H}$ is isomorphic to the group of the n-th roots of 1 in \mathbf{C}; any isomorphism of $\mathfrak{G}/\mathfrak{H}$ onto the latter group may be regarded as a character χ of \mathfrak{G}, with the kernel \mathfrak{H}; such a character, which is of order n, will be said to be *attached* to L. If α is a representative in \mathfrak{G} of a generator of $\mathfrak{G}/\mathfrak{H}$, there is one and only one character χ of \mathfrak{G}, attached to L, such that $\chi(\alpha) = e(1/n)$.

Conversely, let χ be any homomorphism of \mathfrak{G} into \mathbf{C}^{\times}; by lemmas 3 and 4 of Chap. VII-3, it is a character of \mathfrak{G} of finite order n; its kernel \mathfrak{H}

is then an open subgroup of \mathfrak{G}, with a cyclic factor-group of order n, and the subfield L of K_{sep} corresponding to \mathfrak{H} is cyclic of degree n over K; we will then say that L is *attached to* χ.

Let notations be as above; as χ is locally constant on \mathfrak{G}, one can choose, in infinitely many ways, a locally constant mapping Φ of \mathfrak{G} into \mathbf{R} such that $\chi(\sigma) = \mathbf{e}(\Phi(\sigma))$ for all $\sigma \in \mathfrak{G}$. For instance, one may choose Φ so that $0 \leqslant \Phi(\sigma) < 1$ for all σ; if Φ is chosen according to this condition, it is determined uniquely, and it is \mathfrak{H}-regular, since χ is so. In any case, since χ is of order n, Φ maps \mathfrak{G} into $(1/n)\mathbf{Z}$. Consider now the mapping

$$(5) \qquad (\rho, \sigma, \tau) \to e(\rho, \sigma, \tau) = \Phi(\sigma \rho^{-1}) + \Phi(\tau \sigma^{-1}) - \Phi(\tau \rho^{-1})$$

of $\mathfrak{G} \times \mathfrak{G} \times \mathfrak{G}$ into \mathbf{R}; as Φ is locally constant, this is so; as χ is a character, one sees at once that e maps $\mathfrak{G} \times \mathfrak{G} \times \mathfrak{G}$ into \mathbf{Z}. Put now, for any $\theta \in K^{\times}$:

$$(6) \qquad f(\rho, \sigma, \tau) = \theta^{e(\rho, \sigma, \tau)};$$

obviously, f is a covariant mapping of $\mathfrak{G} \times \mathfrak{G} \times \mathfrak{G}$ into $K_{\mathrm{sep}}^{\times}$ (more precisely, into K^{\times}), and one verifies at once that it satisfies condition (2) in th. 2, § 3, i.e. that it is a factor-set. Any factor-set f defined in this manner will be called a *cyclic factor-set*. Let Φ' be another locally constant mapping of \mathfrak{G} into \mathbf{R} such that $\chi(\sigma) = \mathbf{e}(\Phi'(\sigma))$ for all σ; let f' be the factor-set defined by Φ' and θ, just as f has been defined by Φ and θ. Put $\Psi = \Phi' - \Phi$; clearly Ψ maps \mathfrak{G} into \mathbf{Z}; putting $z(\rho, \sigma) = \theta^{\Psi(\sigma \rho^{-1})}$, one sees at once that $f'f^{-1}$ is the coboundary of z. This shows that the class of the factor-set f, modulo the group $\beta(K)$ of trivial factor-sets, is uniquely determined by χ and θ; it will be denoted by $\{\chi, \theta\}$, and every such factor-class will be called *cyclic*.

PROPOSITION 8. *For each* $\theta \in K^{\times}$, $\chi \to \{\chi, \theta\}$ *is a morphism of the group of characters of* \mathfrak{G} *into the group* $H(K)$ *of factor-classes of* K; *for each character* χ *of* \mathfrak{G}, $\theta \to \{\chi, \theta\}$ *is a morphism of* K^{\times} *into* $H(K)$.

This is obvious in view of our definitions.

Let K' be a field containing K; as in § 3, we assume that \bar{K} is contained in \bar{K}', and we denote by ρ the restriction morphism of \mathfrak{G}' into \mathfrak{G}, as well as the morphisms for factor-sets and factor-classes derived from this in the manner explained in § 3. If χ is any character of \mathfrak{G}, $\chi' = \chi \circ \rho$ is a character of \mathfrak{G}'; if χ is of order n, the order n' of χ' divides n; if \mathfrak{H} is the kernel of χ, the kernel of χ' is $\mathfrak{H}' = \rho^{-1}(\mathfrak{H})$, and ρ determines an injective morphism of $\mathfrak{G}'/\mathfrak{H}'$ into $\mathfrak{G}/\mathfrak{H}$; if L is the cyclic extension of K attached to χ, the cyclic extension of K' attached to χ' is the compositum of L and K'; it is cyclic of degree n'. Then, for every $\theta \in K^{\times}$, we have:

$$(7) \qquad \{\chi, \theta\} \circ \rho = \{\chi \circ \rho, \theta\}.$$

PROPOSITION 9. *Let χ be a character of \mathfrak{G}, L the cyclic extension of K attached to χ, and A a simple algebra over K. Then A_L is trivial if and only if the factor-class attached to A can be written in the form $\{\chi, \theta\}$ with $\theta \in K^\times$.*

Call \mathfrak{H} the kernel of χ; it is the subgroup of \mathfrak{G} corresponding to L. If the factor-class attached to A is $\{\chi, \theta\}$, the one attached to A_L is given by (7) when one takes for ρ the restriction morphism of \mathfrak{H} into \mathfrak{G}; then $\chi \circ \rho$, being the character induced by χ on \mathfrak{H}, is trivial, so that A_L is trivial. Conversely, assume that A_L is trivial; then, if n is the degree of L over K, corollary 3 of th. 3, § 3, shows that, after replacing A by an algebra similar to A if necessary, we may assume A to be of dimension n^2 over K. Let F be an L-representation of A, induced on A by an isomorphism of A_L onto $M_n(L)$. As χ is of order n, we may choose $\alpha \in \mathfrak{G}$ such that $\chi(\alpha) = \mathbf{e}(1/n)$; then $\mathfrak{G}/\mathfrak{H}$ is generated by the image of α in that group. There is $X \in M_n(L)^\times$ such that $F^\alpha = X^{-1} F X$, hence, by induction on i, $F^{\alpha^i} = X_i^{-1} F X_i$ if we put

$$X_i = X X^\alpha \ldots X^{\alpha^{i-1}}$$

for all $i > 0$. Take $i = n$; as α^n induces the identity on L, $F^{\alpha^n} = F$; therefore X_n must be of the form $\theta \cdot 1_n$ with $\theta \in L^\times$. Applying α to both sides of the formula defining X_n, we get $X_n^\alpha = X^{-1} X_n X$, hence $\theta^\alpha = \theta$, so that θ is in K^\times. Take any $i \in \mathbf{Z}$ and write it as $i = nv + j$ with v, j in \mathbf{Z} and $1 \leqslant j \leqslant n$; if then we write $X_i = \theta^v X_j$, one verifies easily that, for $i > 0$, this coincides with X_i as above defined, that $X_{nv} = \theta^v \cdot 1_n$ for $v \in \mathbf{Z}$, and that $X_{i+j} = X_i X_j^{\alpha^i}$ for all i, j in \mathbf{Z}. Take now a locally constant function Φ on \mathfrak{G} such that $\chi(\sigma) = \mathbf{e}(\Phi(\sigma))$, hence $n\Phi(\sigma) \in \mathbf{Z}$, for all σ, and put $Y(\rho, \sigma) = (X_{n\Phi(\sigma\rho^{-1})})^\rho$ for all ρ, σ in \mathfrak{G}. One verifies easily that Y, in relation to A and F, has the properties required by th. 2 of § 3, and that the factor-set f determined in terms of Y by (1) of th. 2 is the one given by (5) and (6).

PROPOSITION 10. *Let χ and L be as in proposition 9. Then the kernel of the morphism $\theta \to \{\chi, \theta\}$ of K^\times into $H(K)$ is $N_{L/K}(L^\times)$.*

In the proof of proposition 9, take $A = M_n(K)$; then we may take for F the identity, and, as $F^\alpha = F$, we may take $X = \xi \cdot 1_n$ with any $\xi \in L^\times$. Then $\theta = N_{L/K}(\xi)$, and $\{\chi, \theta\}$ is trivial, since A is so. Conversely, assume $\theta \in K^\times$ to be such that $\{\chi, \theta\}$ is trivial. Take Φ such that $\chi(\sigma) = \mathbf{e}(\Phi(\sigma))$ and $0 \leqslant \Phi(\sigma) < 1$ for all σ; then $\Phi(\alpha^i) = i/n$ for $0 \leqslant i \leqslant n-1$, and Φ is \mathfrak{H}-regular, so that, if we define f by (5) and (6), f is an \mathfrak{H}-regular factor-set. As f is trivial, corollary 2 of th. 3, § 3, shows that it is the coboundary of an \mathfrak{H}-regular covariant mapping z of $\mathfrak{G} \times \mathfrak{G}$ into K_{sep}^\times. As \mathfrak{H} is a normal subgroup of \mathfrak{G}, left cosets and right cosets of \mathfrak{H} in \mathfrak{G} are the same; this implies that, for all ρ, σ in \mathfrak{G}, $z(\rho, \sigma)$ is invariant under all $\lambda \in \mathfrak{H}$ and is therefore in L^\times. For all $\sigma \in \mathfrak{G}$, put $w(\sigma) = z(\varepsilon, \sigma)$, and put $w_i = w(\alpha^i)$ for all i.

Then $z(\rho,\sigma) = w(\sigma\rho^{-1})^\rho$. Write now that $f(\rho,\sigma,\tau)$, as given by (6), is equal to the coboundary of z, as given by (3) of § 3, for $\rho = \varepsilon$, $\sigma = \alpha^i$, $\tau = \alpha^{i+1}$; for $0 \leqslant i \leqslant n-2$, we get $1 = w_i(w_1)^{\alpha^i} w_{i+1}^{-1}$, and, for $i = n-1$, we get $\theta = w_{n-1}(w_1)^{\alpha^{n-1}} w_0^{-1}$. Therefore $\theta = N_{L/K}(w_1)$, which completes the proof.

Let χ and L be as in propositions 9 and 10, \mathfrak{H} and α as in the proofs of these propositions. If Φ is chosen as in the proof of prop. 10, we have $f(\alpha^{-j}, \alpha^{-i}, \varepsilon) = 1$ or θ according as $i \leqslant j$ or $i > j$, and in particular $f(\varepsilon, \varepsilon, \varepsilon) = 1$. We now apply to this factor-set the construction described in the proof of lemma 4, § 3, and define the algebra A as has been explained there. As indicated above, the fact that here \mathfrak{H} is a normal subgroup of \mathfrak{G} implies that every \mathfrak{H}-regular covariant mapping of $\mathfrak{G} \times \mathfrak{G}$ into K_{sep} maps $\mathfrak{G} \times \mathfrak{G}$ into L. For $i \in \mathbb{Z}$, define u_i as the mapping of $\mathfrak{G} \times \mathfrak{G}$ into L given by $u_i(\rho,\sigma) = 1$ or 0 according as $\sigma\rho^{-1}$ is in $\mathfrak{H}\alpha^{-i}$ or not; clearly $u_i \in A$ and $u_{n+i} = u_i$ for all i, and u_0 is the same as the unit $e = 1_A$ of A. One finds at once that, for $0 \leqslant i, j \leqslant n-1$, $u_i u_j = u_{i+j}$ when $i+j \leqslant n-1$ and $u_i u_j = \theta u_{i+j}$ when $i+j \geqslant n$. As in the proof of lemma 4, define ξx, for $\xi \in L$, $x \in A$, as given by the mapping $(\rho,\sigma) \to \xi^\rho x(\rho,\sigma)$; one finds at once that $\xi x = (\xi \cdot 1_A)x$. Similarly, define $x\xi$ as given by the mapping $(\rho,\sigma) \to x(\rho,\sigma)\xi^\sigma$; then $x\xi = x(\xi \cdot 1_A)$. Clearly $\xi u_i = u_i \xi^{\alpha^i}$ for all $\xi \in L$ and all i. As A has the dimension n^2 over K, it has the dimension n over L when considered either as a left vector-space, by $(\xi, x) \to \xi x$, or as a right vector-space, by $(\xi, x) \to x\xi$. Moreover, $\{u_0, u_1, \ldots, u_{n-1}\}$ is a basis for both of these spaces; in fact, if we put $x = \sum \xi_i u_i$ with $\xi_i \in L$ for $0 \leqslant i \leqslant n-1$, we have $x(\varepsilon, \alpha^{-i}) = \xi_i$, so that $x = 0$ implies $\xi_i = 0$ for all i, and a similar proof holds for A as a right vector-space. Finally, as has been observed at the end of the proof of lemma 4, § 3, one can use the isomorphism Φ_ε of A into $M_n(K_{\text{sep}})$ which was defined there, and which is now an L-representation of A, for the calculation of the reduced trace τ and of the reduced norm ν in A. Taking $\{\varepsilon, \alpha^{-1}, \ldots, \alpha^{-n+1}\}$ as the full set \mathfrak{a} of representatives of $\mathfrak{G}/\mathfrak{H}$ in \mathfrak{G} used in the definition of Φ_ε, we get at once, for all $\xi \in L$:

$$(8) \qquad \tau(\xi \cdot 1_A) = Tr_{L/K}(\xi); \quad \tau(\xi u_i) = 0 \qquad (1 \leqslant i \leqslant n-1),$$

$$(9) \qquad \nu(\xi \cdot 1_A) = N_{L/K}(\xi); \quad \nu(u_i) = (-1)^{i(n-i)}\theta^i \qquad (1 \leqslant i \leqslant n-1).$$

PROPOSITION 11. *Let L be a cyclic extension of K of degree n, and α a generator of its Galois group over K. Let X be a left vector-space of dimension n over L, with the basis $\{u_0, u_1, \ldots, u_{n-1}\}$. Then, for each $\theta \in K^\times$, there is one and only one K-bilinear and associative mapping $(x,y) \to xy$ of $X \times X$ into X such that: (i) for all $\xi \in L$ and all $x \in X$, $\xi x = (\xi u_0)x$ and $x u_0 = x$; (ii) $u_i = (u_1)^i$ for $1 \leqslant i \leqslant n-1$; (iii) $(u_1)^n = \theta u_0$; (iv) $\xi u_1 = u_1(\xi^\alpha u_0)$. This makes X into a simple algebra A over K, in which the reduced trace τ and the reduced norm ν satisfy (8) and (9), and the factor-class of K attached to A is $\{\chi, \theta\}$ if χ is the character of \mathfrak{G} attached to L, such that $\chi(\alpha) = \mathbf{e}(1/n)$.*

As all this has been proved above for the algebra A which we constructed there, it only remains for us to show that the conditions (i) to (iv), together with the associativity, determine the multiplication uniquely. In fact, by induction on i, (iv) gives $\xi u_i = u_i(\xi^{\alpha^i} u_0)$ for $0 \leqslant i \leqslant n-1$. Then, using (i) and the associativity of the multiplication, we get, for $0 \leqslant i, j \leqslant n-1$ and for all ξ, η in L:

$$(\xi u_i)(\eta u_j) = (\xi u_0)(u_i(\eta u_0)) u_j = (\xi u_0)(\eta^{\alpha^{-i}} u_i) u_j$$

$$= (\xi u_0)(\eta^{\alpha^{-i}} u_0) u_i u_j = \xi \eta^{\alpha^{-i}} u_i u_j.$$

By (ii), $u_i u_j = u_{i+j}$ if $i+j \leqslant n-1$; by (ii) and (iii), $u_i u_j = \theta u_{i+j-n}$ if $i+j \geqslant n$. This shows that, using (i) to (iv) and the associativity, one can write uniquely $(\xi u_i)(\eta u_j)$ in the form ζu_k with $\zeta \in L$, which completes the proof of our proposition.

DEFINITION 6. *Assumptions and notations being as in proposition 11, the algebra A defined there will be called the cyclic algebra* $[L/K; \chi, \theta]$.

An illustration for the above concepts, which will be considered more closely in the following chapters, is provided by the division algebras over a commutative p-field K. In fact, prop. 5 of Chap. I-4 may now be interpreted as saying precisely that every such algebra D can be written as a cyclic algebra $[K_1/K; \chi, \pi]$, where K_1 is an unramified extension of K, χ a character attached to K_1, and π a suitable prime element of K. But now we can say more; prop. 10, combined with prop. 3 of Chap. VIII-1, shows that $\{\chi, \xi\}$ is trivial for $\xi \in R^\times$, so that $\{\chi, \pi\}$ is independent of the choice of π; so is $[K_1/K; \chi, \pi]$, since there can be only one algebra of given dimension over K in a given class, up to an isomorphism.

As a further illustration for the above theory, we will apply it to the field $K = \mathbf{R}$. We may then identify \bar{K} with \mathbf{C}, and \mathfrak{G} has only two elements, the identity ε and the automorphism σ of \mathbf{C} given by $z \to \bar{z}$, and only one non-trivial character χ, given by $\chi(\sigma) = -1$. The cyclic extension of \mathbf{R}, attached to χ, is \mathbf{C}. Combining now corollary 2 of prop. 3, § 1, corollary 3 of th. 3, § 3, and propositions 9 and 11, we see that every class of simple algebras over \mathbf{R} contains a cyclic algebra $[\mathbf{C}/\mathbf{R}; \chi, \theta]$. As the group $N_{\mathbf{C}/\mathbf{R}}(\mathbf{C}^\times)$ is \mathbf{R}_+^\times and is of index 2 in \mathbf{R}^\times, prop. 10 shows that there are exactly two such algebras, up to isomorphism, viz., a trivial one and the algebra $\mathbf{H} = [\mathbf{C}/\mathbf{R}; \chi, -1]$. The latter is a division algebra; in fact, it is of dimension 4 over \mathbf{R}; writing it as $M_n(D)$, where D is a division algebra over \mathbf{R}, and calling d^2 the dimension of D over \mathbf{R}, we get $nd = 2$, hence $n = 1$ since \mathbf{H} is not trivial. Writing \mathbf{H} in the manner described in prop. 11,

we see that it has a basis over C consisting of two elements $u_0 = 1$ and u_1, hence a basis over \mathbf{R} consisting of $1, i, j = u_1$ and $k = iu_1$; it is then trivial to verify that the multiplication table for $1, i, j, k$ is the well-known one for the "quaternion units" in the algebra of "classical" quaternions.

§ 5. Special cyclic factor-sets.

Now we apply the results of §4 to the characters attached to "Kummer extensions" and to "Artin-Schreier extensions" of K.

In the first place, let n be such that K contains n distinct n-th roots of 1; then these make up a cyclic group E of order n; of course, if K is of characteristic $p > 1$, our assumption implies that n is prime to p. Let ψ be an isomorphism of E onto the group of n-th roots of 1 in C; this will be determined uniquely if we choose a generator ε_1 of E and prescribe that $\psi(\varepsilon_1) = e(1/n)$. Take any $\xi \in K^\times$, and let x be any one of the roots of the equation $X^n = \xi$ in \bar{K}; then x is in K_{sep}^\times, and the equation $X^n = \xi$ has the n distinct roots εx, with $\varepsilon \in E$. In particular, for each $\sigma \in \mathfrak{G}$, x^σ must be one of these roots, so that $x^\sigma x^{-1}$ is in E. Now put

$$\chi_{n,\xi}(\sigma) = \psi(x^\sigma x^{-1});$$

as $E \subset K$, the right-hand side does not change if we replace x by εx with $\varepsilon \in E$ and is therefore independent of the choice of a root x for $X^n = \xi$. For a similar reason, we have, for all ρ, σ in \mathfrak{G}:

$$x^{\rho\sigma} x^{-1} = (x^\rho x^{-1})^\sigma (x^\sigma x^{-1}) = (x^\rho x^{-1})(x^\sigma x^{-1}),$$

and therefore

$$\chi_{n,\xi}(\rho\sigma) = \chi_{n,\xi}(\rho)\chi_{n,\xi}(\sigma),$$

which shows that $\chi_{n,\xi}$ is a character of \mathfrak{G}. Take now any $\eta \in K^\times$, and call y a root of $X^n = \eta$; then xy is a root of $X^n = \xi\eta$, and we have, for all $\sigma \in \mathfrak{G}$:

$$(xy)^\sigma (xy)^{-1} = (x^\sigma x^{-1})(y^\sigma y^{-1})$$

and therefore

$$\chi_{n,\xi\eta} = \chi_{n,\xi}\chi_{n,\eta},$$

which shows that $\xi \to \chi_{n,\xi}$ is a morphism of K^\times into the group of characters of \mathfrak{G}. It is obvious that $\chi_{n,\xi}$ is trivial if and only if the equation $X^n = \xi$ has one root, hence all its roots, in K, i.e. if $\xi \in (K^\times)^n$; in other words, $(K^\times)^n$ is the kernel of $\xi \to \chi_{n,\xi}$. It would be easy to show that the image of K^\times under that morphism consists of all the characters of \mathfrak{G} whose order divides n, but this will not be needed.

Now we put, for ξ and θ in K:

$$\{\chi_{n,\xi}, \theta\} = \{\xi, \theta\}_n;$$

this is known as "Hilbert's symbol"; one should note that it depends upon the choice of ψ, or, what amounts to the same, of the generator ε_1 of the group E of n-th roots of 1 in K. By prop. 8, § 4, we have:

(10) $\{\xi\xi',\theta\}_n=\{\xi,\theta\}_n\cdot\{\xi',\theta\}_n, \quad \{\xi,\theta\theta'\}_n=\{\xi,\theta\}_n\cdot\{\xi,\theta'\}_n$

for all ξ,ξ',θ,θ' in K^\times.

Call again x a root of $X^n=\xi$; clearly the kernel \mathfrak{H} of $\chi_{n,\xi}$ consists of the elements σ of \mathfrak{G} such that $x^\sigma=x$, so that the corresponding subfield of K_{sep}, which is the cyclic extension of K attached to $\chi_{n,\xi}$, is $L=K(x)$. Call d the order of $\chi_{n,\xi}$; then $\chi_{n,\xi}$ determines an isomorphism of $\mathfrak{G}/\mathfrak{H}$ onto the group of d-th roots of 1 in \mathbf{C}; d divides n, and it is also the degree of L over K. Therefore the distinct conjugates of x over K, i.e. its images under the d distinct automorphisms of L over K, are the elements εx, where ε runs through the group E' of d-th roots of 1 in K. Write $e=n/d$, and, for any $\zeta\in K$:

$$\omega=\prod_{\nu=0}^{e-1}(\zeta-\varepsilon_1^\nu x),$$

where ε_1, as before, is a generator of E; as the ε_1^ν for $0\leqslant\nu\leqslant e-1$ are a full set of representatives of the cosets modulo E' in E, we have:

$$N_{L/K}(\omega)=\prod_{\varepsilon\in E}(\zeta-\varepsilon x)=\zeta^n-\xi.$$

For $\zeta=0$ and $\zeta=1$, this shows that $-\xi$ and $1-\xi$ are in $N_{L/K}(L)$. By prop. 10 of § 4, this gives:

(11) $\{\xi,-\xi\}_n=1, \quad \{\xi,1-\xi\}_n=1,$

these formulas being valid whenever they are meaningful, i.e. the first one for all $\xi\in K^\times$, and the second one for all $\xi\neq0,1$ in K. In the first one, replace ξ by $\xi\eta$ with ξ,η in K^\times, and apply (10); we get:

$$\{\xi,-\xi\}_n\cdot\{\xi,\eta\}_n\cdot\{\eta,-\xi\}_n\cdot\{\eta,\eta\}_n=1.$$

Here, by (11), the first factor is 1, and the last one is equal to $\{\eta,-1\}_n$; applying (10) again, we get

(12) $\{\xi,\eta\}_n\cdot\{\eta,\xi\}_n=1,$

which is known as "the law of reciprocity" for the symbol $\{\xi,\eta\}_n$. The same could be proved by the explicit construction of the simple algebra corresponding to the latter factor-class; we merely sketch the proof in the case when the equations $X^n=\xi$, $X^n=\eta$ are both irreducible over K. That being assumed, put $L=K(x)$, where x is a root of $X^n=\xi$; let A be the cyclic algebra $[L/K;\chi_{n,\xi},\eta]$. By prop. 11 of § 4, where we write now y instead of u_1, A has a basis over L consisting of the y^j for $0\leqslant j\leqslant n-1$, hence

a basis over K consisting of the $x^i y^j$ with $0 \leqslant i, j \leqslant n-1$, with the relations $x^n = \xi$, $y^n = \eta$, $xy = \varepsilon_1 yx$. If we exchange ξ and η, and also x and y, A is clearly replaced by its inverse A^0; this implies (12).

Now let K be of characteristic $p > 1$; identify the prime field in K with \mathbf{F}_p, and call ψ the character of the additive group of \mathbf{F}_p given by $\psi(1) = e(1/p)$. Take any $\xi \in K$, and let x be any one of the roots of the equation $X - X^p = \xi$; then x is in K_{sep}, and that equation has the p distinct roots $x + a$ with $a \in \mathbf{F}_p$. In particular, for each $\sigma \in \mathfrak{G}$, x^σ must be one of these roots, so that $x^\sigma - x$ is in \mathbf{F}_p. Now put

$$\chi_{p,\xi}(\sigma) = \psi(x^\sigma - x);$$

as the right-hand side does not change if we replace x by $x + a$ with $a \in \mathbf{F}_p$, it is independent of the choice of x. A calculation, quite similar to the one given above for $\chi_{n,\xi}$, shows that $\chi_{p,\xi}$ is a character of \mathfrak{G}, and that $\xi \to \chi_{p,\xi}$ is a morphism of the additive group of K into the (multiplicative) group of characters of \mathfrak{G}; the kernel of that morphism is the image of K under the mapping $\xi \to \xi - \xi^p$ of K into itself, and it would again be easy to show that the image of that morphism consists of $\chi = 1$ and of the characters of \mathfrak{G} of order p. Put now, for all $\xi \in K$ and all $\theta \in K^\times$:

$$\{\chi_{p,\xi}, \theta\} = \{\xi, \theta\}_p.$$

Then we have:

(13) $$\{\xi + \xi', \theta\}_p = \{\xi, \theta\}_p \cdot \{\xi', \theta\}_p, \quad \{\xi, \theta\theta'\}_p = \{\xi, \theta\}_p \cdot \{\xi, \theta'\}_p$$

for all ξ, ξ' in K and all θ, θ' in K^\times. Assume now that x is not in K; then $L = K(x)$ is the cyclic extension of K attached to $\chi_{p,\xi}$, and it is of degree p over K; $X^p - X + \xi = 0$ must then be the irreducible equation for x over K, so that $N_{L/K}(x) = (-1)^p \xi = -\xi$. By prop. 10 of § 4, this gives

(14) $$\{\xi, -\xi\}_p = 1,$$

which is therefore valid whenever x is not in K. If $x \in K$, $\chi_{p,\xi}$ is trivial, so that (14) is still valid provided it is meaningful, i.e. provided $\xi \neq 0$. Therefore (14) is valid for all $\xi \in K^\times$.

Chapter X

Simple algebras over local fields

§ 1. Orders and lattices. Let D be a division algebra of finite dimension over any field K; we will consider left vector-spaces over D, whose dimension will always be assumed finite and >0. If V and W are such spaces, we write $\mathrm{Hom}(V,W)$ for the space of homomorphisms of V into W, and let it operate on the right on V; in other words, if α is such a homomorphism, and $v \in V$, we write $v\alpha$ for the image of v under α. We consider $\mathrm{Hom}(V,W)$, in an obvious manner, as a vector-space over K; as such, it has a finite dimension, since it is a subspace of the space of K-linear mappings of V into W. As usual, we write $\mathrm{End}(V)$ for $\mathrm{Hom}(V,V)$.

If V, V', V'' are left vector-spaces over D, and $\alpha \in \mathrm{Hom}(V,V')$ and $\beta \in \mathrm{Hom}(V',V'')$, we write $\alpha\beta$ for the morphism $v \to (v\alpha)\beta$ of V into V''. For $V=V'=V''$, this makes $\mathrm{End}(V)$ into a ring; as before, we write $\mathrm{Aut}(V)$ for $\mathrm{End}(V)^{\times}$, this being the group of automorphisms of V. For $V=V'$, $V''=W$, we get for $\mathrm{Hom}(V,W)$ a structure of left $\mathrm{End}(V)$-module; for $V'=V''=W$, we get for $\mathrm{Hom}(V,W)$ a structure of right $\mathrm{End}(W)$-module.

Let D and V be as above; let d^2 be the dimension of D over K, and m that of V over D. Take a basis $\{v_1, \ldots, v_m\}$ of V over D; for each $\xi \in \mathrm{End}(V)$, write $v_i \xi = \sum_j x_{ij} v_j$, with $x_{ij} \in D$ for $1 \leqslant i,j \leqslant m$; this defines a mapping $\xi \to (x_{ij})$ of $\mathrm{End}(V)$ into $M_m(D)$, which is obviously an isomorphism of $\mathrm{End}(V)$ onto $M_m(D)$; in particular, this shows that $\mathrm{End}(V)$ is a simple algebra of dimension $m^2 d^2$ over K. Obviously V, considered as a right $\mathrm{End}(V)$-module, is simple; therefore, by prop. 1 of Chap. IX-1, every such module is a direct sum of modules isomorphic to V.

Let V and W be left vector-spaces over D; call m, n their dimensions; put $A = \mathrm{End}(V)$, $B = \mathrm{End}(W)$, $H = \mathrm{Hom}(V,W)$. As H is a right B-module, it is a direct sum of modules isomorphic to W; comparing dimensions over K, one sees at once that it is the direct sum of m such modules. Similarly, as a left A-module, H is the direct sum of n modules isomorphic to the dual space $V' = \mathrm{Hom}(V,D)$ of V, this being a simple left A-module and a right vector-space of dimension m over D. One could easily see that every endomorphism of H for its structure as a left A-module is of the form $\lambda \to \lambda\beta$ with $\beta \in B$, and that every endomorphism of H for its structure as a right B-module is of the form $\lambda \to \alpha\lambda$ with $\alpha \in A$.

Let D, V and $A = \mathrm{End}(V)$ be as above, and let v be the reduced norm in A; by corollary 1 of prop. 6, Chap. IX-2, the determinant of the endomorphisms $x \rightarrow \alpha x$ and $x \rightarrow x\alpha$ of the underlying vector-space of A over K, for any $\alpha \in A$, is $v(\alpha)^{md}$; in particular, α is in A^{\times} if and only if $v(\alpha) \neq 0$.

If K is a local field, all the above spaces, being vector-spaces of finite dimension over K, can be topologized as such in one and only one way, according to corollary 1 of th. 3, Chap. I-2; conversely, by corollary 2 of the same theorem, the requirement of finite dimensionality over K could everywhere be replaced by that of local compactness. If we write again $A = \mathrm{End}(V)$, the group $A^{\times} = \mathrm{Aut}(V)$ is the open subset of A determined by $v(\alpha) \neq 0$; as such, it is a locally compact group. Moreover, the Haar measure in it is right-invariant as well as left-invariant; this is contained in the following lemma, which generalizes lemma 5 of Chap. VII-4:

LEMMA 1. *Let K be a local field; let D, V and $A = \mathrm{End}(V)$ be as above, and let α be a Haar measure on A. Then the measure μ on A^{\times}, given by*

$$d\mu(x) = \mathrm{mod}_K(N_{A/K}(x))^{-1} \, d\alpha(x) = \mathrm{mod}_K(v(x))^{-md} \, d\alpha(x)$$

is both left-invariant and right-invariant on A^{\times}.

This follows at once from corollary 1 of prop. 6, Chap. IX-2, combined with corollary 3 of th. 3, Chap. I-2.

In the rest of this §, we assume that K is a commutative p-field; D being a division algebra over K, hence also a p-field (a non-commutative one, unless $d = 1$), we write R and R_D for the maximal compact subrings of K and of D, and P and P_D for the maximal ideals in R and in R_D, respectively.

Let V and W be as above; let L be a D-lattice in V, and M a D-lattice in W; then we write $\mathrm{Hom}(V, L; W, M)$ for the set of all morphisms of V into W which map L into M. Choose bases $\{v_1, \ldots, v_m\}$, $\{w_1, \ldots, w_n\}$ for V and W according to th. 1 of Chap. II-2, i.e. so that $L = \sum R_D v_i$ and $M = \sum R_D w_j$. For each $\lambda \in \mathrm{Hom}(V, W)$, we can write $v_i \lambda = \sum x_{ij} w_j$ with $x_{ij} \in D$ for $1 \leqslant i \leqslant m$, $1 \leqslant j \leqslant n$, and this determines a bijection $\lambda \rightarrow (x_{ij})$ of $\mathrm{Hom}(V, W)$ onto the space $M_{m,n}(D)$ of the matrices with m rows and n columns over D; clearly λ is then in $\mathrm{Hom}(V, L; W, M)$ if and only if the matrix (x_{ij}) is in $M_{m,n}(R_D)$. In particular, this shows that $\mathrm{Hom}(V, L; W, M)$ is a K-lattice in the space $\mathrm{Hom}(V, W)$ considered as a vector-space over K, and also that it can be identified with the space of all morphisms of L into M for their structures as R_D-modules. We write $\mathrm{End}(V, L)$ for $\mathrm{Hom}(V, L; V, L)$; this is a K-lattice and an open compact subring of $\mathrm{End}(V)$, which may be identified with $\mathrm{End}(L)$. We write $\mathrm{Aut}(V, L)$ for $\mathrm{End}(V, L)^{\times}$; it is the group of automorphisms of L.

PROPOSITION 1. *Let K be a p-field, D a division algebra over K, V a left vector-space over D, and L a D-lattice in V. Let v be the reduced norm in the algebra $A = \mathrm{End}(V)$ over K. Then $\mathrm{Aut}(V,L)$ consists of the elements ξ of $\mathrm{End}(V,L)$ such that $\mathrm{mod}_K(v(\xi)) = 1$.*

Take $\xi \in A$; it is in A^\times if and only if $v(\xi) \ne 0$. If m, d are as before, the module of the automorphism $x \to x\xi$ of A is $\mathrm{mod}_K(v(\xi))^{md}$; as A, for its structure as a right A-module, is the direct sum of m modules isomorphic to V, this implies that the module of the automorphism $v \to v\xi$ of V is $\mathrm{mod}_K(v(\xi))^d$. Now assume that ξ is in $\mathrm{End}(V,L)$; then it maps L onto a D-lattice $L' = L\xi$ contained in L, so that the module of $v \to v\xi$ is equal to $[L:L']^{-1}$. This shows that $L = L'$ if and only if $\mathrm{mod}_K(v(\xi)) = 1$, which proves our proposition.

COROLLARY. *Notations being as in proposition 1, $\mathrm{Aut}(V,L)$ is a compact open subset of $\mathrm{End}(V,L)$ and of $\mathrm{End}(V)$ and a compact open subgroup of $\mathrm{Aut}(V)$.*

This is now obvious.

PROPOSITION 2. *Let V be as in proposition 1; let X be a multiplicatively closed subset of $\mathrm{End}(V)$. Then X is relatively compact in $\mathrm{End}(V)$ if and only if there is a D-lattice L in V such that $X \subset \mathrm{End}(V,L)$.*

Let X be relatively compact in $\mathrm{End}(V)$; we may assume that it is compact, since otherwise we replace it by its closure. Let L be any D-lattice in V. Call L' the set of the vectors $v \in L$ such that $v\xi \in L$ for all $\xi \in X$; clearly this is an R_D-module, hence closed, by prop. 5 of Chap. II-2; being contained in L, it is compact. As X is compact and L is open, L' is open. Therefore L' is a D-lattice. As X is multiplicatively closed, $v\xi$ is in L' for all $v \in L'$ and all $\xi \in X$, so that $X \subset \mathrm{End}(V,L')$. The converse is obvious.

PROPOSITION 3. *Let V be as above, and let L, L' be two D-lattices in V. Then either $\mathrm{Aut}(V,L)$ is not contained in $\mathrm{End}(V,L')$, or there is $x \in D^\times$ such that $L' = xL$.*

By th. 2 of Chap. II-2, there is a basis $\{v_1, \ldots, v_m\}$ of V, and there are integers v_i, such that $L = \sum R_D v_i$ and $L' = \sum P_D^{v_i} v_i$. Every permutation of the v_i determines an automorphism of V which belongs to $\mathrm{Aut}(V,L)$; if all these are in $\mathrm{End}(V,L)$, all the v_i must be equal; if v is their common value, we have $L' = \pi_D^v L$ for any prime element π_D of D.

THEOREM 1. *Let D be a division algebra over a p-field K, and let V be a left vector-space over D. Then the maximal compact subrings of the*

algebra $A = \mathrm{End}(V)$ *are the rings* $\mathrm{End}(V, L)$, *and the maximal compact subgroups of* A^{\times} *are the groups* $\mathrm{Aut}(V, L)$, *when one takes for* L *all the D-lattices in* V.

By prop. 2, a compact subring of $\mathrm{End}(V)$ must be contained in some $\mathrm{End}(V, L)$, hence must be equal to it if it is maximal. Now assume that, for some L, $\mathrm{End}(V, L)$ is contained in a compact subring X of $\mathrm{End}(V)$; this, in turn, must be contained in some $\mathrm{End}(V, L')$. Then, by prop. 3, $L' = xL$ with some $x \in D^{\times}$; this gives $\mathrm{End}(V, L') = \mathrm{End}(V, L)$, hence $X = \mathrm{End}(V, L)$. Similarly, a compact subgroup of $A^{\times} = \mathrm{Aut}(V)$ must be contained in some $\mathrm{End}(V, L)$, hence in $\mathrm{End}(V, L)^{\times}$, i.e. in $\mathrm{Aut}(V, L)$. If this is contained in a compact subgroup X of $\mathrm{Aut}(V)$, X must be contained in some $\mathrm{End}(V, L')$, and we get $L' = xL$ and $X = \mathrm{Aut}(V, L)$, just as before. In the conclusion of theorem 1, one might take for L, instead of all the D-lattices in V, a full set of representatives for the equivalence relation among D-lattices defined by $L' = xL$, $x \in D^{\times}$.

Compact open subrings of a simple algebra over a p-field are also called *orders*; thus, the first part of theorem 1 states the existence of *maximal orders* in the algebra $A = \mathrm{End}(V)$, viz., all the rings $\mathrm{End}(V, L)$. As we have seen above, these are all isomorphic to $M_m(R_D)$ if m is the dimension of V over D; clearly they are the transforms of one another under the automorphisms of V, since any basis of V over D can be transformed into any other basis by such an automorphism. It amounts to the same to say that they are the transforms of one another under the inner automorphisms of A.

PROPOSITION 4. *Let D be as above, and let V, W be two left vector-spaces over D. Let M, M' be compact open subgroups of* $\mathrm{Hom}(V, W)$, *and let X be the set of the elements ξ of* $\mathrm{End}(V)$ *such that $\xi M \subset M'$. Then X is a compact open subgroup of* $\mathrm{End}(V)$; *if $M = M'$, it is a subring of* $\mathrm{End}(V)$.

Obviously X is a subgroup of $\mathrm{End}(V)$, and a subring if $M = M'$. As M is compact and M' is open, X is open. Now put $H = \mathrm{Hom}(V, W)$. As M is open, it contains a basis $\{\mu_1, \dots, \mu_r\}$ for H regarded as a vector-space over K. If now we regard H as a left $\mathrm{End}(V)$-module, the annihilator of that basis in $\mathrm{End}(V)$ is the same as that of H, hence $\{0\}$ since $\mathrm{End}(V)$ is simple and W is not $\{0\}$. Therefore the mapping $\xi \to (\xi \mu_1, \dots, \xi \mu_r)$ of $\mathrm{End}(V)$ into $H^r = H \times \cdots \times H$ is injective, hence an isomorphism of $\mathrm{End}(V)$ onto its image in H^r, for their structures as vector-spaces over K, hence also for their topological structures. This implies that the set X' of the elements ξ of $\mathrm{End}(V)$ such that $\xi \mu_i \in M'$ for $1 \leqslant i \leqslant r$ is compact. As X is a subgroup of X' and is open in $\mathrm{End}(V)$, it is an open subgroup of X', hence closed in X', hence compact.

For $M=M'$, the ring X defined by proposition 4 is called *the left order* of M. Exchanging right and left, we see that the set Y of the elements η of $\text{End}(W)$ such that $M\eta \subset M$ is a compact open subring of $\text{End}(W)$; this is called *the right order* of M. Now we show that, if one of these orders is maximal, the other is also maximal. This is contained in the following:

THEOREM 2. *Let K and D be as in theorem 1; let V, W be two left vector-spaces over D, and let L be a D-lattice in V. Let N be a compact open subgroup of $\text{Hom}(V,W)$ such that $\xi N \subset N$ for all ξ in $\text{End}(V,L)$. Then there is a D-lattice M in W such that $N=\text{Hom}(V,L;W,M)$, and the left and right orders of N are $\text{End}(V,L)$ and $\text{End}(W,M)$, respectively.*

By th. 1 of Chap. II-2, we can choose a basis $\{v_1,\ldots,v_m\}$ of V over D, so that $L=\sum R_D v_i$; then, as explained above, we can use this basis for identifying $\text{End}(V)$ with $M_m(D)$ and $\text{End}(V,L)$ with $M_m(R_D)$, by assigning to each element ξ of $\text{End}(V)$ the matrix (x_{ij}) given by $v_i\xi=\sum x_{ij}v_j$. Now consider the mapping $\alpha \to (v_1\alpha,\ldots,v_m\alpha)$ of $\text{Hom}(V,W)$ into the direct sum W^m of m spaces isomorphic to W; clearly it is a bijection of $\text{Hom}(V,W)$ onto W^m; call it φ, and put $N'=\varphi(N)$, where N is the set in theorem 2. If α is in $\text{Hom}(V,W)$ and $\varphi(\alpha)=(w_1,\ldots,w_m)$, and if ξ and (x_{ij}) are as above, then $\varphi(\xi\alpha)=(w'_1,\ldots,w'_m)$, with $w'_i=\sum x_{ij}w_j$ for $1\leqslant i\leqslant m$; by our assumption on N, this must be in N' whenever (w_1,\ldots,w_m) is in N' and all the x_{ij} are in R_D. Writing e_{ij} for the "matrix units" in $M_m(D)$, as defined in the proof of th. 1, Chap. IX-1, take first for (x_{ij}) the matrix unit e_{hh}; then we see that, if (w_1,\ldots,w_m) is in N', every one of the elements $(0,\ldots,0,w_h,0,\ldots,0)$, for $1\leqslant h\leqslant m$, must also be in N'. This is the same as to say that, if we call W_1,\ldots,W_m the m summands of W^m, and if we put $N'_h=N'\cap W_h$ for $1\leqslant h\leqslant m$, we have $N'=\sum N'_h$. Similarly, taking for (x_{ij}) the matrix unit e_{hk}, we see now that $N'_h=N'_k$ for all h and k; put $M=N'_h$ for any h. Finally, taking for (x_{ij}) the matrix $x\cdot 1_m$ with $x\in R_D$, we see that M is an R_D-module. As N is open and compact in $\text{Hom}(V,W)$, N' is so in W^m, hence N'_h in W_h, and M in W; therefore M is a D-lattice in W. Now we see that an element α of $\text{Hom}(V,W)$ is in N if and only if $v_i\alpha$ is in M for $1\leqslant i\leqslant m$; this is the same as to say that $N=\text{Hom}(V,L;W,M)$. Then the left order and the right order of N contain $\text{End}(V,L)$ and $\text{End}(W,M)$, respectively; as the latter are maximal orders, this completes our proof.

COROLLARY 1. *Let A be a simple algebra over K, R_A a maximal compact subring of A, and I a left ideal in R_A. Then I is open in A if and only if it can be written as $I=R_A\alpha$ with $\alpha\in R_A\cap A^\times$.*

We may assume that $A = \text{End}(V)$, where V is as in theorem 2; then, by th. 1, we may assume that $R_A = \text{End}(V,L)$, where L is as in theorem 2. If I is open, we may apply theorem 2 to it, taking $W = V$ and $N = I$; this gives $I = \text{Hom}(V,L; V,M)$, where M is a D-lattice in V. Take bases $\{v_1, \ldots, v_m\}$, $\{w_1, \ldots, w_m\}$ of V, so that $L = \sum R_D v_i$ and $M = \sum R_D w_i$, and call ξ the automorphism of V which maps the former basis onto the latter one. Then $M = L\xi$ and $I = R_A \xi$; as ξ is in I, it is in R_A. The converse is obvious.

COROLLARY 2. *Let A and R_A be as in corollary 1, and let J be a compact two-sided R_A-module in A, other than $\{0\}$; then J is open in A. If $A = \text{End}(V)$ and $R_A = \text{End}(V,L)$, with V and L as in theorem 2, then J can be written as $J = \text{Hom}(V,L; V, \pi_D^v L)$, where $v \in \mathbf{Z}$ and π_D is a prime element of D.*

As R_A is a K-lattice in A, we can choose a basis $\{\alpha_1, \ldots, \alpha_N\}$ of A over K, consisting of elements of R_A. If ξ is in A and not 0, the two-sided ideal generated by ξ in A is A, since A is simple; therefore the elements $\alpha_i \xi \alpha_j$, for $1 \leqslant i, j \leqslant N$, generate A as a vector-space over K, so that the R-module they generate in A is a K-lattice, hence open. This implies that the set J in our corollary must be open; then, by theorem 2, we can write it as $J = \text{Hom}(V,L; V,M)$, where M is a D-lattice in V such that $\text{End}(V,M)$ contains $\text{End}(V,L)$. By prop. 3, this gives $M = xL$ with $x \in D^\times$; we have then $M = \pi_D^v L$ for $v = \text{ord}_D(x)$.

If V and W are as in theorem 2, any set N with the properties described there, i.e. any set which can be written as $N = \text{Hom}(V,L; W,M)$ for suitably chosen D-lattices L in V and M in W, will be called a *normal lattice* in $\text{Hom}(V,W)$.

§ 2. Traces and norms.

As before, we consider a local field K, a division algebra D of dimension d^2 over K, and a simple algebra A over K, isomorphic to $M_m(D)$ for some $m \geqslant 1$; we denote by τ and τ_D the reduced traces in A and in D, and by v and v_D the reduced norms in A and in D, respectively. We begin by considering the case of a p-field.

PROPOSITION 5. *Let K be a p-field; let D be a division algebra of dimension d^2 over K; let R_D be the maximal compact subring of D, and let π_D be a prime element of D. For any $m \geqslant 1$, put $A = M_m(D)$ and $R_A = M_m(R_D)$, and let τ be the reduced trace in A. Then the set of the elements x of A, such that $\tau(xy) \in R$ for all $y \in R_A$, is $\omega R_A = R_A \omega$, with $\omega = \pi_D^{1-d} \cdot 1_m$.*

Consider first the case $m = 1$, $A = D$, $R_A = R_D$. As has already been observed in Chap. IX-4, we may use prop. 5 of Chap. I-4 to describe D as a cyclic algebra $[K_1/K; \chi, \pi]$ over K, where K_1 is an unramified

extension of K of degree d, χ a character attached to that extension, and π a prime element of K; then a comparison of that proposition with the definition of a cyclic algebra in prop. 11 of Chap. IX-4 shows that u_1, in the latter proposition, is a prime element of D. We may now substitute u_1 for π_D, as this does not affect the statement in our proposition, and write therefore $\pi_D = u_1$, hence, with the notations of prop. 11 of Chap. IX-4, $u_i = \pi_D^i$ for $0 \leqslant i \leqslant d-1$ and $\pi = \pi_D^d$. Call R_1 the maximal compact subring of K_1; by (b) in prop. 5 of Chap. I-4, R_D is the left R_1-module generated by u_0, \dots, u_{d-1}; therefore any $x \in D$ has the property stated in our proposition if and only if $\tau_D(x \eta \pi_D^j) \in R$ for all $\eta \in R_1$ and for $0 \leqslant j \leqslant d-1$. Again by (b) in prop. 5 of Chap. I-4, we may write $x = \sum_i \xi_i \pi_D^i$ with $\xi_i \in K_1$ for $0 \leqslant i \leqslant d-1$. Using formula (8) of Chap. IX-4 for the reduced trace τ_D in D, we get, for $j = 0$, $Tr_{K_1/K}(\xi_0 \eta) \in R$ for all $\eta \in R_1$; by prop. 3 of Chap. VIII-1, this is so if and only if $\xi_0 \in R_1$. Similarly, for $1 \leqslant j \leqslant d-1$, our condition can be written as $Tr_{K_1/K}(\xi_{d-j} \eta' \pi) \in R$ with $\eta' = \eta^\beta$, $\beta = \alpha^{j-d}$; as every automorphism β of K_1 maps R_1 onto itself, this must be so for all $\eta' \in R_1$, and this, just as before, is equivalent to $\xi_{d-j} \in \pi^{-1} R_1$. Therefore the set defined in our proposition is the R_1-module generated by $1, \pi^{-1}\pi_D, \dots, \pi^{-1}\pi_D^{d-1}$, i.e. by the elements $\omega_D \pi_D^i$ for $0 \leqslant i \leqslant d-1$ if $\omega_D = \pi_D^{1-d}$. In view of (b) in prop. 5, Chap. I-4, this completes our proof for the case $m=1$. For $m > 1$, our conclusion follows immediately from this and from corollary 2 of prop. 6, Chap. IX-2, which says that $\tau(x) = \sum \tau_D(x_{ii})$ for $x = (x_{ij})$ in A.

COROLLARY 1. *Assumptions and notations being as in proposition 5, let χ be a character of K of order 0, and identify A with its topological dual by putting $\langle x,y \rangle = \chi(\tau(xy))$. Then the dual K-lattice to R_A is ωR_A.*

In fact, this dual lattice is defined as the set of the elements x of A such that $\chi(\tau(xy)) = 1$ for all $y \in R_A$. As τ is K-linear, this is the same as to say that $\chi(\tau(xy)z) = 1$ for all $y \in R_A$ and all $z \in R$, hence, by prop. 12 of Chap. II-5, the same as $\tau(xy) \in R$ for all $y \in R_A$.

COROLLARY 2. *Let A be a simple algebra over K; let τ be the reduced trace in A, χ a character of K of order 0, and identify A with its topological dual by $\langle x,y \rangle = \chi(\tau(xy))$. Let M and M' be two K-lattices in A, dual to each other in A. Assume that both M and M' are subrings of A. Then A is trivial over K, M is a maximal compact subring of A, and $M = M'$.*

By th. 1 of § 1, M is contained in some maximal compact subring R_A of A, and we may, by using a suitable isomorphism of A with an algebra of the form $M_m(D)$, identify A with $M_m(D)$ and R_A with $M_m(R_D)$, where notations are as in proposition 5. As $M \subset R_A$, corollary 1 shows now that $M' \supset \omega R_A \supset R_A$, hence, by th. 1 of § 1, $M' = \omega R_A = R_A$. Clearly this

implies that $d=1$, i.e. that A is trivial, and then, again by corollary 1, that $M=R_A$.

PROPOSITION 6. *Let K be a p-field, A a simple algebra over K, and v the reduced norm in A. Then $v(A^\times)=K^\times$.*

By corollary 3 of prop. 6, Chap. IX-2, it is enough to consider the case $A=D$; then, just as above, we can write D as a cyclic algebra and use for v_D the formula (9) in Chap. IX-4. With the same notations as above in the proof of prop. 5, this shows, firstly, that $v_D(D^\times)$ contains $N_{K_1/K}(R_1^\times)$, which is the same as R^\times by prop. 3 of Chap. VIII-1, and secondly that it contains $v_D(u_1)=\pi$. As K^\times is generated by R^\times and π, this proves our proposition.

In the case of **R**-fields, the conclusion of proposition 6 is of course valid for $A=M_m(K)$ with $K=\mathbf{R}$ or **C**, but not for $K=\mathbf{R}$ and $A=M_m(\mathbf{H})$. In fact, as we have seen in Chap. IX-4, the algebra **H** of "classical" quaternions has a basis over **R**, consisting of the "quaternion units" $1, i, j, k$, with the relations $i^2=-1$, $j^2=-1$, $k=ij=-ji$, which imply $k^2=-1$, $i=jk=-kj$, $j=ki=-ik$. Clearly the **R**-linear bijection $x \to \bar{x}$ of **H** onto itself which maps $1, i, j, k$ onto $1, -i, -j, -k$ is an antiautomorphism, i.e. it maps xy onto $\bar{y}\bar{x}$ for all x, y. In order to determine the reduced trace τ and the reduced norm v in **H**, one needs a **C**-representation F of **H**. By applying some of the results in Chap. IX, or by a direct computation, one finds that such a representation is given by

$$F(1)=1_2, \quad F(i)=\begin{pmatrix} i & 0 \\ 0 & -i \end{pmatrix}, \quad F(j)=\begin{pmatrix} 0 & 1 \\ -1 & 0 \end{pmatrix}, \quad F(k)=\begin{pmatrix} 0 & i \\ i & 0 \end{pmatrix}.$$

Then one finds at once, for $x=t+ui+vj+wk$, with t, u, v, w in **R**:

$$\tau(x)=x+\bar{x}=2t, \quad v(x)=x\bar{x}=\bar{x}x=t^2+u^2+v^2+w^2.$$

This shows that v maps \mathbf{H}^\times onto \mathbf{R}_+^\times; by corollary 3 of prop. 6, Chap. IX-2, the same is therefore true for $A=M_m(\mathbf{H})$ for any $m \geqslant 1$.

§ 3. **Computation of some integrals.** Here, in preparation for the computation of the zeta-function of a simple algebra in Chapter XI, we carry out some local calculations, generalizing the results of prop. 11, Chap. VII-4, and of lemma 8, Chap. VII-6.

Take first a p-field K and a division algebra D over K. Call R_D the maximal compact subring of D, P_D the maximal ideal in R_D, and π_D a prime element of D. For each $e \geqslant 0$, choose a full set $A(e)$ of representatives of the classes modulo P_D^e in R_D. Now, for a given $m \geqslant 1$, we define three subsets \mathfrak{X}, \mathfrak{X}', \mathfrak{X}'' of $M_m(D)^\times$, as follows. By \mathfrak{X}, we understand the

group of "triangular" matrices in $M_m(D)^\times$, consisting of the matrices $t=(t_{ij})$ such that $t_{ij}=0$ for $1\leqslant j<i\leqslant m$, and $t_{ii}\neq 0$ for $1\leqslant i\leqslant m$. By \mathfrak{T}', we understand the subset of \mathfrak{T}, consisting of the matrices $t=(t_{ij})$ in \mathfrak{T}, such that $t_{ij}\in R_D$ for all i, j, and that each t_{ii}, for $1\leqslant i\leqslant m$, is of the form $\pi_D^{e_i}$ with $e_i\geqslant 0$. By \mathfrak{T}'', we understand the subset of \mathfrak{T}', consisting of the matrices $t=(t_{ij})$ in \mathfrak{T}', such that $t_{ij}\in A(e_j)$ for $1\leqslant i<j\leqslant m$, e_j being given by $t_{jj}=\pi_D^{e_j}$. With these notations, we have:

LEMMA 2. *Let V be a left vector-space of dimension m over D; let L be a D-lattice in V, and let $\{v_1,\ldots,v_m\}$ be a basis of V such that $L=\sum R_D v_i$. Let L' be a D-lattice in V, contained in L. Then there is one and only one basis $\{v_1',\ldots,v_m'\}$ of V, such that $L'=\sum R_D v_i'$ and that $v_i'=\sum_j t_{ij}v_j$ for $1\leqslant i\leqslant m$, with a matrix $t=(t_{ij})$ belonging to the set \mathfrak{T}''.*

For $1\leqslant i\leqslant m$, call W_i the subspace of V generated by v_i,\ldots,v_m. Let $\{w_1,\ldots,w_m\}$ be any basis of V, and write it as $w_i=\sum x_{ij}v_j$. Clearly the matrix $x=(x_{ij})$ is in \mathfrak{T}, i.e. it is triangular, if and only if, for each i, $\{w_i,\ldots,w_m\}$ is a basis of W_i. By th. 1 of Chap. II-2, one can choose such a basis for which $L'=\sum R_D w_i$; then, since $L'\subset L$, all the x_{ij} are in R_D. Write $x_{ii}=y_i\pi_D^{e_i}$, with $y_i\in R_D^\times$ and $e_i\in\mathbf{Z}$, for $1\leqslant i\leqslant m$, and replace the w_i by the vectors $y_i^{-1}w_i$, which obviously have the same properties; after that is done, the matrix (x_{ij}) is in \mathfrak{T}'. Assume now that there are vectors v_1',\ldots,v_m' such as required by the lemma, and write $v_i'=\sum z_{ij}w_j$; clearly the matrix (z_{ij}) must then be triangular, and, as $L'=\sum R_D v_i'=\sum R_D w_i$, it must be in $M_m(R_D)^\times$; for a triangular matrix (z_{ij}), the latter condition is fulfilled if and only if $z_{ii}\in R_D^\times$ and $z_{ij}\in R_D$ for all i, j. Then the coefficient of v_i in v_i' is $z_{ii}\pi_D^{e_i}$, and, as this must be of the form $\pi_D^{e_i}$, we must have $z_{ii}=1$. Now, for $1\leqslant i<j\leqslant m$, the coefficient t_{ij} of v_j in v_i' is given by

$$t_{ij}=x_{ij}+\sum_{h=i+1}^{j-1}z_{ih}x_{hj}+z_{ij}\pi_D^{e_j},$$

and the proof of the lemma will be complete if one shows that the z_{ij}, for $1\leqslant i<j\leqslant m$, can be uniquely chosen in R_D so that $t_{ij}\in A(e_j)$ for $1\leqslant i<j\leqslant m$. For each value of i, this can be verified at once by induction on j for $i+1\leqslant j\leqslant m$.

LEMMA 3. *The set \mathfrak{T}'' is a full set of representatives for those left cosets of $M_m(R_D)^\times$ in $M_m(D)^\times$ which are contained in $M_m(R_D)$.*

Take a left vector-space V of dimension m over D, a D-lattice L in V, and a basis $\{v_1,\ldots,v_m\}$ of V such that $L=\sum R_D v_i$; as before, identify $\mathrm{End}(V)$ with $M_m(D)$ by assigning, to each $\xi\in\mathrm{End}(V)$, the matrix (x_{ij}) given by $v_i\xi=\sum x_{ij}v_j$: put $A=\mathrm{End}(V)$ and $R_A=\mathrm{End}(V,L)$; then $R_A=$

$=M_m(R_D)$, and R_A^\times, i.e. $M_m(R_D)^\times$, consists of the automorphisms ξ of V such that $L\xi=L$. Therefore two elements α, β of A^\times belong to the same left coset of R_A^\times if and only if $L\alpha=L\beta$; that left coset is contained in R_A if and only if $L\alpha \subset L$. At the same time, by lemma 2, every D-lattice L' of V, contained in L, can be written in one and only one way as $\sum R_D v_i'$ with $v_i'=\sum t_{ij} v_j$ and $(t_{ij})\in\mathfrak{T}''$; this is the same as to say that it can be written in one and only one way as $L\tau$ with $\tau\in\mathfrak{T}''$, which proves our lemma.

PROPOSITION 7. *Let K be a p-field with the module q; let D be a division algebra of dimension d^2 over K. Let A be a simple algebra over K, isomorphic to $M_m(D)$. Let R_A be a maximal order in A, and φ its characteristic function. Let v be the reduced norm in A, and let μ be the Haar measure in A^\times such that $\mu(R_A^\times)=1$. Then the integral*

$$I(s)=\int_{A^\times} \varphi(x)\operatorname{mod}_K(v(x))^s d\mu(x),$$

where $s\in\mathbf{C}$, is absolutely convergent for $\operatorname{Re}(s)>d(m-1)$ and has then the value

$$I(s)=\prod_{i=0}^{m-1}(1-q^{di-s})^{-1}.$$

As before, identify A with $\operatorname{End}(V)$ and R_A with $\operatorname{End}(V,L)$, where V is a left vector-space of dimension m over D, and L a D-lattice in V. By prop. 1 of § 1, the integrand in $I(s)$ is constant on left cosets of R_A^\times in A^\times; in view of the definition of μ, this gives:

$$I(s)=\sum_{\tau}\operatorname{mod}_K(v(\tau))^s,$$

the sum being taken over any full set of representatives of those left cosets of R_A^\times in A^\times which are contained in R_A, and for instance over the set \mathfrak{T}'' supplied by lemma 3. If we identify now A with $M_m(D)$ and R_A with $M_m(R_D)$ as before, \mathfrak{T}'' consists of the triangular matrices $t=(t_{ij})$ such that $t_{ii}=\pi_D^{e_i}$ and $t_{ij}\in A(e_j)$ for all i,j, the e_i being integers $\geqslant 0$. By corollary 2 of prop. 6, Chap. IX-2, we have then $v(t)=\prod v_D(t_{ii})=v_D(\pi_D)^E$ with $E=\sum e_i$, v_D being the reduced norm in D; as we have seen in § 2, $v_D(\pi_D)$ is a prime element of K if π_D has been suitably chosen, and this implies that the same is true for any choice of π_D. This gives $\operatorname{mod}_K(v(t))=$ $=q^{-E}$. On the other hand, prop. 5 of Chap. I-4 shows that the module of D is q^d, so that, for each $e\geqslant0$, the set $A(e)$ consists of q^{de} elements. Therefore, for a given set of integers e_1,\dots,e_m, there are q^{dN} matrices $t\in\mathfrak{T}''$, with $N=\sum_i(i-1)e_i$. Thus we get:

$$I(s)=\prod_{i=1}^{m}\left(\sum_{e=0}^{+\infty}(q^{d(i-1)-s})^e\right).$$

Clearly this is absolutely convergent for $\text{Re}(s) > d(m-1)$ and has then the value stated in our proposition.

COROLLARY. *Let $I(s)$ be as in proposition 7, and let $I_0(s)$ be similarly defined for the algebra $A_0 = M_n(K)$ with $n = dm$. Then we have, for $\text{Re}(s) > n-1$:*

$$I(s)I_0(s)^{-1} = \prod_{\substack{0 < h < n \\ h \not\equiv 0(d)}} (1 - q^{h-s}).$$

This follows at once from proposition 7.

We will also need the corresponding results for algebras over **R**-fields. Here we have either $K = \mathbf{R}$, and $D = \mathbf{R}$ or **H**, or $K = D = \mathbf{C}$. In all three cases, $x \to \bar{x}$ is an antiautomorphism of D such that $x\bar{x} > 0$ for all $x \in D^\times$; it is the identity if $D = \mathbf{R}$, the non-trivial automorphism of **C** over **R** if $D = \mathbf{C}$, and it is as defined at the end of § 2 if $D = \mathbf{H}$. As usual, if $x = (x_{ij})$ is any matrix in $M_m(D)$, we write ${}^t x$ for its transpose, and \bar{x} for the matrix (\bar{x}_{ij}); then $x \to {}^t\bar{x}$ is an antiautomorphism of $M_m(D)$. We will write \mathfrak{T} for the set of all triangular matrices (t_{ij}) in $M_m(D)$ such that $t_{ii} \in \mathbf{R}_+^\times$ for $1 \leqslant i \leqslant m$; clearly this is a subgroup of $M_m(D)^\times$. Now let V be a left vector-space of dimension m over D. For the sake of brevity, and although this does not quite agree with the established usage, we will say that a mapping f of $V \times V$ into D is a *hermitian form* on V if there is a basis $\{v_1, ..., v_m\}$ of V such that, for all x_i, y_i in D:

$$f(\textstyle\sum x_i v_i, \sum y_i v_i) = \sum x_i \bar{y}_i;$$

every basis of V with that property will then be called *orthonormal for f*. One sees at once that a basis $\{w_1, ..., w_m\}$ is orthonormal for f if and only if $f(w_i, w_j) = \delta_{ij}$ for all i, j, or even if this is so merely for $1 \leqslant i \leqslant j \leqslant m$. We topologize the space of all hermitian forms on V by the topology of "uniform convergence on compact subsets" of $V \times V$; in other words, for each compact subset C of $V \times V$, and each $\varepsilon > 0$, the set of the hermitian forms f' such that $\text{mod}_D(f'-f) \leqslant \varepsilon$ on C is to be a neighborhood of f, and these make up a fundamental system of neighborhoods of f in the space of hermitian forms.

LEMMA 4. *For $D = \mathbf{R}$, **H** or **C**, let V be a left vector-space over D, with the basis $\{v_1, ..., v_m\}$, and let f be a hermitian form on V. Then there is one and only one orthonormal basis $\{v_1', ..., v_m'\}$ for f such that $v_i' = \sum t_{ij} v_j$ with $(t_{ij}) \in \mathfrak{T}$, and it depends continuously upon f.*

The proof is straightforward, and so well known that it may be omitted here.

Let V be as above; let f be a hermitian form on V, with the orthonormal basis $\{v_1, ..., v_m\}$. Let $\{w_1, ..., w_m\}$ be a basis of V, given by $w_i = \sum u_{ij} v_j$ with $u = (u_{ij}) \in M_m(D)$; a trivial calculation shows at once that this is orthonormal for f if and only if $u \cdot {}^t\bar{u} = 1_m$; clearly the matrices u with that property make up a compact subgroup of $M_m(D)^\times$, which we will denote by \mathfrak{U}. Now let α be any automorphism of V; we will write f^α for the transform of f by α, i.e. for the mapping defined by $f^\alpha(v, w) = {} = f(v\alpha^{-1}, w\alpha^{-1})$ for all v, w in V; this is a hermitian form, with the orthonormal basis $\{v_1 \alpha, ..., v_m \alpha\}$. Clearly, when we identify $A = \mathrm{End}(V)$ with $M_m(D)$, as before, by means of the basis $\{v_1, ..., v_m\}$, \mathfrak{U} is the subgroup of $A^\times = M_m(D)^\times$ consisting of the automorphisms ξ of V such that $f^\xi = f$.

LEMMA 5. *The subgroups \mathfrak{T} and \mathfrak{U} of $A^\times = M_m(D)^\times$ being as defined above, the mapping $(u, t) \to ut$ is a homeomorphism of $\mathfrak{U} \times \mathfrak{T}$ onto A^\times.*

Let V, f and the orthonormal basis $\{v_1, ..., v_m\}$ be as above; use that basis again to identify $A = \mathrm{End}(V)$ with $M_m(D)$, hence A^\times with $M_m(D)^\times$. Let α, β be in A^\times; we have $f^\alpha = f^\beta$ if and only if $f = f^{\beta \alpha^{-1}}$, i.e. if and only if $\beta \alpha^{-1} \in \mathfrak{U}$, or $\beta \in \mathfrak{U} \alpha$. Now, for any $\alpha \in A^\times$, apply lemma 4 to f^α; it shows that there is one and only one matrix $(t_{ij}) \in \mathfrak{T}$ such that the vectors $v_i' = \sum t_{ij} v_j$ make up an orthonormal basis for f^α. This is the same as to say that the automorphism τ of V which corresponds to that matrix, i.e. which maps $\{v_1, ..., v_m\}$ onto $\{v_1', ..., v_m'\}$, transforms f into $f^\tau = f^\alpha$. Moreover, by lemma 4, the matrix (t_{ij}) depends continuously upon f^α, hence upon α. Expressing this in terms of the matrices x, u, t in $M_m(D)^\times$ which correspond respectively to α, $\alpha \tau^{-1}$ and τ, we get the assertion in our lemma.

LEMMA 6. *Notations being as in lemma 5, let μ be a Haar measure on A^\times. For every continuous function F with compact support on \mathfrak{T}, call F' the function on A^\times such that $F'(ut) = F(t)$ for all $u \in \mathfrak{U}$ and all $t \in \mathfrak{T}$. Then F' is continuous with compact support on A^\times, and there is a right-invariant measure θ on \mathfrak{T} such that $\int F' d\mu = \int F d\theta$ for all F.*

The first assertion follows at once from lemma 5 and the compacity of the group \mathfrak{U}. Then, as μ is right-invariant on A^\times by lemma 1, § 1, it is obvious that the mapping $F \to \int F' d\mu$ is invariant under right translations in \mathfrak{T}. By the theory of the Haar measure, this shows that θ is the image of a Haar measure, i.e. of a left-invariant measure on \mathfrak{T}, under the homeomorphism $t \to t^{-1}$ of \mathfrak{T} onto itself.

LEMMA 7. *Let α be a Haar measure on D; put $\delta = [K : \mathbf{R}]$, and write, for $t = (t_{ij}) \in \mathfrak{T}$:*

$$d\theta(t) = \prod_{i=1}^{m} (t_{ii}^{-\frac{1}{2}d^2(i-1)-1} dt_{ii}) \cdot \prod_{1 \leq i < j \leq m} d\alpha(t_{ij}).$$

Then this defines a right-invariant measure on \mathfrak{T}.

As δd^2 is the dimension of D over \mathbf{R}, corollary 2 of th. 3, Chap. I-2, shows that, for every $a \in \mathbf{R}_+^\times$, the module of the automorphism $x \to xa$ of D is $a^{\delta d^2}$. A straightforward computation shows then at once, firstly that the measure θ in lemma 7 is invariant under $t \to tt'$ for every diagonal matrix $t' \in \mathfrak{T}$, and secondly that it is invariant under $t \to tt''$ for every matrix $t'' = (t_{ij}'') \in \mathfrak{T}$ such that $t_{ii}'' = 1$ for $1 \leqslant i \leqslant m$. As every matrix in \mathfrak{T} can be written in the form $t't''$, this proves our lemma.

PROPOSITION 8. *Take $K = \mathbf{R}$ and $D = \mathbf{R}$ or \mathbf{H}, or $K = D = \mathbf{C}$; call δ the dimension of K over \mathbf{R}, and d^2 that of D over K. Call τ the reduced trace and ν the reduced norm in the algebra $A = M_m(D)$ over K. Let μ be a Haar measure in A^\times. Then the integral*

$$I(s) = \int_{A^\times} \exp(-\pi \delta \tau({}^t\bar{x} \cdot x)) \operatorname{mod}_K(\nu(x))^s \, d\mu(x)$$

is absolutely convergent for $\operatorname{Re}(s) > d(m-1)$, *and, for a suitable choice of μ, it has then the value*

$$I(s) = (\pi \delta d)^{-m\delta ds/2} \prod_{i=0}^{m-1} \Gamma(\delta d(s-di)/2).$$

Clearly the first factor in the integrand of $I(s)$ is constant on left cosets of \mathfrak{U} in A^\times. Now, for any $u \in \mathfrak{U}$, put $z = \nu(u)$. If $D = K$, this means that $z = \det(u)$, so that $u \in \mathfrak{U}$ implies $z\bar{z} = 1$, hence $\operatorname{mod}_K(z) = 1$. If $K = \mathbf{R}$ and $D = \mathbf{H}$, we have, for all $x \in A$, $\nu(x) = \det(F(x))$, where F is an isomorphism of A into $M_{2m}(\mathbf{C})$; but then $x \to {}^tF({}^t\bar{x})$ is also such an isomorphism, so that $\nu(\bar{x}) = \nu(x)$; this implies that, for $u \in \mathfrak{U}$, $\nu(u)^2 = 1$, and therefore $\nu(u) = 1$ since we have seen in § 2 that ν maps $M_m(\mathbf{H})^\times$ into \mathbf{R}_+^\times. Therefore, in all cases, the second factor in the integrand of $I(s)$ is also constant on left cosets of \mathfrak{U} in A^\times. That being so, lemmas 6 and 7 show that, for a suitable choice of μ, $I(s)$ is the same as the integral with the same integrand, but taken on \mathfrak{T} with the measure $d\theta(t)$. The reduced trace τ_D in D is given by $\tau_D(x) = x$ if $D = K$, and $\tau_D(x) = x + \bar{x}$ if $K = \mathbf{R}$ and $D = \mathbf{H}$; in view of corollary 2 of prop. 6, Chap. IX-2, we have now, for $t = (t_{ij}) \in \mathfrak{T}$:

$$\tau({}^t\bar{t} \cdot t) = d \cdot \sum_{1 \leqslant i \leqslant j \leqslant m} \bar{t}_{ij} t_{ij}, \quad \nu(t) = \prod_{1 \leqslant i \leqslant m} (t_{ii})^d.$$

This gives:

$$I(s) = \prod_{i=1}^{m} \left(\int_0^{+\infty} \exp(-\pi \delta d t^2) \, t^{s_i - 1} \, dt \right) \cdot \left(\int_D \exp(-\pi \delta d \bar{t} t) \, d\alpha(t) \right)^{\frac{m(m-1)}{2}}$$

with $s_i = \delta d(s - di + d)$ for $1 \leqslant i \leqslant m$. The last factor is independent of s and is > 0. The other factors can be transformed into the usual integral for the gamma function by an obvious change of variable. Up to a

constant factor >0, which can be rendered equal to 1 by changing the Haar measure μ, the result is that stated in our proposition.

COROLLARY. *Assume, in proposition* 8, *that* $K=\mathbf{R}$ *and* $D=\mathbf{H}$; *let* $I(s)$ *be as defined there, and let* $I_0(s)$ *be the similarly defined integral for the algebra* $A_0 = M_n(\mathbf{R})$ *with* $n=2m$. *Then we have, for* $\mathrm{Re}(s)>n-1$:

$$I(s)I_0(s)^{-1} = \gamma \prod_{\substack{0<h<n \\ h \not\equiv 0(2)}} (s-h)$$

with a constant $\gamma > 0$.

This is an immediate consequence of proposition 8 and of the identities

$$\Gamma(s+1)=s\Gamma(s), \qquad \Gamma(s)=\pi^{-1/2}2^{s-1}\Gamma\left(\frac{s}{2}\right)\Gamma\left(\frac{s+1}{2}\right)$$

from the theory of the gamma function.

Simple algebras over A-fields

§ 1. Ramification. In this Chapter, k will be an A-field; we use all the notations introduced for such fields in earlier Chapters, such as k_A, k_v, r_v, etc. We shall be principally concerned with a simple algebra A over k; as stipulated in Chapter IX, it is always understood that A is central, i.e. that its center is k, and that it has a finite dimension over k; by corollary 3 of prop. 3, Chap. IX-1, this dimension can then be written as n^2, where n is an integer $\geqslant 1$. We use A_v, as explained in Chapters III and IV, for the algebra $A_v = A \otimes k_v$ over k_v, where, in agreement with Chapter IX, it is understood that the tensor-product is taken over k. By corollary 1 of prop. 3, Chap. IX-1, this is a simple algebra over k_v; therefore, by th. 1 of Chap. IX-1, it is isomorphic to an algebra $M_{m(v)}(D(v))$, where $D(v)$ is a division algebra over k_v; the dimension of $D(v)$ over k_v can then be written as $d(v)^2$, and we have $m(v)d(v) = n$; the algebra $D(v)$ is uniquely determined up to an isomorphism, and $m(v)$ and $d(v)$ are uniquely determined. One says that A is *unramified* or *ramified* at v according as A_v is trivial over k_v or not, i.e. according as $d(v) = 1$ or $d(v) > 1$.

THEOREM 1. *Let A be a simple algebra over an A-field k; let α be a finite subset of A, containing a basis of A over k. For each finite place v of k, call α_v the r_v-module generated by α in A_v. Then, for almost all v, A_v is trivial over k_v, and α_v is a maximal compact subring of A_v.*

By corollary 1 of th. 3, Chap. III-1, we may assume that α is a basis of A over k, and that 1_A belongs to it. Call τ the reduced trace in A; by prop. 6 of Chap. IX-2, it is not 0, and its k_v-linear extension to A_v is the reduced trace in A_v. By lemma 3 of Chap. III-3, we may identify the underlying vector-space of A over k with its algebraic dual by putting $[x, y] = \tau(xy)$. Now, as in th. 3 of Chap. IV-2, take a "basic character" χ of k_A. By corollary 1 of that theorem, χ_v is of order 0 for almost all v; by corollary 3 of the same theorem, the k_v-lattice α_v is its own dual for almost all v, when A_v is identified with its topological dual by putting $\langle x, y \rangle = \chi_v(\tau(xy))$. By corollary 2 of th. 3, Chap. III-1, α_v is a compact subring of A_v for almost all v. Therefore, at almost all places v of k, the assumptions of corollary 2 of prop. 5, Chap. X-2, are valid, the conclusion being as stated in our theorem.

The first part of theorem 1 can be expressed by saying that A is unramified at almost all places of k. The object of § 2 will be to show that it cannot be unramified at all places of k unless it is trivial.

§ 2. **The zeta-function of a simple algebra.** Let all notations be as in § 1, and let α be a basis of A over k. By th. 1 of § 1, α_v is a maximal compact subring of A_v for almost all v; therefore we may, for each finite place v of k, choose a maximal compact subring R_v of A_v, in such a way that $R_v = \alpha_v$ for almost all v; that being done, call Φ_v the characteristic function of R_v. For each infinite place v of k, choose an isomorphism of A_v with $M_{m(v)}(D(v))$, where $D(v)$ is **R**, **H** or **C**, as the case may be; identifying A_v with the latter algebra by means of that isomorphism, define Φ_v on A_v by putting, for all $x \in A_v$, $\Phi_v(x) = \exp(-\pi \delta \tau({}^t\bar{x} \cdot x))$, where notations are the same as in prop. 8 of Chap. X-3. Then $\Phi = \prod \Phi_v$ is a standard function on A_A. Taking now a Haar measure μ on A_A^\times, we have:

PROPOSITION 1. *The integral*

$$Z_A(s) = \int\limits_{A_A^\times} \Phi(z)|\nu(z)|_A^s \, d\mu(z)$$

is absolutely convergent for $\mathrm{Re}(s) > n$ *and is then given by the formula*

$$Z_A(s) = C \prod_{i=0}^{n-1} Z_k(s-i) \cdot \prod_v \left(\prod_{\substack{0 < h < n \\ h \not\equiv 0(d(v))}} (1 - q_v^{h-s}) \right) \cdot \left(\prod_{\substack{0 < h < n \\ h \not\equiv 0(2)}} (s-h) \right)^\rho,$$

where Z_k *is the function defined in theorem 3 of Chap. VII -6, or the zeta-function of* k, *according as* k *is of characteristic 0 or not, where* ρ *is the number of real places* v *of* k *for which* $D(v) =$ **H**, *and* C *is a constant* > 0.

For each v, choose a Haar measure μ_v on A_v^\times, so that $\mu_v(R_v^\times) = 1$ for all finite places v of k; we may then assume that we have taken $\mu = \prod \mu_v$, in the same sense as has been explained in Chap. VII-4 for the case $A = k$. By following step by step the proof of prop. 10, Chap. VII-4, one finds that the integral $Z_A(s)$ is absolutely convergent, and equal to the infinite product

$$\prod_v (\int\limits_{A_v^\times} \Phi_v(x)|\nu(x)|_v^s \, d\mu_v(x)),$$

whenever the factors in that product, and the product itself, are absolutely convergent. Those factors have been calculated in propositions 7 and 8 of Chap. X-3; the absolute convergence of $Z_A(s)$ for $\mathrm{Re}(s) > n$ is then an immediate consequence of the latter results, combined with prop. 1 of Chap. VII-1. The same results, combined with the definitions in Chap. VII-6, give now the final formula in our proposition for the case $A = M_n(k)$; combining this with the corollaries of propositions 7 and 8 of Chap. X-3, one obtains at once the general case of the same formula.

One should note that the middle product, in the formula for $Z_A(s)$ in proposition 1, is a finite one by th. 1 of § 1, since that theorem shows that $d(v)=1$ for almost all the places of k. It should also be pointed out that the computation of the constant C in that formula, for an explicitly given Haar measure μ on A_A^\times, offers no difficulty, and that it is important for some purposes, e.g. for the determination of the "Tamagawa number" of the subgroup $A^{(1)}$ of A^\times given by $v(x)=1$. As this lies beyond the scope of the present volume, it will not be pursued any further.

PROPOSITION 2. *Let* D *be a division algebra of dimension* d^2 *over* k, *and let* $Z_D(s)$ *be defined as in proposition* 1. *Then, if* k *is of characteristic* 0, $Z_D(s)$ *has no other pole than* $s=0$ *and* $s=d$; *if* k *is of characteristic* $p>1$ *and has the field of constants* \mathbf{F}_q, $Z_D(s)$ *has no other pole than the zeros of* $(1-q^{-s})(1-q^{d-s})$.

This will be proved by following step by step the proof of th. 2, Chap. VII-5. In analogy with that proof, it will be convenient to adopt the following notations. For $z \in D_A^\times$ and $s \in \mathbf{C}$, write $\omega_s(z)=|v(z)|_A^s$; ω_1 is then a morphism of D_A^\times into \mathbf{R}_+^\times. For $\xi \in D^\times$, $v(\xi)$ is in k^\times; therefore, by th. 5 of Chap. IV-4, ω_1 is trivial on D^\times. Consider first the case where k is of characteristic $p>1$; by prop. 6 of Chap. X-2, v maps D_v^\times onto k_v^\times for all v, so that ω_1 maps D_v^\times onto the subgroup of \mathbf{R}_+^\times generated by q_v; by corollary 6 of th. 2, Chap. VII-5, this implies that ω_1 maps D_A^\times onto the subgroup of \mathbf{R}_+^\times generated by q, if \mathbf{F}_q is the field of constants of k. In that case, take $z_1 \in D_A^\times$ such that $\omega_1(z_1)=q$, and call M the subgroup of D_A^\times generated by z_1; then D_A^\times is the product of M and of the kernel D_A^1 of ω_1. On the other hand, if k is of characteristic 0, call M the subgroup of k_A^\times defined in corollary 2 of th. 5, Chap. IV-4; k being identified with the center of D, k_A^\times is to be considered as a subgroup of D_A^\times; as $v(z)=z^d$ for $z \in k$, the corollary we have just quoted shows that ω_1 maps M onto \mathbf{R}_+^\times, so that D_A^\times is again the product of M and of the kernel D_A^1 of ω_1. In both cases, th. 4 of Chap. IV-3 shows that D_A^1/D^\times is compact.

As in the proof of th. 1, § 1, take a basic character χ of k_A; identify D_A with its topological dual by $\langle x,y \rangle = \chi(\tau(xy))$, and, for each v, identify D_v with its topological dual by $\langle x,y \rangle = \chi_v(\tau(xy))$. For each v, call α_v the self-dual Haar measure on D_v; then, by corollary 1 of th. 1, Chap. VII-2, the measures α_v are coherent, and $\alpha = \prod \alpha_v$ is the Tamagawa measure on D_A.

Let again $\Phi = \prod \Phi_v$ be the standard function on D_A which was used above in the construction of $Z_D(s)$, and let $\Psi = \prod \Psi_v$ be any standard function on D_A. Call $Z(\Psi,s)$ the integral obtained by replacing Φ by Ψ in the definition of $Z_D(s)$; with the present notations, this can be written:

$$Z(\Psi,s)=\int_{D_A^\times} \Psi(z)\omega_s(z)d\mu(z).$$

It will now be shown that this is absolutely convergent for $\mathrm{Re}(s)>d$ and that it can be continued as a meromorphic function in the whole s-plane, with no other poles than those mentioned in proposition 2; this will then contain that proposition as a special case. As to the convergence, we have $\Psi_v=\Phi_v$ for almost all v, by the definition of a standard function. For every finite place v of k, the support R_v of Φ_v is an open subgroup of D_v, and the support S_v of Ψ_v is compact; then, choosing $a_v\in k_v^\times$ such that $a_vS_v\subset R_v$, and putting $\gamma_v=\sup|\Psi_v|$, we have $|\Psi_v(x)|\leqslant\gamma_v\Phi_v(a_vx)$ for all $x\in D_v$. Similarly, the definition of standard functions shows at once that, for any infinite place v, one can find ε_v and γ_v in \mathbf{R}_+^\times such that $|\Psi_v(x)|\leqslant$ $\leqslant\gamma_v\Phi_v(\varepsilon_vx)$ for all $x\in D_v$. This shows that there is $a\in k_\mathbf{A}^\times$ and $\gamma\in\mathbf{R}_+^\times$ such that $|\Psi(x)|\leqslant\gamma\Phi(ax)$ for all $x\in D_\mathbf{A}$. Therefore the integral $Z(\Psi,s)$, for $\mathrm{Re}(s)=\sigma$, is majorized by

$$\gamma\int_{D_\mathbf{A}^\times}\Phi(az)\omega_\sigma(z)d\mu(z)=\gamma\omega_\sigma(a^{-1})Z_D(\sigma),$$

which, by prop. 1, is convergent for $\sigma>d$.

Now take the same two functions F_0, F_1 as in the proof of th. 2 of Chap. VII-5; $Z(\Psi,s)$ is then the sum of the two integrals

$$Z_i=\int_{D_\mathbf{A}^\times}\Phi(z)\omega_s(z)F_i(\omega_1(z))d\mu(z).$$

Exactly as in that proof (but taking now $B>d$), we see that Z_0 is absolutely convergent for all s and defines therefore an entire function of s, and that the same is true for the integral Z_0' obtained by replacing Ψ by its Fourier transform Ψ', s by $d-s$ and F_0 by $t\to F_1(t^{-1})$ in the definition of Z_0. Just as there, one can also apply Poisson's summation formula (i.e. formula (1) of Chap. VII-2), in combination with lemma 1 of Chap. VII-2, to the function $x\to\Psi(zx)$ on $D_\mathbf{A}$; in applying the latter lemma, one has to use the fact that the module of the automorphism $x\to z^{-1}x$ of $D_\mathbf{A}$, for $z\in D_\mathbf{A}^\times$, is

$$|N_{D/k}(z^{-1})|_\mathbf{A}=|\nu(z)|_\mathbf{A}^{-d}=\omega_{-d}(z).$$

Then, proceeding exactly as in the proof in question, one finds that $Z(\Psi,s)$ is the sum of the entire function Z_0+Z_0' and of the integral

$$f(s)=\int_{D_\mathbf{A}^\times/D}(\Psi'(0)-\omega_d(z)\Psi(0))\omega_{s-d}(z)F_1(\omega_1(z))d\mu(z).$$

Here the integrand is constant on the cosets of the compact subgroup $G_1=D_\mathbf{A}^1/D^\times$ of the group $G=D_\mathbf{A}^\times/D^\times$. As G_1 is the kernel of the morphism of G into \mathbf{R}_+^\times determined by ω_1, we may identify G/G_1 with the image N of G in \mathbf{R}_+^\times under that morphism, which is \mathbf{R}_+^\times or the group generated by

q, according to the characteristic of k. In view of lemma 6 of Chap. VII-5, and taking for v the measure occurring in that lemma, we have therefore (up to a constant factor which may be made equal to 1 by a suitable choice of μ):

$$f(s) = \int_N (\Psi'(0) - n^d \Psi(0)) n^{s-d} F_1(n) dv(n) = \Psi'(0) \lambda(s-d) - \Psi(0) \lambda(s)$$

where λ is as defined there. In view of the statement about the poles of λ in that lemma, this completes our proof.

Now the comparison between propositions 1 and 2 will give us the main result of this Chapter.

THEOREM 2. *A simple algebra A over an A-field k is trivial if and only if it is everywhere unramified, i.e. if and only if A_v is trivial over k_v for every place v of k.*

It is clearly enough to prove this for a division algebra D. If D_v is trivial for all v, prop. 1 shows that its zeta-function $Z_D(s)$, up to a constant factor, is given by

$$Z_D(s) = \prod_{i=0}^{d-1} Z_k(s-i).$$

In view of theorems 3 and 4 of Chap. VII-6, this has poles of order 2 at $s = 1, 2, \ldots, d-1$ if $d > 1$. By prop. 2, this cannot be. Therefore $d = 1$, and $D = k$.

Actually the combination of propositions 1 and 2 allows one to draw stronger conclusions than theorem 2; for instance, it shows at once that, if $d > 1$, D must be ramified at least at two places of k. This need not be pursued any further now, since much stronger results will be obtained in Chapter XIII.

§ 3. Norms in simple algebras. As a first application of theorem 2, we will now reproduce Eichler's proof for the following:

PROPOSITION 3. *Let A be a simple algebra over an A-field k, and let v be the reduced norm in A. Then $v(A^\times)$ is the subgroup γ of k^\times, consisting of the elements whose image in k_v is > 0 for every real place v of k where A is ramified.*

That proof depends upon the following lemmas.

LEMMA 1. *Let K be a commutative p-field, $L = K(\xi)$ a separably algebraic extension of K of degree n, and put*

$$F(X) = N_{L/K}(X - \xi) = X^n + \sum_{i=1}^{n} a_i X^{n-i}.$$

Let $G(X) = \sum_{i=1}^{n} b_i X^{n-i}$ be a polynomial of degree $n-1$ in $K[X]$. Then, if all the coefficients of G are sufficiently close to 0 in K, the polynomial $F + G$ is irreducible over K and has a root in L.

It will be convenient to extend mod_K to a mapping $x \to |x|$ of an algebraic closure \bar{K} of L into \mathbf{R}_+ by putting $|x| = \text{mod}_{K'}(x)^{1/\nu}$ whenever $K(x) \subset K' \subset \bar{K}$, and K' has the finite degree ν over K; by corollary 2 of th. 3, Chap. I-2, this is independent of the choice of K' when x is given in \bar{K}. Take $A \in \mathbf{R}_+^\times$ such that $|a_i| \leqslant A^i$ for $1 \leqslant i \leqslant n$, and assume, for some $B < A$, that $|b_i| \leqslant B^i$ for $1 \leqslant i \leqslant n$. Let η be any root of $F + G$ in \bar{K}; then we have

$$\eta^n = - \sum_{i=1}^{n} (a_i + b_i) \eta^{n-i}$$

and therefore

$$|\eta|^n \leqslant \sup_i (A^i |\eta|^{n-i}),$$

hence $|\eta| \leqslant A$. Call ξ_1, \dots, ξ_n the roots of F; they are all distinct, since L is separable over K, and they are the images of ξ under the automorphisms of \bar{K} over K. What we have proved for η can be applied to the ξ_ν, by taking $G = 0$, so that $|\xi_\nu| \leqslant A$ for $1 \leqslant \nu \leqslant n$. Now put

$$\alpha = \inf_{1 \leqslant \mu < \nu \leqslant n} |\xi_\nu - \xi_\mu|;$$

we have $0 < \alpha \leqslant A$. Assume now that we have taken $B < A(\alpha/A)^n$. As η is a root of $F + G$, we have

$$\prod_{\nu=1}^{n} (\eta - \xi_\nu) = - \sum_{i=1}^{n} b_i \eta^{n-i}$$

and therefore

$$\inf_\nu |\eta - \xi_\nu|^n \leqslant \sup_i (B^i A^{n-i}) \leqslant B A^{n-1} < \alpha^n,$$

so that there is ν such that $|\eta - \xi_\nu| < \alpha$. Clearly this implies that $|\eta - \xi_\mu| \geqslant \alpha$ for all $\mu \neq \nu$. Let σ be an automorphism of \bar{K} over K, mapping ξ_ν onto ξ. After replacing η by η^σ, which is also a root of $F + G$, we see that $|\eta - \xi| < \alpha$ and $|\eta - \xi_\nu| \geqslant \alpha$ for all $\xi_\nu \neq \xi$. Assume that L is not contained in $K(\eta)$; then there is an automorphism τ of \bar{K} over $K(\eta)$ such that $\xi^\tau \neq \xi$; as this must leave $|\eta - \xi|$ invariant, we get a contradiction. Therefore $K(\eta) \supset L$; as η is at most of degree n over K, this implies that $K(\eta) = L$ and that $F + G$ is irreducible.

Incidentally, since every extension of K of degree n can obviously be generated by a root of a monic polynomial F of degree n with coefficients in the maximal compact subring of K, lemma 1 shows that K has at most

* finitely many separable extensions of given degree, hence also (by corollary 2 of prop. 4, Chap. I-4) finitely many algebraic extensions of given degree.

LEMMA 2. *Let K be a commutative p-field, R its maximal compact subring, and L an unramified extension of K. Then, for every $x \in R^\times$, there is $y \in L$ such that $N_{L/K}(y) = x$ and that $K(y) = L$.*

Call n the degree of L over K; call δ the number of divisors of n; since L is cyclic over K, δ is also the number of distinct fields between K and L. We will first construct $\varepsilon \in L$ such that $N_{L/K}(\varepsilon) = 1$ and that $L = K(\varepsilon^i)$ for $1 \leqslant i \leqslant \delta$. Take a common multiple D of the integers $1, 2, \ldots, \delta$, e.g. $D = \delta!$. Call α a generator of the Galois group of L over K. For $1 \leqslant h \leqslant n-1$, consider the mapping

$$\xi \to P_h(\xi) = (\xi^{\alpha^{h+1}} \xi)^D - (\xi^{\alpha^h} \xi^\alpha)^D$$

of L into itself. This is a polynomial mapping, when L is regarded as a vector-space over K, as one sees at once by choosing a basis for L over K and expressing ξ in terms of that basis. Taking again for \bar{K} an algebraic closure of L, we can extend the mappings P_h to the algebra $\mathscr{L} = L \otimes_K \bar{K}$ over \bar{K}. Now apply prop. 3 of Chap. III-2 to that algebra and to the n distinct isomorphisms α^i of L into \bar{K}, for $0 \leqslant i \leqslant n-1$. As in that proposition, call μ_i the \bar{K}-linear extension of α^i to \mathscr{L}, and put $\varphi = (\mu_0, \ldots, \mu_{n-1})$; that proposition shows that φ is an isomorphism of \mathscr{L} onto \bar{K}^n. Then the mapping $\mu_0 \circ P_h \circ \varphi^{-1}$ of \bar{K}^n into \bar{K} is given by

$$(x_0, \ldots, x_{n-1}) \to (x_{h+1} x_0)^D - (x_h x_1)^D,$$

where it should be understood, for $h = n-1$, that $x_n = x_0$. As this is not 0. and as K is an infinite field, we see now that none of the P_h is 0 and that one can choose $\xi \in L$ such that $P_h(\xi) \neq 0$ for $1 \leqslant h \leqslant n-1$. Let ξ be so chosen, and put $\varepsilon = \xi^\alpha \xi^{-1}$. Then $N_{L/K}(\varepsilon) = 1$, and the images $(\varepsilon^{\alpha^h})^D$ of ε^D, under the automorphisms α^h with $1 \leqslant h \leqslant n-1$, are all $\neq \varepsilon^D$, so that $L = K(\varepsilon^D)$. As D is a multiple of i for $1 \leqslant i \leqslant \delta$, we have, for each such i, $\varepsilon^D = (\varepsilon^i)^{D/i}$, hence $K(\varepsilon^D) \subset K(\varepsilon^i)$, so that $L = K(\varepsilon^i)$. Now take any $x \in R^\times$; by prop. 3 of Chap. VIII-1, we can write it in the form $x = N_{L/K}(y_1)$ with $y_1 \in L^\times$. Consider the infinite sequence of fields $K_i = K(\varepsilon^i y_1)$, for all $i \geqslant 0$. At most δ of them can be distinct; therefore there are pairs (i,j) of integers such that $0 \leqslant i < j$ and $K_i = K_j$, and, if we take such a pair for which $j - i$ has the smallest value, we have $0 < j - i \leqslant \delta$. As $\varepsilon^i y_1$ and $\varepsilon^j y_1$ are both in K_i, ε^{j-i} is in K_i. In view of our choice of ε, this implies $K_i = L$. Thus $y = \varepsilon^i y_1$ satisfies the requirements in our lemma.

We can now proceed to prove proposition 3. Call n^2 the dimension of the given algebra A over k, and R_∞ the set of the infinite places of k where

it is ramified. If v is in R_∞, it must be real, and A_v must be isomorphic to an algebra $M_m(\mathbf{H})$; as this implies $n = 2m$, this can only happen if n is even, and, of course, if k is of characteristic 0. We have seen in Chap. X-2 that the reduced norm ν maps $M_m(\mathbf{H})^\times$ onto \mathbf{R}_+^\times; therefore $\nu(A^\times)$ is contained in the group γ defined in proposition 3. Choose now a non-empty finite set R' of finite places of k, containing all the finite places of k where A is ramified, and put $R = R' \cup R_\infty$. Take any $v \in R'$, and a prime element π_v of k_v; by prop. 6 of Chap. X-2, there is $x_v \in A_v$ such that $v(x_v) = = \pi_v$. Apply corollary 2 of th. 3, Chap. IV-2, to A and to some place v_0 of k, not in R'; it shows that we can choose $\alpha \in A$ so that its image in A_v is arbitrarily close to x_v, and that, for all $w \neq v$ in R', its image in A_w is arbitrarily close to 1. In view of the continuity of ν, this can be done so that the image of $\nu(\alpha)$ in k_v is so close to π_v as to be a prime element of k_v, and that its image in k_w, for every $w \neq v$ in R', is so close to 1 as to be in r_v^\times; then $\alpha \in A^\times$, since $\nu(\alpha) \neq 0$. For each $v \in R'$, choose an element α_v of A^\times in this manner. Now take any ξ in the subgroup γ of k^\times defined in our proposition; we have to show that it is in $\nu(A^\times)$. For each $v \in R'$, put $n(v) = = \mathrm{ord}_v(\xi)$, and put $\alpha = \prod \alpha_v^{n(v)}$; after replacing ξ by $\xi \nu(\alpha)^{-1}$, we see that it is enough to prove our assertion under the additional assumption that $\mathrm{ord}_v(\xi) = 0$ for all $v \in R'$. For each place $v \in R'$, take an unramified extension k_v' of k_v, of degree n over k_v. By lemma 2, there is $y_v \in k_v'$ such that $\xi = = N_{k_v'/k_v}(y_v)$ and $k_v' = k_v(y_v)$. As y_v is then of degree n over k_v, it is the root of an irreducible polynomial F_v of degree n over k_v, given by:

$$F_v(X) = N_{k_v'/k_v}(X - y_v) = X^n + \sum_{i=1}^{n-1} a_{i,v} X^{n-i} + (-1)^n \xi,$$

with $a_{i,v} \in k_v$ for $1 \leqslant i \leqslant n-1$. For each $v \in R_\infty$, put $a_{i,v} = 0$ for $1 \leqslant i \leqslant n-1$, and consequently, since the existence of such a place implies that n is even:

$$F_v(X) = X^n + (-1)^n \xi = X^n + \xi;$$

then, because of our assumption $\xi \in \gamma$, F_v has no root in $k_v = \mathbf{R}$, so that the same is true of every monic polynomial of degree n over \mathbf{R} whose coefficients are close enough to those of F_v. Applying corollary 2 of th. 3, Chap. IV-2, to k and to some place v_0 of k, not in R, we see that we can choose $\omega_i \in k$, for $1 \leqslant i \leqslant n-1$, so that its image in k_v is arbitrarily close to $a_{i,v}$ for every $v \in R$. In view of lemma 1 and what has just been said, this can be done so that the polynomial

$$F(X) = X^n + \sum_{i=1}^{n-1} \omega_i X^{n-i} + (-1)^n \xi$$

has the following properties: (a) for every $v \in R'$, F is irreducible over k_v and has a root in k_v'; (b) for every $v \in R_\infty$, F has no root in $k_v = \mathbf{R}$. As R' is

not empty, (a) implies that F is irreducible over k and has no multiple roots. Call ζ a root of F in some algebraic closure of k, and put $k' = k(\zeta)$. Take any $v \in R'$, and a place w of k', lying above v; as the completion of k' at w must then be generated over k_v by a root of F, this completion, by (a), is isomorphic to k'_v, with which we can identify it; as it is of degree n over k_v, corollary 1 of th. 4, Chap. III-4, shows that w is the only place of k' lying above v, and that theorem shows then that we may identify $k'_w = k'_v$ with $k' \otimes_k k_v$. Similarly, (b) shows that, if v is in R_∞, all the places of k' lying above v are imaginary.

Now consider the algebra $A' = A_{k'}$ over k'. Take any place w of k', and call v the place of k lying below w. By the elementary properties of tensor-products, A'_w, which is the algebra $A' \otimes_{k'} k'_w$ over k', may be identified in an obvious manner with $A_v \otimes_{k_v} k'_w$. As A_v is trivial over k_v for v not in R, this shows that also A'_w must then be trivial. If v is in R_∞, w is imaginary, so that $k'_w = C$ and that A'_w is trivial. Finally, let v be in R', and write A_v as $M_{m(v)}(D(v))$, where $D(v)$ is a division algebra over k_v; if its dimension over k_v is $d(v)^2$, we have $n = m(v)d(v)$, so that $d(v)$ divides n. Then k'_w, which is unramified, hence cyclic, and of degree n over k_v, contains a field k'' which is of degree $d(v)$ over k_v, and is of course unramified over k_v. By prop. 5 of Chap. I-4, $D(v)$ contains a field isomorphic to k''; therefore, by corollary 6 of prop. 3, Chap. IX-1, $D(v)_{k''}$ is trivial over k''; obviously this implies that $(A_v)_{k''}$ is trivial over k'', hence that $A'_w = (A_v)_{k'_w}$ is so over k'_w.

Having thus shown that A' is unramified at all places of k', we can conclude, by th. 2 of §2, that it is trivial over k', which is the same as to say that A has a k'-representation into $M_n(k')$. Therefore, by th. 2 of Chap. IX-3, A has an \mathfrak{H}-regular factor-set, if \mathfrak{H} is the Galois group over k' of the separable algebraic closure k_{sep} of k'. Then, by lemma 4 of Chap. IX-3, we can construct an algebra of dimension n^2 over k, containing a field isomorphic to k', with the same factor-set as A; as this implies that it is similar to A, and as it has the same dimension as A over k, it is isomorphic to A and may be identified with it. As shown there, we have then $v(\zeta \cdot 1_A) = N_{k'/k}(\zeta) = \xi$.

§ 4. Simple algebras over algebraic number-fields.

We will now combine the results of § 1 with some of those of Chapter V in order to obtain a few basic results in the theory of ideals in simple algebras over algebraic number-fields.

In this §, k will be an algebraic number-field, \mathfrak{r} its maximal order, and all algebras will be simple algebras over k. We recall that, by prop. 4 of Chap. V-2, if L is any k-lattice in a vector-space E over k, and if v is a finite place of k, the closure L_v of L in E_v is the \mathfrak{r}_v-module generated by L in E_v.

Let D be a division algebra over k. As in Chap. X-1, let V, V', V'' be left vector-spaces of finite dimension over D, other than $\{0\}$; put $H = \mathrm{Hom}(V, V')$, $H' = \mathrm{Hom}(V', V'')$, $H'' = \mathrm{Hom}(V, V'')$. If X, X' are sub-groups of the additive groups of H and of H', respectively, we write XX', as usual, for the subgroup of H'' generated by the elements $\xi\xi'$ for $\xi \in X$, $\xi' \in X'$; it is easily seen, e. g. by taking bases for V, V', V'' over D, that $HH' = H''$. Now let L, L' be k-lattices, in H and in H', respectively, when these are regarded as vector-spaces over k; then LL' is obviously a finitely generated \mathfrak{r}-module in H'', and, as $H'' = HH'$, it is a k-lattice in H''.

PROPOSITION 4. *Let A be a simple algebra over k. Then there are maximal orders in A; these are k-lattices in A, and a k-lattice R in A is a maximal order if and only if its closure R_v in A_v is a maximal order in A_v for every finite place v of k. Every order in A is contained in a maximal order.*

If R is any order in A, the \mathfrak{r}-module generated by R in A is also an order, and it is a k-lattice; this shows that, unless R is a k-lattice, it cannot be maximal. Let X be any k-lattice in A; the last part of th. 1, § 1, may be expressed by saying that X_v is a maximal order in A_v for almost all v. Then th. 2 of Chap. V-2 shows that there is a one-to-one correspondence between the orders R in A which are k-lattices, and the possible choices of an order R_v in A_v for every finite place v of k, subject to the condition that R_v be a k_v-lattice for all v, and $R_v = X_v$ for almost all v; if R is given, R_v is the closure of R in A_v, and, if the R_v are given, R is defined by $R = \bigcap(A \cap R_v)$. In view of th. 1, Chap. X-1, all our assertions are now obvious.

PROPOSITION 5. *Let D be a division algebra over k. Let V, W be two left vector-spaces of finite dimension over D; put $H = \mathrm{Hom}(V, W)$ and $A = \mathrm{End}(V)$. Let M, M' be two k-lattices in H. Then the set X of the elements ξ of A such that $\xi M \subset M'$ is a k-lattice in A, whose closure in A_v, for every finite place v of k, is the set X_v of the elements x of A_v such that $xM_v \subset M'_v$. If $M = M'$, X is an order of A.*

For every v, by prop. 4 of Chap. X-1, X_v is a k_v-lattice in A_v, and it is an order if $M_v = M'_v$. Let L be any k-lattice in A; as we have seen above, LM is a k-lattice in H, whose closure in H_v, for every v, is clearly $L_v M_v$. Therefore, for almost all v, we have $L_v M_v = M_v = M'_v$; this implies that, for almost all v, X_v is an order and contains L_v. As L_v is a maximal order in A_v for almost all v, we see that $X_v = L_v$ for almost all v. Therefore, by th. 2 of Chap. V-2, there is a k-lattice $X' = \bigcap(A \cap X_v)$ in A with the closure X_v for all v. Clearly $X \subset X'$; conversely, since

$X'M \subset M'_v$ for every v, and $M' = \bigcap (H \cap M'_v)$, we have $X' \subset X$. This completes the proof, except for the last assertion, which is now obvious.

If notations and assumptions are as in proposition 5, and if $M = M'$, the set X is called *the left order* of M. Exchanging right and left, we see that the set of the elements η of $B = \mathrm{End}(W)$, such that $M\eta \subset M$, is an order of B; this is called *the right order* of M.

PROPOSITION 6. *Let V, W and M be as in proposition 5; assume that there is a maximal order R of $A = \mathrm{End}(V)$, such that M is a left R-module. Then R is the left order of M, and its right order is a maximal order of $B = \mathrm{End}(W)$.*

This is an immediate consequence of prop. 5, combined with th. 2 of Chap. X-1.

With the same notations and assumptions as in propositions 5 and 6, a k-lattice M in $\mathrm{Hom}(V, W)$ with the left order R and the right order S will be called an (R, S)-lattice; it is called a *normal lattice* if R or S, and consequently both R and S, are maximal orders. If $V = W$, hence $H = A = B$, a normal lattice is also known as a "normal fractional ideal". Clearly, in that case, the three relations $M \cdot M \subset M$, $M \subset R$, $M \subset S$ are equivalent; when they hold, M is a left ideal in the ring R and a right ideal in the ring S; it is then called a normal ideal and an (R, S)-ideal. By using the above results and those of Chap. X, one can see at once that, if R and S are any two maximal orders in A, there are always (R, S)-ideals. Furthermore, if a normal (R, S)-lattice M is a maximal left ideal in R, i. e. if $M \subset R$, $M \neq R$, and if there is no left ideal other than R and M between R and M, then it is a maximal right ideal in S, in the same sense; when that is so, one must have $M_v = R_v = S_v$ for all finite places v of k except one. If the multiplication law $(M, M') \to MM'$ is restricted to those pairs (M, M') of normal lattices in A for which the right order of M is the same as the left order of M', the normal lattices, for this law, make up a so-called "groupoid" whose units are the maximal orders of A. It is also easily seen that, for this law, every normal ideal can be written, although in general not uniquely, as a product of maximal ideals. For two-sided ideals and (R, R)-lattices, one has a more precise result:

PROPOSITION 7. *Let R be a maximal order in A. Then, for the law $(M, M') \to MM'$, the (R, R)-lattices in A make up a commutative group; it is the free group generated by the maximal two-sided ideals in R; for every prime ideal \mathfrak{p} in \mathfrak{r}, there is one such ideal, and only one, between R and $\mathfrak{p}R$.*

This follows in a quite straightforward manner from the above results and corollary 2 of th. 2, Chap. X-1.

Chapter XII

Local classfield theory

§ 1. The formalism of classfield theory. The purpose of classfield theory is to give a description of the abelian extensions of the types of fields studied in this book, viz., local fields and A-fields. Here we assemble part of the formal machinery common to both types.

LEMMA 1. *Let* $G = G_1 \times N$ *be a quasicompact group,* G_1 *being compact and* N *isomorphic to* **R** *or* **Z***; let* H *be an open subgroup of* G. *Then, if* H *is contained in* G_1 *(i.e. if it is compact),* N *is isomorphic to* **Z***, and* H *is of finite index in* G_1*; otherwise it is of finite index in* G.

Put $H_1 = H \cap G_1$; as this is open in G_1, and G_1 is compact, it is of finite index in G_1; this proves the first assertion. As $H \cap N$ is an open subgroup of N, it is N if N is isomorphic to **R**; therefore $H = H_1 \times N$ in that case, and G/H is isomorphic to G_1/H_1. If N is isomorphic to **Z**, let n_1 be a generator of N; if H is not contained in G_1, it has an element of the form $g_1 n_1^\mu$ with $g_1 \in G_1$, $\mu \in$ **Z**, $\mu \neq 0$. As G_1/H_1 is finite, there is $\nu \neq 0$ such that $g_1^\nu \in H_1$. Then $n_1^{\mu\nu}$ is in H, so that H contains the group H' generated by H_1 and $n_1^{\mu\nu}$. As H' is obviously of finite index in G, this proves the lemma. Theorem 1 of Chap. IV-4 may be regarded as the special case where $G = k_A^\times / k^\times$, H being the image of $\Omega(P)$ in k_A^\times / k^\times.

LEMMA 2. *Let* $G = G_1 \times N$, $G' = G_1' \times N'$ *be quasicompact groups,* G_1 *and* G_1' *being compact and* N, N' *isomorphic to* **R** *or* **Z***. Let* F *be a morphism of* G' *into* G *but not into* G_1*. Then* $F^{-1}(G_1) = G_1'$*; the kernel of* F *is compact;* $F(G')$ *is closed in* G, *and* $G/F(G')$ *is compact.*

As G_1 is the maximal compact subgroup of G, $F(G_1')$ is contained in G_1. For $n' \in N'$, call $f(n')$ the projection of $F(n')$ onto N in G; f is then a non-trivial morphism of N' into N, hence, obviously, an isomorphism of N' onto a closed subgroup of N with compact quotient; our first and second assertions follow from this at once. We also see now that F induces on N' an isomorphism of N' onto $F(N')$, and that $F(N') \cap G_1 = \{1\}$; therefore $F(G')$ is the direct product of $F(G_1')$ and $F(N')$ and is closed. Finally, $G/G_1 F(N')$ is clearly isomorphic to $N/f(N')$, hence compact; as the kernel of the obvious morphism of $G/F(G')$ onto $G/G_1 F(N')$ is the image of G_1 in $G/F(G')$, hence compact, $G/F(G')$ must also be compact.

From now on, in this §, we will consider a field K; later on, this will be either a local field or an A-field. As in Chapter IX, we write \bar{K} for an algebraic closure of K, K_{sep} for the union of all separable extensions of K contained in \bar{K}, and \mathfrak{G} for the Galois group of K_{sep} over K, topologized as usual. We will write K_{ab} for the maximal abelian extension of K contained in \bar{K}; this is the same as the union of all abelian extensions of K of finite degree, contained in \bar{K}, i.e. of all the Galois extensions of K of finite degree, contained in \bar{K}, whose Galois group is commutative; by definition, this is contained in K_{sep}. We denote by $\mathfrak{G}^{(1)}$ the subgroup of \mathfrak{G} corresponding to K_{ab}; this is the smallest closed normal subgroup of \mathfrak{G} such that $\mathfrak{G}/\mathfrak{G}^{(1)}$ is commutative; it is therefore the same as the "topological commutator-group" of \mathfrak{G}, i.e. the closure of the subgroup of \mathfrak{G} generated by the commutators of elements of \mathfrak{G}. We write \mathfrak{A} for the Galois group of K_{ab} over K; this may be identified with $\mathfrak{G}/\mathfrak{G}^{(1)}$; it is a compact commutative group. Let χ be any character of \mathfrak{G}; as in Chap. IX-4, call \mathfrak{H} its kernel and L the subfield of K_{sep} corresponding to \mathfrak{H}, which is the cyclic extension of K attached to χ; clearly $L \subset K_{\text{ab}}$ and $\mathfrak{H} \supset \mathfrak{G}^{(1)}$, so that we may identify χ with a character of \mathfrak{A}, for which we will also write χ. Conversely, every character of \mathfrak{A} determines in an obvious manner a character of \mathfrak{G}, with which we identify it. Thus the group of characters of \mathfrak{G}, for which we will write X_K, is identified with the group of characters of \mathfrak{A}; the latter is the same as the dual \mathfrak{A}^* of \mathfrak{A}, except that we will always write the group X_K multiplicatively; we put on X_K the discrete topology, this being in agreement with the fact that the dual of a compact commutative group is always discrete. By the duality theory, the intersection of the kernels of all the characters of \mathfrak{A} is the neutral element; this is the same as to say that the intersection of the kernels \mathfrak{H} of all the characters χ of \mathfrak{G} is $\mathfrak{G}^{(1)}$, or also that K_{ab} is generated by all the cyclic extensions L of K; this is of course well-known.

Let K' be any field containing K; as in Chap. IX-3, we take an algebraic closure \bar{K}' of K' and assume at the same time that we have taken for \bar{K} the algebraic closure of K in \bar{K}'; then, as we have seen there, K_{sep} is contained in K'_{sep}, and, if \mathfrak{G}' is the Galois group of K'_{sep} over K', the restriction morphism ρ of \mathfrak{G}' into \mathfrak{G} is the one which maps every automorphism of K'_{sep} over K' onto its restriction to K_{sep}. Obviously ρ maps $\mathfrak{G}'^{(1)}$ into $\mathfrak{G}^{(1)}$, so that it determines a morphism of $\mathfrak{A}' = \mathfrak{G}'/\mathfrak{G}'^{(1)}$ into $\mathfrak{A} = \mathfrak{G}/\mathfrak{G}^{(1)}$, which we also denote by ρ and call the restriction morphism of \mathfrak{A}' into \mathfrak{A}. It amounts to the same to say that K_{ab} is contained in K'_{ab}, and that ρ maps an element α' of \mathfrak{A}', i.e. an automorphism of K'_{ab} over K', onto its restriction to K_{ab}. Correspondingly, $\chi \to \chi \circ \rho$ is a morphism of X_K into $X_{K'}$.

In classfield theory, one defines a "pairing" of the group X_K of the characters of \mathfrak{G} (or, what amounts to the same, of \mathfrak{A}) with a locally

compact commutative group G_K, invariantly associated with K. In this Chapter, where K will be a local field, we will take $G_K = K^\times$; in the next one, K will be an A-field, and we will take at first $G_K = K_A^\times$ and later on $G_K = K_A^\times / K^\times$. This pairing, which will be called *the canonical pairing*, is a mapping of $X_K \times G_K$ into \mathbf{C}^\times, whose value, for $\chi \in X_K$ and $g \in G_K$, will be written as $(\chi, g)_K$; to begin with, we assume that it satisfies the following condition:

[I] (i) *For all* χ, χ' *in* X_K, *and all* g, g' *in* G_K:

$$(\chi\chi', g)_K = (\chi, g)_K \cdot (\chi', g)_K, \ (\chi, gg')_K = (\chi, g)_K \cdot (\chi, g')_K;$$

(ii) $(\chi, g) \to (\chi, g)_K$ *is a continuous mapping of* $X_K \times G_K$ *into* \mathbf{C}^\times.

As X_K is discrete, the pairing is continuous, i.e. [I(ii)] is satisfied, if and only if $g \to (\chi, g)_K$ is a continuous mapping of G_K into \mathbf{C}^\times for every $\chi \in X_K$; then [I(i)] implies that it is a character of G_K, of an order dividing that of χ. Consequently, if [I(i)] is assumed, [I(ii)] is equivalent to the following:

[I(ii')] *For every* $\chi \in X_K$, *the kernel of* $g \to (\chi, g)_K$ *is an open subgroup of* G_K.

Assume that such a pairing has been given. Then, for each $g \in G_K$, $\chi \to (\chi, g)_K$ is a character of X_K. As X_K is the same as the dual of \mathfrak{A}, the duality theory shows that this can be uniquely written as $\chi \to \chi(\alpha)$ with $\alpha \in \mathfrak{A}$. We will write \mathfrak{a}, or, when necessary, \mathfrak{a}_K for the mapping $g \to \alpha$ of G_K into \mathfrak{A} determined in this manner. Obviously we have $\mathfrak{a}(gg') = \mathfrak{a}(g)\mathfrak{a}(g')$ for all g, g' in G_K, and the continuity of our pairing, i.e. condition [I(ii)], implies at once that \mathfrak{a} is continuous. Thus \mathfrak{a} is a morphism of G_K into \mathfrak{A}, determined by the relation

$$(1) \qquad\qquad (\chi, g)_K = \chi(\mathfrak{a}(g)),$$

which is valid for all $\chi \in X_K$ and all $g \in G_K$. We will call \mathfrak{a} *the canonical morphism* of G_K into \mathfrak{A}.

Now we assume the additional condition:

[II] *If* $(\chi, g)_K = 1$ *for all* $g \in G_K$, *then* $\chi = 1$.

In view of [I], this is clearly equivalent to either one of the following conditions:

[II'] $\chi \to \chi \circ \mathfrak{a}$ *is an injective morphism of* X_K *into the group of characters of* G_K.

[II''] *The image* $\mathfrak{a}(G_K)$ *of* G_K *by* \mathfrak{a} *is dense in* \mathfrak{A}.

We also add the assumption that G_K should be quasicompact, or rather a more precise one, which is as follows:

[III] *Either* (a) G_K *is the direct product of a compact group* G_K^1 *and of a group* N *isomorphic to* **R**, *or* (b) G_K *is the direct product of a compact group* G_K^1 *and of a group* N *isomorphic to* **Z**, *and there is, for each integer* $n \geqslant 1$, *a character* $\chi \in X_K$ *of order n such that* $(\chi, g)_K = 1$ *for all* $g \in G_K^1$.

The two cases in [III] will be referred to as case [III(a)] and case [III(b)], respectively. In both cases, as has been observed in Chap. VII-3, G_K^1 may be characterized as the unique maximal compact subgroup of G_K.

From now on, we will write U_K for the kernel of the canonical morphism a of G_K into \mathfrak{A}; it is the intersection of the kernels of the characters $\chi \circ a$ of G_K, i.e. of the characters $g \to (\chi, g)_K$, for all $\chi \in X_K$.

PROPOSITION 1. *In case* [III(a)], *the canonical morphism* a *determines an isomorphism of* G_K/U_K *onto* \mathfrak{A}; *every character of* G_K, *trivial on* U_K, *can be uniquely written as* $\chi \circ a$ *with* $\chi \in X_K$; *and* $\chi \to \chi \circ a$ *is an injective morphism of* X_K *into the group of characters of finite order of* G_K.

As every character χ of \mathfrak{A} is of finite order, the last assertion is no more than a restatement of [II']. For every $\chi \in X_K$, $\chi \circ a$ induces on the subgroup N of G_K a character of N of finite order; as N is isomorphic to **R**, there is no such character except the trivial one. Therefore $N \subset U_K$; if we put $U_K^1 = U_K \cap G_K^1$, we have $U_K = U_K^1 \times N$, and G_K/U_K may be identified with G_K^1/U_K^1; as this is compact, a determines an isomorphism of that group onto a closed subgroup of \mathfrak{A}, hence onto \mathfrak{A} itself, by [II'']. Then, by the duality theory, $\chi \to \chi \circ a$ is the "dual" or "transpose" of a, hence an isomorphism of X_K onto the subgroup of the group of characters of G_K which is associated by duality with U_K; this subgroup consists of the characters of G_K, trivial on U_K.

COROLLARY. *In case* [III(a)], *every character of* G_K^1, *trivial on* $U_K^1 = U_K \cap G_K^1$, *can be uniquely extended to a character of* G_K *of the form* $\chi \circ a$.

In fact, it can be uniquely extended to a character of G_K, trivial on N; this is then trivial on U_K and is as required.

PROPOSITION 2. *In case* [III(b)], *call* X_0 *the subgroup of* X_K *consisting of the characters* χ *such that* $(\chi, g)_K = 1$ *for all* $g \in G_K^1$; *call* n_1 *a generator of the subgroup* N *of* G_K. *Then* $\chi \to (\chi, n_1)_K$ *is an isomorphism of* X_0 *onto the group of all roots of* 1 *in* **C**.

As every $\chi \in X_K$ is of finite order, $(\chi, g)_K$ is always a root of 1, for all χ and all g. As G_K is generated by G_K^1 and n_1, a character of G_K which is trivial on G_K^1 is uniquely determined by its value at n_1; in view of [II'], this shows that $\chi \to (\chi, n_1)_K$ is an injective morphism of X_0 into the group of roots of 1 in **C**; in particular, it maps every character χ of order n, belonging to X_0, onto a primitive n-th root of 1 in **C**. By [III(b)], there are such characters for every $n \geqslant 1$; therefore the image of X_0 by that morphism contains all the roots of 1 in **C**.

COROLLARY 1. *Assumptions and notations being as in proposition 2,* G_K^1 *is the set of the elements g of* G_K *such that* $(\chi,g)_K=1$ *for all* $\chi\in X_0$.

Let v be any integer other than 0; by proposition 2, there is $\chi\in X_0$ such that $(\chi,n_1^v)_K\ne1$, hence $(\chi,n_1^v g)_K\ne1$ for all $g\in G_K^1$. As G_K is the union of the cosets $n_1^v G_K^1$ for all $v\in\mathbf{Z}$, this proves our assertion.

COROLLARY 2. *In case* [III(b)], *the kernel* U_K *of the canonical morphism* \mathfrak{a} *is contained in* G_K^1; \mathfrak{a} *determines an isomorphism of* G_K^1/U_K *onto the intersection* \mathfrak{A}_0 *of the kernels in* \mathfrak{A} *of the characters* $\chi\in X_0$; *and* $\mathfrak{a}^{-1}(\mathfrak{A}_0)=G_K^1$.

The first and last assertion follow at once from corollary 1. Put $\mathfrak{B}=\mathfrak{a}(G_K^1)$; clearly \mathfrak{B} is compact, and \mathfrak{a} determines an isomorphism of G_K^1/U_K onto \mathfrak{B}; moreover, by the definition of X_0, a character χ of \mathfrak{A} belongs to X_0 if and only if it is trivial on \mathfrak{B}, so that $\mathfrak{B}=\mathfrak{A}_0$.

COROLLARY 3. *In case* [III(b)], *every character of* G_K^1, *trivial on* U_K, *is of finite order and can be extended to a character of* G_K *of the form* $\chi\circ\mathfrak{a}$, *where* χ *is a character of* \mathfrak{A}.

By corollary 2, every character of G_K^1, trivial on U_K, can be written as $\chi_1\circ\mathfrak{a}_1$, where χ_1 is a character of \mathfrak{A}_0 and \mathfrak{a}_1 is the morphism of G_K^1 onto \mathfrak{A}_0 induced by \mathfrak{a}. As χ_1 can be extended (although not uniquely) to a character χ of \mathfrak{A}, and as every character of \mathfrak{A} is of finite order, this proves our assertions.

COROLLARY 4. *In case* [III(b)], *the mapping* $\chi\to\chi\circ\mathfrak{a}$ *is a bijective morphism of* X_K *onto the group of the characters of* G_K *of finite order, trivial on* U_K; *it maps* X_0 *onto the group of the characters of* G_K *of finite order, trivial on* G_K^1.

All we need show is that the mappings in question are surjective. Take first a character ψ of G_K of finite order, trivial on G_K^1. As $\psi(n_1)$ is then a root of 1 in \mathbf{C}, proposition 2 shows that there is $\chi\in X_0$ such that $(\chi,n_1)_K=$ $=\psi(n_1)$; then $\chi\circ\mathfrak{a}$ coincides with ψ on G_K^1 and at n_1, hence on G_K. Now take any character ψ of G_K of finite order, trivial on U_K; by corollary 3, we can find $\chi\in X_K$ such that ψ coincides with $\chi\circ\mathfrak{a}$ on G_K^1; then $\psi'=\psi\cdot(\chi\circ\mathfrak{a})^{-1}$ is trivial on G_K^1 and of finite order, so that, by what we have just proved, it can be written as $\chi'\circ\mathfrak{a}$. This completes our proof.

PROPOSITION 3. *Assume* [I], [II] *and* [III], *and call* \mathfrak{a} *the canonical morphism of* G_K *into* \mathfrak{A}. *For every extension* L *of* K *of finite degree, contained in* K_{ab}, *call* $\mathfrak{B}(L)$ *the subgroup of* \mathfrak{A} *corresponding to* L, *and put* $N(L)=\mathfrak{a}^{-1}(\mathfrak{B}(L))$. *Then* $\mathfrak{B}(L)$ *is the closure of* $\mathfrak{a}(N(L))$ *in* \mathfrak{A}; L *consists of the elements of* K_{ab} *which are invariant under* $\mathfrak{a}(g)$ *for all* $g\in N(L)$; \mathfrak{a} *determines an isomorphism of* $G_K/N(L)$ *onto the Galois group of* L *over* K; *and* $L\to N(L)$ *is a one-to-one correspondence between subfields* L *of* K_{ab}

of finite degree over K, and the open subgroups of G_K of finite index in G_K, containing U_K.

As $\mathfrak{B}(L)$ is open in \mathfrak{A}, $N(L)$ is open in G_K. By [II''], $\mathfrak{a}(G_K)$ is dense in \mathfrak{A}; this implies that $\mathfrak{a}(N(L))$ is dense in $\mathfrak{B}(L)$ and that \mathfrak{a} determines an isomorphism of $G_K/N(L)$ onto $\mathfrak{A}/\mathfrak{B}(L)$, which is the same as the Galois group of L over K. As the operation of \mathfrak{A} on K_{ab} is continuous, every element of K_{ab} which is invariant under $\mathfrak{a}(N(L))$ is invariant under its closure $\mathfrak{B}(L)$, so that it is in L. Finally, let H be any open subgroup of G_K of finite index n, containing U_K; call ψ_i, for $1 \leqslant i \leqslant n$, all the distinct characters of G_K, trivial on H; then H is the intersection of their kernels. By prop. 1 in case [III(a)] and by corollary 4 of prop. 2 in case [III(b)], we can write $\psi_i = \chi_i \circ \mathfrak{a}$ for $1 \leqslant i \leqslant n$, the χ_i being characters of \mathfrak{A}; by [II'], the χ_i are uniquely determined and make up a finite subgroup of X_K, since the ψ_i make up a finite subgroup of the group of characters of G_K. Call \mathfrak{B} the intersection of the kernels of the χ_i in \mathfrak{A}; it is an open subgroup of \mathfrak{A}, of index n; therefore the subfield L of K_{ab}, corresponding to \mathfrak{B}, is of degree n over K. Clearly $H = \mathfrak{a}^{-1}(\mathfrak{B})$, hence $H = N(L)$. This completes our proof.

COROLLARY. *In case* [III(b)], *call K_0 the subfield of K_{ab} corresponding to the subgroup $\mathfrak{A}_0 = \mathfrak{a}(G_K^1)$ of \mathfrak{A}. Then, for each integer $v \geqslant 1$, K_0 contains one and only one extension K_v of K of degree v; this is the cyclic extension of K, attached to any one of the characters of order v, belonging to X_0; and $N(K_v)$ is the subgroup of G_K generated by G_K^1 and n_1^v.*

By corollary 2 of prop. 2, we have $G_K^1 = \mathfrak{a}^{-1}(\mathfrak{A}_0)$; therefore, if L and $N(L)$ are as in proposition 3, we have $L \subset K_0$ if and only if $N(L) \supset G_K^1$; this implies that $N(L)$ is generated by G_K^1 and n_1^v if v is the index of $N(L)$ in G_K. Then, by proposition 3, L is cyclic of degree v over K, and, if χ is a character of \mathfrak{A} attached to L, $N(L)$ is the kernel of $\chi \circ \mathfrak{a}$, so that χ belongs to X_0 and has the order v. Conversely, if χ is such, the kernel of $\chi \circ \mathfrak{a}$ is generated by G_K^1 and n_1^v, so that the cyclic extension attached to χ is L.

Now we consider a cyclic extension K' of K, contained in K_{sep}. We use the notations \mathfrak{G}', $\mathfrak{G}'^{(1)}$, $\mathfrak{A}' = \mathfrak{G}'/\mathfrak{G}'^{(1)}$ as explained above, and write ρ for the restriction morphism of \mathfrak{G}' into \mathfrak{G} and also for that of \mathfrak{A}' into \mathfrak{A}. As K' is cyclic over K, \mathfrak{G}' is an open normal subgroup of \mathfrak{G}, with cyclic factor-group; consequently, we have $\mathfrak{G} \supset \mathfrak{G}' \supset \mathfrak{G}^{(1)} \supset \mathfrak{G}'^{(1)}$, and $\mathfrak{G}'^{(1)}$ is a normal subgroup of \mathfrak{G}. For every $\lambda \in \mathfrak{G}$, the inner automorphism $\sigma \to \lambda \sigma \lambda^{-1}$ induces on \mathfrak{G}' an automorphism of \mathfrak{G}'; therefore, if χ' is any character of \mathfrak{G}', we can define a character χ'^λ of \mathfrak{G}' by putting, for every $\sigma' \in \mathfrak{G}'$, $\chi'^\lambda(\sigma') = \chi'(\lambda \sigma' \lambda^{-1})$. Clearly $\chi'^\lambda = \chi'$ if $\lambda \in \mathfrak{G}'$, so that $\chi' \to \chi'^\lambda$ deter-

mines an operation of the Galois group $\mathfrak{G}/\mathfrak{G}'$ of K' over K on the group $X_{K'}$ of the characters of \mathfrak{G}'.

Furthermore, we assume that we have been given canonical pairings $(\chi, g)_K$, $(\chi', g')_{K'}$ of X_K with a group G_K, and of $X_{K'}$ with a group $G_{K'}$, both of them satisfying [I], [II], [III]; to simplify notations, put $G = G_K$, $G' = G_{K'}$, $G_1 = G_K^1$, $G_1' = G_{K'}^1$; call \mathfrak{a}, \mathfrak{a}' the canonical morphisms of G into \mathfrak{A}, and of G' into \mathfrak{A}', respectively defined by these pairings. Assume also that \mathfrak{G} operates on G', the action of any $\lambda \in \mathfrak{G}$ on G' being written as $g' \to g'^\lambda$ and satisfying the following condition:

[IV] (i) *For* $\lambda \in \mathfrak{G}'$, $g' \to g'^\lambda$ *is the identity on* G'; (ii) *For each* $\lambda \in \mathfrak{G}$, $g' \to g'^\lambda$ *is an automorphism of* G', *and* $g'^\lambda g'^{-1} \in G_1'$ *for all* $g' \in G'$; (iii) *For all* $\chi' \in X_{K'}$, $g' \in G'$ *and* $\lambda \in \mathfrak{G}$, *we have*:

$$(\chi'^\lambda, g'^\lambda)_{K'} = (\chi', g')_{K'}.$$

Finally, assume that we have been given a morphism F of G' into G, satisfying the following condition:

[V] (i) *For all* $g' \in G'$ *and all* $\lambda \in \mathfrak{G}$, *we have* $F(g'^\lambda) = F(g')$; (ii) *For all* $\chi \in X_K$ *and all* $g' \in G'$, *we have*:

$$(\chi \circ \rho, g')_{K'} = (\chi, F(g'))_K.$$

Clearly [V(ii)] may also be written as $\rho \circ \mathfrak{a}' = \mathfrak{a} \circ F$.

PROPOSITION 4. *Let* K' *be a cyclic extension of* K; *let* \mathfrak{a}, \mathfrak{a}' *be the canonical morphisms respectively defined by canonical pairings of* X_K *with a group* G, *and of* $X_{K'}$ *with a group* G', *both satisfying* [I], [II], [III]. *Assume that the Galois group* \mathfrak{G} *of* K_{sep} *acts on* G', *that* F *is a morphism of* G' *into* G, *and that* [IV] *and* [V] *are satisfied. Then* $U \cap F(G_1') = F(U' \cap G_1')$, *where* U, U' *are the kernels of* \mathfrak{a} *and of* \mathfrak{a}'; *moreover, we have* $U \cap F(G') = = F(U')$ *if* G' *satisfies* [III(a)], *or if* G *and* G' *satisfy* [III(b)] *and* F *does not map* G' *into* G_1.

By [V(ii)], we have $\rho \circ \mathfrak{a}' = \mathfrak{a} \circ F$; therefore $F(U')$ is contained in U, hence in $U \cap F(G')$, and, if we put $U_1' = U' \cap G_1'$, $F(U_1')$ is contained in $U \cap F(G_1')$. Let ψ be a character of G, trivial on $F(U_1')$; then $\psi \circ F$ is a character of G', trivial on U_1'. Apply now the corollary of prop. 1, in the case [III(a)], and corollary 3 of prop. 2, in the case [III(b)], to the character induced on G_1' by $\psi \circ F$; this shows that $\psi \circ F$ coincides on G_1' with a character of the form $\chi' \circ \mathfrak{a}'$, with $\chi' \in X_{K'}$. In other words, we have, for all $g' \in G_1'$:

$$\psi(F(g')) = (\chi', g')_{K'}.$$

By [IV(ii)], this must hold if we substitute $g'^\lambda g'^{-1}$ for g', with any $g' \in G'$ and any $\lambda \in \mathfrak{G}$. In view of [V(i)], this gives

$$1 = (\chi', g'^\lambda g'^{-1})_{K'} = (\chi', g'^\lambda)_{K'} \cdot (\chi', g')_{K'}^{-1},$$

and therefore, by [IV(iii)]:

$$(\chi',g')_{K'} = (\chi',g'^{\lambda})_{K'} = (\chi'^{\lambda^{-1}},g')_{K'}.$$

By [II], this shows that χ' is invariant under λ for every $\lambda \in \mathfrak{G}$; more explicitly, it is invariant under all the automorphisms of \mathfrak{G}' induced on \mathfrak{G}' by inner automorphisms of \mathfrak{G}. Therefore the same must be true of the kernel \mathfrak{H}' of χ', so that \mathfrak{H}', which is an open subgroup of \mathfrak{G}' with cyclic factor-group, is a normal subgroup of \mathfrak{G}. Let α be a representative in \mathfrak{G} of a generator of the cyclic group $\mathfrak{G}/\mathfrak{G}'$; let β be a representative in \mathfrak{G}' of a generator of $\mathfrak{G}'/\mathfrak{H}'$; then \mathfrak{G}' is generated by \mathfrak{H}' and β, and \mathfrak{G} is generated by \mathfrak{G}' and α, hence by \mathfrak{H}', β and α. Consequently $\mathfrak{G}/\mathfrak{H}'$ is generated by the images α', β' of α, β in $\mathfrak{G}/\mathfrak{H}'$. As χ' is invariant under $\sigma' \rightarrow \alpha \sigma' \alpha^{-1}$, we get, for $\sigma' = \beta, \chi'(\beta) = \chi'(\alpha\beta\alpha^{-1})$. This shows that $\alpha\beta\alpha^{-1}\beta^{-1}$ is in the kernel \mathfrak{H}' of χ', so that α' commutes with β' in $\mathfrak{G}/\mathfrak{H}'$. Consequently $\mathfrak{G}/\mathfrak{H}'$ is commutative. Therefore the character of $\mathfrak{G}'/\mathfrak{H}'$ determined by χ' can be extended to a character of $\mathfrak{G}/\mathfrak{H}'$. This is the same as to say that χ' can be extended to a character χ of \mathfrak{G}, so that we have $\chi' = \chi \circ \rho$. In view of [V(ii)], the definition of χ' gives now, for all $g' \in G_1'$:

$$\psi(F(g')) = (\chi \circ \rho, g')_{K'} = (\chi, F(g'))_K.$$

This is the same as to say that ψ coincides with $\chi \circ \mathfrak{a}$ on $F(G_1')$, so that it is trivial on $U \cap F(G_1')$. As $F(U_1')$ is a compact subgroup of G, and as we have proved that every character ψ of G, trivial on $F(U_1')$, is trivial on $U \cap F(G_1')$, we see that $F(U_1') \supset U \cap F(G_1')$; in view of what we had proved before, this completes the proof of the first part of our proposition. If G and G' satisfy [III(b)], we have $U' \subset G_1'$ and $U \subset G_1$, by corollary 2 of prop. 2; if F does not map G' into G_1, we have $F^{-1}(G_1) = G_1'$, by lemma 2; $U \cap F(G_1')$ is then the same as $U \cap F(G')$, which completes the proof of the second part in that case. Now assume that $G' = G_1' \times N'$ with N' isomorphic to \mathbf{R}. As we have seen before, for every $\chi' \in X_{K'}$, the character of N' induced on N' by $\chi' \circ \mathfrak{a}'$, being of finite order, is trivial, so that $N' \subset U'$, hence $U' = U_1' \times N'$; the same argument, applied to the character induced on N' by $\chi \circ \mathfrak{a} \circ F$ for $\chi \in X_K$, gives now $F(N') \subset U$, and therefore:

$$U \cap F(G') = (U \cap F(G_1')) \cdot F(N') = F(U_1')F(N') = F(U').$$

§ 2. **The Brauer group of a local field.** From now on, K will be a local field. As in Chapter IX, we write $B(K)$ for its Brauer group, $H(K)$ for the group of its factor-classes, and we identify these groups with each other by means of th. 3, Chap. IX-3. In Chap. IX-4, we have already determined these groups in the cases $K = \mathbf{R}$, $K = \mathbf{C}$, and we begin by recalling the results found there, and introducing some additional notations which will be useful in the next Chapter. As $B(\mathbf{R})$ has two elements, it

has a unique isomorphism η onto the subgroup $\{\pm 1\}$ of \mathbf{C}^{\times}; for any simple algebra A over \mathbf{R}, we write $h(A) = \eta(\mathrm{Cl}(A))$, and call this *the Hasse invariant* of A; it is $+1$ or -1 according as A is trivial or not. As $B(\mathbf{C})$ has only one element, we write η for the mapping which maps it onto $\{+1\}$, and, for every simple algebra A over \mathbf{C}, we write $h(A) = \eta(\mathrm{Cl}(A)) = +1$ and call this *the Hasse invariant* of A. For $K = \mathbf{R}$, the Galois group \mathfrak{G} of K_{sep} over K consists of the identity ε and of the automorphism $x \to \bar{x}$ of \mathbf{C} over \mathbf{R}; for $K = \mathbf{C}$, $\mathfrak{G} = \{\varepsilon\}$. For every character χ of \mathfrak{G}, and every $\theta \in K^{\times}$, we have defined in Chap. IX-4 the factor-class $\{\chi, \theta\}$; identifying $H(K)$ with $B(K)$ as we have said, we may now write, for $K = \mathbf{R}$ or \mathbf{C}:

$$(\chi, \theta)_K = \eta(\{\chi, \theta\}).$$

Clearly this is 1 if $K = \mathbf{C}$, or if $K = \mathbf{R}$ and χ is the trivial character of \mathfrak{G}; if $K = \mathbf{R}$ and χ is the non-trivial character of \mathfrak{G}, our results of Chap. IX-4 show that it is $+1$ or -1 according as $\theta > 0$ or $\theta < 0$. One verifies immediately that this is a canonical pairing of X_K with K^{\times}, in the sense of § 1, and that it satisfies conditions [I], [II], [III(a)]; the kernel U_K of the canonical morphism is \mathbf{C}^{\times} if $K = \mathbf{C}$, and \mathbf{R}_+^{\times} if $K = \mathbf{R}$.

From now on, K will always denote a commutative p-field, except that occasionally we will point out the validity of some of our results for $K = \mathbf{R}$ or \mathbf{C}. As usual, we write R for the maximal compact subring of K, q for its module, P for the maximal ideal of R, and π for a prime element of K. We use the notations \bar{K}, K_{sep}, \mathfrak{G}, K_{ab}, \mathfrak{A}, as in § 1. Write \mathfrak{M} for the set of all roots of 1 of order prime to p in \bar{K}; this is clearly a subgroup of $K_{\mathrm{sep}}^{\times}$. Put $K_0 = K(\mathfrak{M})$, and call \mathfrak{H}_0 the closed subgroup of \mathfrak{G} corresponding to K_0, i.e. consisting of the automorphisms of K_{sep} over K which leave invariant all the elements of K_0, or, what amounts to the same, all those of \mathfrak{M}. By corollary 2 of th. 7, Chap. I-4, every finite subset of \mathfrak{M} generates over K an unramified extension of K. Conversely, every extension of K, contained in an unramified extension, is itself unramified, so that, by corollary 3 of th. 7, Chap. I-4, it is generated by a finite subset of \mathfrak{M}; moreover, by the same corollary, there is one and only one such extension K_n of degree n over K, for every $n \geq 1$. Consequently, K_0 is the union of the fields K_n for all $n \geq 1$. Again by corollary 2 of the same theorem, the mapping $\mu \to \mu^q$ of \mathfrak{M} into itself is an automorphism of \mathfrak{M}, and, for every $n \geq 1$, there is one and only one automorphism of K_n over K, viz., the Frobenius automorphism, which coincides with that mapping on $\mathfrak{M} \cap K_n$. Clearly this implies that there is one and only one automorphism φ_0 of K_0 over K which induces $\mu \to \mu^q$ on \mathfrak{M}; this will be called *the Frobenius automorphism* of K_0 over K, and every automorphism φ of K_{sep} over K which induces φ_0 on K_0 will be called *a Frobenius automorphism* of K_{sep}

over K; then the Frobenius automorphisms of K_{sep} over K make up the coset $\mathfrak{H}_0 \varphi$ in \mathfrak{G}.

DEFINITION 1. *A character χ of \mathfrak{G} will be called unramified if the cyclic extension of K attached to χ is unramified; we will write X_0 for the set of all unramified characters of \mathfrak{G}.*

In view of what has been said above, it is clear that χ is unramified if and only if the cyclic extension attached to χ is contained in K_0, or, what amounts to the same, if and only if χ is trivial on the subgroup \mathfrak{H}_0 of \mathfrak{G} corresponding to K_0; therefore X_0 is a subgroup of the group X_K of all characters of \mathfrak{G}.

PROPOSITION 5. *Let φ be a Frobenius automorphism of K_{sep} over K. Then $\chi \to \chi(\varphi)$ is an isomorphism of the group X_0 of the unramified characters of \mathfrak{G} onto the group of all roots of 1 in \mathbf{C}; it is independent of the choice of φ.*

Clearly that mapping is a morphism of X_0 into the group of the roots of 1 in \mathbf{C}. With the notations explained above, the cyclic extension of K attached to an unramified character χ of order n is K_n. As φ induces on K_n the Frobenius automorphism of K_n over K, and this generates the Galois group of K_n over K, $\chi(\varphi)$ is a primitive n-th root of 1; therefore the morphism in our proposition is both injective and surjective. The last assertion follows from the fact that two Frobenius automorphisms can differ only by an element of \mathfrak{H}_0, and every unramified character is trivial on \mathfrak{H}_0.

THEOREM 1. *Let K be a commutative p-field, and π a prime element of K. Let X_0 be the group of the unramified characters of \mathfrak{G}. Then $\chi \to \{\chi, \pi\}$ is an isomorphism of X_0 onto the group $H(K)$ of factor-classes of K; it is independent of the choice of π.*

We can identify $H(K)$ with the Brauer group $B(K)$ of K. Every element of $B(K)$, i.e. every class of simple algebras over K, contains one and only one division algebra over K. As has already been pointed out in Chap. IX-4, and again in Chap. X-2, prop. 5 of Chap. I-4 shows that such an algebra, if it is of dimension n^2 over K, can be written as $[K_n/K; \chi, \pi_1]$, where χ is a character attached to K_n and π_1 is a suitable prime element of K; therefore the factor-class belonging to that algebra is $\{\chi, \pi_1\}$. Combining prop. 10, Chap. IX-4, with prop. 3 of Chap. VIII-1, we see that this is independent of π_1, so that it is the same as $\{\chi, \pi\}$. Consequently, $\chi \to \{\chi, \pi\}$ is a surjective morphism of X_0 onto $H(K)$. As K_n is unramified of degree n over K, its modular degree over K is n; therefore π cannot be in $N_{K_n/K}(K_n^\times)$ unless $n=1$; again by prop. 10, Chap. IX-4, this shows that,

if χ is attached to K_n, $\{\chi,\pi\}\neq 1$ unless $n=1$, i.e. unless $\chi=1$. This completes our proof.

COROLLARY 1. *Let K and π be as in theorem 1; let K_n be the unramified extension of K of degree n, and let χ be a character attached to K_n. Then $[K_n/K;\chi,\pi]$ is a division algebra over K.*

At any rate, it is of the form $M_m(D)$, D being a division algebra over K; if d^2 is the dimension of D over K, D can be written as $[K_d/K;\chi',\pi]$, where χ' is a character attached to K_d. By theorem 1, this implies $\chi'=\chi$, hence $n=d$ and $m=1$.

COROLLARY 2. *Let φ be a Frobenius automorphism of K_{sep} over K. There is one and only one isomorphism η of $H(K)$ onto the group of all roots of 1 in \mathbf{C}, such that $\eta(\{\chi,\pi\})=\chi(\varphi)$ for all $\chi\in X_0$; it is independent of the choice of π and of φ.*

This follows at once from theorem 1, combined with prop. 5.

COROLLARY 3. *Notations being as above, let X_K be the group of all characters of \mathfrak{G}; for all $\chi\in X_K$ and all $\theta\in K^\times$, put*

$$(\chi,\theta)_K=\eta(\{\chi,\theta\}).$$

Then this defines a pairing between X_K and K^\times which satisfies conditions [I] and [III(b)] of § 1.

By prop. 8 of Chap. IX-4, it satisfies [I(i)]. By prop. 10 of Chap. IX-4 and prop. 5 of Chap. VIII-1, it satisfies [I(ii')]. As to [III(b)], we have here to take $G_K=K^\times$, $G_K^1=R^\times$, and we can take for N the subgroup of K^\times generated by π. Then [III(b)] is satisfied by taking for χ any character attached to the unramified extension K_n of K of degree n, as follows at once from prop. 10 of Chap. IX-4 and prop. 3 of Chap. VIII-1.

COROLLARY 4. *For all $\chi\in X_0$, and all $\theta\in K^\times$, we have $(\chi,\theta)_K=\chi(\varphi)^{ord(\theta)}$; if π is any prime element of K, $(\chi,\pi)_K=\chi(\varphi)$.*

The latter assertion is a restatement of corollary 2. Then the former holds for $\theta=\pi$, and also, as proved in the proof of corollary 3, for $\theta\in R^\times$; the general case follows from this at once.

COROLLARY 5. *Let K_1 be a field isomorphic to K; let \bar{K}_1 be an algebraic closure of K_1, and λ an isomorphism of \bar{K} onto \bar{K}_1, mapping K onto K_1. For every character χ of \mathfrak{G}, write χ^λ for its transform by λ, i.e. for the character of the Galois group \mathfrak{G}_1 of $(K_1)_{sep}$ over K_1 given by $\chi^\lambda(\sigma_1)=\chi(\lambda\sigma_1\lambda^{-1})$ for all $\sigma_1\in\mathfrak{G}_1$. Then $(\chi,\theta)_K=(\chi^\lambda,\theta^\lambda)_{K_1}$ for all $\chi\in X_K$ and all $\theta\in K^\times$.*

This follows at once from corollary 2, since obviously λ maps a prime element of K onto a prime element of K_1, and transforms a Frobenius automorphism of K_{sep} over K into one of $(K_1)_{\text{sep}}$ over K_1.

From now on, the pairing of X_K with K^\times, defined in corollary 3, will be called *the canonical pairing* for K. As explained in § 1, we derive from this a morphism \mathfrak{a} of K^\times into the Galois group \mathfrak{A} of K_{ab} over K which will be called *the canonical morphism* for K; it is defined by $(\chi, \theta)_K = \chi(\mathfrak{a}(\theta))$, this being valid for all $\chi \in X_K$ and all $\theta \in K^\times$. Corollary 4 of th. 1 shows that $\mathfrak{a}(\pi)$ induces on K_0 the Frobenius automorphism of K_0 over K whenever π is a prime element of K.

As we identify the Brauer group $B(K)$ with the group $H(K)$ considered in theorem 1 and its corollaries, we may consider the mapping η defined in corollary 2 of th. 1 as an isomorphism of $B(K)$ onto the group of roots of 1 in \mathbf{C}; for every simple algebra A over K, we will write $h(A) = \eta(\text{Cl}(A))$, and will call this *the Hasse invariant of* A; it is 1 if and only if A is trivial.

THEOREM 2. *Let K' be an extension of K of finite degree, contained in \bar{K}; let $\mathfrak{G}, \mathfrak{G}'$ be the Galois groups of K_{sep} over K, and of K'_{sep} over K', respectively, and let ρ be the restriction morphism of \mathfrak{G}' into \mathfrak{G}. Then, for every $\chi \in X_K$, and every $\theta' \in K'^\times$, we have:*

(2) $(\chi \circ \rho, \theta')_{K'} = (\chi, N_{K'/K}(\theta'))_K.$

Let f be the modular degree of K' over K; then the module of K' is q^f, and, if φ, φ' are Frobenius automorphisms of K_{sep} over K, and of K'_{sep} over K', respectively, φ' coincides with φ^f on the group \mathfrak{M} of the roots of 1 of order prime to p in \bar{K}, hence on $K_0 = K(\mathfrak{M})$, so that $\rho(\varphi')\varphi^{-f}$ is in the subgroup \mathfrak{H}_0 of \mathfrak{G} which corresponds to K_0. Now assume first that the character χ in (2) is unramified, hence trivial on \mathfrak{H}_0; this implies that $\chi(\rho(\varphi')) = \chi(\varphi)^f$. As we have observed in Chap. IX-4, the cyclic extension of K' attached to $\chi \circ \rho$ is the compositum of K' and of the cyclic extension of K attached to χ; as the latter is unramified, hence generated by elements of \mathfrak{M}, the same is true of the former, so that $\chi \circ \rho$ is unramified. We can now apply corollary 4 of th. 1 to both sides of (2); it shows that the left-hand side is $\chi(\rho(\varphi'))^r$ with $r = \text{ord}_{K'}(\theta')$, and that the right-hand side is $\chi(\varphi)^s$ with $s = \text{ord}_K(N_{K'/K}(\theta'))$, hence $s = fr$ by formula (2) of Chap. VIII-1. This proves (2) when χ is unramified. In the general case, call n the order of χ; as neither side of (2) is changed if we replace θ' by $\theta' \eta'^n$ with $\eta' \in K'^\times$, we may assume that $r = \text{ord}_{K'}(\theta') \neq 0$. As we have just shown, if χ_1 is any unramified character of \mathfrak{G}, $(\chi_1 \circ \rho, \theta')_{K'}$ is equal to $\chi_1(\varphi)^{fr}$; in view of prop. 5, we can choose χ_1 so that this is equal to any given root of 1 in \mathbf{C}, and in particular to the left-hand side of (2); as (2) has already been proved for unramified characters, it will

therefore be enough, after replacing χ by $\chi\chi_1^{-1}$, to prove our result under the additional assumption that the left-hand side has the value 1. That being now assumed, call L the cyclic extension of K attached to χ; the cyclic extension of K' attached to $\chi \circ \rho$ is then the compositum L' of K' and L. As the left-hand side of (2) is 1, prop. 10 of Chap. IX-4 shows that there is $\eta' \in L'$ such that $\theta' = N_{L'/K'}(\eta')$. This gives, by Chap. III-3: $N_{K'/K}(\theta') = N_{L'/K}(\eta') = N_{L/K}(N_{L'/L}(\eta'))$; the same proposition shows then that the right-hand side of (2) is 1, which completes the proof.

COROLLARY 1. *If \mathfrak{a}, \mathfrak{a}' are the canonical morphisms for K and for K', respectively, we have $\rho \circ \mathfrak{a}' = \mathfrak{a} \circ N_{K'/K}$.*

In view of our definitions, this is just another way of writing (2).

COROLLARY 2. *Let K and K' be as in theorem 2; call n the degree of K' over K. Then, for every simple algebra A over K, we have $h(A_{K'}) = h(A)^n$.*

By th. 1, the factor-class belonging to A can be written as $\{\chi, \pi\}$. By formula (7) of Chap. IX-4, the restriction morphism of $H(K)$ into $H(K')$ maps the class $\{\chi, \theta\}$ onto the class $\{\chi \circ \rho, \theta\}$ for every $\chi \in X_K$ and every $\theta \in K^\times$; moreover, for $\theta \in K^\times$, we have $N_{K'/K}(\theta) = \theta^n$. By th. 2, this gives:

$$h(A_{K'}) = (\chi \circ \rho, \pi)_{K'} = (\chi, \pi^n)_K = h(A)^n.$$

COROLLARY 3. *If χ is a non-trivial character of \mathfrak{G}, $\theta \to (\chi, \theta)_K$ is a non-trivial character of K^\times.*

Call n and d the orders of these two characters; clearly d divides n. Call L the cyclic extension of K attached to χ; call χ_1 an unramified character of \mathfrak{G} of order n, K_n the unramified extension of K of degree n, and put $D = [K_n/K; \chi_1, \pi]$. By corollary 2 of th. 1, we have $h(D) = \chi_1(\varphi)$, so that $h(D)$ is a primitive n-th root of 1. By corollary 2, we have then $h(D_L) = h(D)^n = 1$, so that D_L is trivial; this is the same as to say that D has an L-representation into $M_n(L)$; by prop. 9 of Chap. IX-4, the factor-class attached to D can then be written in the form $\{\chi, \theta\}$, with some $\theta \in K^\times$, and we have $h(D) = (\chi, \theta)_K$. Therefore $d = n$. This shows that our canonical pairing satisfies condition [II] of § 1.

COROLLARY 4. *If L is any cyclic extension of K of degree n, $N_{L/K}(L^\times)$ is an open subgroup of K^\times of index n.*

In fact, by prop. 10 of Chap. IX-4, it is the kernel of $\theta \to (\chi, \theta)_K$, where χ is a character of \mathfrak{G} attached to L, and we have just proved that this is of order n.

If $K' = K$, or if $\chi = 1$, the conclusion (2) of theorem 2 is trivial; if K' is the cyclic extension of K attached to χ, (2) is equivalent to prop. 10

of Chap. IX-4, since $\chi \circ \rho = 1$ in that case. No other case than these can occur if K is an **R**-field, as one sees at once. Therefore theorem 2 remains valid for $K = \mathbf{R}$ or \mathbf{C}; so do its corollaries.

PROPOSITION 6. *A character χ of \mathfrak{G} is unramified if and only if $(\chi,\theta)_K = 1$ for all $\theta \in R^\times$.*

Call X_0' the group of the characters with the latter property; as before, we call X_0 the group of the unramified characters of \mathfrak{G}. By corollary 4 of th. 1, $X_0 \subset X_0'$. By prop. 2 of §1, $\chi \to (\chi,\pi)_K$ is an isomorphism of X_0' onto the group of all roots of 1 in \mathbf{C}; by th. 1, combined with prop. 5, this induces on X_0 an isomorphism of X_0 onto the same group. Therefore $X_0' = X_0$.

COROLLARY. *A cyclic extension L of K is unramified if and only if $N_{L/K}(L^\times)$ contains R^\times.*

In view of prop. 10, Chap. IX-4, this follows at once from the application of proposition 6 to a character of \mathfrak{G} attached to L.

§ 3. **The canonical morphism.** We have now verified conditions [I], [II], [III(b)] of §1 for the canonical pairing $(\chi,\theta)_K$, and we have also shown that the subgroup X_0 of X_K defined by means of such a pairing in § 1 is here the same as the group X_0 of the unramified characters of \mathfrak{G}. As in § 1, we will now call U_K the kernel of the canonical morphism \mathfrak{a} of K^\times into \mathfrak{A}; our main result in this § will be that $U_K = \{1\}$. In applying the results of § 1, we have to keep in mind that here G_K^1 must be replaced by R^\times, n_1 by a prime element π of K, and N by the subgroup of K^\times generated by π. Corollary 2 of prop. 2, § 1, shows that U_K is contained in R^\times, and that \mathfrak{a} determines a morphism of R^\times onto the intersection \mathfrak{A}_0 of the kernels of the characters $\chi \in X_0$, when these are considered as characters of \mathfrak{A}. Here, by prop. 6 of § 2, X_0 consists of the characters of \mathfrak{G} which are trivial on the subgroup \mathfrak{H}_0 of \mathfrak{G} corresponding to the union K_0 of all unramified extensions of K. Therefore \mathfrak{A}_0 is the image of \mathfrak{H}_0 in \mathfrak{A}, i.e. the subgroup of \mathfrak{A} corresponding to the subfield K_0 of K_{ab}, or in other words the Galois group of K_{ab} over K_0.

PROPOSITION 7. *Let K_0 be the union of all the unramified extensions of K, contained in K_{sep}; let φ_0 be the Frobenius automorphism of K_0 over K, and let \mathfrak{a} be the canonical morphism of K^\times into the Galois group \mathfrak{A} of K_{ab} over K. Then, for every $\theta \in K^\times$, $\mathfrak{a}(\theta)$ induces on K_0 the automorphism φ_0^r with $r = \mathrm{ord}(\theta)$.*

In fact, corollary 4 of th. 1, § 2, can be expressed by saying that $\chi(\mathfrak{a}(\theta)) = \chi(\varphi)^r$ for every $\chi \in X_0$, if φ is an automorphism of K_{sep} over K which induces φ_0 on K_0. This is the same as to say that, if φ induces φ'

on K_{ab}, $a(\theta)\varphi'^{-r}$ is in the intersection of the kernels of all the characters $\chi \in X_0$, or again, in view of the definition of \mathfrak{A}_0 and K_0, that $a(\theta)\varphi'^{-r}$ induces the identity on K_0, as was to be proved.

COROLLARY. *Notations being as in proposition 7, call φ' an automorphism of K_{ab} over K, inducing φ_0 on K_0. Then a maps R^\times onto \mathfrak{A}_0; it maps K^\times onto the union of the cosets $\mathfrak{A}_0 \varphi'^n$ for $n \in \mathbf{Z}$, and this union is dense in \mathfrak{A}.*

This follows at once from proposition 7 and from [II″] in § 1.

Now we consider the kernel U_K of a. By definition, it is the intersection of the kernels of the characters $\theta \to (\chi, \theta)_K$ of K^\times, when one takes for χ all the characters of \mathfrak{G}. By prop. 10 of Chap. IX-4, this is the same as to say that it is the intersection of the groups $N_{L/K}(L^\times)$ when one takes for L all the cyclic extensions of K.

PROPOSITION 8. *Let K' be an abelian extension of K of finite degree. Then $U_K = N_{K'/K}(U_{K'})$.*

Assume first that K' is cyclic over K. Then we can apply prop. 4 of § 1, by taking $F = N_{K'/K}$; in fact, [IV(i)] and [IV(ii)] are obviously satisfied by the automorphisms $x \to x^\lambda$ of K'^\times, for all $\lambda \in \mathfrak{G}$; so is [IV(iii)], by corollary 5 of th. 1, § 2; [V(i)] is obviously satisfied, and so is [V(ii)], by th. 2 of § 2. In the conclusion of prop. 4, U and U' are here the same, respectively, as U_K and $U_{K'}$; moreover, as we have seen, U_K is contained in $N_{K'/K}(K'^\times)$, which, in the notation of prop. 4, is the same as $F(G')$. This proves our assertion when K' is cyclic over K. Otherwise we can find a sequence $K, K_1, \ldots, K_m = K'$ of fields between K and K', such that each one is cyclic over the preceding one. If we use induction on m, the induction assumption gives $U_{K_1} = N_{K'/K_1}(U_{K'})$, and what we have proved gives $U_K = N_{K_1/K}(U_{K_1})$; putting these together, we get our conclusion. The same proof would be valid for any solvable extension, but this will not be needed.

PROPOSITION 9. *Assume that K contains n distinct n-th roots of 1. Then the intersection of the kernels of the characters $\theta \to (\chi_{n,\xi}, \theta)_K$ of K^\times, for all $\xi \in K^\times$, is $(K^\times)^n$.*

Here the assumption on K implies that n is not a multiple of the characteristic of K, and $\chi_{n,\xi}$ is as defined in Chap. IX-5. By definition, the set in question is the intersection of the kernels of all the morphisms $\theta \to \{\xi, \theta\}_n$ of K^\times into $H(K)$. By formula (12) of Chap. IX-5 (the "reciprocity law"), this consists of the elements θ of K^\times such that $\{\theta, \xi\}_n = 1$, i.e. $\{\chi_{n,\theta}, \xi\} = 1$, i.e. $(\chi_{n,\theta}, \xi)_K = 1$, for all $\xi \in K^\times$. By corollary 3 of th. 2, § 2,

this is equivalent to $\chi_{n,\theta}=1$; as we have observed in Chap. IX-5, this is so if and only if $\theta\in(K^\times)^n$.

COROLLARY. *Let K be any p-field; if n is not a multiple of the characteristic of K, $U_K\subset(K^\times)^n$.*

The assumption on n implies that there are n distinct n-th roots of 1 in K_{sep}; then they generate an abelian extension K' of K. By proposition 9, we have $U_{K'}\subset(K'^\times)^n$. By prop. 8, this gives

$$U_K=N_{K'/K}(U_{K'})\subset N_{K'/K}((K'^\times)^n)\subset(K^\times)^n.$$

PROPOSITION 10. *Assume that K is of characteristic p. Then the intersection of the kernels of the characters $\theta\to(\chi_{p,\xi},\theta)_K$ of K^\times, for all $\xi\in K$, is $(K^\times)^p$.*

Call Z that intersection; as all the characters $\chi_{p,\xi}$ are of order p or 1, Z is a subgroup of K^\times, containing $(K^\times)^p$; as $\chi_{p,\xi}=1$ for $\xi=0$, Z may be defined as consisting of the elements θ of K^\times such that $\{\xi,\theta\}_p=1$ for all $\xi\in K^\times$, or, what amounts to the same, such that $\{\xi\theta,\theta\}_p=1$ for all $\xi\in K^\times$. By formulas (13) and (14) of Chap. IX-5, we have, for all $\xi\in K^\times$, $\theta\in K^\times$:

$$1=\{\xi\theta,-\xi\theta\}_p=\{\xi\theta,-\xi\}_p\cdot\{\xi\theta,\theta\}_p,$$

so that Z is also the set of the elements θ of K^\times such that $\{\xi\theta,-\xi\}_p=1$ for all $\xi\in K^\times$. Then, by the first formula (13) of Chap. IX-5, $Z\cup\{0\}$ is an additive subgroup of K. As Z is a subgroup of K^\times, containing $(K^\times)^p$, we see now that $Z\cup\{0\}$ is a subfield of K, containing K^p; therefore, by corollary 1 of prop. 4, Chap. I-4, it is either K or K^p. If it was K, all the characters of the form $\chi_{p,\xi}$ would be trivial. As we have observed in Chap. IX-5, the kernel of the morphism $\xi\to\chi_{p,\xi}$ is the image of K under the mapping $x\to x-x^p$; in view of th. 8 of Chap. I-4, one sees at once that this image cannot contain π^{-1}, if π is any prime element of K; therefore $\chi_{p,\xi}$ is not trivial for $\xi=\pi^{-1}$. This proves that $Z\cup\{0\}=K^p$, hence $Z=(K^\times)^p$.

COROLLARY. *If K is of characteristic p, $U_K\subset(U_K)^p$.*

By proposition 10, $U_K\subset(K^\times)^p$, so that, if $\theta\in U_K$, it can be written as η^p with $\eta\in K^\times$. Take any cyclic extension L of K; by prop. 8, $U_K=N_{L/K}(U_L)$, and, by prop. 10, $U_L\subset(L^\times)^p$; therefore we can write θ as $N_{L/K}(\zeta^p)$ with $\zeta\in L^\times$. This gives $\eta^p=N_{L/K}(\zeta)^p$; as p is the characteristic, this implies $\eta=N_{L/K}(\zeta)$. We have thus shown that η is in the intersection of the groups $N_{L/K}(L^\times)$ for all cyclic extensions L of K; as this intersection is U_K, this proves our corollary.

THEOREM 3. *The mapping $\chi \to \chi \circ \mathfrak{a}$ is a bijective morphism of the group X_K of characters of \mathfrak{A} onto the group of the characters of finite order of K^\times.*

Take any integer $n \geqslant 1$. If K is not of characteristic p, we have $U_K \subset (K^\times)^n$, by the corollary of prop. 9. If K is of characteristic p, write $n = n'p^i$ with n' prime to p, and $i \geqslant 0$, and take any $\theta \in U_K$; by the same corollary, we can write $\theta = \zeta^{n'}$ with $\zeta \in K^\times$. By the corollary of prop. 10, and using induction on i, we see at once that $U_K \subset (U_K)^{p^i}$, so that we can write $\theta = \eta^{p^i}$ with $\eta \in U_K$. Take integers a, b such that $n'a + p^i b = 1$; then $\theta = (\zeta^b \eta^a)^n$. This shows that, in all cases, $U_K \subset (K^\times)^n$, so that every character of K^\times, of order dividing n, is trivial on U_K. As this is so for all n, our conclusion follows now at once from corollary 4 of prop. 2, § 1.

COROLLARY 1. *The canonical morphism \mathfrak{a} of K^\times into the Galois group \mathfrak{A} of K_{ab} over K is injective.*

By lemma 2 of § 1, applied to the endomorphism $x \to x^n$ of K^\times, $(K^\times)^n$ is a closed subgroup of K^\times for every $n \geqslant 1$; this implies that it is the intersection of the kernels of all the characters of K^\times whose order divides n; that being so, theorem 3 shows that the kernel U_K of \mathfrak{a} is the same as the intersection U' of the groups $(K^\times)^n$ for all $n \geqslant 1$. Clearly U' is contained in R^\times. As it is obvious that the compact group R^\times is totally disconnected, lemma 4 of Chap. VII-3 shows that all its characters are of finite order. If π is a prime element of K, every character of R^\times can be uniquely extended to a character ω of K^\times such that $\omega(\pi) = 1$, which then must also be of finite order. This implies that U' is contained in the kernel of all the characters of R^\times, so that it is $\{1\}$.

COROLLARY 2. *The canonical morphism \mathfrak{a} induces on R^\times an isomorphism of R^\times onto the Galois group \mathfrak{A}_0 of K_{ab} over the union K_0 of all unramified extensions of K in \bar{K}.*

This is now obvious, by corollary 1 and the corollary of prop. 7.

THEOREM 4. *Let K' be an extension of K of finite degree, contained in \bar{K}; put $L = K' \cap K_{ab}$. Then, for $\theta \in K^\times$, $\mathfrak{a}(\theta)$ induces the identity on L if and only if θ is in $N_{K'/K}(K'^\times)$.*

Call ρ the restriction morphism of \mathfrak{A}' into \mathfrak{A}, and put $\mathfrak{B} = \rho(\mathfrak{A}')$. An element of K_{ab} is invariant under \mathfrak{B} if and only if it is in K'; then it is in L; therefore \mathfrak{B} is the subgroup of \mathfrak{A} corresponding to L. Put $X = \mathfrak{a}^{-1}(\mathfrak{B})$ and $X' = N_{K'/K}(K'^\times)$; what we have to prove is that $X = X'$. By lemma 2 of § 1, X' is closed in K^\times. If n is the degree of K' over K, we have $N_{K'/K}(\theta) = = \theta^n$ for $\theta \in K^\times$, so that $X' \supset (K^\times)^n$; therefore, if ψ is a character of K^\times, trivial on X', it is trivial on $(K^\times)^n$, hence of a finite order dividing n, so

that, by th. 3, it can be written as $\chi \circ \mathfrak{a}$ with $\chi \in X_K$. Then $\chi \circ \mathfrak{a} \circ N_{K'/K}$ is trivial on K'^\times; by corollary 1 of th. 2, § 2, it is the same as $\chi \circ \rho \circ \mathfrak{a}'$, so that $\chi \circ \rho$ must be trivial on \mathfrak{A}', hence χ on $\rho(\mathfrak{A}') = \mathfrak{B}$, hence ψ on X. This shows that $X' \supset X$. Conversely, if $\theta = N_{K'/K}(\theta')$ with $\theta' \in K'^\times$, corollary 1 of th. 2, § 2, gives $\mathfrak{a}(\theta) = \rho(\mathfrak{a}'(\theta'))$; as this is in \mathfrak{B}, we see that $X' \subset X$, which completes the proof.

COROLLARY 1. *Assumptions and notations being as in theorem 4, call* \mathfrak{B} *the subgroup of* \mathfrak{A} *corresponding to L. Then* $N_{L/K}(L^\times) = N_{K'/K}(K'^\times) = = \mathfrak{a}^{-1}(\mathfrak{B})$.

The latter equality is just a restatement of theorem 4. Applying theorem 4 to $K' = L$, we get $N_{L/K}(L^\times) = \mathfrak{a}^{-1}(\mathfrak{B})$.

COROLLARY 2. *For every extension L of K of finite degree, contained in* K_{ab}, *call* $\mathfrak{B}(L)$ *the subgroup of* \mathfrak{A} *corresponding to L, and put* $N(L) = N_{L/K}(L^\times)$. *Then* $N(L) = \mathfrak{a}^{-1}(\mathfrak{B}(L))$; $\mathfrak{B}(L)$ *is the closure of* $\mathfrak{a}(N(L))$ *in* \mathfrak{A}; *L consists of the elements of* K_{ab}, *invariant under* $\mathfrak{a}(\theta)$ *for all* $\theta \in N(L)$, *and* \mathfrak{a} *determines an isomorphism of* $K^\times / N(L)$ *onto the Galois group* $\mathfrak{A}/\mathfrak{B}(L)$ *of L over K. Moreover,* $L \to N(L)$ *maps the subfields of* K_{ab}, *of finite degree over K, bijectively onto the open subgroups of* K^\times, *of finite index in* K^\times.

All this is a restatement of prop. 3 of § 1, once theorems 3 and 4 are taken into account. Traditionally, when L and $N(L)$ are as in our corollary, one says that L is "the classfield" to the subgroup $N(L)$ of K^\times. In applying our corollary, it is frequently useful to keep in mind that, by lemma 1 of § 1, an open subgroup of K^\times is of finite index in K^\times if and only if it is not contained in R^\times.

COROLLARY 3. *Let K and K' be as in theorem 4; let M be a subfield of* K_{ab}, *of finite degree over K, and call M' its compositum with K'. Then* $N_{M'/K'}(M'^\times) = N_{K'/K}^{-1}(N_{M/K}(M^\times))$.

By corollary 2, $N_{M/K}(M^\times)$, which is the same as $N(M)$, consists of the elements θ of K^\times such that $\mathfrak{a}(\theta)$ leaves every element of M invariant. Similarly, $N_{M'/K'}(M'^\times)$ consists of the elements θ' of K'^\times such that $\mathfrak{a}'(\theta')$ leaves every element of M' invariant; the latter condition is fulfilled if and only if $\rho(\mathfrak{a}'(\theta'))$ leaves every element of M invariant; in view of corollary 1 of th. 2, § 2, this is the same as to say that $\mathfrak{a}(N_{K'/K}(\theta'))$ leaves every element of M invariant, i.e. that $N_{K'/K}(\theta')$ is in $N_{M/K}(M^\times)$.

It is easily seen that theorem 4 and its corollaries retain their validity for R-fields; so does theorem 3.

§ 4. **Ramification of abelian extensions.** The above theory would be incomplete without the knowledge of the ramification properties of the abelian extensions of K, and in particular of their differents and discrimi-

nants. As shown in Chap. VIII-3, these properties can be fully expressed by a description of the Herbrand distribution on the Galois group \mathfrak{A} of K_{ab} over K. We begin with some preliminary results, the first one of which has no reference to abelian extensions and may be regarded as supplementing Chap. VIII-3. We adopt the same notations as there, e.g. in prop. 9 of that Chapter, calling K' a Galois extension of K of degree n with the Galois group $\mathfrak{g} = \mathfrak{g}_0$, and calling \mathfrak{g}_v, for $v \geqslant 1$, the higher ramification groups of K' over K. We also call R, R' the maximal compact subrings of K, K', and P, P' the maximal ideals of R, R', respectively. We denote by ε the neutral element of \mathfrak{g}.

PROPOSITION 11. *Let e be the order of ramification of K' over K, and let P'^d be its different. Take $h \geqslant 1$, $z \in P'^h$ and put*:

$$N_{K'/K}(X - z) = X^n + a_1 X^{n-1} + \cdots + a_n,$$

where X is an indeterminate.

(i) *If $v(\lambda) \leqslant h + 1$ for all $\lambda \neq \varepsilon$, then, for $1 \leqslant i \leqslant n$:*

$$e \cdot \mathrm{ord}_K(a_i) \geqslant h + d - e + 1.$$

(ii) *If $v(\lambda) \leqslant h$ for all $\lambda \neq \varepsilon$, then, for $2 \leqslant i \leqslant n$:*

$$e \cdot \mathrm{ord}_K(a_i) > h + d - e + 1.$$

(iii) *If $v(\lambda) \geqslant h + 1$ for all $\lambda \neq \varepsilon$, then, for $1 \leqslant i \leqslant n$:*

$$\mathrm{ord}_K(a_i) \geqslant h.$$

(iv) *If $v(\lambda) \geqslant h + 2$ for all $\lambda \neq \varepsilon$, then, for $1 \leqslant i \leqslant n-1$:*

$$\mathrm{ord}_K(a_i) > h.$$

As $-a_1 = \mathit{Tr}_{K'/K}(z)$, the inequality in (i), for $i = 1$, is nothing else than corollary 1 of prop. 4, Chap. VIII-1, and does not depend upon the assumption about $v(\lambda)$ in (i). In any case, we have

$$(3) \qquad\qquad (-1)^i a_i = \sum z^{\lambda_1} z^{\lambda_2} \ldots z^{\lambda_i},$$

where the sum is taken over all combinations of i distinct elements of \mathfrak{g}, or, what amounts to the same, over all subsets $\mathfrak{s} = \{\lambda_1, \ldots, \lambda_i\}$ of \mathfrak{g} of cardinal number i. For each such subset \mathfrak{s}, write:

$$z(\mathfrak{s}) = z^{\lambda_1} z^{\lambda_2} \ldots z^{\lambda_i}.$$

Take such a subset \mathfrak{s}; for each $\sigma \in \mathfrak{g}$, write $\mathfrak{s}\sigma$ for the image of \mathfrak{s} under the translation $\lambda \rightarrow \lambda\sigma$ in \mathfrak{g}; call \mathfrak{h} the subgroup of \mathfrak{g}, consisting of the elements σ such that $\mathfrak{s}\sigma = \mathfrak{s}$; call l the order of \mathfrak{h}, and take a full set $\{\rho_1, \ldots, \rho_r\}$ of representatives of the left cosets $\mathfrak{h}\rho$ of \mathfrak{h} in \mathfrak{g}. Clearly \mathfrak{s} is a disjoint union of right cosets $\mu\mathfrak{h}$ of \mathfrak{h} in \mathfrak{g}; take a full set $\mathfrak{m} = \{\mu_1, \ldots, \mu_m\}$ of representatives

for these cosets, so that \mathfrak{s} is the disjoint union of the cosets $\mu_1\mathfrak{h},\ldots,\mu_m\mathfrak{h}$; we have $i=ml$. Put $w=z(m)$, and call K'' the subfield of K' corresponding to the subgroup \mathfrak{h} of \mathfrak{g}; we have now:

$$z(\mathfrak{s})=\prod_{\sigma\in\mathfrak{h}}w^\sigma=N_{K'/K''}(w).$$

In view of the definition of \mathfrak{h}, the sets $\mathfrak{s}\,\rho_1,\ldots,\mathfrak{s}\,\rho_r$ are all distinct; as they have the same cardinal number i as \mathfrak{s}, all the terms $z(\mathfrak{s}\rho_j)$, for $1\leqslant j\leqslant r$, occur in the right-hand side of (3). As the ρ_j induce on K'' all the distinct isomorphisms of K'' into \bar{K}, the sum of these terms can be written as

$$(4)\qquad \sum_j z(\mathfrak{s}\rho_j)=\sum_j z(\mathfrak{s})^{\rho_j}=\sum_j N_{K'/K''}(w)^{\rho_j}=\mathit{Tr}_{K''/K}(N_{K'/K''}(w)).$$

Consequently, the right-hand side of (3) can be written as a sum of terms, each of which has the form shown in the right-hand side of (4); moreover, for each one of these terms, we have $ml=i$, where l is the order of \mathfrak{h}, i.e. the degree of K' over K''. All we need do now is to prove the inequalities in our proposition for each term of that form, with $\operatorname{ord}_{K'}(w)\geqslant mh$ in view of the assumption on z and of the definition of w. Call e' the order of ramification and f' the modular degree of K' over K'', so that $l=e'f'$ by corollary 6 of th. 6, Chap. I-4; then, by (2) of Chap. VIII-1, the order of $N_{K'/K''}(w)$ in K'' is $\geqslant f'mh$. Call e'' the order of ramification and d'' the differential exponent of K'' over K. If ω is the order in K of the right-hand side of (4), we have, by corollary 1 of prop. 4, Chap. VIII-1:

$$e''\omega\geqslant f'mh+d''-e''+1.$$

As $e=e'e''$, $i=ml$, and $e'f'=l$, this gives

$$e\omega\geqslant ih+e'd''-e+e'.$$

If now we call d' the differential exponent of K' over K'', corollary 4 of prop. 4, Chap. VIII-1, gives $d=e'd''+d'$, so that our last inequality can be written as

$$e\omega\geqslant ih+(d-e+1)-(d'-e'+1).$$

Now formula (9) of Chap. VIII-3, applied to K' and K'', gives:

$$d'-e'+1=\sum_\lambda(v(\lambda)-1)^+,$$

where the sum is taken over all $\lambda\neq\varepsilon$ in \mathfrak{h}; moreover, as pointed out there, the number of terms >0 in that sum is $\leqslant e'-1$. If the assumption in (i) is satisfied, every one of these terms is $\leqslant h$; this gives

$$d'-e'+1\leqslant(e'-1)h,$$

and therefore
$$e\omega-(h+d-e+1)\geqslant(i-e')h;$$

as $e'\leqslant l\leqslant i$, this proves part (i) of our proposition. If the assumption in (ii) is satisfied, we get in the same way:
$$d'-e'+1\leqslant(e'-1)(h-1)$$
and therefore
$$e\omega-(h+d-e+1)\geqslant i-1+(i-e')(h-1);$$

this cannot be 0 unless $i=1$, which proves (ii). On the other hand, if we apply formula (9) of Chap. VIII-3 to K' and K, we get
$$(d-e+1)-(d'-e'+1)=\sum(v(\lambda)-1)^+$$

where the sum is now taken over all $\lambda\in\mathfrak{g}-\mathfrak{h}$ and consists of $n-l$ terms, so that it is $\geqslant(n-l)h$ if the assumption in (iii) is satisfied. Then we get:
$$e(\omega-h)\geqslant(i+n-l-e)h,$$

which proves (iii), since $l\leqslant i$ and $e\leqslant n$. Similarly, the assumption in (iv) gives:
$$e(\omega-h)\geqslant(i+n-l-e)h+n-l;$$

as $l\leqslant i$, $e\leqslant n$, $l\leqslant n$, the right-hand side cannot be 0 unless $l=i$, $e=n$, $l=n$, hence $i=n$. This proves (iv).

COROLLARY 1. *Notations being as in proposition 11, take again $h\geqslant1$, $z\in P'^h$. That being so:*

(i) *If $v(\lambda)\leqslant h+1$ for all $\lambda\neq\varepsilon$, then:*
$$e\cdot\mathrm{ord}_K(N_{K'/K}(1+z)-1)\geqslant h+d-e+1.$$

(ii) *If $v(\lambda)\leqslant h$ for all $\lambda\neq\varepsilon$, and $h=\rho e-(d-e+1)$ with $\rho\in\mathbf{Z}$, then:*
$$N_{K'/K}(1+z)\equiv1\ (P^\rho),\quad N_{K'/K}(1+z)\equiv1+\mathit{Tr}_{K'/K}(z)\ (P^{\rho+1}).$$

(iii) *If $v(\lambda)\geqslant h+1$ for all $\lambda\neq\varepsilon$, then:*
$$N_{K'/K}(1+z)\equiv1\ (P^h).$$

(iv) *If $v(\lambda)\geqslant h+2$ for all $\lambda\neq\varepsilon$, then:*
$$N_{K'/K}(1+z)\equiv1+N_{K'/K}(z)\ (P^{h+1}).$$

In fact, with the notations of proposition 11, we have:
$$N_{K'/K}(1+z)=1+\sum_{i=1}^{n}(-1)^i a_i;$$

the four assertions in our corollary follow now immediately from the corresponding ones in proposition 11.

COROLLARY 2. *Assumptions being as in corollary* 1 (ii), *we have* $N_{K'/K}(1+P'^h)=1+P^\rho$.

As $d-e+1\geqslant 0$, these assumptions imply $h\leqslant \rho e$, hence $\rho \geqslant 1$. By corollary 1 (ii), $N_{K'/K}(1+P'^h)$ is contained in $1+P^\rho$. Conversely, take any $x_0 \in P^\rho$; then we can define by induction two sequences (x_0, x_1, \ldots) and (z_0, z_1, \ldots), with $x_i \in P^{i+\rho}$ and $z_i \in P'^{ie+h}$ for all $i \geqslant 0$, by taking, for each $i \geqslant 0$, $z_i \in P'^{ie+h}$ such that $Tr_{K'/K}(z_i)=x_i$, as may be done by prop. 4, Chap. VIII-1, and then putting

$$x_{i+1}=(1+x_i)N_{K'/K}(1+z_i)^{-1}-1;$$

corollary 1 (ii) shows at once that this is in $P^{i+1+\rho}$, as it should. Then, obviously, $1+x_0 = N_{K'/K}(y)$ with y given by the convergent product

$$y=\prod_{i=0}^{\infty}(1+z_i);$$ as y is in $1+P'^h$, this proves our corollary.

PROPOSITION 12. *Let K and K' be as above, and assume that $v(\lambda)$ has the same value $i \geqslant 2$ for all $\lambda \neq \varepsilon$ in* \mathfrak{g}. *Then, for $1 \leqslant h \leqslant i$, $N_{K'/K}(1+P'^h)$ is contained in $1+P^h$; and $N_{K'/K}(1+P'^{i-1})$ is contained in $1+P^i$ if and only if the degree n of K' over K is equal to the module q of K.*

Here the higher groups of ramification of K' over K are given by $\mathfrak{g}_v = \mathfrak{g}$ for $v \leqslant i$ and $\mathfrak{g}_v = \{\varepsilon\}$ for $v \geqslant i+1$. As $\mathfrak{g}_1 = \mathfrak{g}$, we have $e=n$; K' has the modular degree $f=1$ over K and has the same module q as K. By formula (9) of Chap. VIII-3, we have $d=(n-1)i$. Taking $h=i$ in corollary 1(i) of prop. 11, we get our first assertion for that case; in the case $h<i$, it follows at once from corollary 1 (iii) of the same proposition. By corollary 3 of prop. 9, Chap. VIII-3, the degree n of K' over K divides q; by that proposition, if we take a prime element π' of K' and put $y_\lambda = \pi'^\lambda \pi'^{-1}$ for all $\lambda \in \mathfrak{g}$, the mapping $\lambda \to y_\lambda$ maps \mathfrak{g} onto a set Y of elements of $1+P'^{i-1}$ which are all incongruent to each other modulo P'^i. In particular, Y makes up a full set of representatives of the cosets of $1+P'^i$ in $1+P'^{i-1}$ if and only if $n=q$. Since obviously $N_{K'/K}(y_\lambda)=1$ for all λ, this shows that, if $n=q$, $N_{K'/K}(1+P'^{i-1})$ is the same as $N_{K'/K}(1+P'^i)$, hence contained in $1+P^i$. In order to prove the converse, take $z \in K'^\times$ such that $\mathrm{ord}_{K'}(z)=i-1$. As in prop. 11, write

$$N_{K'/K}(X-z)=X^n+\sum_{j=1}^{n}a_j X^{n-j}.$$

Then $a_n = N_{K'/K}(-z)$, so that, by (2) of Chap. VIII-1, $\mathrm{ord}_K(a_n)=i-1$. Taking $h=i-1$ in prop. 11 (i), we get $\mathrm{ord}_K(a_j) \geqslant i-1$ for $1 \leqslant j \leqslant n$, so that, if we put $b_j = a_j/a_n$, all the b_j are in R. Now take any $y \in 1+P'^{i-1}$; as

$(1-y)/z$ is in R', and as K' has the same module as K, there is $\alpha \in R$ such that $(1-y)/z \equiv \alpha \ (P')$, or, what amounts to the same, $y=(1-\alpha z)u$ with $u \in 1+P'^i$. Then $N_{K'/K}(u)$ is in $1+P^i$, so that we have:

$$N_{K'/K}(y) \equiv N_{K'/K}(1-\alpha z) = 1+ \sum_{j=1}^{n} a_j \alpha^j = 1+a_n(\alpha^n + \sum_{j=1}^{n-1} b_j \alpha^j) \quad (P^i).$$

For $1 \leqslant j \leqslant n\text{-}1$, call \bar{b}_j the image of b_j in the field R/P, under the canonical homomorphism of R onto that field; then the above formula shows that $N_{K'/K}(y)$ is in $1+P^i$ if and only if the image of α in the same field is a root of the polynomial $T^n + \sum \bar{b}_j T^j$; in particular, if this is so for all y, all the elements of R/P must be roots of that polynomial, so that $n \geqslant q$. This completes the proof.

PROPOSITION 13. *Let π be a prime element of K; for each $v \geqslant 1$, call N_v the subgroup of K^\times generated by π and $1+P^v$, and call K_v the subfield of K_{ab} such that $N(K_v)=N_v$, in the sense of corollary 2 of theorem 4, § 3. Call $\mathfrak{g}^{(v)}$ the Galois group of K_v over K, and \mathfrak{a}_v the morphism of K^\times onto $\mathfrak{g}^{(v)}$, with the kernel N_v, determined by the canonical morphism \mathfrak{a} of K. Call $\mathfrak{g}_i^{(v)}$, for $i \geqslant 1$, the higher groups of ramification of K_v over K. Then $\mathfrak{g}_1^{(v)} = \mathfrak{g}^{(v)}$; for $1 \leqslant \rho \leqslant v$ and $q^{\rho-1} < i \leqslant q^\rho$, $\mathfrak{g}_i^{(v)} = \mathfrak{a}_v(N_\rho)$.*

Choose some $v \geqslant 1$, and then, to simplify notations, write N instead of N_v, L instead of K_v, \mathfrak{g} instead of $\mathfrak{g}^{(v)}$, \mathfrak{g}_i instead of $\mathfrak{g}_i^{(v)}$; as in Chap. VIII-3, call g_i the number of elements of \mathfrak{g}_i, for all $i \geqslant 1$; then g_i divides g_j for $i \geqslant j$. As \mathfrak{g} is isomorphic to K^\times/N, the degree of L over K is the index of N in K^\times, which is $n=(q-1)q^{v-1}$. By the corollary of prop. 6, § 2, the maximal unramified extension L' of K, contained in L, is the one for which $N(L'^\times) = R^\times N$; as $R^\times N = K^\times$, we get $L'=K$; in other words, L is fully ramified over K, its order of ramification is $e=n$, and we have $\mathfrak{g}_1 = \mathfrak{g}$; moreover, L has the same module q as K, and the same must then be true of all fields between K and L, so that, if $K \subset K' \subset K'' \subset L$, K'' is fully ramified over K'. By corollary 1 of prop. 9, Chap. VIII-3, g_i/g_{i+1} divides q for $i \geqslant 2$; therefore $g_1/g_2 = q-1$. Put $r_1=0$ and $r_i=(g_2+\cdots+g_i)/n$ for all $i \geqslant 2$; for each integer $\rho \geqslant 0$, call $i(\rho)$ the largest of the integers i such that $r_i \leqslant \rho$. Assume $r_i < \rho < r_{i+1}$ for any $\rho \geqslant 0$ and any $i \geqslant 1$; then $0 < \rho n-(g_2+\cdots+g_i) < g_{i+1}$; this is clearly a contradiction, since n, g_2, \ldots, g_i are multiples of g_{i+1}. Therefore, for all ρ, we have $r_{i(\rho)}=\rho$. We have $i(0)=1$, and $i(\rho)>1$ for $\rho>0$. Take now any ρ such that $0 \leqslant \rho < v$; put $i=i(\rho)$; call K', K'' the subfields of L consisting of the elements invariant under \mathfrak{g}_i and under \mathfrak{g}_{i+1}, respectively; the Galois group \mathfrak{g}'' of K'' over K' may then be identified with $\mathfrak{g}_i/\mathfrak{g}_{i+1}$. If $\rho=0$, $i=1$, and the degree of K'' over K' is $g_1/g_2=q-1$. From now on, assume that $1 \leqslant \rho < v$; we will show that then K'' is of degree q over K'. We first observe that the higher ramification groups \mathfrak{g}_j''

of K'' over K' are given by applying formula (11) of Chap. VIII-3 to K', K'' and L; as K'' is fully ramified over K', its order of ramification is the same as its degree over K', and then that formula shows at once that $v(\sigma'')=i$ for all $\sigma'' \in g''$ except the identity, so that $g_j''=g''$ for $j \leqslant i$, and $g_j''=\{\varepsilon\}$ for $j \geqslant i+1$. Similarly, write $g'=g/g_i$ for the Galois group of K' over K, and call g_j' its higher ramification groups; in exactly the same manner, we find that g_j' is the image g_j/g_i of g_j in g', for $j < i$, and that it is $\{\varepsilon\}$ for $j \geqslant i$. Call R', R'' the maximal compact subrings of K', K'', and P', P'' the maximal ideals in R', R'', respectively. As K' is fully ramified over K, its order of ramification e' is the same as its degree $n'=n/g_i$; then, if $P'^{d'}$ is the different of K' over K, formula (10) of Chap. VIII-3, gives

$$d'-e'+1=\sum_{j=2}^{i}((g_j/g_i)-1)=r_i n/g_i-i+1=\rho \acute{n}'-i+1.$$

Take any $z \in L$ such that $\mathrm{ord}_L(z) \geqslant i-1$; put $v=N_{L/K''}(z)$, so that $v \in P''^{i-1}$. Applying corollary 1 (iv) of prop. 11 to K'' and L, with $h=i-1$, we get

$$N_{L/K''}(1+z) \equiv 1+v \quad (P''^i).$$

Define now $w \in K'$ by writing

$$1+w=N_{L/K'}(1+z)=N_{K''/K'}(N_{L/K''}(1+z)).$$

Applying to K', K'' the first assertion in prop. 12 with $h=i$, we get

$$1+w \equiv N_{K''/K'}(1+v) \quad (P'^i);$$

the case $h=i-1$ of the same assertion in prop. 12 gives then $w \in P'^{i-1}$. Now, taking $h=i-1$ in corollary 1 (ii) of prop. 11, we can apply it to K and K'; this gives

$$N_{L/K}(1+z)=N_{K'/K}(1+w) \equiv 1+\mathit{Tr}_{K'/K}(w) \quad (P^{\rho+1}).$$

By the definition of K_v, and corollary 2 of th. 4, § 3, this must be in $N_v=N$, hence in $N_v \cap R^\times$, i.e. in $1+P^v$; as $\rho < v$, this implies that $\mathit{Tr}_{K'/K}(w)$ is in $P^{\rho+1}$. In view of the values found above for e' and $d'-e'+1$, prop. 4 of Chap. VIII-1 shows that $\mathit{Tr}_{K'/K}$ maps P'^{i-1} surjectively onto P^ρ, and P'^i onto $P^{\rho+1}$; in particular, there is $w' \in P'^{i-1}$ such that $\mathit{Tr}_{K'/K}(w')$ is not in $P^{\rho+1}$. Then, if we had $\mathrm{ord}_{K'}(w)=i-1$, $w'w^{-1}$ would be in R', so that we could write $w'w^{-1} \equiv \alpha \ (P')$ with $\alpha \in R$, since K' has the same module as K; this can be written $w' \equiv \alpha w \ (P'^i)$, which implies

$$\mathit{Tr}_{K'/K}(w') \equiv \alpha \, \mathit{Tr}_{K'/K}(w) \equiv 0 \quad (P^{\rho+1}),$$

against our assumption. This shows that w is in P'^i, or in other words that $N_{K''/K'}(1+v)$ is in $1+P'^i$ whenever $v=N_{L/K''}(z)$ with $z \in L$, $\mathrm{ord}_L(z) \geqslant$

$\geqslant i-1$. Now choose $z_0 \in L^\times$ so that $\mathrm{ord}_L(z_0) = i-1$; put $v_0 = N_{L/K''}(z_0)$, so that $\mathrm{ord}_{K''}(v_0) = i-1$; call M^\times the group of $(q-1)$-th roots of 1 in K, and take $z = \mu z_0$ with $\mu \in M^\times$. As the degree of L over K'' is g_{i+1}, and as g_j/g_{j+1} divides q for all $j \geqslant 2$, we have $v = \mu^Q v_0$, where $Q = g_{i+1}$ divides a power of q and is therefore prime to $q-1$, so that $\mu \to \mu^Q$ is an automorphism of M^\times. Consequently, when μ runs through the set $M = M^\times \cup \{0\}$, $1+v$ runs through a full set of representatives of the cosets modulo $1 + P''^i$ in $1 + P''^{i-1}$. As $N_{K''/K'}(1+v)$ is in $1+P'^i$ for all these elements v, prop. 12 shows, firstly, that $N_{K''/K'}(1 + P''^{i-1})$ is contained in $1 + P'^i$, and then that the degree of K'' over K' must therefore be q.

In other words, we have shown that $g_{i(\rho)}/g_{i(\rho)+1}$ is equal to q for $1 \leqslant \rho \leqslant \leqslant v-1$; we had already found that it has the value $q-1$ for $\rho = 0$. As we have

$$n = (q-1)q^{v-1} = \prod_{i=1}^{\infty} (g_i/g_{i+1}),$$

this implies that $g_i = g_{i+1}$ whenever i is not one of the integers $i(0) = 1$, $i(1), \ldots, i(v-1)$; therefore $g_j = g_{i(\rho)+1}$ for $i(\rho) < j \leqslant i(\rho+1)$, so that $g_{i(\rho+1)}$ is of index q in $g_{i(\rho)}$ for $1 \leqslant \rho < v$, while this index is $q-1$ for $\rho = 0$. By induction on ρ, we see at once that $g_{i(\rho)} = q^{v-\rho}$ for $1 \leqslant \rho \leqslant v$. The definition of the integers r_i gives now:

$$r_{i(\rho+1)} - r_{i(\rho)} = (g_{i(\rho)+1} + \cdots + g_{i(\rho+1)})/n = (i(\rho+1) - i(\rho))q^{v-\rho-1}n^{-1},$$

for $0 \leqslant \rho < v$. As $r_{i(\rho)} = \rho$, the left-hand side is 1; this gives

$$i(\rho+1) - i(\rho) = (q-1)q^\rho,$$

and therefore $i(\rho) = q^\rho$ by induction on ρ.

To complete the proof, observe that, in view of the values found above for e' and $d' - e' + 1$, we may apply corollary 2 of prop. 11, with $h = i-1$, to K and the same field K' as above; it shows that $1 + P^\rho$ is the same as $N_{K'/K}(1 + P'^{i-1})$ and is therefore contained in the group $N' = N_{K'/K}(K'^\times)$ associated with K' according to corollary 2 of th. 4, § 3. As N' contains the group $N = N_v$ associated with $L = K_v$, it contains π, hence the group N_ρ generated by π and $1 + P^\rho$. Let \mathfrak{a}_v be the morphism of K^\times onto $\mathfrak{g} = \mathfrak{g}^{(v)}$, with the kernel N_v, defined in our proposition; by corollary 2 of th. 4, § 3, this maps N' onto the subgroup $\mathfrak{g}_{i(\rho)}$ of \mathfrak{g} corresponding to K'; therefore $\mathfrak{g}_{i(\rho)}$ contains $\mathfrak{a}_v(N_\rho)$ for $1 \leqslant \rho < v$. In view of our definitions, the same is obviously true for $\rho = 0$ if we define N_0 by $N_0 = K^\times$, and also for $\rho = v$. Now we prove by induction on ρ that $\mathfrak{g}_{i(\rho)} = \mathfrak{a}_v(N_\rho)$ for $0 \leqslant \rho \leqslant v$. It is true for $\rho = 0$. Assume $\mathfrak{g}_{i(\rho-1)} = \mathfrak{a}_v(N_{\rho-1})$, and let N' be as above; N' contains N_ρ, as we have seen, and it is contained in $N_{\rho-1}$, since $\mathfrak{g}_{i(\rho)}$ is contained in $\mathfrak{g}_{i(\rho-1)}$; its index in $N_{\rho-1}$ is the same as that of $\mathfrak{g}_{i(\rho)}$ in $\mathfrak{g}_{i(\rho-1)}$,

which is $q-1$ if $\rho=1$ and q if $\rho>1$. As this is the same as the index of N_ρ in $N_{\rho-1}$, we get $N'=N_\rho$. In view of what has already been proved above, this completes the proof of our proposition.

COROLLARY. *Notations being as in proposition 13, the order of ramification of K_v over K is the same as its degree and is given by $e_v=(q-1)q^{v-1}$; if d_v is the differental exponent of K_v over K, we have $d_v/e_v=v-(q-1)^{-1}$.*

The value of e_v has already been given above; that of d_v can be obtained at once by applying formula (10) of Chap. VIII-3 to the results stated in proposition 13; this gives the formula in our corollary.

As will presently be seen, proposition 13 contains in substance the determination of the Herbrand distribution on the Galois group \mathfrak{A} of K_{ab} over K, which was our main object in this §. We recall that this has been defined in Chap. VIII-3 as a certain linear form $f \to H(f)$ on the space of all locally constant functions on \mathfrak{A}. As explained there, if X is any open and closed subset of \mathfrak{A}, its characteristic function f_X is locally constant, and then we write $H(X)$ instead of $H(f_X)$; $X \to H(X)$ is thus a finitely additive function of X.

LEMMA 3. *Let H be the Herbrand distribution on \mathfrak{A}. Then there is a unique distribution H_0 on \mathfrak{A}_0 such that $H(f)=H_0(f_0)$ whenever f is a locally constant function on \mathfrak{A}, and f_0 is the function induced on \mathfrak{A}_0 by f.*

Let \mathfrak{B} be any open subgroup of \mathfrak{A}; let L be the subfield of K_{ab} corresponding to \mathfrak{B}. Let K_0 be as in § 2, i.e. the union of all unramified extensions of K, so that \mathfrak{A}_0 is the subgroup of \mathfrak{A} corresponding to K_0. Then the maximal unramified extension L_0 of K, contained in L, is $K_0 \cap L$ and corresponds to the subgroup $\mathfrak{B}\mathfrak{A}_0$ of \mathfrak{A}. If $\mathfrak{B}\alpha$ is any coset of \mathfrak{B} in \mathfrak{A}, other than \mathfrak{B}, and α induces on L the automorphism λ, $H(\mathfrak{B}\alpha)$ is by definition equal to $-v(\lambda)/e$, where e is the order of ramification of L over K; this is 0 if λ does not induce the identity on L_0, i.e. if $\mathfrak{B}\alpha$ is not contained in $\mathfrak{B}\mathfrak{A}_0$, or in other words if $\mathfrak{B}\alpha \cap \mathfrak{A}_0=\emptyset$. As H is finitely additive, this implies that $H(X)=0$ whenever $X \cap \mathfrak{A}_0=\emptyset$, and $H(f)=0$ whenever the locally constant function f is 0 on \mathfrak{A}_0. On the other hand, take any locally constant function f_0 on \mathfrak{A}_0. As \mathfrak{A}_0 is compact, f_0 is uniformly continuous, so that there is an open subgroup \mathfrak{B} of \mathfrak{A} such that f_0 is constant on the cosets of $\mathfrak{B}\cap\mathfrak{A}_0$ in \mathfrak{A}_0. Then f_0 can be uniquely continued to a function f on \mathfrak{A}, constant on the cosets of \mathfrak{B} in \mathfrak{A} and 0 outside $\mathfrak{B}\mathfrak{A}_0$. If then we put $H_0(f_0)=H(f)$, it is clear that H_0 is as required by our lemma. Except for obvious notational changes, the lemma and its proof remain valid for the Herbrand distribution on the Galois group of any Galois extension of K, abelian or not. This will not be needed.

As the canonical morphism \mathfrak{a} of K^\times into \mathfrak{A} maps R^\times isomorphically onto \mathfrak{A}_0, we can transport to R^\times the distribution H_0 of lemma 3 by means of the inverse to that isomorphism. This defines a distribution H_R on R^\times, which we extend to a distribution H_K on K^\times by prescribing that $H_K(X) = H_R(X \cap R^\times)$ for every open and closed subset X of K^\times. We will call H_K the Herbrand distribution on K^\times; in an obvious sense, its support is contained in R^\times. In view of the definition of H_0, we have $H(f) = H_K(f \circ \mathfrak{a})$ for every locally constant function f on \mathfrak{A}, and $H(X) = H_K(\mathfrak{a}^{-1}(X))$ for every open and closed subset X of \mathfrak{A}. The distribution H_K is given by the following theorem:

THEOREM 5. *Let H_K be the Herbrand distribution on K^\times. Then its support is R^\times; $H_K(R^\times) = 0$; $H_K(1 + P^\nu) = \nu - (q-1)^{-1}$ for all $\nu \geqslant 1$; if $0 \leqslant \rho < \nu$, $\xi \in R^\times$, and $\mathrm{ord}_K(1 - \xi) = \rho$, then*

$$H_K((1 + P^\nu)\xi) = -q^{\rho + 1 - \nu}(q-1)^{-1}.$$

By the definition of the Herbrand distribution, we have $H(\mathfrak{A}) = 0$; this gives $H_K(K^\times) = 0$, hence $H_K(R^\times) = 0$. Let K_ν, N_ν, d_ν, e_ν be as in prop. 13 and its corollary; if \mathfrak{B}_ν is the subgroup of \mathfrak{A} corresponding to K_ν, we have $H(\mathfrak{B}_\nu) = d_\nu/e_\nu$ by the definition of the Herbrand distribution; as $N_\nu = \mathfrak{a}^{-1}(\mathfrak{B}_\nu)$ by corollary 2 of th. 4, § 3, and as $N_\nu \cap R^\times = 1 + P^\nu$, this, together with the corollary of prop. 13, gives for $H_K(1 + P^\nu)$ the value given in our theorem. Finally, let ξ be as in our theorem; call λ the automorphism of K_ν induced by $\mathfrak{a}(\xi)$. By the definition of the Herbrand distribution, $H(\mathfrak{B}_\nu \mathfrak{a}(\xi))$ is $-\nu(\lambda)/e_\nu$, or, what amounts to the same, it is $-i/e_\nu$ if i is the largest integer such that $\lambda \in \mathfrak{g}_i^{(\nu)}$. By prop. 13, this is $i = q^\rho$ if ρ is the largest integer such that $\lambda \in \mathfrak{a}_\nu(N_\rho)$, or, what amounts to the same, such that $\xi \in N_\rho$; this is given by $\rho = \mathrm{ord}_K(1 - \xi)$. On the other hand, $H(\mathfrak{B}_\nu \mathfrak{a}(\xi))$ is the same as $H_K(N_\nu \xi)$ and as $H_K((1 + P^\nu)\xi)$. This completes our proof.

COROLLARY 1. *Let χ be a character of \mathfrak{A}, and P^f the conductor of the character $\chi \circ \mathfrak{a}$ of K^\times. Then $f = H(\chi) = H_K(\chi \circ \mathfrak{a})$.*

Put $\omega = \chi \circ \mathfrak{a}$. If $f = 0$, ω is trivial on R^\times, so that $H_K(\omega) = H_K(R^\times) = 0$ by theorem 5. Assume now that $f \geqslant 1$; call φ_0 the characteristic function of R^\times, and φ_i that of $1 + P^i$ for $1 \leqslant i \leqslant f$. We have:

$$H_K(\omega) = \sum_{i=0}^{f-1} H_K((\varphi_i - \varphi_{i+1})\omega) + H_K(\varphi_f \omega).$$

By the definition of the conductor, ω is trivial on $1 + P^f$, so that the last term is equal to $H_K(1 + P^f)$, hence to $f - (q-1)^{-1}$ by theorem 5. Also by theorem 5, and in view of the fact that ω is constant on cosets modulo $1 + P^f$ in R^\times, we have, for $0 \leqslant i \leqslant f-1$:

$$H_K((\varphi_i - \varphi_{i+1})\omega) = -q^{i+1-f}(q-1)^{-1}(S_i - S_{i+1}),$$

where S_i is the sum $\sum \omega(\xi)$ taken over a full set of representatives of the cosets modulo $1 + P^f$ in R^\times if $i=0$ and in $1 + P^i$ if $i \geqslant 1$. By the definition of the conductor, ω is not trivial on R^\times, nor on $1 + P^i$ for any $i < f$; therefore $S_i = 0$ for $i < f$, and $S_f = 1$. Our conclusion follows from this at once.

COROLLARY 2. *Let L be an abelian extension of K of finite degree. Let $\omega_1, \ldots, \omega_n$ be all the distinct characters of K^\times, trivial on the subgroup $N(L) = N_{L/K}(L^\times)$ of K^\times associated with L; for each i, call P^{f_i} the conductor of ω_i. Then the discriminant of L over K is P^δ with $\delta = \sum_i f_i$.*

Call \mathfrak{B} the closure of $\mathfrak{a}(N(L))$ in \mathfrak{A}; by corollary 2 of th. 4, § 3, it is the subgroup of \mathfrak{A} corresponding to L, and \mathfrak{a} determines an isomorphism of $K^\times/N(L)$ onto $\mathfrak{A}/\mathfrak{B}$. Therefore we can write, for each i, $\omega_i = \chi_i \circ \mathfrak{a}$, where χ_i is a character of \mathfrak{A}, trivial on \mathfrak{B}, and the χ_i, for $1 \leqslant i \leqslant n$, are then all the characters of \mathfrak{A}, trivial on \mathfrak{B}, so that the characteristic function of \mathfrak{B} on \mathfrak{A} can be written as $n^{-1}\sum_i \chi_i$. Call e, f and d the order of ramification, the modular degree and the differential exponent of L over K; then $n = ef$, and, by the corollary of prop. 6, Chap. VIII-2, the discriminant of L over K is P^{fd}. By the definition of the Herbrand distribution, we have $H(\mathfrak{B}) = d/e$; this can be written as

$$d/e = H\left(n^{-1}\sum_i \chi_i\right) = n^{-1}\sum_i H(\chi_i),$$

hence $fd = \sum_i f_i$, in view of corollary 1. This completes the proof.

§ 5. **The transfer.** Notations being as before, let K' be an extension of K of finite degree; call \mathfrak{a}, \mathfrak{a}' the canonical morphisms of K^\times into \mathfrak{A}, and of K'^\times into \mathfrak{A}', respectively. As \mathfrak{a} is injective, there is a mapping t of the image $\mathfrak{a}(K^\times)$ of K^\times in \mathfrak{A} into the image $\mathfrak{a}'(K'^\times)$ of K'^\times in \mathfrak{A}', defined by $t(\mathfrak{a}(\theta)) = \mathfrak{a}'(\theta)$ for every $\theta \in K^\times$, or in other words by $t \circ \mathfrak{a} = \mathfrak{a}' \circ j$ if j is the natural injection of K^\times into K'^\times. The question arises whether this can be characterized in group-theoretical terms, and extended by continuity to a morphism t of \mathfrak{A} into \mathfrak{A}'; this will now be answered affirmatively. For simplicity, we will assume K' to be separable over K; a consideration of the general case would complicate our statements without adding to them anything of value.

Consequently, let K' be a subfield of K_{sep}, of finite degree n over K. As before, we call \mathfrak{G}' the subgroup of \mathfrak{G} corresponding to K', and identify \mathfrak{A} with $\mathfrak{G}/\mathfrak{G}^{(1)}$ and \mathfrak{A}' with $\mathfrak{G}'/\mathfrak{G}'^{(1)}$. It will be shown that the morphism we are looking for is none other than the so-called "transfer homo-

morphism" t of \mathfrak{A} into \mathfrak{A}'; we recall that this is defined as follows. Take a full set $\{\sigma_1,...,\sigma_n\}$ of representatives of the right cosets $\sigma\mathfrak{G}'$ of \mathfrak{G}' in \mathfrak{G}. For every $\sigma\in\mathfrak{G}$, the mapping $\sigma_i\mathfrak{G}'\to\sigma\sigma_i\mathfrak{G}'$ is a permutation of these cosets, so that one may write, for each i, $\sigma\sigma_i\mathfrak{G}'=\sigma_{j(i)}\mathfrak{G}'$, where $i\to j(i)$ is a permutation of $\{1,...,n\}$; this can be written as $\sigma\sigma_i=\sigma_{j(i)}\tau_i$ with $\tau_i\in\mathfrak{G}'$. It is now easily seen that the image of $\tau_1\cdots\tau_n$ in $\mathfrak{A}'=\mathfrak{G}'/\mathfrak{G}'^{(1)}$, under the canonical morphism of \mathfrak{G}' onto \mathfrak{A}', depends neither upon the choice of the representatives σ_i nor upon their ordering, so that, if we call that image $\alpha(\sigma)$, the mapping $\sigma\to\alpha(\sigma)$ of \mathfrak{G} into \mathfrak{A}' depends only upon \mathfrak{G} and \mathfrak{G}'. One sees then at once that $\alpha(\sigma\sigma')=\alpha(\sigma)\alpha(\sigma')$ for all σ, σ' in \mathfrak{G}. Moreover, the subgroup \mathfrak{G}'' of \mathfrak{G}, consisting of the elements σ such that $\sigma\sigma_i\in\sigma_i\mathfrak{G}'$ for all i, is the intersection of the open subgroups $\sigma_i\mathfrak{G}'\sigma_i^{-1}$ for $1\leqslant i\leqslant n$, hence itself an open subgroup of \mathfrak{G}; as it is obvious that $\sigma\to\alpha(\sigma)$ is continuous on \mathfrak{G}'', it is continuous on \mathfrak{G}, hence a morphism of \mathfrak{G} into \mathfrak{A}'. As \mathfrak{A}' is commutative, the kernel of this morphism must contain $\mathfrak{G}^{(1)}$, so that it determines a morphism t of $\mathfrak{A}=\mathfrak{G}/\mathfrak{G}^{(1)}$ into \mathfrak{A}'; by definition, this is the *transfer homomorphism* of \mathfrak{A} into \mathfrak{A}'.

THEOREM 6. *Let K' be an extension of K of finite degree, contained in K_{sep}; let \mathfrak{a}, \mathfrak{a}' be the canonical morphisms of K^\times into \mathfrak{A}, and of K'^\times into \mathfrak{A}', respectively. Let t be the transfer homomorphism of \mathfrak{A} into \mathfrak{A}', and j the natural injection of K^\times into K'^\times. Then $t\circ\mathfrak{a}=\mathfrak{a}'\circ j$.*

Let \mathfrak{G}, \mathfrak{G}' be as above; let L be any Galois extension of K of finite degree, containing K' and contained in K_{sep}, and let \mathfrak{H} be the subgroup of \mathfrak{G} corresponding to L. Then \mathfrak{H} is a normal open subgroup of \mathfrak{G}, contained in \mathfrak{G}'; the Galois group of L over K is $\mathfrak{g}=\mathfrak{G}/\mathfrak{H}$, and the subgroup of \mathfrak{g} corresponding to K' is $\mathfrak{g}'=\mathfrak{G}'/\mathfrak{H}$. Let K'' be any field between K and L; let \mathfrak{G}'' and $\mathfrak{g}''=\mathfrak{G}''/\mathfrak{H}$ be the subgroups of \mathfrak{G} and of \mathfrak{g}, respectively, corresponding to K''. The canonical morphism \mathfrak{a}'' for K'' is then a morphism of K''^\times into $\mathfrak{A}''=\mathfrak{G}''/\mathfrak{G}''^{(1)}$, which, to every $\xi\in K''^\times$, assigns an automorphism $\mathfrak{a}''(\xi)$ of K''_{ab} over K''; we will write $\mathfrak{b}(K'';\xi)$ for the automorphism of $L\cap K''_{\text{ab}}$ over K'' induced on that field by $\mathfrak{a}''(\xi)$. As the subgroup of \mathfrak{G} corresponding to $L\cap K''_{\text{ab}}$ is $\mathfrak{H}\mathfrak{G}''^{(1)}$, $\xi\to\mathfrak{b}(K'';\xi)$ is a morphism of K''^\times into the group $\mathfrak{G}''/\mathfrak{H}\mathfrak{G}''^{(1)}$; clearly the latter group may be identified with $\mathfrak{g}''/\mathfrak{g}''^{(1)}$, where $\mathfrak{g}''^{(1)}$ is the commutator subgroup of \mathfrak{g}''. In particular, $\theta\to\mathfrak{b}(K;\theta)$ is a morphism of K^\times into $\mathfrak{g}/\mathfrak{g}^{(1)}$. We will denote by t_0 the transfer homomorphism defined for \mathfrak{g} and \mathfrak{g}' just as t has been defined above for \mathfrak{G} and \mathfrak{G}'; it is a morphism of $\mathfrak{g}/\mathfrak{g}^{(1)}$ into $\mathfrak{g}'/\mathfrak{g}'^{(1)}$. Our theorem will be proved if we show that, for all $\theta\in K^\times$, $\mathfrak{b}(K';\theta)=t_0(\mathfrak{b}(K;\theta))$; for this implies that $\mathfrak{a}'(\theta)$ can differ from $t(\mathfrak{a}(\theta))$ only by an element of the image of $\mathfrak{H}\mathfrak{G}'^{(1)}$ in $\mathfrak{G}'/\mathfrak{G}'^{(1)}$, i.e. by an element which is arbitrarily close to the identity, since we can take for \mathfrak{H} an arbitrarily small open subgroup of \mathfrak{G}', normal in \mathfrak{G}.

We will denote by h, h' the canonical morphisms of \mathfrak{g} onto $\mathfrak{g}/\mathfrak{g}^{(1)}$ and of \mathfrak{g}' onto $\mathfrak{g}'/\mathfrak{g}'^{(1)}$, respectively; h is the same as the "restriction morphism" which, to every automorphism of L over K, assigns its restriction to $L \cap K_{ab}$, and h' can be similarly interpreted. If now K'' is a field between K and L, corresponding to the subgroup \mathfrak{g}'' of \mathfrak{g}, $K'' \cap K_{ab}$ is the subfield of L corresponding to the subgroup $\mathfrak{g}'' \mathfrak{g}^{(1)}$ of \mathfrak{g}, or, what amounts to the same, it is the subfield of $L \cap K_{ab}$ corresponding to the subgroup $h(\mathfrak{g}'')$ of the Galois group $\mathfrak{g}/\mathfrak{g}^{(1)}$ of $L \cap K_{ab}$ over K. Consequently, by th. 4 of § 3, and in view of our definition of $\mathfrak{b}(K;\theta)$, the subgroup $N_{K''/K}(K''^{\times})$ of K^{\times} consists of the elements θ of K^{\times} such that $\mathfrak{b}(K;\theta)$ is in $h(\mathfrak{g}'')$. Now assume that \mathfrak{g}'' is commutative, so that $\xi \to \mathfrak{b}(K'';\xi)$ maps K''^{\times} into \mathfrak{g}''; then we see in a similar manner, by applying corollary 1 of th. 2, § 2, to K and K'', that we have, for all $\xi \in K''^{\times}$:

$$(5) \qquad h(\mathfrak{b}(K'';\xi)) = \mathfrak{b}(K; N_{K''/K}(\xi)).$$

If $K'' \supset K'$, i.e. if $\mathfrak{g}'' \subset \mathfrak{g}'$, we have a similar formula with K', h' replacing K, h.

Now, for a given $\theta \in K^{\times}$, we can choose a cyclic subgroup Γ of \mathfrak{g} such that $\mathfrak{b}(K;\theta)$ is in $h(\Gamma)$; for instance, we may take for Γ the group generated by any $\gamma \in \mathfrak{g}$ such that $h(\gamma) = \mathfrak{b}(K;\theta)$. Then, as we have seen above, if Z is the subfield of L corresponding to Γ, θ may be written as $N_{Z/K}(\zeta)$ with $\zeta \in Z^{\times}$. Take a full set of representatives $\{\lambda_1,...,\lambda_r\}$ for the double cosets $\Gamma \lambda \mathfrak{g}'$ of Γ and \mathfrak{g}' in \mathfrak{g}, and call γ_1 a generator of Γ. For each i, $\Gamma \lambda_i \mathfrak{g}'$ is a union of right cosets $\gamma \lambda_i \mathfrak{g}'$ of \mathfrak{g}', with $\gamma \in \Gamma$. If γ, γ' are in Γ, $\gamma \lambda_i \mathfrak{g}'$ is the same as $\gamma' \lambda_i \mathfrak{g}'$ if and only if $\gamma^{-1}\gamma'$ is in the group $\Gamma_i = \Gamma \cap \lambda_i \mathfrak{g}' \lambda_i^{-1}$. Call d_i the index of Γ_i in Γ; then Γ_i is generated by $\gamma_1^{d_i}$, and d_i is also the smallest of the integers d such that $\lambda_i^{-1} \gamma_1^{d} \lambda_i$ is in \mathfrak{g}'. That being so, $\Gamma \lambda_i \mathfrak{g}'$ is the disjoint union of the cosets $\gamma_1^{j} \lambda_i \mathfrak{g}'$ for $0 \leqslant j < d_i$. Consequently, the elements $\gamma_1^{j} \lambda_i$, for $1 \leqslant i \leqslant r$, $0 \leqslant j < d_i$, make up a full set of representatives of the right cosets of \mathfrak{g}' in \mathfrak{g}, and we can use it for computing the transfer $t_0(\gamma)$ of any element γ of Γ. Taking at first $\gamma = \gamma_1$, we find at once, in that case:

$$(6) \qquad t_0(\gamma) = h'\left(\prod_{i=1}^{r} (\lambda_i^{-1} \gamma^{d_i} \lambda_i) \right) = \prod_{i=1}^{r} h'(\lambda_i^{-1} \gamma^{d_i} \lambda_i);$$

this being true for $\gamma = \gamma_1$, it is obvious that it remains so for $\gamma = \gamma_1^{j}$ for all j, or in other words for all $\gamma \in \Gamma$.

For each i, put $Z_i = Z^{\lambda_i}$, and call Z_i' the compositum of Z_i and K'; obviously (λ_i, Z_i') is a proper embedding of Z above K' in the sense of Chap. III-2. Let (λ, Z') be any such embedding; after replacing it if necessary by an equivalent one, we may assume that Z' is contained in K_{sep}, hence in L, so that the isomorphism λ of Z onto Z' can be extended to an automorphism λ of L over K. Then (λ, Z') is equivalent to (λ_i, Z_i')

if and only if there is a K'-linear isomorphism of Z' onto Z_i', which we can then extend to an automorphism σ of L over K', such that λ coincides with $\lambda_i \sigma$ on Z. Then $\lambda = \gamma \lambda_i \sigma$ with $\gamma \in \Gamma$ and $\sigma \in \mathfrak{g}'$. Consequently (λ, Z') is equivalent to one and only one of the embeddings (λ_i, Z_i'). Now prop. 4 of Chap. III-3 gives:

$$\theta = N_{Z/K}(\zeta) = \prod_{i=1}^{r} N_{Z_i/K'}(\zeta^{\lambda_i}).$$

As we may apply (5) to K' and to $K'' = Z_i'$ for each i, we get:

(7)
$$\mathfrak{b}(K'; \theta) = \prod_{i=1}^{r} h'(\mathfrak{b}(Z_i'; \zeta^{\lambda_i})).$$

Put $\gamma = \mathfrak{b}(Z; \zeta)$; by the definition of \mathfrak{b}, this is in Γ. By corollary 5 of th. 1, § 2, we have then:

$$\mathfrak{b}(Z_i; \zeta^{\lambda_i}) = \lambda_i^{-1} \gamma \lambda_i.$$

We may apply (5) to the fields Z_i, Z_i' instead of K, K'', replacing at the same time h by the identity since the Galois group of L over Z_i is the commutative group $\lambda_i^{-1} \Gamma \lambda_i$. The Galois group of L over Z_i' is the intersection of the latter group with \mathfrak{g}'; with the same notations as before, this is $\lambda_i^{-1} \Gamma_i \lambda_i$; it is of index d_i in $\lambda_i^{-1} \Gamma \lambda_i$, so that d_i is the degree of Z_i' over Z_i. As ζ^{λ_i} is in Z_i, we have then $N_{Z_i'/Z_i}(\zeta^{\lambda_i}) = (\zeta^{\lambda_i})^{d_i}$. Therefore (5), applied to Z_i, Z_i' and ζ^{λ_i}, gives:

$$\mathfrak{b}(Z_i'; \zeta^{\lambda_i}) = \mathfrak{b}(Z_i; (\zeta^{\lambda_i})^{d_i}) = (\lambda_i^{-1} \gamma \lambda_i)^{d_i}.$$

In view of (6) and (7), our conclusion follows from this at once.

Chapter XIII

Global classfield theory

§ 1. **The canonical pairing.** In this Chapter, k will be an A-field; we use the same notations as in earlier chapters, e.g. k_v, r_v, q_v, k_A, etc. We choose an algebraic closure \bar{k} of k, and, for each place v of k, an algebraic closure K_v of k_v, containing \bar{k}. We write k_{sep}, $k_{v,\text{sep}}$ for the maximal separable extensions of k in \bar{k}, and of k_v in K_v, respectively. We write k_{ab}, $k_{v,ab}$ for the maximal abelian extensions of k in k_{sep}, and of k_v in $k_{v,\text{sep}}$, respectively. One could easily deduce from lemma 1, Chap. XI-3, that $k_{v,\text{sep}}$ is generated over k_v by k_{sep}, and therefore K_v by \bar{k}, and we shall see in § 9 of this Chapter that $k_{v,ab}$ is generated over k_v by k_{ab}; no use will be made of these facts. We write \mathfrak{G} and $\mathfrak{A} = \mathfrak{G}/\mathfrak{G}^{(1)}$ for the Galois groups of k_{sep} and of k_{ab}, respectively, over k; we write \mathfrak{G}_v and $\mathfrak{A}_v = \mathfrak{G}_v/\mathfrak{G}_v^{(1)}$ for those of $k_{v,\text{sep}}$ and of $k_{v,ab}$, respectively, over k_v. We write ρ_v for the restriction morphism of \mathfrak{G}_v into \mathfrak{G}, and also, as explained in Chap. XII-1, for that of \mathfrak{A}_v into \mathfrak{A}. We write X_k for the group of characters of \mathfrak{G}, or, what amounts to the same, of \mathfrak{A}; for each $\chi \in X_k$, we write $\chi_v = \chi \circ \rho_v$; this is a character of \mathfrak{G}_v, or, what amounts to the same, of \mathfrak{A}_v.

PROPOSITION 1. *Take any* $\chi \in X_k$; *call* L *the cyclic extension of* k *attached to* χ. *Let* v *be any place of* k; *let* L' *be the cyclic extension of* k_v *attached to* $\chi_v = \chi \circ \rho_v$, *and let* w *be any place of* L *lying above* v. *Then there is a* k_v-*linear isomorphism of* L' *onto* L_w.

As observed in Chap. IX-4, L' is the compositum of L and k_v in K_v. As it is of finite degree over k_v, it is a local field, and prop. 1 of Chap. III-1 shows that L is dense in it; therefore it is the completion of L at a place lying above v. Our conclusion follows now from corollary 4 of th. 4, Chap. III-4.

COROLLARY. *Notations being as in proposition 1,* χ_v *is unramified for almost all* v; *if* χ_v *is trivial for almost all* v, χ *is trivial.*

The first assertion follows at once from th. 1 of Chap. VIII-4, combined with proposition 1; the second one follows similarly from corollary 4 of th. 2, Chap. VII-5, when one takes there for V the set of all the finite places of k where χ_v is trivial.

We can now apply to k_v and χ_v the definitions and results of Chap. XII-2. For any $z \in k_v^\times$, we will write $(\chi_v, z)_v$ instead of $(\chi_v, z)_{k_v}$. The canonical

morphism of k_v^\times into \mathfrak{A}_v will be denoted by \mathfrak{a}_v; then we have $(\chi_v, z)_v = \chi_v(\mathfrak{a}_v(z))$ for all $z \in k_v^\times$. For every $z \in k_v^\times$, $\rho_v(\mathfrak{a}_v(z))$ is the automorphism of k_{ab} over k induced on k_{ab} by the automorphism $\mathfrak{a}_v(z)$ of $k_{v,ab}$ over k_v.

Take now $z = (z_v) \in k_A^\times$. For almost all v, z_v is in r_v^\times, and, by the corollary of prop. 1, χ_v is unramified, so that we have $(\chi_v, z_v)_v = 1$ by corollary 4 of th. 1, Chap. XII-2. Therefore, in the product

$$(1) \qquad (\chi, z)_k = \prod_v (\chi_v, z_v)_v,$$

taken over all the places of k, almost all the factors are equal to 1, so that the product is well defined. The continuity of $z_v \to (\chi_v, z_v)_v$ for each v, together with the facts mentioned above, implies that $z \to (\chi, z)_k$ is continuous on k_A^\times; therefore it is a character of k_A^\times, whose order is finite since it divides that of χ. The pairing of X_k with k_A^\times, given by (1), will be called *the canonical pairing* for k; it is clear that it satisfies condition [I] of Chap. XII-1. As to condition [II], assume that $z \to (\chi, z)_k$ is trivial on k_A^\times; then $z_v \to (\chi_v, z_v)_v$ must be trivial for every v. As [II] is satisfied for local fields, this implies that all χ_v are trivial, hence that χ is so, by the corollary of prop. 1. This proves [II] for the pairing (1).

As explained in Chap. XII-1, we can now define *the canonical morphism* \mathfrak{a} of k_A^\times into \mathfrak{A} by writing, for all $\chi \in X_k$ and all $z = (z_v) \in k_A^\times$:

$$(2) \qquad \chi(\mathfrak{a}(z)) = (\chi, z)_k = \prod_v (\chi_v, z_v)_v.$$

Then, by [II''] of Chap. XII-1, \mathfrak{a} maps k_A^\times onto a dense subgroup of \mathfrak{A}.

PROPOSITION 2. *Let j_v be the natural injection of k_v^\times into k_A^\times, mapping k_v^\times onto the quasifactor k_v^\times of k_A^\times. Then $\mathfrak{a} \circ j_v = \rho_v \circ \mathfrak{a}_v$.*

In fact, if $z_v \in k_v^\times$, $z = j_v(z_v)$ is the idele whose coordinates are all 1 except the one corresponding to v which is z_v. Put $\alpha = \mathfrak{a}_v(z_v)$; (2) gives here:

$$\chi(\mathfrak{a}(z)) = (\chi_v, z_v)_v = \chi_v(\alpha) = \chi(\rho_v(\alpha)).$$

As this is so for all characters χ of \mathfrak{A}, it implies $\mathfrak{a}(z) = \rho_v(\alpha)$, as was to be proved.

THEOREM 1. *Let k' be an extension of k of finite degree, contained in \bar{k}; let \mathfrak{G}, \mathfrak{G}' be the Galois groups of k_{sep} over k, and of k'_{sep} over k', respectively, and let ρ be the restriction morphism of \mathfrak{G}' into \mathfrak{G}. Then, for every character $\chi \in X_k$, and for every $z' \in k_A'^\times$, we have:*

$$(\chi \circ \rho, z')_{k'} = (\chi, N_{k'/k}(z'))_k.$$

In view of our definitions, this is an immediate consequence of th. 2 of Chap. XII-2, combined with corollary 3 of th. 1, Chap. IV-1.

COROLLARY 1. *If* \mathfrak{a}, \mathfrak{a}' *are the canonical morphisms for* k *and for* k', *respectively, we have* $\rho \circ \mathfrak{a}' = \mathfrak{a} \circ N_{k'/k}$.

This is just a restatement of theorem 1.

COROLLARY 2. *Assumptions and notations being as in theorem* 1, $N_{k'/k}(k_A^\times)$ *is contained in the kernel of* $\chi \circ \mathfrak{a}$ *if and only if* k' *contains the cyclic extension of* k *attached to* χ.

In fact, theorem 1 shows that it is contained in that kernel if and only if $\chi \circ \rho \circ \mathfrak{a}'$ is trivial, hence, by [II], if and only if $\chi \circ \rho$ is trivial. Let L be the cyclic extension of k attached to χ; then the cyclic extension of k' attached to $\chi \circ \rho$ is the compositum L' of k' and L, and $\chi \circ \rho$ is trivial if and only if $L' = k'$, i.e. $k' \supset L$.

Our main business in this chapter will be to determine the kernel of the canonical morphism \mathfrak{a}. For the time being, we merely observe that it must contain the kernel of \mathfrak{a}_v for every v; this is $\{1\}$ if v is finite, but it is \mathbf{R}_+^\times if v is real, and \mathbf{C}^\times if v is imaginary. We will write $k_{\infty+}^\times$ for the product of the latter kernels in k_A^\times, i.e. for the group of the ideles (z_v) such that $z_v = 1$ for every finite place v, and $z_v > 0$ for every real place v; then this group is contained in the kernel of \mathfrak{a}; of course it is $\{1\}$ if k is of characteristic $p > 1$.

We will now give explicit formulas for $(\chi, z)_k$ in some special cases, and begin by considering a field k of characteristic $p > 1$. Let then F be the field of constants of k; call q the number of elements of F, and \bar{F} the algebraic closure of F in \bar{k}. By th. 2 of Chap. I-1, \bar{F}^\times is the group of the roots of 1 in \bar{k}, and all these roots have an order prime to p. We will call k_0 the compositum of k and \bar{F}, \mathfrak{H}_0 the subgroup of \mathfrak{G} corresponding to k_0, and X_0 the subgroup of X_k consisting of the characters of \mathfrak{G}, trivial on \mathfrak{H}_0. Clearly every extension of k of finite degree, contained in k_0, is generated over k by finitely many elements of \bar{F}, hence by an extension F' of F of finite degree. More precisely, we have the following:

LEMMA 1. *Let* F' *be the extension of* F *of degree* n, *contained in* \bar{k}; *then the compositum* k' *of* k *and* F' *is cyclic of degree* n *over* k; *its field of constants is* F'; *and the restriction morphism of the Galois group of* k' *over* k *into that of* F' *over* F *is an isomorphism of the former onto the latter group.*

Call F'' the field of constants of k', n' the degree of k' over k, and n'' that of F'' over F; clearly $n' \leqslant n \leqslant n''$. Take $\zeta \in F''$ such that $F'' = F(\zeta)$, and call P the irreducible monic polynomial in $F[X]$ with the root ζ. If Q is a monic polynomial in $k[X]$, dividing P in $k[X]$, all its roots are in \bar{F}, so that its coefficients are in $\bar{F} \cap k$, i.e. in F. Therefore P is irreducible in $k[X]$, so

that $n' \geqslant n''$, hence $n' = n = n''$. The assertion about the Galois groups may now be regarded as a special case of corollary 1 of prop. 3, Chap. III-2, or also as following from $k' = F' \otimes_F k$, which is an immediate consequence of prop. 2, Chap. III-2.

Whenever k and k' are as in lemma 1, we will say that k' is a *constant-field extension* of k. In view of th. 2, Chap. I-1, there is, for every integer $n \geqslant 1$, one and only one such extension of k of degree n; this will be denoted by k_n. Then k_0 is the union of the cyclic extensions k_n for all $n \geqslant 1$; in particular, it is contained in k_{ab}; we will denote by \mathfrak{A}_0 the subgroup of \mathfrak{A} corresponding to k_0, i.e. the group of automorphisms of k_{ab} over k_0. We may then consider X_0 as being the group of characters of \mathfrak{A}, trivial on \mathfrak{A}_0; a character $\chi \in X_k$ belongs to X_0 if and only if the cyclic extension of k attached to χ is contained in k_0, hence if and only if it is one of the fields k_n.

By corollary 2 of th. 2, Chap. I-1, combined with lemma 1, there is, for every $n \geqslant 1$, one and only one automorphism of k_n over k, inducing on the field of constants $F_n = \bar{F} \cap k_n$ of k_n the automorphism $x \to x^q$, where q, as before, is the number of elements of F; moreover, this generates the Galois group of k_n over k. Consequently there is one and only one automorphism φ_0 of k_0 over k, inducing on \bar{F} the automorphism $x \to x^q$; this will be called *the Frobenius automorphism* of k_0 over k. Every automorphism φ of k_{sep} over k, inducing φ_0 on k_0, will be called *a Frobenius automorphism* of k_{sep} over k; then the Frobenius automorphisms of k_{sep} over k make up the coset $\mathfrak{H}_0 \varphi$ in \mathfrak{G}.

PROPOSITION 3. *Let k be an A-field of characteristic $p > 1$ with the field of constants $F = \mathbf{F}_q$. Let χ be a character of \mathfrak{G} belonging to X_0, i.e. such that the cyclic extension of k attached to χ is a constant-field extension of k. Let φ be any Frobenius automorphism of k_{sep} over k. Take $z \in k_A^\times$, and put $|z|_A = q^{-r}$. Then $(\chi, z)_k = \chi(\varphi)^r$.*

Put $z = (z_v)$. Let v be a place of k of degree d, i.e. such that the module of k_v is $q_v = q^d$. Let L be the cyclic extension of k attached to χ; this is generated over k by some extension F' of F, hence by roots of 1 of order prime to p. Therefore the extension of k_v attached to χ_v, being generated by F' over k_v, is unramified, so that χ_v is unramified. Moreover, a Frobenius automorphism of $k_{v, sep}$ over k_v induces on \bar{F} the automorphism $x \to x^{q^d}$ and therefore coincides with φ^d on \bar{F}, hence on k_0. By corollary 4 of th. 1, Chap. XII-2, this gives $(\chi_v, z_v)_v = \chi(\varphi^d)^\nu$ with $\nu = \mathrm{ord}_v(z_v)$. As $|z_v|_v = q^{-d\nu}$ and $|z|_A = \prod |z_v|_v$, our conclusion follows from this at once.

COROLLARY 1. *Assumptions and notations being as in proposition 3, and \mathfrak{a} being the canonical morphism for k, $\mathfrak{a}(z)$ coincides with φ^r on k_0.*

In fact, they can differ only by an element belonging to the kernels of all the characters $\chi \in X_0$, and the intersection of these kernels is \mathfrak{H}_0, which is the group leaving k_0 invariant.

COROLLARY 2. *If* $\chi \in X_0$ *and* $\theta \in k^\times$, $(\chi, \theta)_k = 1$.

This follows at once from proposition 3 and from the fact that $|\theta|_A = 1$ by th. 5 of Chap. IV-4.

COROLLARY 3. *A character* χ *of* \mathfrak{G} *belongs to* X_0 *if and only if* $(\chi, z)_k = 1$ *for all* $z \in k_A^1$.

If $\chi \in X_0$, proposition 3 shows that χ has the latter property. Now assume that χ has that property. By corollary 6 of th. 2, Chap. VII-5, there is $z_1 \in k_A^\times$ such that $|z_1|_A = q$, and then k_A^\times is generated by k_A^1 and z_1. Let n be the order of χ; then $(\chi, z_1)_k$ is a primitive n-th root of 1 in C. As φ induces on k_n a generator of the Galois group of k_n over k, there is a character χ' attached to k_n, such that $\chi'(\varphi) = (\chi, z_1)_k$. By proposition 3, we have then $(\chi', z_1)_k = \chi'(\varphi)^{-1}$, hence $(\chi\chi', z_1)_k = 1$, and therefore, in view of proposition 3 and our assumption on χ, $(\chi\chi', z)_k = 1$ for all $z \in k_A^\times$. This implies that $\chi = \chi'^{-1}$, so that $\chi \in X_0$.

When k is of characteristic 0, there is no such convenient tool as the one supplied by the constant-field extensions in the case of characteristic $p > 1$; the nearest substitute is provided by the "cyclotomic" extensions; here we merely consider the case $k = \mathbf{Q}$; then \mathfrak{G} is the Galois group of $\bar{\mathbf{Q}} = \mathbf{Q}_{\text{sep}}$ over \mathbf{Q}. For $m \geqslant 1$, let ε be a primitive m-th root of 1 in $\bar{\mathbf{Q}}$; call \mathfrak{H}_m the subgroup of \mathfrak{G} corresponding to $\mathbf{Q}(\varepsilon)$, so that the Galois group of $\mathbf{Q}(\varepsilon)$ over \mathbf{Q} is $\mathfrak{g} = \mathfrak{G}/\mathfrak{H}_m$. It is well known that \mathfrak{g} consists of the automorphisms determined by $\varepsilon \to \varepsilon^x$, when one takes for x all the integers prime to m modulo m; it may thus be identified with $(\mathbf{Z}/m\mathbf{Z})^\times$, i.e. with the multiplicative group of the ring $\mathbf{Z}/m\mathbf{Z}$. Let χ be any character of \mathfrak{g}, with the kernel \mathfrak{h}; we identify this in an obvious manner with a character of \mathfrak{G}, which we also call χ; this has a kernel $\mathfrak{H} \supset \mathfrak{H}_m$. On the other hand, when we identify \mathfrak{g} with $(\mathbf{Z}/m\mathbf{Z})^\times$, we also identify χ with a function on the latter group and therefore with a function on the set of all integers prime to m, which we also call χ, and which is then such that $\chi(ab) = \chi(a)\chi(b)$ whenever a, b are two such integers. This can then be uniquely extended to a character of the subgroup of \mathbf{Q}^\times consisting of the fractions a/b, with a, b in \mathbf{Z} and prime to m; also the latter character will be denoted by χ. With these notations, we have:

PROPOSITION 4. *Let* χ *be as above, and let* Z *be the cyclic extension of* \mathbf{Q} *attached to* χ. *Then, for every rational prime* p *not dividing* m, χ_p *is unramified, and, for every* $z \in \mathbf{Q}_p^\times$, $(\chi_p, z)_p = \chi(|z|_p^{-1})$; *for every* $z \in \mathbf{R}^\times$, $(\chi_\infty, z)_\infty = \chi(\text{sgn } z)$.

Let p be any prime not dividing m; let w be a place of Z, lying above p, and let u be a place of the field $L = \mathbf{Q}(\varepsilon)$, lying above w. By prop. 1 of Chap. III-1, L_u is generated over \mathbf{Q}_p by ε; as this is of order m prime to p, L_u is unramified, and so is Z_w, hence also χ_p, by prop. 1. A Frobenius automorphism φ, over \mathbf{Q}_p, of the algebraic closure of \mathbf{Q}_p induces on L_u, hence on L, the automorphism determined by $\varepsilon \to \varepsilon^p$; therefore $\chi(\varphi)$, according to the notations explained above, is the same as $\chi(p)$. In view of this, our assertion about $(\chi_p, z)_p$ is an immediate consequence of corollary 4 of th. 1, Chap. XII-2, and of $|z|_p = p^{-\operatorname{ord}(z)}$. Similarly, let w be a place of Z lying above the place ∞ of \mathbf{Q}, and u a place of L lying above w. If $m = 1$ or 2, χ is trivial, and our last assertion is obvious. If $m > 2$, $L_u = \mathbf{R}(\varepsilon) = \mathbf{C}$ has the non-trivial automorphism $x \to \bar{x}$ over \mathbf{R}; this is the one determined by $\varepsilon \to \varepsilon^{-1}$, so that, if \mathfrak{g} and \mathfrak{h} are as explained above, it induces on Z_w the automorphism corresponding to the image of -1 in $\mathfrak{g}/\mathfrak{h}$. If $\chi(-1) = 1$, -1 is in \mathfrak{h}, $Z_w = \mathbf{R}$, and χ_∞ is trivial; if $\chi(-1) = -1$, -1 is not in \mathfrak{h}, $Z_w = \mathbf{C}$, and χ_∞ is non-trivial. The last assertion in proposition 4 follows at once from this and from the results stated at the beginning of Chap. XII-2.

COROLLARY 1. *Assumptions and notations being as above, let w be a place of Z. If w lies above a rational prime p, not dividing m, the degree of Z_w over \mathbf{Q}_p is the order of $\chi(p)$ in the group \mathbf{C}^\times; if w lies above ∞, the degree of Z_w over \mathbf{R} is the order of $\chi(-1)$ in \mathbf{C}^\times.*

The latter assertion was proved above; as to the former, prop. 1 shows that the degree in question is equal to the order of the character $z \to (\chi_p, z)_p$ of \mathbf{Q}_p; proposition 4 shows that this is as stated in our corollary.

COROLLARY 2. *The character χ being as above, take $z = (z_v) \in \mathbf{Q}_A^\times$ such that, for every prime p dividing m, $\operatorname{ord}_p(z_p) = 0$ and $(\chi_p, z_p)_p = 1$. Then $(\chi, z)_\mathbf{Q} = \chi(r(z))$, with $r(z)$ given by*

$$r(z) = \operatorname{sgn}(z_\infty) \prod_p |z_p|_p^{-1}.$$

In the latter formula, the product is taken over all the rational primes, or (what amounts to the same, in view of the assumption on z) over all the primes not dividing m; then $\chi(r(z))$ is well defined. Our assertion follows now at once from proposition 4 and the definitions.

COROLLARY 3. *The character χ being as above, one can choose, for every prime p dividing m, an open subgroup g_p of \mathbf{Q}_p^\times such that $(\chi, \xi)_\mathbf{Q} = 1$ for all $\xi \in \bigcap (\mathbf{Q}^\times \cap g_p)$.*

For each p dividing m, let p^μ be the highest power of p dividing m; then $1 + m\mathbf{Z}_p$ is the same as the subgroup $1 + p^\mu \mathbf{Z}_p$ of \mathbf{Q}_p^\times. Take now for g_p, for each p dividing m, the intersection of $1 + m\mathbf{Z}_p$ with the kernel

of $z \to (\chi_p, z)_p$ in \mathbf{Q}_p^\times. Then, if ξ is as in our corollary, corollary 2 shows that $(\chi, \xi)_{\mathbf{Q}}$ is equal to $\chi(r(\xi))$; by th. 5 of Chap. IV-4, $r(\xi)$ is equal to $\operatorname{sgn}(\xi) \cdot |\xi|_\infty$, i.e. to ξ. Write $\xi = a/b$, with a, b in \mathbf{Z} and $(a,b)=1$. If p is any prime dividing m, ξ is in \mathbf{Z}_p, so that b is prime to p; if p^μ is as above, ξ is in $1 + p^\mu \mathbf{Z}_p$, so that $a \equiv b \ (p^\mu)$. Therefore a and b are prime to m, and $a \equiv b \ (m)$; this implies $\chi(a) = \chi(b)$, hence, in view of our definitions, $\chi(\xi) = 1$, which completes the proof.

§ 2. **An elementary lemma.** As above, let χ be a character of $(\mathbf{Z}/m\mathbf{Z})^\times$; considering it again as a function on the set of all integers prime to m, we now associate with it the function ψ on \mathbf{Z} such that $\psi(x) = \chi(x)$ whenever x is prime to m, and $\psi(x) = 0$ otherwise. It is customary, by abuse of language, to call such a function ψ "a multiplicative character modulo m", or, more briefly, "a character modulo m" on \mathbf{Z}. Obviously a function ψ on \mathbf{Z} is such a character if and only if $\psi(x+m) = \psi(x)$ for all $x \in \mathbf{Z}$, $\psi(x) = 0$ for $(x,m) \neq 1$, $\psi(1) = 1$, and $\psi(ab) = \psi(a)\psi(b)$ for all a, b in \mathbf{Z}; it will be called trivial if $\psi(x)$ takes no other values than 0 and 1, and of order n if ψ^n is trivial. If ψ, ψ' are such characters, modulo m and modulo m' respectively, $\psi \psi'$ is a character modulo mm'.

The object of this § is to prove lemma 3; the lemma and the proof are due to van der Waerden. We begin with a special case.

LEMMA 2. *Let l be a rational prime, n an integer $\geqslant 1$, and a_1, \ldots, a_r integers > 1. Then there is a multiplicative character ψ on \mathbf{Z}, such that, for every i, $\psi(a_i)$ is a root of 1 whose order is a multiple of l^n; moreover, there is such a ψ whose order is a power of l, and, if $l=2$, there is such a ψ for which $\psi(-1) = -1$.*

Clearly the order of $\psi(a_i)$ is a multiple of that of $\psi(a_i^2)$. If $l=2$, we replace each a_i by a_i^2; after doing this, we may therefore assume, in that case, that $a_i \equiv 0$ or 1 (4) for all i. For each i, we will now define a sequence of primes $p_{i,\nu}(\nu = 0, 1, \ldots)$ as follows. If $a_i \not\equiv 1 \ (l)$, we take for $p_{i,\nu}$ any prime divisor of the integer

$$(3) \qquad \frac{a_i^{l^{\nu+1}} - 1}{a_i^{l^\nu} - 1} = 1 + a_i^{l^\nu} + \cdots + a_i^{l^\nu(l-1)}.$$

Clearly, as $l^{\nu+1} \equiv 1 \ (l-1)$, the numerator of the left-hand side is $\equiv a_i - 1 \ (l)$, hence not a multiple of l; therefore $p_{i,\nu} \neq l$. On the other hand, if $a_i \equiv 1 \ (l)$, write $a_i^{l^\nu}$ in the form $a_i^{l^\nu} = 1 + l^\alpha b$ with $\alpha \geqslant 1$ and $b \not\equiv 0 \ (l)$; if $l=2$, we have $\alpha \geqslant 2$, in view of our assumption on the a_i in that case. Then:

$$a_i^{l^{\nu+1}} = (1 + l^\alpha b)^l \equiv 1 + l^{\alpha+1} b \qquad (l^{\alpha+2}).$$

This shows that the left-hand side of (3) is then a multiple of l and not of l^2; as the right-hand side shows, it is $>l$; we take then for $p_{i,v}$ any prime divisor of that left-hand side, other than l. Now we show that, in all cases, $p_{i,v}$ cannot divide the denominator of the left-hand side of (3). In fact, assume that it does; then all the terms in the right-hand side are $\equiv 1\ (p_{i,v})$, so that the right-hand side itself is $\equiv l\ (p_{i,v})$; as it is a multiple of $p_{i,v}$, and as $p_{i,v} \neq l$, this cannot be. This shows that the image of a_i in the group $(\mathbf{Z}/p_{i,v}\mathbf{Z})^{\times}$ is exactly of order l^{v+1}; in particular, for each i, all the $p_{i,v}$ are distinct. Therefore, choosing an integer ρ such that $l^{\rho}>r$, we may, for each i, choose an integer $v_i \geqslant n+\rho-2$ such that the prime $p_i = p_{i,v_i}$ does not divide any of the integers a_1,\ldots,a_r. For each i, the group $(\mathbf{Z}/p_i\mathbf{Z})^{\times}$ is cyclic of order p_i-1, and the image of a_i in that group has the order l^{v_i+1}; call χ_i a generator of the group of characters of that group; call l^{λ_i} the highest power of l dividing p_i-1; put $\mu = \lambda_i - v_i + n + \rho - 2$, $m = l^{-\mu}(p_i-1)$ and $\chi'_i = \chi_i^m$; clearly $\lambda_i > v_i$, so that $\mu > 0$; then χ'_i is a character of order l^{μ}, and it is easily verified that $\chi'_i(a_i)$ is a root of 1 of order $l^{n+\rho-1}$. For each i, extend χ'_i to a multiplicative character ψ_i modulo p_i on \mathbf{Z}, as explained above; let M be an integer such that l^M is a multiple of the order of all the χ'_i, hence of all the ψ_i. As each p_i is prime to all the a_j, we can then write

$$\psi_i(a_j) = \mathbf{e}(l^{-M} b_{ij})$$

with $b_{ij} \in \mathbf{Z}$, for $1 \leqslant i,j \leqslant r$; moreover, for each i, the highest power of l dividing b_{ii} is $l^{M-n-\rho+1}$. Now consider the l^{Mr} characters

$$\omega_x = \prod_{i=1}^{r} (\psi_i)^{x_i}$$

with $0 \leqslant x_i < l^M$ for $1 \leqslant i \leqslant r$; of course they need not all be distinct. For each j, we have

$$\omega_x(a_j) = \mathbf{e}\left(l^{-M} \sum_{i=1}^{r} b_{ij} x_i \right).$$

This is a root of 1, of order dividing l^M; that order is a multiple of l^n unless it divides l^{n-1}, i.e. unless we have

$$b_{jj} x_j \equiv -\sum_{i \neq j} b_{ij} x_i \qquad (l^{M-n+1}).$$

For a given j, and for each set of values for the x_i for $i \neq j$, this congruence has either no solution x_j at all, or exactly $l^{M-\rho}$ solutions modulo l^M; therefore, for each j, there are at most $l^{Mr-\rho}$ sets of values for the x_i, satisfying $0 \leqslant x_i < l^M$ for $1 \leqslant i \leqslant r$, such that $\omega_x(a_j)$ has an order dividing l^{n-1}. Consequently there are at most $r l^{Mr-\rho}$ such sets for which at least

one of the $\omega_x(a_j)$ has an order dividing l^{n-1}; as $r < l^p$, the number of such sets is $< l^{Mr}$. This proves that one can choose x so that the order of $\omega_x(a_j)$ is a multiple of l^n for all j. Then $\psi = \omega_x$ is a solution of our problem, except perhaps in the case $l = 2$, since then we also want ψ to be such that $\psi(-1) = -1$. In that case, if $\omega_x(-1) = -1$, we take $\psi = \omega_x$. If not, take a prime p_0 dividing $4a_1 a_2 \ldots a_r - 1$, and $\equiv -1$ (4); clearly there is at least one such prime. Then the group $(\mathbf{Z}/p_0\mathbf{Z})^\times$ is cyclic of order $2m_0$ with $m_0 = (p_0 - 1)/2 \equiv 1$ (2), so that it has exactly one character χ_0 of order 2; this satisfies $\chi_0(-1) = -1$. Extending χ_0 to a multiplicative character ψ_0 modulo p_0 on \mathbf{Z}, one sees at once that $\psi = \psi_0 \omega_x$ is a solution of our problem, provided one has taken $n \geqslant 2$, as may of course always be assumed. This completes the proof.

LEMMA 3. *Let* a_1, \ldots, a_r, n_1, \ldots, n_r *be integers* > 1. *Then there is a multiplicative character* ψ *on* \mathbf{Z} *such that* $\psi(-1) = -1$ *and that, for every* i, $\psi(a_i)$ *is a root of* 1 *whose order is a multiple of* n_i.

Put $N = 2\prod n_i$; for every prime l dividing N, let l^n be the highest power of l dividing N, and let ψ_l be chosen according to lemma 2, so that its order is a power of l, the order of $\psi_l(a_i)$ is a multiple of l^n for every i, and $\psi_l(-1) = -1$ if $l = 2$. When l is odd, $\psi_l(-1)$, being ± 1 and of odd order, is 1. That being so, it is clear that $\psi = \prod \psi_l$ solves the problem.

§ 3. **Hasse's "law of reciprocity".** As in Chap. XI, if A is a simple algebra over k, and v any place of k, we write A_v for the algebra $A \otimes_k k_v$ over k_v; we have seen in Chap. IX-3 that the mapping $\mathrm{Cl}(A) \to \mathrm{Cl}(A_v)$ is then a morphism of the Brauer group $B(k)$ of k into the Brauer group $B(k_v)$ of k_v. It has been shown in Chap. XII-2 that the Hasse invariant h determines an isomorphism of $B(k_v)$ onto a group H_v consisting of all the roots of 1 in \mathbf{C} if v is a finite place, of ± 1 if v is real and of 1 if v is imaginary. From now on, for any simple algebra A over k, we will write $h_v(A) = h(A_v)$; this will be called *the Hasse invariant of* A *at* v. By th. 1 of Chap. XI-1, we have $h_v(A) = 1$ for almost all v; therefore the mapping $A \to (h_v(A))$ determines a morphism \mathbf{h} of $B(k)$ into the "direct sum" of the groups H_v for all v, i.e. into the subset H of $\prod_v H_v$ consisting of the elements (η_v) of that product such that $\eta_v = 1$ for almost all v. By th. 2 of Chap. XI-2, the kernel of \mathbf{h} is the class of trivial algebras over k, so that \mathbf{h} is injective. It will be shown in § 6 that $\mathbf{h}(B(k))$ consists of the elements (η_v) of H such that $\prod_v \eta_v = 1$. In this §, we show that $\prod_v h_v(A) = 1$ for every simple algebra A over k.

As in § 1, let χ be a character of \mathfrak{G}, and L the cyclic extension of k attached to χ; for any $\theta \in k^\times$, consider the cyclic algebra $A = [L/k; \chi, \theta]$, corresponding to the factor-class $\{\chi, \theta\}$. As we have seen in Chap. IX-4, the restriction morphism maps the factor-class $\{\chi, \theta\}$ of k onto the factor-class $\{\chi_v, \theta\}$ of k_v, so that A_v belongs to the latter class. Therefore, by the definition of the Hasse invariant, we have, for all v:

(4) $$h_v(A) = (\chi_v, \theta)_v, \quad \prod_v h_v(A) = (\chi, \theta)_k.$$

On the other hand, let k' be an extension of k of finite degree; A being any simple algebra over k, put $A' = A \otimes_k k'$; let w be a place of k', and v the place of k lying below w. The transitivity properties of tensor-products show at once that the algebra $(A')_w = A' \otimes_{k'} k'_w$ over k'_w may be identified with $A_v \otimes_{k_v} k'_w$; therefore, by corollary 2 of th. 2, Chap. XII-2, we have $h_w(A') = h_v(A)^{n(w)}$ if $n(w)$ is the degree of k'_w over k_v. In particular, in view of what has been said above, A' is trivial if and only if $h_v(A)^{n(w)} = 1$ for every place w of k'.

PROPOSITION 5. *For any $\chi \in X_k$, let L be the cyclic extension of k attached to χ. Let A be a simple algebra over k. Then the following assertions are equivalent: (i) A_L is trivial; (ii) for every place v of k, and every place w of L above v, the degree of L_w over k_v is a multiple of the order of $h_v(A)$ in the group \mathbf{C}^\times; (iii) A is similar to a cyclic algebra $[L/k; \chi, \theta]$ with some $\theta \in k^\times$; (iv) there is $z = (z_v)$ in $k_\mathbf{A}^\times$, such that $h_v(A) = (\chi_v, z_v)_v$ for every place v of k. Moreover, if θ is as in (iii) and z as in (iv), $\theta^{-1} z$ is in $N_{L/k}(L_\mathbf{A}^\times)$.*

The equivalence of (i) and (ii) is a special case of what has just been proved above; that of (i) and (iii) is contained in prop. 9 of Chap. IX-4. Assume (iii); then, by (4), (iv) is satisfied if we take $z = \theta$. Now assume (iv); then the order of $h_v(A)$ divides that of χ_v, which, by prop. 1 of § 1, is equal to the degree of L_w over k_v for every place w of L lying above v, so that (ii) is satisfied. Finally, let θ be as in (iii), z as in (iv), and put $z' = \theta^{-1} z$; by (4), we have then $(\chi_v, z'_v)_v = 1$ for all v; by prop. 10 of Chap. IX-4, and prop. 1 of § 1, this implies that, if w is any place of L above v, z'_v is in $N_{L_w/k_v}(L_w^\times)$. For each place v of k, choose $t_w \in L_w^\times$, for all the places w of L lying above v, so that $z'_v = N_{L_w/k_v}(t_w)$ for one of these places, and $t_w = 1$ for all the others; as $|z'_v|_v = 1$ for almost all v, this implies that $|t_w|_w = 1$ for almost all w, so that $t = (t_w)$ is in $L_\mathbf{A}^\times$; then we have $z' = N_{L/k}(t)$.

We will now use proposition 5 in order to show that every simple algebra A over k is similar to one of a very special type. For this, we require two lemmas.

LEMMA 4. *Let k be of characteristic $p > 1$. For every place v of k, let $v(v)$ be an integer ≥ 1, such that $v(v) = 1$ for almost all v. Then there is a*

constant-field extension k' of k such that, if v is any place of k and w a place of k' above v, the degree of k'_w over k_v is a multiple of $v(v)$.

Let $F = \mathbf{F}_q$ be the field of constants of k. For a place v of k of degree $d(v)$, the module of k_v is $q^{d(v)}$; therefore, by corollary 3 of th. 7, Chap. I-4, if k'_w contains a primitive root of 1 of order $q^{d(v)f} - 1$, its degree over k_v must be a multiple of f. The condition in lemma 4 will therefore be satisfied if we take for k' a constant-field extension of k whose degree over k is a multiple of all the integers $d(v)v(v)$ corresponding to the finitely many places v where $v(v) > 1$.

LEMMA 5. *Let k be of characteristic 0. For every place v of k, let $v(v)$ be an integer ≥ 1, such that $v(v) = 1$ for almost all v, $v(v) = 1$ or 2 whenever v is real, and $v(v) = 1$ whenever v is imaginary. Then there is an integer $m \geq 1$ and a cyclic extension Z of \mathbf{Q}, contained in the extension $\mathbf{Q}(\varepsilon)$ generated by a primitive m-th root ε of 1, with the following properties: (a) if v is any place of k, and w a place, lying above v, of the compositum k' of k and Z, the degree of k'_w over k_v is a multiple of $v(v)$; (b) $|m|_v = 1$ whenever v is a finite place of k and $v(v) > 1$.*

To begin with, let Z be any extension of \mathbf{Q}, and let k' be its compositum with k. Let v be any place of k, w a place of k' lying above v, u the place of Z lying below w, and t the place of \mathbf{Q} lying below u. Then k_v, Z_u and \mathbf{Q}_t are respectively the closures of k, Z and \mathbf{Q} in k'_w, so that t also lies below v. We have:

$$[k'_w : k_v] = [k'_w : Z_u] \cdot [Z_u : \mathbf{Q}_t] \cdot [k_v : \mathbf{Q}_t]^{-1};$$

therefore, if we put

$$v'(v) = v(v) \cdot [k_v : \mathbf{Q}_t],$$

and if $[Z_u : \mathbf{Q}_t]$ is a multiple of $v'(v)$, $[k'_w : k_v]$ will be a multiple of $v(v)$. Now, for every finite place t of \mathbf{Q} above which there lies some place v of k where $v(v) > 1$, call $n(t)$ some common multiple of the integers $v'(v)$ for all the places v of k above t; for all other finite places t of \mathbf{Q}, put $n(t) = 1$; put $n(\infty) = 2$, this being a multiple of $v'(v)$ for every infinite place v of k, as one sees at once. Then m and Z, in our lemma, will satisfy our requirements if $[Z_u : \mathbf{Q}_t]$ is always a multiple of $n(t)$ and if $|m|_t = 1$ when $t \neq \infty$, $n(t) > 1$; in other words, it is enough to prove our lemma for $k = \mathbf{Q}$. Call then p_1, \ldots, p_r the rational primes p for which $n(p) > 1$; apply lemma 3 of § 2 to the integers $a_i = p_i$, $n_i = n(p_i)$; we get a multiplicative character ψ on \mathbf{Z}, modulo some integer m, such that $\psi(-1) = -1$ and that, for each i, $\psi(p_i)$ is a root of 1 whose order is a multiple of $n(p_i)$. As $\psi(x) = 0$ when x is not prime to m, m is then prime to all the p_i, which is the same as to say that $|m|_p = 1$ when $n(p) > 1$. Let then χ be the character of $(\mathbf{Z}/m\mathbf{Z})^\times$ deter-

mined by ψ; consider this as a character of the Galois group of $\mathbf{Q}(\varepsilon)$ over \mathbf{Q}, ε being a primitive m-th root of 1, and call Z the cyclic extension of \mathbf{Q} attached to χ; then corollary 1 of prop. 4, § 1, shows that m and Z satisfy all the requirements in our lemma.

THEOREM 2. *If A is any simple algebra over k, we have $\prod_v h_v(A) = 1$, the product being taken over all the places v of k.*

If k is of characteristic $p > 1$, prop. 5 and lemma 4 show that A is similar to a cyclic algebra $[k'/k; \chi, \theta]$, where k' is a constant-field extension of k, χ a character attached to k', and $\theta \in k^\times$. Then χ is in X_0, where X_0 is as defined in § 1, and our conclusion follows at once from (4) and corollary 2 of prop. 3, § 1. If k is of characteristic 0, we apply prop. 5 and lemma 5, taking for $\nu(v)$, in the latter lemma, the order of $h_v(A)$ in \mathbf{C}^\times; this shows that A is similar to a cyclic algebra $[k'/k; \chi', \theta]$, where k' is as in lemma 5, χ' is any character attached to k', and $\theta \in k^\times$. By (4), what we have to prove is that $(\chi', \theta)_k = 1$. Let m and Z be as in lemma 5; then we can take $\chi' = \chi \circ \rho$, where ρ is the restriction morphism of the Galois group of $\bar{\mathbf{Q}}$ over k into that of $\bar{\mathbf{Q}}$ over \mathbf{Q}, and χ is a character of the latter group attached to Z. Call v_1, \ldots, v_M all the places of k lying above some rational prime dividing m; for each i, choose a place w_i of k' lying above v_i; call w'_1, \ldots, w'_N all the places of k', other than the w_i, lying above some v_i; for each i, call k_i, k'_i the completions of k at v_i, and of k' at w_i, respectively; for each j, call k''_j the completion of k' at w'_j. By condition (b) in lemma 5, and in view of our choice of the $\nu(v)$, we have $h_{v_i}(A) = 1$ for all i; by (4), prop. 1 of § 1, and prop. 10 of Chap. IX-4, this implies that, for each i, we can write $\theta = N_{k_i/k_i}(z_i)$ with $z_i \in k_i'^\times$. By corollary 2 of th. 3, Chap. IV-2, there is an element ζ of k' whose image in k'_i, for $1 \leqslant i \leqslant M$, is arbitrarily close to z_i, and whose image in k''_j, for $1 \leqslant j \leqslant N$, is arbitrarily close to 1. In view of corollary 3 of th. 1, Chap. IV-1, this implies that we can choose $\zeta \in k'^\times$ so that the image of $\theta_1 = \theta N_{k'/k}(\zeta)^{-1}$ in k_i is arbitrarily close to 1 for $1 \leqslant i \leqslant M$. By prop. 10 of Chap. IX-4, one does not change the factor-class $\{\chi', \theta\}$ if one replaces θ by θ_1; consequently, this does not change the invariants $h_v(A) = (\chi'_v, \theta)_v$ of A. Therefore it is enough if we prove our assertion $(\chi', \theta)_k = 1$ under the additional assumption that the image of θ in k_i, for $1 \leqslant i \leqslant M$, is in a prescribed neighborhood of 1. By corollary 3 of th. 1, Chap. IV-1, these neighborhoods can be so chosen that the image of $N_{k/\mathbf{Q}}(\theta)$ in \mathbf{Q}_p, for every prime p dividing m, is arbitrarily close to 1. As $\chi' = \chi \circ \rho$, and as we have, by th. 1 of § 1:

$$(\chi \circ \rho, \theta)_k = (\chi, N_{k/\mathbf{Q}}(\theta))_\mathbf{Q},$$

our assertion follows now from corollary 3 of prop. 4, § 1.

COROLLARY. *For every* $\chi \in X_k$, *and every* $\theta \in k^\times$, *we have* $(\chi, \theta)_k = 1$.

This follows at once from (4) and theorem 2.

The corollary of theorem 2 is known as "Artin's law of reciprocity" because Artin discovered it (in substance) and pointed out that the "laws of reciprocity" of classical number-theory can easily be derived from it and from purely local considerations. Theorem 2 is due to Hasse; its close connection with "Artin's law" accounts for the name of "Hasse's law of reciprocity" which is usually given to it.

The corollary of theorem 2 may be expressed by saying that, for every $\chi \in X_k$, the character $z \to (\chi, z)_k$ of k_A^\times is trivial on k^\times, or that k^\times is contained in the kernel of the canonical morphism. Consequently, we may now regard $(\chi, z)_k$ as defining a pairing between X_k and the "idele-class group" $G_k = k_A^\times / k^\times$ of k; in order not to complicate notations, we do not introduce any new symbol for this pairing, but we will apply to it the results of Chap. XII-1 in an obvious manner. It is clear that it satisfies conditions [I] and [II] of Chap. XII-1, since these have been verified in § 1 for $(\chi, z)_k$ considered as a pairing of X_k and k_A^\times. By th. 6 of Chap. IV-4, G_k is quasi-compact, and we have $G_k^1 = k_A^1 / k^\times$. If k is of characteristic 0, condition [III(a)] of Chap. XII-1 is satisfied. If k is of characteristic $p > 1$, corollary 3 of prop. 3, § 1, together with lemma 1 of § 1, shows that condition [III(b)] of Chap. XII-1 is satisfied by taking for χ, in that condition, a character attached to the constant-field extension of degree n of k; it also shows that the group denoted by X_0 in § 1, and consisting of the characters attached to the constant-field extensions of k, is now the same as the group which was so denoted in Chap. XII-1. In that case, we can now apply corollary 2 of prop. 2, Chap. XII-1, which shows that the canonical morphism \mathfrak{a} maps k_A^1 onto the subgroup \mathfrak{A}_0 of \mathfrak{A} corresponding to the union k_0 of the constant-field extensions of k; similarly, if k is of characteristic 0, prop. 1 of Chap. XII-1 shows that \mathfrak{a} maps k_A^\times onto \mathfrak{A}. If we call again U_k the kernel of \mathfrak{a}, it contains k^\times, and, if k is of characteristic $p > 1$, corollary 2 of prop. 2, Chap. XII-1, shows now that $U_k \subset k_A^1$; on the other hand, if k is of characteristic 0 and if the subgroup $k_{\infty+}^\times$ of k_A^\times is defined as in § 1, U_k contains the closure of $k^\times k_{\infty+}^\times$. In § 8, it will be shown that U_k is that closure if k is of characteristic 0, and that otherwise $U_k = k^\times$.

We will now reformulate prop. 4 of Chap. XII-1 for the pairing between X_k and G_k defined above. Let k' be a cyclic extension of k. Then $\xi \to \xi^\lambda$, for every $\lambda \in \mathfrak{G}$, and $\xi \to N_{k'/k}(\xi)$ are polynomial mappings of k', into k' and into k respectively, when k' is regarded as a vector-space over k, and we have $N_{k'/k}(\xi^\lambda) = N_{k'/k}(\xi)$ for all ξ. As explained in Chap. IV-1, $N_{k'/k}$, as a mapping of k'_A into k_A, is the extension to these spaces of the polynomial mapping $N_{k'/k}$ of k' into k, and we now extend to k'_A the

k-linear mapping $\xi \to \xi^\lambda$ of k' onto k' in the same manner. Then we have $N_{k'/k}(x'^\lambda) = N_{k'/k}(x')$ for all $x' \in k'_A$, and in particular for all $x' \in k'^\times_A$. At the same time, by the corollary of prop. 3, Chap. IV-3, we have $|z'|_A = = |N_{k'/k}(z')|_A$ for all $z' \in k'^\times_A$, hence $|z'^\lambda|_A = |z'|_A$ for all $z' \in k'^\times_A$ and all $\lambda \in \mathfrak{G}$. As the morphisms $z' \to z'^\lambda$, $z' \to N_{k'/k}(z')$ of k'^\times_A onto k'^\times_A and of k'^\times_A into k^\times_A map k'^\times onto k'^\times and into k^\times respectively, they determine morphisms of $G_{k'}$ onto $G_{k'}$ and into G_k respectively; we take these as the mappings $g' \to g'^\lambda$, $g' \to F(g')$, in prop. 4, Chap. XII-1. It is now clear that these mappings satisfy conditions [IV(i)–(ii)] and [V(i)]; [IV(iii)] is an immediate consequence of the definitions and of corollary 5 of th. 1, Chap. XII-2 (it admits of an obvious generalization, quite similar to the latter corollary), while [V(ii)] is here the same as the assertion in th. 1, § 1. All conditions for the application of prop. 4, Chap. XII-1, being thus fulfilled, we conclude that we have, in our present notation:

$$(5) \qquad U_k \cap k^\times N_{k'/k}(k'^\times_A) = k^\times N_{k'/k}(U_{k'}),$$

where k is any A-field, k' any cyclic extension of k, and U_k, $U_{k'}$ are the kernels of the canonical morphisms for k and for k', respectively.

§ 4. Classfield theory for Q. The results already obtained make it easy to conclude our investigation in the special case $k = \mathbf{Q}$; this is due to the fact expressed in the following lemma:

LEMMA 6. *We have the direct product decomposition:*

$$\mathbf{Q}^\times_A = \mathbf{Q}^\times \times \mathbf{R}^\times_+ \times \prod_p \mathbf{Z}^\times_p$$

where the last product is taken over all the rational primes p.

Here \mathbf{R}^\times_+ and the \mathbf{Z}^\times_p are to be understood as subgroups of the quasi-factors $\mathbf{R}^\times = \mathbf{Q}^\times_\infty$ and \mathbf{Q}^\times_p of \mathbf{Q}^\times_A. As in corollary 2 of prop. 4, § 1, define a morphism r of \mathbf{Q}^\times_A into \mathbf{Q}^\times by putting, for $z = (z_v)$ in \mathbf{Q}^\times_A:

$$r(z) = \mathrm{sgn}(z_\infty) \prod_p |z_p|^{-1}_p;$$

as we have already observed (in the proof of corollary 3 of prop. 4, § 1), r induces the identity on \mathbf{Q}^\times, as follows at once from th. 5 of Chap. IV-4. Therefore, if R is the kernel of r, r determines a direct product decomposition $\mathbf{Q}^\times_A = \mathbf{Q}^\times \times R$ and is the projection from that product onto its first factor. Clearly $R = \mathbf{R}^\times_+ \times \prod_p \mathbf{Z}^\times_p$. This proves the lemma.

Clearly the subgroup $\prod \mathbf{Z}^\times_p$ of \mathbf{Q}^\times_A is totally disconnected, so that, by lemma 4 of Chap. VII-3, all its characters are of finite order. It follows now at once from lemma 6, combined with corollary 2 of prop. 7, Chap.

VII-3, that every quasicharacter of Q_A^\times, trivial on Q^\times, is of the form $\omega_s \psi$, where ψ is trivial on $Q^\times \times R_+^\times$ and where ω_s, as in Chap. VII, denotes the quasicharacter $z \to |z|_A^s$ and is trivial on Q^\times and on $\prod Z_p^\times$; ψ is then a character of finite order. We recall that, if ω is any quasicharacter of Q_A^\times, trivial on Q^\times, the conductor of ω, according to the definition given in Chap. VII-7 for an arbitrary number-field, is the ideal $\prod p^{f(p)}$ of Z, where, for each rational prime p, $p^{f(p)}$ is the conductor of the quasi-character ω_p induced by ω on Q_p^\times; here, as usual, we identify a non-zero ideal in Z with the integer >0 which generates it.

As explained in § 1, if ε is a primitive m-th root of 1 in \bar{Q}, we identify the Galois group \mathfrak{g} of $Q(\varepsilon)$ over Q with $(Z/mZ)^\times$, and every character χ of \mathfrak{g} with a character of the Galois group \mathfrak{G} of \bar{Q} over Q, or, what amounts to the same, with a character of the Galois group \mathfrak{A} of Q_{ab} over Q. Of course $Q(\varepsilon) \subset Q_{ab}$ for all m.

THEOREM 3. *For any* $m > 1$, *let* ε *be a primitive* m-th root of 1 in \bar{Q}, *and let* $\mathfrak{g} = (Z/mZ)^\times$ *be the Galois group of* $Q(\varepsilon)$ *over* Q. *Then* $\chi \to \chi \circ \mathfrak{a}$ *is an isomorphism of the group of the characters* χ *of* \mathfrak{g} *onto the group of the characters of* Q_A^\times, *trivial on* $Q^\times \times R_+^\times$, *whose conductor divides* m.

Call Γ the latter group. Call P the set consisting of ∞ and of the primes p dividing m; for each prime $p \in P$, put $g_p = 1 + p^\mu Z_p$, where p^μ is the highest power of p dividing m; call H the subgroup of Q_A^\times consisting of the ideles (z_v) such that $z_\infty > 0$, $z_p \in g_p$ for every prime $p \in P$, and $z_p \in Z_p^\times$ for p not in P. Then Γ is the group of the characters ω of Q_A^\times which are trivial on Q^\times and on H. Put $g_\infty = R^\times$ and $g = \prod g_v$, the latter product being taken over all $v \in P$; as in Chap. VII-8, call G_P the subgroup of Q_A^\times consisting of the ideles (z_v) such that $z_v = 1$ for all $v \in P$; as $g \times G_P$ is an open subgroup of Q_A^\times, and as $Q^\times G_P$ is dense in Q_A^\times by prop. 15, Chap. VII-8, we have $Q_A^\times = Q^\times \cdot (g \times G_P)$. The morphism r of Q_A^\times onto Q^\times defined in the proof of lemma 6 maps $g \times G_P$ onto the subgroup $Q^{(m)}$ of Q^\times consisting of the fractions a/b, where a, b are in Z and are prime to m; the kernel of the morphism of $g \times G_P$ onto $Q^{(m)}$ induced by r is the group H defined above. As every character in Γ is trivial on H, this implies that, for any $\omega \in \Gamma$, there is a character χ of $Q^{(m)}$ such that $\chi \circ r$ coincides with ω on $g \times G_P$. Then, if $a \in Z$ and $a \equiv 1 \ (m)$, we have $a \in g \times G_P$ and $r(a) = a$, hence $\chi(a) = \omega(a) = 1$. Therefore χ determines a character of $(Z/mZ)^\times$; this being also denoted by χ, and being regarded as a character of \mathfrak{g}, hence of \mathfrak{G}, corollary 2 of prop. 4, § 1, shows that $(\chi, z)_Q = \chi(r(z))$ for all $z \in g' \times G_P$, if g' is a suitable open subgroup of g. This means that $\chi \circ \mathfrak{a}$ coincides with $\chi \circ r$, hence with ω, on $g' \times G_P$. As $Q_A^\times = Q^\times \cdot (g' \times G_P)$ by prop. 15, Chap. VII-8, and as $\chi \circ \mathfrak{a}$ and ω are both trivial on Q^\times, this proves that $\chi \circ \mathfrak{a} = \omega$. Conversely, let now

χ be any character of $\mathfrak{g} = (\mathbf{Z}/m\mathbf{Z})^{\times}$; as in § 1, consider this as a function on the set of all integers prime to m, and extend this to a character χ of $\mathbf{Q}^{(m)}$; then $\chi \circ r$ is a character of $g \times G_P$. Take $\xi \in \mathbf{Q}^{\times} \cap (g \times G_P)$; then $r(\xi) = \xi$, and one sees at once, as in the proof of corollary 3 of prop. 4, § 1, that $\xi \in \mathbf{Q}^{(m)}$ and $\chi(\xi) = 1$. Therefore $\chi \circ r$ is trivial on $\mathbf{Q}^{\times} \cap (g \times G_P)$, so that it can be uniquely extended to a character ω of $\mathbf{Q}_A^{\times} = \mathbf{Q}^{\times} \cdot (g \times G_P)$, trivial on \mathbf{Q}^{\times}; as r is trivial on H, ω is also trivial on H, so that it belongs to Γ. As above, corollary 2 of prop. 4, § 1, shows now that $\chi \circ \mathfrak{a}$ coincides with $\chi \circ r$, hence with ω, on $g' \times G_P$, if g' is a suitable open subgroup of g; as above, this gives $\chi \circ \mathfrak{a} = \omega$, which completes our proof. We see also that $\chi \circ \mathfrak{a}$ coincides with $\chi \circ r$, not only on $g' \times G_P$, but even on $g \times G_P$; in other words, the conclusion of corollary 2 of prop. 4, § 1, is valid provided $z_p \in g_p$ for every prime $p \in P$; we will not formulate this as a separate result, but will use it in the proof of our next corollary.

COROLLARY 1. *Let ε be as in theorem 3; take any $z = (z_r)$ in $\prod \mathbf{Z}_p^{\times}$ and put $\alpha = \mathfrak{a}(z)^{-1}$. Then there is an integer a such that $a \in z_p + m\mathbf{Z}_p$ for every prime p, and, for every such a, we have $\varepsilon^{\alpha} = \varepsilon^a$.*

The condition on a can also be written as $a \equiv z_p \ (p^{\mu})$ for every prime p dividing m, p^{μ} being the highest power of p dividing m; it is well known that these congruences have a unique solution modulo m (this may also be regarded as a special case of corollary 1 of th. 1, Chap. V-2). As $z_p \in \mathbf{Z}_p^{\times}$ for all p, a is then prime to m; in particular, it is not 0. Put then $z' = a^{-1}z$; then $z'_p \in g_p$ for all primes $p \in P$; therefore, as shown at the end of the proof of theorem 3, we have $\chi(\mathfrak{a}(z')) = \chi(r(z'))$. As \mathfrak{a} is trivial on \mathbf{Q}^{\times}, $\mathfrak{a}(z') = \mathfrak{a}(z) = \alpha^{-1}$; as $r(a) = a$ and $r(z) = 1$, we get $\chi(\alpha) = \chi(a)$. As this is so for all characters χ of \mathfrak{g}, it shows that the automorphism of $\mathbf{Q}(\varepsilon)$ induced by α is the one determined by $\varepsilon \to \varepsilon^a$.

COROLLARY 2. *The kernel of the canonical morphism \mathfrak{a} for \mathbf{Q} is $\mathbf{Q}^{\times} \times \mathbf{R}_+^{\times}$, and \mathfrak{a} determines an isomorphism of $\prod \mathbf{Z}_p^{\times}$ onto the Galois group \mathfrak{A} of \mathbf{Q}_{ab} over \mathbf{Q}.*

In fact, we already knew that the kernel of \mathfrak{a} contains $\mathbf{Q}^{\times} \times \mathbf{R}_+^{\times}$, and theorem 3 shows that it is contained in it. The last assertion follows now at once from lemma 6, and prop. 1 of Chap. XII-1.

COROLLARY 3. *\mathbf{Q}_{ab} is generated over \mathbf{Q} by the roots of 1 in the algebraic closure $\bar{\mathbf{Q}}$ of \mathbf{Q}.*

Let K be the extension of \mathbf{Q} generated by these roots, which is the same as the union of the fields $\mathbf{Q}(\varepsilon)$ for all $m > 1$, where ε is as in theorem 3. Let \mathfrak{B} be the subgroup of \mathfrak{A} corresponding to K. Then, if χ is as in theorem 3, it is trivial on \mathfrak{B}, so that $\chi \circ \mathfrak{a}$ is trivial on $\mathfrak{a}^{-1}(\mathfrak{B})$. By theo-

rem 3, $a^{-1}(\mathfrak{B})$ must therefore be contained in $\mathbf{Q}^{\times} \times \mathbf{R}^{\times}_+$; as this, by corollary 2, is the kernel of a, we must have $\mathfrak{B} = \{1\}$, hence $K = \mathbf{Q}_{ab}$.

§ 5. The Hilbert symbol. The determination of the kernel of the canonical morphism in the general case depends on two results, corresponding to propositions 9 and 10 of Chap. XII-3. In this §, we deal with the former one; this will require some preparations.

By n, we will understand any integer > 1.

LEMMA 7. *Let G be a quasicompact group. Let γ be a group of characters of G, all of order dividing n, and let X be the intersection of their kernels. Then every character of G, trivial on X, is in γ.*

By lemma 2 of Chap. XII-1, applied to the endomorphism $x \to x^n$ of G, G^n is a closed subgroup of G, and G/G^n is compact; therefore the subgroup of the dual of G, associated by duality with G^n, is discrete; it consists of all the characters of G which are trivial on G^n, i.e. whose order divides n. Consequently γ is discrete, hence closed, in the dual of G. Our assertion follows now from the duality theory.

PROPOSITION 6. *Let K be a local field containing n distinct n-th roots of 1. For x, y in K^{\times}, put $(x,y)_{n,K} = (\chi_{n,x}, y)_K$. Then*

$$(y,x)_{n,K} = (x,y)_{n,K}^{-1}$$

for all x, y in K^{\times}; $(K^{\times})^n$ is the set of the elements y of K^{\times} such that $(x,y)_{n,K} = 1$ for all $x \in K^{\times}$; if $\operatorname{mod}_K(n) = 1$, and if R is the maximal compact subring of K, the set of the elements y of K^{\times} such that $(x,y)_{n,K} = 1$ for all $x \in R^{\times}$ is $(K^{\times})^n R^{\times}$.

In view of our definitions in Chap. IX-5 and in Chap. XII-2, $(x,y)_{n,K}$ is the same as $\eta(\{x,y\}_n)$, where η is as defined in corollary 2 of th. 1, Chap. XII-2; our first assertion is then nothing else than formula (12) of Chap. IX-5. The second one is identical with prop. 9 of Chap. XII-3 if K is a p-field; it is trivial if $K = \mathbf{C}$; it can be verified at once if $K = \mathbf{R}$, since in that case our assumption, about the n-th roots of 1 being in K, implies that $n = 2$. As to the last assertion, the assumption $\operatorname{mod}_K(n) = 1$ implies that K is a p-field, with p prime to n. In view of our first formula, and of prop. 6 of Chap. XII-2, our assertion amounts to saying that $\chi_{n,y}$ is unramified if and only if y is in $(K^{\times})^n R^{\times}$. Call q the module of K; our assumption about the n-th roots of 1 implies that n divides $q - 1$. In an algebraic closure \bar{K} of K, take a primitive root ζ of 1 of order $n(q-1)$. For any $f \geq 1$, let K_f be the unramified extension of K of degree f, contained in \bar{K}; then ζ is in K_f if and only if $n(q-1)$ divides $q^f - 1$, i.e. if and only if $1 + q + \cdots + q^{f-1} \equiv 0 \ (n)$; as $q \equiv 1 \ (n)$, this is so if and only

if $f \equiv 0$ (n). This shows that $K(\zeta) = K_n$. Put $\varepsilon = \zeta^n$; as this is a primitive $(q-1)$-th root of 1, it is in K. In view of the definitions of Chap. IX-5, we have thus shown that $\chi_{n,\varepsilon}$ is an unramified character of order n, attached to K_n; therefore, by prop. 5 of Chap. XII-2, it generates the group of the unramified characters of order dividing n. In particular, for $y \in K^\times$, $\chi_{n,y}$ is unramified if and only if it is equal to $(\chi_{n,\varepsilon})^\nu$ for some $\nu \in \mathbf{Z}$, i.e. if $y \varepsilon^{-\nu}$ is in the kernel of the morphism $x \to \chi_{n,x}$; as we have seen in Chap. IX-5, that kernel is $(K^\times)^n$. Consequently $\chi_{n,y}$ is unramified if and only if y is in the subgroup of K^\times generated by $(K^\times)^n$ and ε. By prop. 8 of Chap. II-3, $(K^\times)^n$ contains $1 + P$; as R^\times is generated by $1 + P$ and ε, our assertion is now obvious.

COROLLARY. *For every local field K containing n distinct n-th roots of 1, $(x,y)_{n,K}$ defines a locally constant mapping of $K^\times \times K^\times$ into the group of the n-th roots of 1 in \mathbf{C}.*

This is obvious if $K = \mathbf{R}$ or \mathbf{C}; if K is a p-field, it is an immediate consequence of proposition 6, and of the fact (contained in prop. 8 of Chap. II-3 if K is of characteristic p, since then n must be prime to p, and otherwise in the corollary of prop. 9, Chap. II-3) that $(K^\times)^n$ is an open subgroup of K^\times, of finite index in K^\times. The symbol $(x,y)_{n,K}$ may be said to determine a duality between the finite group $K^\times/(K^\times)^n$ and itself, by means of which that group can be identified with its own dual.

PROPOSITION 7. *Let k be an A-field containing n distinct n-th roots of 1. Then, for all $z = (z_v)$, $z' = (z'_v)$ in $k_\mathbf{A}^\times$, almost all factors of the product*

$$(z,z')_n = \prod_v (z_v, z'_v)_{n,k_v}$$

taken over all the places v of k, are equal to 1; it defines a locally constant mapping of $k_\mathbf{A}^\times \times k_\mathbf{A}^\times$ into the group of the n-th roots of 1 in \mathbf{C}, and satisfies $(z,z')_n = (z',z)_n^{-1}$ for all z, z'. Moreover, $(k_\mathbf{A}^\times)^n$ is the set of the elements z of $k_\mathbf{A}^\times$ such that $(z,z')_n = 1$ for all $z' \in k_\mathbf{A}^\times$.

If k is of characteristic $p > 1$, our assumption about k implies that n is prime to p; consequently, in all cases, we have $|n|_v = 1$ for almost all v. As z_v, z'_v are in r_v^\times for almost all v, our first assertion follows now at once from prop. 6; the same facts, combined with the corollary of prop. 6, show that $(z,z')_n$ is locally constant. By prop. 6, if z is in the kernel of all the characters $z \to (z,z')_n$, we must have $z_v \in (k_v^\times)^n$ for all v; then, if we write $z_v = t_v^n$ with $t_v \in k_v^\times$, the fact that z_v is in r_v^\times for almost all v implies the same for t_v, so that $t = (t_v)$ is in $k_\mathbf{A}^\times$ and that $z = t^n$.

COROLLARY 1. *For every finite set P of places of k, containing all the places v for which $|n|_v \neq 1$, put*

$$\Omega(P)=\prod_{v\in P} k_v^\times \times \prod_{v\notin P} r_v^\times, \quad \Omega'(P)=\prod_{v\in P} (k_v^\times)^n \times \prod_{v\notin P} r_v^\times.$$

Then these are open subgroups of k_A^\times, and the set of the elements z of k_A^\times such that $(z,z')_n=1$ for all $z'\in\Omega(P)$ (resp. for all $z'\in\Omega'(P)$) is $(k_A^\times)^n\,\Omega'(P)$ (resp. $(k_A^\times)^n\,\Omega(P)$).

Concerning the definition of P, one should observe that $|n|_v>1$ for every infinite place of k, so that P contains all these places. Then $\Omega(P)$ is the same as the open subgroup of k_A^\times which was so denoted in Chap. IV-4; as we have seen above, $(k_v^\times)^n$ is open in k_v^\times for all v, so that $\Omega'(P)$ is open in $\Omega(P)$. The first set considered in our corollary consists of the ideles (z_v) such that $(z_v,z_v')_{n,k_v}=1$ for all $z_v\in k_v^\times$ if $v\in P$, and for all $z_v'\in r_v^\times$ if v is not in P. Our assertion follows now at once from prop. 6. The other set can be treated in the same manner.

COROLLARY 2. *Let P be as in corollary 1, and assume also that $k_A^\times = =k^\times\Omega(P)$. Then $(k^\times)^n=k^\times\cap(k_A^\times)^n\Omega'(P)$.*

In this last relation, $(k^\times)^n$ is clearly contained in the right-hand side. Conversely, let ξ be an element of this right-hand side. Then, by corollary 1, $(\xi,z)_n=1$ for all $z\in\Omega(P)$; by definition, this is the same as to say that $\Omega(P)$ is in the kernel of the character $z\to(\chi_{n,\xi},z)_k$ of k_A^\times. As that kernel contains k^\times, by the corollary of th. 2, § 3, and as $k_A^\times=k^\times\Omega(P)$, this implies that $\chi_{n,\xi}$ is trivial, hence that $\xi\in(k^\times)^n$.

The symbol $(z,z')_n$ defined in prop. 7 may be called *the Hilbert symbol* for k. As the last assertion of prop. 7 implies that $(k_A^\times)^n$ is a closed subgroup of k_A^\times, the main content of that proposition may be expressed by saying that the Hilbert symbol determines a duality between the group $k_A^\times/(k_A^\times)^n$ and itself, by means of which it can be identified with its own dual. As observed above, we have, for $\xi\in k^\times$, $z\in k_A^\times$:

$$(\xi,z)_n=(\chi_{n,\xi},z)_k=\chi_{n,\xi}(a(z))$$

and therefore, by the corollary of th. 2, § 3, $(\xi,\eta)_n=1$ for all ξ,η in k^\times.

PROPOSITION 8. *Let k contain n distinct n-th roots of 1. Then $k^\times(k_A^\times)^n$ is the set of the elements z of k_A^\times such that $(\xi,z)_n=1$ for all $\xi\in k^\times$, and it contains the kernel U_k of a.*

Call X_n the set in question; it may also be described as the intersection of the kernels of the characters $\chi_{n,\xi}\circ a$ of k_A^\times for all $\xi\in k^\times$; clearly it contains U_k. As before, put $G_k=k_A^\times/k^\times$; applying lemma 2 of Chap. XII-1 to G_k and to the endomorphism $x\to x^n$ of G_k, we see that $k^\times(k_A^\times)^n$ is a closed subgroup of k_A^\times with compact factor-group. Applying lemma 7 to G_k, and to the group of the characters of G_k determined by characters

of k_A^\times of the form $\chi_{n,\xi} \circ a$ with $\xi \in k^\times$, we see that every character of k_A^\times, trivial on X_n, is of that form. Clearly X_n contains $k^\times (k_A^\times)^n$; as they are both closed in k_A^\times, our proposition will be proved if we show that there are arbitrarily small neighborhoods U of 1 in k_A^\times such that X_n is contained in $k^\times (k_A^\times)^n U$; we will choose U as follows. Let P_0 be a finite set of places of k, containing all the places v where $|n|_v \neq 1$, and satisfying the condition in the corollary of th. 7, Chap. IV-4, i.e. such that $k_A^\times = k^\times \Omega(P_0)$; then every finite set of places $P \supset P_0$ has these same properties. Take any such set P; take $U = \prod U_v$, where U_v is an arbitrary neighborhood of 1 in $(k_v^\times)^n$ for $v \in P$, and $U_v = r_v^\times$ for v not in P; clearly U is a neighborhood of 1 in k_A^\times and can be made arbitrarily small by suitable choices of P and the neighborhoods U_v for $v \in P$. One sees at once that $(k_A^\times)^n U$ is the same as $(k_A^\times)^n \Omega'(P)$, where $\Omega'(P)$ is as defined in corollary 1 of prop. 7. What we have to prove is that X_n is contained in the group $W(P) = k^\times (k_A^\times)^n \Omega'(P)$, or in other words that $X_n W(P) = W(P)$. By lemma 1 of Chap. XII-1, applied to $G_k = k_A^\times / k^\times$ and to the image of $W(P)$ in G_k, we see that $W(P)$ has a finite index in k_A^\times; it will thus be enough to show that $W(P)$ and $X_n W(P)$ have the same index in k_A^\times.

The index of $X_n W(P)$ in k_A^\times is equal to the number of distinct characters of k_A^\times, trivial on X_n and on $W(P)$. Being trivial on X_n, such a character must be of the form $\chi_{n,\xi} \circ a$ with $\xi \in k^\times$. As X_n contains $k^\times (k_A^\times)^n$, this is trivial on $W(P)$ if and only if it is trivial on $\Omega'(P)$, hence, by corollary 1 of prop. 7, if and only if ξ is in $(k_A^\times)^n \Omega(P)$. In view of our assumptions on P, we can write

$$(k_A^\times)^n \Omega(P) = (k^\times \Omega(P))^n \Omega(P) = (k^\times)^n \Omega(P).$$

As in Chap. IV-4, put $E(P) = k^\times \cap \Omega(P)$; we see now that the characters in question are those of the form $\chi_{n,\xi} \circ a$ with $\xi \in (k^\times)^n E(P)$, and we must compute the number of distinct ones among these, which is the index in $(k^\times)^n E(P)$ of the kernel of the morphism $\xi \to \chi_{n,\xi} \circ a$. That kernel is the same as that of $\xi \to \chi_{n,\xi}$, which is $(k^\times)^n$; consequently that index is the same as that of $E(P)^n$ in $E(P)$. In view of th. 9 of Chap. IV-4, and of the fact that n divides the order of the group of the roots of 1 in k, that index is n^c with $c = \operatorname{card}(P)$.

Now we have to compute the index of $W(P)$ in k_A^\times. Consider the groups $G = k^\times \times \Omega(P)$, $G' = k^\times \times \Omega'(P)$, and the morphism f of G into k_A^\times given by $f(\xi, u) = \xi u$ for $\xi \in k^\times$, $u \in \Omega(P)$. Call H the kernel of f; this consists of the elements (ξ, ξ^{-1}) of G with $\xi \in E(P)$. In view of our assumption on P, f maps G onto k_A^\times; it maps G' onto $W(P)$, as appears from the formula

$$W(P) = k^\times (k_A^\times)^n \Omega'(P) = k^\times (k^\times \Omega(P))^n \Omega'(P) = k^\times (\Omega(P)^n \Omega'(P)) = k^\times \Omega'(P).$$

This gives $f^{-1}(W(P)) = HG'$, and therefore:

$$[k_A^\times : W(P)] = [G : HG'] = [G : G'] \cdot [HG' : G']^{-1}.$$

Here $[G : G']$ is given by

$$[G : G'] = [\Omega(P) : \Omega'(P)] = \prod_{v \in P} [k_v^\times : (k_v^\times)^n].$$

In the right-hand side, each factor corresponding to an imaginary place v is equal to 1, which can also be written as $n^2 |n|_v^{-1}$, since in that case $|n|_v = n^2$. If v is real, n must be 2, since $k_v = \mathbf{R}$ must contain a primitive n-th root of 1; then the corresponding factor is 2, which can again be written as $n^2 |n|_v^{-1}$. The factors corresponding to the finite places $v \in P$ are given by the corollary of prop. 9, Chap. II-3, if k is of characteristic 0, and by prop. 8 of Chap. II-3 otherwise; here one has to take into account the fact that n divides the order of the group of roots of 1 in k, hence also in k_v, and that consequently it is prime to p if k is of characteristic p. Then one sees that the factors in question are again respectively equal to $n^2 |n|_v^{-1}$. This gives

$$[G : G'] = \prod_{v \in P} (n^2 |n|_v^{-1}) = n^{2c} \prod_v |n|_v^{-1} = n^{2c}$$

since $|n|_v = 1$ for all places v not in P.

Our proof will now be complete if we show that $[HG' : G'] = n^c$. This is the same as the index of $H \cap G'$ in H, or, in view of the definition of H and G', as that of $E(P) \cap \Omega'(P)$ in $E(P)$. By corollary 2 of prop. 7, $E(P) \cap \Omega'(P)$ is contained in $E(P) \cap (k^\times)^n$, i.e. in $E(P)^n$, and it is obvious that it contains $E(P)^n$. Therefore the index is question is that of $E(P)^n$ in $E(P)$; we have already found above that this is n^c; this completes our proof.

§ 6. **The Brauer group of an A-field.** In § 3, we have seen that a class of simple algebras A over k is uniquely determined by its local invariants $h_v(A)$, with $h_v(A) = 1$ for almost all v, $h_v(A) = 1$ for all imaginary places, and $h_v(A) = 1$ or 2 for all real places; and we have proved that $\prod h_v(A) = 1$. Therefore the Brauer group $H(k)$ will be known if we prove the following:

THEOREM 4. *Let k be an A-field. For each place v of k, let η_v be a root of 1 in \mathbf{C}. Assume that $\eta_v = 1$ for almost all v, $\eta_v = 1$ for every imaginary place v, $\eta_v = 1$ or 2 for every real place v, and that $\prod_v \eta_v = 1$. Then there is a simple algebra A over k with the invariants $h_v(A) = \eta_v$.*

The proof of this, for a field k of characteristic 0, will be postponed until the end of this §; we proceed to prove it now for a field k of characteristic $p > 1$. As in Chap. VI, write $D(k)$ for the group of divisors of k,

$D_0(k)$ for the group of divisors of degree 0, and $P(k)$ for the group of principal divisors; call h the number of divisor-classes of degree 0, i.e. the index of $P(k)$ in $D_0(k)$. Let v_1, \ldots, v_N be all the places of k where $\eta_v \neq 1$; taking for n an integer $\geqslant 1$ such that $(\eta_{v_i})^n = 1$ for all i, we can write $\eta_{v_i} = e(a_i/n)$ with $a_i \in \mathbf{Z}$ for $1 \leqslant i \leqslant N$. As $\prod \eta_v = 1$, we have $\sum a_i = na$ with $a \in \mathbf{Z}$; after replacing a_1 by $a_1 - na$, we may assume that $\sum a_i = 0$. For each i, call d_i the degree of the place v_i; put $d = \prod d_i$ and $\mathfrak{m} = \sum (a_i d/d_i) v_i$. Then $\deg(\mathfrak{m}) = \sum a_i d = 0$, \mathfrak{m} is in $D_0(k)$, hence $h\mathfrak{m}$ in $P(k)$, so that there is $\theta \in k^\times$ such that $\operatorname{div}(\theta) = h\mathfrak{m}$, i.e. $\operatorname{ord}_{v_i}(\theta) = ha_i d/d_i$ for $1 \leqslant i \leqslant N$, and $\operatorname{ord}_v(\theta) = 0$ when $\eta_v = 1$. Now consider the constant-field extension k' of k of degree hnd; let φ be the Frobenius automorphism of k' over k, and χ the character of the Galois group of k' over k such that $\chi(\varphi) = = e(1/hnd)$. Just as in the proof of prop. 3 of § 1, one sees at once, by applying corollary 4 of th. 1, Chap. XII-2, that, if v is any place of k, and δ its degree, $(\chi_v, \theta)_v = \chi(\varphi^\delta)^\nu$ with $\nu = \operatorname{ord}_v(\theta)$. In view of our choice of θ and χ, this gives $(\chi_v, \theta)_v = \eta_v$ for all v. Therefore, by formula (4) of § 3, the cyclic algebra $A = [k'/k; \chi, \theta]$ solves our problem.

PROPOSITION 9. *For each* $\chi \in X_k$, *call* $U(\chi)$ *the kernel of the character* $\chi \circ \mathfrak{a}$ *of* $k_{\mathbf{A}}^\times$. *Then:* (a) *if* k' *is the cyclic extension of* k *attached to* χ, $U(\chi) = = k^\times N_{k'/k}(k_{\mathbf{A}}'^\times)$; (b) *for every* $n \geqslant 1$, *prime to* p *if* k *is of characteristic* $p > 1$, *the intersection* U_n *of the kernels* $U(\chi)$, *for all the characters* $\chi \in X_k$ *of order dividing* n, *is* $k^\times (k_{\mathbf{A}}^\times)^n$.

Put $U'(\chi) = k^\times N_{k'/k}(k_{\mathbf{A}}'^\times)$, when χ and k' are as above, and put $U_n' = k^\times (k_{\mathbf{A}}^\times)^n$. Applying lemma 2 of Chap. XII-1 to the endomorphism $x \to x^n$ of the group $G_k = k_{\mathbf{A}}^\times / k^\times$, we see that U_n' is closed in $k_{\mathbf{A}}^\times$; applying the same lemma to the morphism of $G_{k'}$ into G_k determined by the morphism $N_{k'/k}$ of $k_{\mathbf{A}}'^\times$ into $k_{\mathbf{A}}^\times$, we see that $U'(\chi)$ is also closed in $k_{\mathbf{A}}^\times$. If χ is of order n, and k' is as in (a), n is the degree of k' over k, so that $N_{k'/k}(z) = z^n$ for $z \in k$, hence also for $z \in k_{\mathbf{A}}$; therefore we have then $U'(\chi) \supset U_n'$. A character of $k_{\mathbf{A}}^\times$ is trivial on $(k_{\mathbf{A}}^\times)^n$ if and only if its order divides n; therefore a character of $k_{\mathbf{A}}^\times$, trivial on k^\times, has an order dividing n if and only if it is trivial on U_n'. As before, call U_k the kernel of \mathfrak{a}; we know that it contains k^\times. If k is of characteristic 0, apply prop. 1 of Chap. XII-1 to the pairing between X_k and G_k determined by $(\chi, z)_k$; otherwise apply corollary 4 of prop. 2, Chap. XII-2; in both cases we see that every character of $k_{\mathbf{A}}^\times$ of finite order, trivial on U_k, can be uniquely written as $\chi \circ \mathfrak{a}$ with $\chi \in X_k$. This implies that U_n is the intersection of the characters of $k_{\mathbf{A}}^\times$, trivial on U_n' and on U_k; therefore it is the closure of $U_n' U_k$, and we have $U_n = U_n'$ if and only if $U_n' \supset U_k$. We also see now that every character of $k_{\mathbf{A}}^\times$, trivial on $U'(\chi)$ and on U_k, must be of the form $\chi' \circ \mathfrak{a}$ with $\chi' \in X_k$; by corollary 2 of th. 1, § 1, this is trivial on $U'(\chi)$ if

and only if the cyclic extension of k attached to χ' is contained in k', i.e. if and only if $\chi' = \chi^v$ with some $v \in \mathbb{Z}$; as the intersection of the kernels $U(\chi^v)$ for $v \in \mathbb{Z}$ is obviously $U(\chi)$, this shows that $U(\chi)$ is the closure of $U'(\chi) U_k$, and that we have $U(\chi) = U'(\chi)$ if and only if $U'(\chi) \supset U_k$. Now consider first the case of characteristic 0. Proceeding by induction on n, we assume that, for all fields k of characteristic 0, (a) holds for every χ of order $< n$. Then $U_k \subset k^\times N_{k'/k}(k_A^\times)$ whenever k is such a field and k' is a cyclic extension of k of degree $< n$; by formula (5) at the end of § 3, this implies $U_k = k^\times N_{k'/k}(U_{k'})$. Let L be the extension of k generated by a primitive n-th root of 1; as this is abelian of degree $< n$ over k, we can find a sequence $k_0 = L$, $k_1, \ldots, k_r = k$ of fields between L and k, such that, for $1 \leqslant i \leqslant r$, k_{i-1} is cyclic of degree $< n$ over k_i; therefore we have, for $1 \leqslant i \leqslant r$:

$$U_{k_i} = k_i^\times N_{k_{i-1}/k_i}(U_{k_{i-1}}).$$

By induction on i for $1 \leqslant i \leqslant r$, we see now at once that U_{k_i} is contained in $k_i^\times((k_i)_A^\times)^n$, since this is so for $i = 0$ by prop. 8. For $i = r$, we get $U_k \subset U_n'$; as we have seen above, this proves (b), and it implies $U_k \subset U'(\chi)$ for every χ of order n, which proves (a) for such characters and completes the induction. Now let k be of characteristic $p > 1$. Take χ and k' as in (a); take $z \in U(\chi)$, so that $(\chi, z)_k = 1$; then, if we put $z = (z_v)$ and $\eta_v = (\chi_v, z_v)_v$, these satisfy the conditions in th. 4. As th. 4 has been proved for characteristic $p > 1$, we conclude that there is a simple algebra A over k with the invariants $h_v(A) = \eta_v$; as this is condition (iv) in prop. 5 of § 3, we can apply that proposition, the last assertion in which shows now that $z \in U'(\chi)$. This proves (a). Just as above, we can now conclude, by applying formula (5) at the end of § 3, that $U_k = k^\times N_{k'/k}(U_{k'})$ for all cyclic extensions k' of k. Assuming that n is prime to p, take for k' the constant-field extension of k generated by a primitive n-th root of 1. Then prop. 8 gives $U_{k'} \subset k'^\times (k_A'^\times)^n$; therefore the same is true for k.

We can now prove theorem 4 in the case of characteristic 0. For every place v of k, call $v(v)$ the order of η_v in \mathbb{C}^\times. One can construct a character χ such that, for every v, the order of χ_v is a multiple of $v(v)$; for instance, this will be so if we take for χ a character attached to the cyclic extension k' of k described in lemma 5 of § 3. Then, for every v, $z \to (\chi_v, z)_v$ is a character of k_v^\times whose order, being equal to that of χ_v, is a multiple of $v(v)$; therefore we can choose $z_v \in k_v^\times$ so that $(\chi_v, z_v)_v = \eta_v$. If, in doing so, we take $z_v = 1$ whenever $\eta_v = 1$, $z = (z_v)$ is in k_A^\times, and the assumption $\prod \eta_v = 1$ implies that z is in the kernel of $\chi \circ a$; therefore, by prop. 9, it is in $k^\times N_{k'/k}(k_A'^\times)$, k' being the cyclic extension of k attached to χ. Writing $z = \theta N_{k'/k}(z')$ with $z' \in k_A'^\times$, one sees at once, by combining prop. 10 of Chap. IX-4 with corollary 3 of th. 1, Chap. IV-1, and with

prop. 1 of § 1, that the cyclic algebra $A = [k'/k; \chi, \theta]$ has the required local invariants $h_v(A) = \eta_v$, so that it solves our problem.

§ 7. The Hilbert p-symbol.

By now, as will be seen in the next §, our investigation is essentially complete, so far as only algebraic number-fields are concerned. For the case of characteristic $p > 1$, we still need a symbol, similar to the Hilbert symbol studied in § 5 but based on the factor-classes $\{\xi, \theta\}_p$ of Chap. IX-5.

In any field K of characteristic $p > 1$, we will denote by Φ the endo-morphism $x \to x - x^p$ of the additive group of K; its kernel is the prime field \mathbf{F}_p. We begin by considering a local p-field K of characteristic p; as usual, we write R for its maximal compact subring, P for the maximal ideal of R, and q for the module of K. Obviously, Φ maps R into R, P into P, and, if $\mathrm{ord}(x) = v < 0$, we have $\mathrm{ord}(\Phi(x)) = pv < 0$, so that $\Phi^{-1}(R) = R$.

PROPOSITION 10. *Let K, R, P and Φ be as above; for $x \in K$, $z \in K^\times$, put $(x, z)_{p, K} = (\chi_{p, x}, z)_K$. Then $\Phi(K)$ contains P and not R; the set of the elements x of K, such that $(x, z)_{p, K} = 1$ for all $z \in K^\times$ (resp. for all $z \in R^\times$) is $\Phi(K)$ (resp, $R + \Phi(K)$); the set of the elements z of K^\times such that $(x, z)_{p, K} = 1$ for all $x \in K$ (resp. for all $x \in R$) is $(K^\times)^p$ (resp. $(K^\times)^p R^\times$).*

For $x \in P$, put $\Psi(x) = \sum_{n=0}^{\infty} x^{p^n}$; clearly this is convergent and defines an endomorphism of P, and one sees at once that both $\Phi \circ \Psi$ and $\Psi \circ \Phi$ induce the identity on P. Therefore Φ induces on P an automorphism of the additive group of P, so that $P \subset \Phi(K)$. Call F the algebraic closure of the prime field \mathbf{F}_p in K; by th. 7 of Chap. I-4, F is a field with q elements, and $R = F + P$, so that $\Phi(R) = \Phi(F) + P$. As the endomorphism induced by Φ on the finite field F has the kernel \mathbf{F}_p, it is not surjective; therefore $\Phi(R) \ne R$; as $\Phi^{-1}(R) = R$, this shows that R is not contained in $\Phi(K)$. If $(x, z)_{p, K} = 1$ for all $z \in K^\times$, $\chi_{p, x}$ must be trivial; as we have seen in Chap. IX-5, this is so if and only if $x \in \Phi(K)$. Take any $x \in R + \Phi(K)$; as $R = F + P$ and $P \subset \Phi(K)$, we can write $x = a + \Phi(u)$ with $a \in F$, $u \in K$; then $\chi_{p, x}$ is the same as $\chi_{p, a}$ and is a character attached to the cyclic extension of K generated by any root α of $X - X^p = a$. As α is algebraic over F, it is 0 or a root of 1 of order prime to p, so that $K(\alpha)$, hence also $\chi_{p, x}$, are unramified over K; therefore $(x, z)_{p, K} = 1$ for all $z \in R^\times$. Now take a root ζ of $X - X^q = 1$ in some algebraic closure of K, and put $\varepsilon = \zeta - \zeta^p$. The Galois group of $K(\zeta)$ over K is generated by the Frobenius automorphism; as ζ is algebraic over F, this maps ζ onto $\zeta^q = \zeta - 1$, so that it leaves ε invariant; therefore ε is in F, $K(\zeta)$ is the unramified extension of K of degree p, and $\chi_{p, \varepsilon}$ is a character attached to that extension.

Then the unramified characters over K, of order p or 1, are those of the form $(\chi_{p,\varepsilon})^v$ with $v\in\mathbf{Z}$; consequently a character $\chi_{p,x}$ is unramified if and only if it can be so written, i.e. if and only if $x=v\varepsilon+\Phi(u)$ with $u\in K$; then x is in $R+\Phi(K)$. As to the last two assertions, one is nothing else than prop. 10 of Chap. XII-3. Finally, if ε is as above, the kernel of $z\to(\varepsilon,z)_{p,K}$ is the subgroup of K^\times of index p containing R^\times; this is $(K^\times)^p R^\times$; for every $x\in R$, we have seen that $\chi_{p,x}$ is unramified, and it is of order p or 1, so that the kernel of $z\to(x,z)_{p,K}$ contains $(K^\times)^p R^\times$. This completes our proof.

COROLLARY. *For each integer* $m\geqslant0$, *call* $\Omega'(m,K)$ *the set of the elements* z *of* K^\times *such that* $(x,z)_{p,K}=1$ *for all* $x\in P^{-m}$. *Then this is an open subgroup of* K^\times, *containing* $(K^\times)^p$; *its index in* K^\times *is* $p\cdot q^{m-m'}$ *if* m' *is the largest integer* $\leqslant m/p$. *For every neighborhood* U *of 1 in* K^\times, *there is* $m\geqslant0$ *such that* $\Omega'(m,K)\subset(K^\times)^p U$. *Moreover,* $\Omega'(0,K)=(K^\times)^p R^\times$; *and, for every* $m\geqslant0$, *the set of the elements* x *of* K *such that* $(x,z)_{p,K}=1$ *for all* $z\in\Omega'(m,K)$ *is* $P^{-m}+\Phi(K)$.

Let the finite field F be as above, and let π be a prime element of K; by th. 8 of Chap. I-4, K may be identified with the field of formal power-series in π with coefficients in F; therefore, if we call V_m the space of polynomials of degree $\leqslant m$ in π^{-1} with coefficients in F, P^{-m} is the direct sum of V_m and P, and V_m is a vector-space of dimension $m+1$ over F; moreover, one verifies at once that $V_m\cap\Phi(K)=\Phi(V_{m'})$, with m' as in our corollary. By prop. 10, $\Omega'(m,K)$ is the intersection of the kernels of the characters $z\to(x,z)_{p,K}$ of K^\times for $x\in V_m$; therefore it is open and contains $(K^\times)^p$, and, by lemma 7 of §5, all the characters of K^\times, trivial on $\Omega'(m,K)$, are of that form. This implies that the index of $\Omega'(m,K)$ in K^\times is equal to the number of distinct characters of that form, which is the index of $V_m\cap\Phi(K)$ in V_m; as V_m, $V_{m'}$ have respectively q^{m+1} and $q^{m'+1}$ elements, and as the morphism Φ of $V_{m'}$ onto $V_m\cap\Phi(K)$ has the kernel \mathbf{F}_p, that index is $p\cdot q^{m-m'}$. By lemma 2 of Chap. XII-1, and lemma 7 of §5, the group $G'=K^\times/(K^\times)^p$ is compact, and its characters are those determined by the characters $z\to(x,z)_{p,K}$ of K^\times for $x\in K$. Therefore the intersections of the kernels of finitely many such characters make up a fundamental system of neighborhoods of 1 in G'. This is the same as to say that, if U is any neighborhood of 1 in K^\times, one can find finitely many characters $z\to(x_i,z)_{p,K}$ such that the intersection W of their kernels is contained in $(K^\times)^p U$; then W contains $\Omega'(m,K)$ if we take $m\geqslant0$ such that $-m\leqslant\mathrm{ord}(x_i)$ for all i. By prop. 10, $\Omega'(0,K)=(K^\times)^p R^\times$. Finally, if $z\to(y,z)_{p,K}$ is trivial on $\Omega'(m,K)$, it must coincide with a character $z\to(x,z)_{p,K}$ with some $x\in V_m$; by prop. 10, this is the same as to say that $y\in V_m+\Phi(K)$; as $\Phi(K)\supset P$, and $V_m+P=P^{-m}$, this proves the last assertion in our corollary.

From now on, in this §, k will be an A-field of characteristic p. We will need the following lemma:

LEMMA 8. *If v is any place of $k, k \cap (k_v)^p = k^p$.*

Clearly $k \cap (k_v)^p$ is a field between k and k^p; by lemma 1 of Chap. VIII-5, it must be k or k^p. As $(k_v)^p$ contains no prime element of k_v, it is not dense in k_v; as k is dense in $k_v, (k_v)^p$ cannot contain k.

PROPOSITION 11. *For all $x = (x_v)$ in k_A, and all $z = (z_v)$ in k_A^\times, almost all the factors of the product*

$$(x, z)_p = \prod_v (x_v, z_v)_{p, k_v}$$

are equal to 1; it defines a locally constant mapping of $k_A \times k_A^\times$ into the group of the p-th roots of 1 in \mathbf{C}; the set of the elements x of k_A (resp. of the elements z of k_A^\times) such that $(x, z)_p = 1$ for all $z \in k_A^\times$ (resp. for all $x \in k_A$) is $\Phi(k_A)$ (resp. $(k_A^\times)^p$).

All this follows at once from prop. 10.

COROLLARY 1. *For every divisor $\mathfrak{m} = \sum m(v) \cdot v \succ 0$ of k, put*

$$\Omega'(\mathfrak{m}) = \prod_v \Omega'(m(v), k_v).$$

Then this is an open subgroup of k_A^\times, containing $(k_A^\times)^p$. For every neighborhood U of 1 in k_A^\times, there is a divisor \mathfrak{m} such that $\Omega'(\mathfrak{m}) \subset (k_A^\times)^p U$. The set of the elements x of k_A such that $(x, z)_p = 1$ for all $z \in \Omega'(\mathfrak{m})$ is

$$\left(\prod_v p_v^{-m(v)} \right) + \Phi(k_A).$$

For all v, $\Omega'(m(v), k_v)$ is an open subgroup of k_v^\times, containing $(k_v^\times)^p$, and, for almost all v, $m(v) = 0$, so that $\Omega'(m(v), k_v)$ contains r_v^\times; this proves the first assertion. In the second assertion, it is enough to consider a neighborhood $U = \prod U_v$, where U_v is a neighborhood of 1 in k_v^\times for all v, and $U_v = r_v^\times$ for almost all v; then our assertion follows at once from the corollary of prop. 10. Assume that $x = (x_v)$ is as in the last assertion; then, by the same corollary, we can write $x_v = y_v + \Phi(u_v)$ with $y_v \in p_v^{-m(v)}$, $u_v \in k_v$ for all v, and $y_v = x_v$, $u_v = 0$ whenever $m(v) = 0$ and $x_v \in r_v$, hence for almost all v. Then $y = (y_v)$ and $u = (u_v)$ are in k_A, and $x = y + \Phi(u)$, $y \in \prod p_v^{-m(v)}$.

COROLLARY 2. *Notations being as in corollary 1, assume also that $\deg(\mathfrak{m}) > 2g - 2$. Then $(k^\times)^p = k^\times \cap \Omega'(\mathfrak{m})$.*

Clearly the right-hand side of this last formula contains $(k^\times)^p$. Now take $\xi \in k^\times$; if $\xi \in \Omega'(\mathfrak{m})$, we have $(x, \xi)_p = 1$ for all $x \in \prod p_v^{-m(v)}$ and also for all $x \in k$, hence for all $x \in k_A$ by corollary 3 of th. 2, Chap. VI. By proposition 11 this implies $\xi \in (k_A^\times)^p$, hence $\xi \in (k^\times)^p$ by lemma 8.

COROLLARY 3. *Notations being as in corollary* 1, *assume that* $\mathfrak{n} = \sum n(v) \cdot v$ *is a divisor of* k, *of degree* $> 2g - 2$, *such that* $\mathfrak{m} \succ p\mathfrak{n}$. *Then the set of the elements* x *of* $k_{\mathbf{A}}$ *such that* $(x, z)_p = 1$ *for all* $z \in \Omega'(\mathfrak{m})$ *is*

$$\left(\prod_v p_v^{-m(v)} \right) + \Phi(k).$$

Again by corollary 3 of th. 2, Chap. VI, we can write

$$\Phi(k_{\mathbf{A}}) = \Phi(k) + \Phi\left(\prod_v p_v^{-n(v)} \right).$$

Our assumptions imply that the second term in the right-hand side is a subgroup of $\prod p_v^{-m(v)}$. Our assertion follows now from the last one in corollary 1.

PROPOSITION 12. *The set of the elements* z *of* $k_{\mathbf{A}}^{\times}$ *such that* $(\xi, z)_p = 1$ *for all* $\xi \in k$ *is* $k^{\times} (k_{\mathbf{A}}^{\times})^p$.

Call that set X_p; it is the intersection of the kernels of the characters $\chi_{p, \xi} \circ \mathfrak{a}$ of $k_{\mathbf{A}}^{\times}$ for all $\xi \in k$; it contains $k^{\times} (k_{\mathbf{A}}^{\times})^p$. By lemma 2 of Chap. XII-1, applied to $G_k = k_{\mathbf{A}}^{\times}/k^{\times}$, $k^{\times} (k_{\mathbf{A}}^{\times})^p$ is a closed subgroup of $k_{\mathbf{A}}^{\times}$ with compact factor-group; by lemma 7 of § 5, every character of $k_{\mathbf{A}}^{\times}$, trivial on X_p, must be of the form $\chi_{p, \xi} \circ \mathfrak{a}$ with $\xi \in k$. Choose a divisor $\mathfrak{n} > 0$ of k, of degree $> 2g - 2$. In view of corollary 1 of prop. 11, it will be enough for us to show that X_p is contained in $W(\mathfrak{m}) = k^{\times} \Omega'(\mathfrak{m})$ for all divisors $\mathfrak{m} = \sum m(v) \cdot v \succ p\mathfrak{n}$. By lemma 1 of Chap. XII-1, $W(\mathfrak{m})$ is of finite index in $k_{\mathbf{A}}^{\times}$; therefore it will be enough to show that $X_p W(\mathfrak{m})$ and $W(\mathfrak{m})$ have the same index in $k_{\mathbf{A}}^{\times}$.

The index N of $X_p W(\mathfrak{m})$ is equal to the number of distinct characters of $k_{\mathbf{A}}^{\times}$ of the form $\chi_{p, \xi} \circ \mathfrak{a}$, or, what amounts to the same, of the form $z \to (\xi, z)_p$, with $\xi \in k$, which are trivial on $\Omega'(\mathfrak{m})$. By corollary 3 of prop. 11, the latter character is trivial on $\Omega'(\mathfrak{m})$ if and only if $\xi \in U(\mathfrak{m}) + \Phi(k)$, with $U(\mathfrak{m}) = \prod p_v^{-m(v)}$. As in Chap. VI, put $\Lambda(\mathfrak{m}) = k \cap U(\mathfrak{m})$. Then we see that N is the index of $\Phi(k)$ in $\Lambda(\mathfrak{m}) + \Phi(k)$, or, what amounts to the same, of $\Lambda(\mathfrak{m}) \cap \Phi(k)$ in $\Lambda(\mathfrak{m})$. Put $\mathfrak{m}' = \sum m'(v) \cdot v$, where $m'(v)$, for each v, is the largest integer $\leqslant m(v)/p$; call m, m', n the degrees of $\mathfrak{m}, \mathfrak{m}', \mathfrak{n}$, respectively; then $m \geqslant m' \geqslant n > 2g - 2$. Clearly $\Lambda(\mathfrak{m}) \cap \Phi(k)$ is the same as $\Phi(\Lambda(\mathfrak{m}'))$; corollary 2 of th. 2, Chap. VI, shows that $\Lambda(\mathfrak{m}), \Lambda(\mathfrak{m}')$ are vector-spaces, of dimension $m - g + 1$ and $m' - g + 1$ respectively, over the field of constants F of k; as Φ maps $\Lambda(\mathfrak{m}')$ onto $\Lambda(\mathfrak{m}) \cap \Phi(k)$ with the kernel \mathbf{F}_p, we see that the latter group has $p^{-1} q^{m' - g + 1}$ elements while $\Lambda(\mathfrak{m})$ has $q^{m - g + 1}$ elements. This gives $N = p \cdot q^{m - m'}$.

Now we have to compute the index of $W(\mathfrak{m})$ in $k_{\mathbf{A}}^{\times}$. Take a finite set P of places of k, containing all the places v for which $m(v) > 0$, and satisfying the condition in the corollary of th. 7, Chap. IV-4, i.e. such that $k_{\mathbf{A}}^{\times} = k^{\times} \Omega(P)$. Put:

$$\Omega'' = \prod_{v \in P} \Omega'(m(v), k_v) \times \prod_{v \notin P} r_v^\times.$$

Clearly this contains $\Omega(P)^p$, and we have $\Omega'(\mathfrak{m}) = (k_\mathbf{A}^\times)^p \Omega''$, hence:

$$W(\mathfrak{m}) = k^\times (k_\mathbf{A}^\times)^p \Omega'' = k^\times (k^\times \Omega(P))^p \Omega'' = k^\times \Omega''.$$

Put now $G = k^\times \times \Omega(P)$, $G' = k^\times \times \Omega''$; call f the morphism of G into $k_\mathbf{A}^\times$ given by $f(\xi, u) = \xi u$ for $\xi \in k^\times$, $u \in \Omega(P)$, and call H the kernel of f. Then f maps G onto $k_\mathbf{A}^\times$, G' onto $W(\mathfrak{m})$, and H consists of the elements (ξ, ξ^{-1}) with ξ in $E(P) = k^\times \cap \Omega(P)$. We have now:

$$[k_\mathbf{A}^\times : W(\mathfrak{m})] = [G : HG'] = [G : G'] \cdot [HG' : G']^{-1}.$$

Here, in view of the corollary of prop. 10, $[G : G']$ is given by

$$[G : G'] = [\Omega(P) : \Omega''] = \prod_{v \in P} [k_v^\times : \Omega'(m(v), k_v)] = p^c q^{m-m'}$$

with $c = \mathrm{card}(P)$. Finally, $[HG' : G']$ is the same as the index of $H \cap G'$ in H, i.e. as that of $E(P) \cap \Omega''$ in $E(P)$. Clearly $E(P) \cap \Omega''$ is contained in $k^\times \cap \Omega'(\mathfrak{m})$, which is $(k^\times)^p$ by corollary 2 of prop. 11, and it contains $E(P)^p$; as $E(P) \cap (k^\times)^p$ is the same as $E(P)^p$, we see that $E(P) \cap \Omega''$ is $E(P)^p$, and it follows at once from th. 9 of Chap. IV-4 that its index in $E(P)$ is p^{c-1}. This completes the proof.

COROLLARY. *If k is as above, and U_k is the kernel of the canonical morphism* \mathfrak{a}, *we have $U_k \subset k^\times (U_k)^p$.*

By proposition 12, U_k is contained in $k^\times (k_\mathbf{A}^\times)^p$, so that, if u is any element of U_k, it can be written as $u = \xi v^p$ with $\xi \in k^\times$, $v \in k_\mathbf{A}^\times$. Take any $\chi \in X_k$; call k' the cyclic extension of k attached to χ. By prop. 9 of §6, the kernel $U(\chi)$ of $\chi \circ \mathfrak{a}$ is $k^\times N_{k'/k}(k_\mathbf{A}^{'\times})$; as $U_k \subset U(\chi)$, this implies, by formula (5) at the end of §3, that $U_k = k^\times N_{k'/k}(U_{k'})$. Again by proposition 12, $U_{k'}$ is contained in $k^{'\times}(k_\mathbf{A}^{'\times})^p$, so that U_k is contained in $k^\times N_{k'/k}(k_\mathbf{A}^{'\times})^p$; therefore, if u is as above, we can write $u = \eta N_{k'/k}(w)^p$ with $\eta \in k^\times$, $w \in k_\mathbf{A}^{'\times}$. This gives $\xi \eta^{-1} = v^{-p} N_{k'/k}(w)^p$. As $\xi \eta^{-1}$ is in k^\times and in $(k_\mathbf{A}^\times)^p$, it is in $(k^\times)^p$, by lemma 8; writing it as ζ^p with $\zeta \in k^\times$, we get $v = \zeta^{-1} N_{k'/k}(w)$, since p is the characteristic. This shows that v is in $U(\chi)$; as this is so for all $\chi \in X_k$, it is in U_k, which completes the proof.

§ 8. The kernel of the canonical morphism. We are now able to determine U_k in all cases.

THEOREM 5. *Let k be an A-field, and \mathfrak{a} the canonical morphism of $k_\mathbf{A}^\times$ into the Galois group \mathfrak{A} of k_{ab} over k. Then $\chi \to \chi \circ \mathfrak{a}$ is a bijective morphism of the group X_k of characters of \mathfrak{A} onto the group of the characters of $k_\mathbf{A}^\times$ of finite order, trivial on k^\times.*

Every character of k_A^\times of order n, trivial on k^\times, is trivial on $k^\times (k_A^\times)^n$, hence on U_k, by prop. 9 of § 6, if k is of characteristic 0; in that case, our conclusion follows from this at once by applying prop. 1 of Chap. XII-1 to $G_k = k_A^\times / k^\times$. Now let k be of characteristic $p > 1$, and let ω be a character of k_A^\times of order n, trivial on k^\times. Write $n = n' p^i$ with n' prime to p and $i \geq 0$; taking integers a, b such that $n' a + p^i b = 1$, we have $\omega = \omega' \omega''$ with $\omega' = \omega^{p^i b}$ of order n', and $\omega'' = \omega^{n' a}$ of order p^i, both being trivial on k^\times. Just as above, we conclude from prop. 9 of § 6 that ω' is trivial on U_k. On the other hand, one concludes at once from the corollary of prop. 12 of § 7, by induction on i, that U_k is contained in $k^\times (U_k)^{p^i}$, hence in $k^\times (k_A^\times)^{p^i}$, and then, just as above, that ω'' is trivial on U_k. This shows that ω is trivial on U_k; our conclusion follows from this at once by applying corollary 4 of prop. 2, Chap. XII-1, to $G_k = k_A^\times / k^\times$.

COROLLARY 1. *The kernel U_k of \mathfrak{a} is the intersection of the closed subgroups $k^\times (k_A^\times)^n$ of k_A^\times for all $n \geq 1$.*

In the proof of prop. 9, § 6, we have already seen that these are closed subgroups; clearly, then, $k^\times (k_A^\times)^n$ is the intersection of the kernels of all the characters of k_A^\times, trivial on k^\times, whose order divides n. Our assertion follows now at once from theorem 5.

COROLLARY 2. *If k is of characteristic $p > 1$, $U_k = k^\times$.*

Write $G_k = G_k^1 \times N$, with $G_k^1 = k_A^1 / k$ and N isomorphic to \mathbf{Z}. As G_k^1 is compact, and as it is obvious that it is totally disconnected, lemma 4 of Chap. VII-3 shows that all its characters are of finite order; every such character can be uniquely extended to one of G_k, trivial on N, which is then also of finite order, hence, by theorem 5, trivial on the image of U_k in G_k. As corollary 2 of prop. 2, Chap. XII-1, shows that this image is contained in G_k^1, it must therefore be $\{1\}$, which is the same as to say that $U_k = k^\times$.

COROLLARY 3. *If k is of characteristic 0, U_k is the closure of $k^\times k_{\infty+}^\times$ in k_A^\times, $k_{\infty+}^\times$ being the group of the ideles (z_v) such that $z_v > 0$ for all real places and $z_v = 1$ for all finite places v of k.*

Write $G_k = G_k^1 \times N$, with $G_k^1 = k_A^1 / k^\times$, N being the image in G_k of the group M defined in corollary 2 of th. 5, Chap. IV-4. Call U' the closure of $k^\times k_{\infty+}^\times$ in k_A^\times, U'' its image in G_k, and put $U_1'' = U'' \cap G_k^1$. Obviously $k_A^\times / k_{\infty+}^\times$ is totally disconnected, so that the same is true of k_A^\times / U', hence of G_k / U''. As M is contained in U', N is contained in U'', so that $U'' = = U_1'' \times N$ and that G_k / U'' may be identified with G_k^1 / U_1'' and therefore is compact. This shows that every character of G_k, trivial on U_1'', or, what amounts to the same, every character of k_A^\times, trivial on U', is of finite

order. Consequently U_k is contained in U'; as we already knew that it contains U', we see that it is U'.

In order to obtain more precise results about U_k in the case of characteristic 0, one needs an algebraic lemma:

LEMMA 9. *Let l be a rational prime, K a field, not of characteristic l, and \bar{K} an algebraic closure of K. For each $n \geqslant 0$, call K_n the extension of K generated by a primitive l^n-th root of 1 in \bar{K}. Then, if $l \neq 2$, or if $l = 2$ and $K = K_2$, $K^{\times} \cap (K_n^{\times})^{l^n} = (K^{\times})^{l^n}$ for all n. If $l = 2$, $K \neq K_2$, $2 \leqslant m < n$ and $K_2 \neq K_{m+1}$, then $K^{\times} \cap (K_n^{\times})^{2^n} \subset (K^{\times})^{2^{n-m}}$.*

Take $a \in K^{\times} \cap (K_n^{\times})^{l^n}$; assume at first that a is not in $(K_1^{\times})^{l^n}$, and * let i be the smallest integer such that a is in $(K_{i+1}^{\times})^{l^n}$ and not in $(K_i^{\times})^{l^n}$. Then $1 \leqslant i < n$, $K_i \neq K_{i+1}$, and we can write $a = x^{l^n}$ with x in K_{i+1} and not in K_i. Call η a primitive l^{i+1}-th root of 1 in K_{i+1}, and put $\varepsilon = \eta^l$, $\zeta = \eta^{l^i}$; ε, ζ are roots of 1 of order l^i and l, respectively, and are in K_i. We have $K_{i+1} = K_i(\eta)$, $\eta^l = \varepsilon$, and ε is in K_i and is not 1; therefore K_{i+1} is cyclic of degree l over K_i, with a Galois group generated by the automorphism σ given by $\eta^{\sigma} = \zeta \eta$. Put $\theta = x^{\sigma} x^{-1}$; then $\theta \in K_{i+1}$ and $\theta^{l^n} = 1$, so that θ is a root of 1 of some order l^{ν} dividing l^n; therefore θ^{σ} must be of the form θ^s with $s \in \mathbf{Z}$; moreover, if $\nu \leqslant i$, θ is in K_i, so that we can take $s = 1$, while, if $\nu > i$, we have $\eta = \theta^r$ with some $r \in \mathbf{Z}$, hence $\eta^{\sigma} = \eta^s$, $\zeta = \eta^{s-1}$, and therefore $s \equiv 1 + l^i$ (l^{i+1}). By induction on h, one sees at once that $x^{\sigma^h} = x \theta^{1 + s + \cdots + s^{h-1}}$; for $h = l$, this gives $1 + s + \cdots + s^{l-1} \equiv 0$ (l^{ν}). If $\nu \leqslant i$, we have $s = 1$, so that the latter congruence implies $\nu \leqslant 1$. If $\nu > i$, we have $s = 1 + al^i$ with $a \equiv 1$ (l); if $l \neq 2$, or if $l = 2$ and $i \geqslant 2$, this implies $s^l = 1 + bl^{i+1}$ with $b \equiv 1$ (l), which shows that $(s^l - 1)/(s - 1)$ cannot be a multiple of l^2, hence also not of l^{ν}; as this contradicts what we have found above, we conclude that $\nu \leqslant 1$ except possibly for $l = 2$, $i = 1$. Therefore, except in that case, we can write $\theta = \zeta^t$ with $t \in \mathbf{Z}$; writing then $x' = \eta^{-t} x$, we have $x'^{\sigma} = x'$, so that x' is in K_i, and $a = x'^{l^n}$, which contradicts the definition of i. This proves that a is in $(K_1^{\times})^{l^n}$ if $l \neq 2$, and, for $l = 2$, it proves that, if it is not in $(K_1^{\times})^{2^n}$, it is in $(K_2^{\times})^{2^n}$. In the former case, write $a = y^{l^n}$ with $y \in K_1$. As the Galois group of K_1 over K is a subgroup of $(\mathbf{Z}/l\mathbf{Z})^{\times}$, the degree d of K_1 over K divides $l - 1$ and is prime to l; write $1 = de + l^n f$; we have $a^d = b^{l^n}$ with $b = N_{K_1/K}(y)$, hence $a = (a^f b^e)^{l^n}$. If $l = 2$, we have $K_1 = K$; if then a is not in $(K^{\times})^{2^n}$, we must have $K \neq K_2$, and we can apply what we have found above, with $i = 1$; if at the same time $K_2 \neq K_{m+1}$, the order 2^{ν} of θ cannot be a multiple of 2^{m+1}, so that it divides 2^m; putting then $b = x^{2^m}$, we have $b^{\sigma} = b$, so that b is in $K_1 = K$, and, if $n > m$, $a = b^{2^{n-m}}$. In the case $l = 2$, $K \neq K_2$, one can easily show, by using similar arguments, that $(K^{\times})^{2^n}$ is a subgroup of index 2 of $K^{\times} \cap (K_n^{\times})^{2^n}$, the latter group being generated by $(K^{\times})^{2^n}$ and $(1 + \omega)^{2^n}$ if ω is a generator of the group

of the roots u of 1 in K_2, of order dividing 2^n, such that $N_{K_2/K}(u)=1$. These facts will not be used here.

PROPOSITION 13. *Let P be a finite set of places of k, containing P_∞; put $H=\prod_{v\in P}k_v^\times$. Then $U_k\cap H$ is $k_{\infty+}^\times$ if k is of characteristic 0, and $\{1\}$ if it is of characteristic $p>1$.*

The latter assertion is obvious, since in that case $U_k=k^\times$; we may therefore assume that k is of characteristic 0. By corollary 1 of th. 5, $U_k\cap H$ is the same as the intersection of the groups $k^\times(k_A^\times)^N\cap H$ for all $N\geqslant 1$. An element of k_A^\times belongs to the latter group if and only if it can be written as ξz^N with $\xi\in k^\times$, $z=(z_v)\in k_A^\times$ and $\xi=z_v^{-N}$ for all v not in P. Take $N=l^n$, where l is a rational prime; let k' be the extension of k generated by a primitive N-th root of 1 in \bar{k}, and k'' the extension of k' generated by any root of $X^N=\xi$ in \bar{k}. Clearly, for all places w of k' which do not lie above a place $v\in P$, we have $\xi\in(k_w'^\times)^N$; therefore, by corollary 4 of th. 2, Chap. VII-5, we have $k''=k'$, so that ξ is in $(k'^\times)^N$. By lemma 9, this implies $\xi\in(k^\times)^N$ if $l\neq 2$. If $l=2$, call k_2 the extension of k generated by a primitive 4-th root of 1; if $k_2=k$, we have again $\xi\in(k^\times)^N$. If $k_2\neq k$, call 2^m the highest power of 2 dividing the order of the group of roots of 1 in k_2; then lemma 9 gives $\xi\in(k^\times)^{N'}$ with $N'=2^{-m}N$, provided $n>m$. Taking $N=2^{m+\mu}$ in the latter case, and otherwise $N=l^\mu$, we see that $k^\times(k_A^\times)^N\cap H$ is then contained in $(k_A^\times)^{l^\mu}\cap H$, which is the same as H^{l^μ}. This shows that $U_k\cap H$ is contained in H^{l^μ} for all primes l and all $\mu>0$. Take any integer $M>1$; for every prime l dividing M, let l^μ be the highest power of l dividing M; we can find integers $a(l)$ such that $1/M=\sum l^{-\mu}a(l)$. Take any $h\in U_k\cap H$, and, for each l, write $h=(h_l)^{l^\mu}$ with $h_l\in H$; then $h=h'^M$ with $h'=\prod(h_l)^{a(l)}$; therefore $U_k\cap H\subset H^M$ for all $M>1$. In corollary 1 of th. 3, Chap. XII-3, we have shown that the intersection of all the groups $(k_v^\times)^M$, for a given finite place v of k, is $\{1\}$; this same intersection is obviously C^\times if $k_v=C$, and R_+^\times if $k_v=R$. Therefore the intersection of all the groups H^M is $k_{\infty+}^\times$, so that $U_k\cap H$ is contained in $k_{\infty+}^\times$: as it obviously contains it, this completes our proof.

COROLLARY. *For every place v of k, $k_{v,ab}$ is generated over k_v by k_{ab}.*

This is trivial if $k_v=C$, and it is obvious if $k_v=R$, since then $k_{v,ab}$ is C and is generated by a primitive 4-th root of 1 in \bar{k}. Assume now that v is a finite place. The union $k_{v,0}$ of all unramified extensions of k_v is generated over k_v by roots of 1; therefore, if k' is the subfield of $k_{v,ab}$ generated over k_v by k_{ab}, it contains $k_{v,0}$. As in § 1, let \mathfrak{A}_v be the Galois group of $k_{v,ab}$ over k_v, and let ρ_v be the restriction morphism of \mathfrak{A}_v into \mathfrak{A}. An automorphism α of $k_{v,ab}$ over k_v induces the identity on k' if and only if it induces the identity on k_{ab}, i.e. if and only if $\rho_v(\alpha)$ is the identity.

Assume that this is so; then, as $k_{v,0} \subset k'$, α is in the Galois group of $k_{v,ab}$ over $k_{v,0}$, so that, by corollary 2 of th. 3, Chap. XII-3, it can be written as $\alpha = a_v(z)$ with $z \in r_v^\times$. Then, by prop. 2 of §1, we have $\rho_v(\alpha) = a(j_v(z))$, where j_v is the natural injection of k_v^\times into k_A^\times; if $\rho_v(\alpha)$ is the identity, $j_v(z)$ must be in U_k; taking for P, in proposition 13, a set containing v, we see now that α itself must then be the identity. This proves our corollary.

As an example for the above corollary, we may apply it to the case $k = \mathbf{Q}$; then, in combination with corollary 3 of th. 3, §4, it shows that, for every rational prime p, the maximal abelian extension of \mathbf{Q}_p, in an algebraic closure of \mathbf{Q}_p, is generated by all the roots of 1. This could also, of course, have been derived directly from the results of Chap. XII.

§ 9. The main theorems. The main results of classfield theory are either immediate consequences of those found above, or can be derived from them by following exactly the proofs given for the corresponding theorems in Chap. XII.

THEOREM 6. *If k is of characteristic 0, the canonical morphism a determines an isomorphism of k_A^\times/U_k onto the Galois group \mathfrak{A} of k_{ab} over k, U_k being the closure of $k^\times k_{\infty+}^\times$ in k_A^\times; if k is of characteristic $p > 1$, a determines a bijective morphism of k_A^\times/k^\times onto a dense subgroup of \mathfrak{A}, and an isomorphism of k_A^1/k^\times onto the Galois group \mathfrak{A}_0 of k_{ab} over the union k_0 of all constant-field extensions of k.*

The first assertion merely repeats part of prop. 1, Chap. XII-1, corollary 3 of th. 5, §8, being taken into account. The other assertions repeat part of corollary 2 of prop. 2, Chap. XII-1, and [II″] of Chap. XII-1, taking into account the fact that $U_k = k^\times$ and that \mathfrak{A}_0 has been determined in §1.

THEOREM 7. *Let k' be an extension of k of finite degree, contained in \bar{k}; put $L = k' \cap k_{ab}$. Then, for $z \in k_A^\times$, $a(z)$ induces the identity on L if and only if z is in $k^\times N_{k'/k}(k_A'^\times)$.*

The proof is identical to that of th. 4, Chap. XII-3, except that of course one must now make use of th. 5 of §8, instead of th. 3 of Chap. XII-3, and corollary 1 of th. 1, §1, instead of corollary 1 of th. 2, Chap. XII-2.

COROLLARY 1. *Assumptions and notations being as in theorem 7, call \mathfrak{B} the subgroup of \mathfrak{A} corresponding to L. Then:*

$$k^\times N_{L/k}(L_A^\times) = k^\times N_{k'/k}(k_A'^\times) = a^{-1}(\mathfrak{B}).$$

The latter equality is a restatement of theorem 7. Applying theorem 7 to $k' = L$, we get $k^\times N_{L/k}(L_A^\times) = a^{-1}(\mathfrak{B})$.

COROLLARY 2. *For every extension L of k of finite degree, contained in k_{ab}, call $\mathfrak{B}(L)$ the subgroup of \mathfrak{A} corresponding to L, and put $N(L) = = k^{\times} N_{L/k}(L_{\mathbf{A}}^{\times})$. Then $N(L) = a^{-1}(\mathfrak{B}(L))$; $\mathfrak{B}(L)$ is the closure of $a(N(L))$ in \mathfrak{A}; L consists of the elements of k_{ab}, invariant under $a(z)$ for all $z \in N(L)$, and a determines an isomorphism of $k_{\mathbf{A}}^{\times}/N(L)$ onto the Galois group $\mathfrak{A}/\mathfrak{B}(L)$ of L over k. Moreover, $L \to N(L)$ maps the subfields of k_{ab}, of finite degree over k, bijectively onto the open subgroups of $k_{\mathbf{A}}^{\times}$, of finite index in $k_{\mathbf{A}}^{\times}$ and containing k^{\times}.*

All this merely repeats prop. 3 of Chap. XII-1, the corollaries of th. 5, §8, being taken into account; one should notice here that, when k is of characteristic 0, the group $k_{\infty+}^{\times}$, being a product of finitely many factors isomorphic to \mathbf{R}_{+}^{\times} or to \mathbf{C}^{\times}, is generated by every neighborhood of 1 in that group, and is therefore contained in every open subgroup of $k_{\mathbf{A}}^{\times}$.

COROLLARY 3. *Notations being as in corollary 2, let Γ be the group of the characters of \mathfrak{A}, trivial on $\mathfrak{B}(L)$. Then the subgroup $N(L)$ of $k_{\mathbf{A}}^{\times}$ associated with L is the intersection of the kernels of the characters $\omega = \chi \circ a$ of $k_{\mathbf{A}}^{\times}$ for $\chi \in \Gamma$, and $\chi \to \chi \circ a$ is an isomorphism of Γ onto the group γ of the characters of $k_{\mathbf{A}}^{\times}$, trivial on $N(L)$.*

The first assertion is merely a restatement, in other terms, of the equality $N(L) = a^{-1}(\mathfrak{B}(L))$; similarly, the second one is a restatement of the fact that a determines an isomorphism of $k_{\mathbf{A}}^{\times}/N(L)$ onto $\mathfrak{A}/\mathfrak{B}(L)$.

COROLLARY 4. *Let χ be any character of \mathfrak{A}; then, if L is the cyclic extension of k attached to χ, the subgroup $N(L)$ associated with L is the kernel of the character $\omega = \chi \circ a$ of $k_{\mathbf{A}}^{\times}$.*

This is a special case of corollary 3, since here the group Γ of that corollary is the one generated by χ.

COROLLARY 5. *Let k and k' be as in theorem 7; let M be a subfield of k_{ab}, of finite degree over k, and call M' its compositum with k'. Let $U = = k^{\times} N_{M/k}(M_{\mathbf{A}}^{\times})$, $U' = k^{\times} N_{M'/k'}(M_{\mathbf{A}}^{\prime \times})$ be the open subgroups of $k_{\mathbf{A}}^{\times}$ and of $k_{\mathbf{A}}^{\prime \times}$ associated with the abelian extensions M of k, and M' of k', respectively, by corollary 2. Then $U' = N_{k'/k}^{-1}(U)$.*

The proof is identical to that of corollary 3 of th. 4, Chap. XII-3.

THEOREM 8. *Let k' be an extension of k of finite degree, contained in k_{sep}; let a, a' be the canonical morphisms of $k_{\mathbf{A}}^{\times}$ into \mathfrak{A}, and of $k_{\mathbf{A}}^{\prime \times}$ into the Galois group \mathfrak{A}' of k_{ab}' over k', respectively. Let t be the transfer homomorphism of \mathfrak{A} into \mathfrak{A}', and j the natural injection of $k_{\mathbf{A}}$ into $k_{\mathbf{A}}'^{\times}$. Then $t \circ a = a' \circ j$.*

The proof is identical to that of th. 6, Chap. XII-5, except that here, of course, one must use th. 7, instead of th. 4 of Chap. XII-3.

§ 10. Local behavior of abelian extensions. Let k be as above; let v be any place of k; as in § 1, we choose an algebraic closure K_v of k_v, containing the algebraic closure \bar{k} of k. If k' is any extension of k of finite degree, contained in \bar{k}, prop. 1 of Chap. III-1 shows that the sub-field of K_v generated by k' over k_v may be identified with the completion k'_w of k' at one of the places w lying above v. If k' is a Galois extension of k, with the Galois group \mathfrak{g}, we can apply corollary 4 of th. 4, Chap. III-4, as we have already done in similar cases on earlier occasions. This shows that k'_w is a Galois extension of k_v; if \mathfrak{h} is its Galois group over k_v, the restriction morphism of \mathfrak{h} into \mathfrak{g} is injective and may be used to identify \mathfrak{h} with a subgroup of \mathfrak{g}; then the completions of k' at the places of k' lying above v are in a one-to-one correspondence with the cosets of \mathfrak{h} in \mathfrak{g} and are all isomorphic to k'_w.

We now apply this to the case when k' is abelian over k. Then, by corollary 2 of th. 7, § 9, its Galois group \mathfrak{g} is isomorphic to k_A^\times/U with $U = N(k') = k^\times N_{k'/k}(k_A'^\times)$, U being then an open subgroup of k_A^\times of finite index. More precisely, if \mathfrak{B} is the subgroup of the Galois group \mathfrak{A} of k_{ab} over k, corresponding to k', the canonical morphism \mathfrak{a} determines an isomorphism of k_A^\times/U onto $\mathfrak{g} = \mathfrak{A}/\mathfrak{B}$. On the other hand, if k_v, k'_w are as above, k'_w is an abelian extension of k_v, with which corollary 2 of th. 4, Chap. XII-3, associates the open subgroup $U_v = N_{k'_w/k_v}(k'_w{}^\times)$ of k_v^\times. Call \mathfrak{A}_v, as before, the Galois group of $k_{v,ab}$ over k_v; call \mathfrak{B}_v the subgroup of \mathfrak{A}_v corresponding to k'_w; then the same corollary shows that the canonical morphism \mathfrak{a}_v of k_v^\times into \mathfrak{A}_v determines an isomorphism of k_v^\times/U_v onto $\mathfrak{h} = \mathfrak{A}_v/\mathfrak{B}_v$. The relation between these various groups is given by the following:

PROPOSITION 14. *Assumptions and notations being as above, the subgroup U_v of k_v^\times, associated with k'_w, is given by $U_v = k_v^\times \cap U$. If \mathfrak{g} is identified with k_A^\times/U by means of \mathfrak{a}, and \mathfrak{h} with k_v/U_v, by means of \mathfrak{a}_v, the restriction morphism of \mathfrak{h} into \mathfrak{g} is the same as the morphism of k_v^\times/U_v into k_A^\times/U determined by the natural injection j_v of k_v^\times into k_A^\times, and the places of k' which lie above v are in a one-to-one correspondence with the cosets of $k_v^\times U$ in k_A^\times.*

Take any $z_v \in k_v^\times$, and put $\alpha = \mathfrak{a}_v(z_v)$. By prop. 2 of § 1, the automorphism of k_{ab} induced by α is $\rho_v(\alpha) = \mathfrak{a}(z)$ with $z = j_v(z_v)$. As k'_w is generated by k' over k_v, α induces the identity on k'_w if and only if $\rho_v(\alpha)$ induces the identity on k'; in view of corollary 2 of th. 4, Chap. XII-3, and of corollary 2 of th. 7, § 9, this amounts to saying that z_v is in U_v if and only if $j_v(z_v)$ is

in U, which we express by $U_v = k_v^\times \cap U$. The second assertion in our proposition follows at once from the same facts; they also imply that the image of \mathfrak{h} in \mathfrak{g} can be identified with that of k_v^\times in k_A^\times/U, which is $k_v^\times U/U$, and that $\mathfrak{g}/\mathfrak{h}$ can be identified with $k_A^\times/k_v^\times U$. As we have recalled above, the places of k' above v correspond bijectively to the cosets of \mathfrak{h} in \mathfrak{g}, hence also to those of $k_v^\times U$ in k_A^\times; this completes the proof. Our proposition and its proof remain valid when v is an infinite place, since theorem 4 of Chap. XII-3, and its corollaries, remain valid for **R** and **C**, as has been observed at the time. The relations between the various groups and morphisms considered above are illustrated by the following diagram.

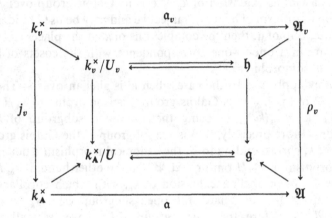

COROLLARY 1. *Let γ be the group of the characters of k_A^\times, trivial on U; let γ_v be the group of the characters of k_v^\times, trivial on U_v. Then the mapping which, to every $\omega \in \gamma$, assigns the character ω_v induced by ω on k_v^\times, is a surjective morphism of γ onto γ_v, and the order of its kernel is equal to the number of places of k' lying above v.*

Clearly $\omega \to \omega_v$ determines a morphism of γ into γ_v. Every character of k_v^\times, trivial on U_v, can be uniquely extended to one of $k_v^\times U$, trivial on U, and this can be extended to one of k_A^\times, which then belongs to γ; therefore the morphism in question is surjective. Its kernel consists of the characters of k_A^\times, trivial on $k_v^\times U$; this is the dual group to $k_A^\times/k_v^\times U$; in view of the last assertion in proposition 14, its order is therefore as stated in our corollary.

COROLLARY 2. *Assumptions being as above, assume also that v is a finite place of k. Then the modular degree f, and the order of ramification*

e, of k'_w over k_v are given by

$$f=[k_v^\times:r_v^\times U_v]=[k_v^\times U:r_v^\times U], \quad e=[r_v^\times U_v:U_v]=[r_v^\times U:U].$$

By the corollary of prop. 6, Chap. XII-2, and corollary 2 of th. 4, Chap. XII-3, the maximal unramified extension of k_v contained in k'_w is the one associated with the subgroup $r_v^\times U_v$ of k_v^\times; the first part of our corollary follows from this at once; the second part is an immediate consequence of the first.

COROLLARY 3. *Assumptions being as in corollary 2, k'_w is unramified over k_v if and only if $U \supset r_v^\times$; when that is so, the automorphism of k' over k, induced by the Frobenius automorphism of k'_w over k_v, is the image in $\mathfrak{g}=k_A^\times/U$ of any prime element π_v of k_v, and it is an element of \mathfrak{g} of order f.*

To say that k'_w is unramified over k_v is to say that $e=1$, so that the first assertion is a special case of corollary 2. The second one follows at once from proposition 14, combined with corollary 4 of th. 1, Chap. XII-2, which says that $a_v(\pi_v)$ is here the Frobenius automorphism of k'_w over k_v.

Notations being as in corollaries 2 and 3, we know from the corollary of prop. 3, Chap. VIII-1, that k'_w is unramified over k_v if and only if its different over k_v is r'_w. In view of the definitions of the different and of the discriminant in Chap. VIII-4, and of the fact that the completions of k', at the places of k' lying above v, are all isomorphic to k'_w, it amounts to the same to say that k'_w is unramified over k_v if and only if v does not occur in the discriminant of k' over k. By corollary 3 of prop. 14, this is so if and only if $U \supset r_v^\times$. This qualitative result can be refined into a more precise one, as follows:

THEOREM 9. *Let k' be an extension of k of finite degree, contained in k_{ab}; let $U = k^\times N_{k'/k}(k'^\times_A)$ be the subgroup of k_A^\times associated with k', and call γ the group of the characters of k_A^\times, trivial on U. For each $\omega \in \gamma$, call $\mathfrak{f}(\omega)$ the conductor of ω. Then the discriminant \mathfrak{D} of k' over k is given by $\mathfrak{D} = \prod_{\omega \in \gamma} \mathfrak{f}(\omega)$, or by $\mathfrak{D} = \sum_{\omega \in \gamma} \mathfrak{f}(\omega)$, according as k is of characteristic 0 or not.*

Let notations be the same as in corollary 2 of prop. 14; let p_v be the maximal ideal in the maximal compact subring r_v of k_v; call p_v^δ the discriminant of k'_w over k_v, and v the number of places of k' lying above v. As the completions of k' at these places are all isomorphic to k'_w, they all make the same contribution to the discriminant \mathfrak{D}, so that their total contribution is $p_v^{\delta v}$ (resp. $\delta v \cdot v$). Let γ_v be defined as in corollary 1 of prop. 14; call ω'_i, for $1 \leqslant i \leqslant d$, the distinct elements of γ_v, and, for

each i, call $p_v^{f(i)}$ the conductor of ω_i'; by corollary 2 of th. 5, Chap. XII-4, we have $\delta = \sum f(i)$. By corollary 1 of prop. 14, each ω_i' is induced on k_v^\times by exactly v characters $\omega \in \gamma$. Our assertion is now obvious.

THEOREM 10. *Assumptions and notations being as in theorem* 9, *the Dedekind zeta-function of* k' *is given by* $\zeta_{k'}(s) = \prod_{\omega \in \gamma} L(s, \omega)$.

It is enough to prove this for $\mathrm{Re}(s) > 1$, when the infinite products for these functions are absolutely convergent; and then it is enough to show that, for each finite place v of k, the contribution of the places of k' above v to $\zeta_{k'}(s)$ is equal to the product of the contributions of v to the products $L(s, \omega)$. If f is as in corollary 2 of prop. 14, the contribution of w to the product $\zeta_{k'}(s)$ is $(1 - q_v^{-fs})^{-1}$; that is also the contribution of each one of the places of k' above v, so that, if v is their number, their total contribution is $(1 - q_v^{-fs})^{-v}$. On the other hand, for $\omega \in \gamma$, the contribution of v to $L(s, \omega)$ is 1 unless ω_v is unramified, and $(1 - \omega_v(\pi_v)q_v^{-s})^{-1}$ if it is unramified. In view of corollary 1 of prop. 14, their product is equal to $\prod(1 - \omega'(\pi_v)q_v^{-s})^{-v}$, where the latter product is taken over all the distinct characters ω' of k_v^\times, trivial on U_v and on r_v^\times, i.e. trivial on $r_v^\times U_v$. By corollary 2 of prop. 14, the group $k_v^\times / r_v^\times U_v$ is of order f; clearly it is generated by the image of π_v in it, hence cyclic; therefore there are f characters ω', and the values they take at π_v are the f-th roots of 1 in \mathbf{C}. This implies that the product $\prod(1 - \omega'(\pi_v)t)$, for every $t \in \mathbf{C}$, is equal to $1 - t^f$, which completes our proof.

COROLLARY. *Assumptions and notations being as in theorems* 9 *and* 10, *assume also that* k *is an algebraic number-field. Then* $Z_{k'}(s) = \pi^{n\rho/2} \prod_{\omega \in \gamma} \Lambda(s, \omega)$, *where* n *is the degree of* k' *over* k, *and* ρ *is the number of real places of* k *such that the places of* k' *above them are imaginary.*

Here $Z_{k'}(s)$ and $\Lambda(s, \omega)$ are the functions defined in theorem 3 of Chap. VII-6, and in theorem 5 of Chap. VII-7, respectively. In view of theorem 10, what we have to show is that each infinite place v of k contributes the same G-factors to both sides of the formula in our corollary. Define v as above; then the total contribution to $Z_{k'}(s)$ of the v places of k' above v is $G_1(s)^v$ or $G_2(s)^v$, according as w is real or not. The contribution of v to $\Lambda(s, \omega)$ is $G_1(s + s_v)$ or $G_2(s + s_v)$ according as v is real or not, s_v depending upon ω_v in the manner described in Chap. VII-7. Here ω_v has to be trivial on U_v, which, being an open subgroup of k_v^\times, is \mathbf{C}^\times if $k_v = \mathbf{C}$, and either \mathbf{R}^\times or \mathbf{R}_+^\times if $k_v = \mathbf{R}$. If ω_v is trivial on k_v^\times, we must put $s_v = 0$; if not, we must have $k_v = \mathbf{R}$, $U_v = \mathbf{R}_+^\times$ and $\omega_v(x) = x^{-1}|x|$, hence $s_v = 1$. As the degree of k_w' over k_v is $[k_v^\times : U_v]$, it is 2 in the latter case, and otherwise 1. Taking now corollary 1 of prop. 14 into account,

we see that the contribution of v to $\prod \varLambda(s,\omega)$ is $G_1(s)^\nu$ if v and w are real, $G_2(s)^\nu$ if they are both imaginary, and $G_1(s)^\nu G_1(s+1)^\nu$ if v is real and w imaginary. In the latter case, we have $\nu = n/2$, e.g. by corollary 1 of th. 4, Chap. III-4. Our corollary follows now at once from these facts and from the identity $G_2(s) = \pi \, G_1(s) G_1(s+1)$, which is the same as the identity between gamma functions already quoted at the end of Chap. X.

§ 11. "Classical" classfield theory.

§ 11. "Classical" classfield theory. The reinterpretation of our results, in the traditional language of this theory, depends upon the following facts:

(a) Let \mathfrak{U} be the set of all the open subgroups of $k_\mathbf{A}^\times$, containing k^\times; let \mathfrak{U}' be the set of those which are of finite index in $k_\mathbf{A}^\times$, and \mathfrak{U}'' the set of those which are contained in $k_\mathbf{A}^1$ and of finite index in $k_\mathbf{A}^1$. Lemma 1 of Chap. XII-1 shows that $\mathfrak{U} = \mathfrak{U}'$ and $\mathfrak{U}'' = \emptyset$ if k is an algebraic number-field, and that $\mathfrak{U} = \mathfrak{U}' \cup \mathfrak{U}''$ if k is of characteristic $p > 1$.

(b) Let \mathfrak{R} be the set of all the fields between k and k_{ab}, of finite degree over k; when k is of characteristic $p > 1$, let \mathfrak{R}_0 be the set of all the fields between k_0 and k_{ab}, of finite degree over k_0. Then corollary 2 of th. 7, § 9, defines a one-to-one correspondence between \mathfrak{U}' and \mathfrak{R}, while, by the last assertion in th. 6, § 9, and Galois theory, there is a one-to-one correspondence between \mathfrak{U}'' and \mathfrak{R}_0 when k is of characteristic $p > 1$.

(c) As the open subgroups of any group are the kernels of its morphisms onto discrete groups, we may regard the open subgroups of $k_\mathbf{A}^\times$ in (a) as kernels of such morphisms, and describe these morphisms in terms of morphisms of the groups $I(P)$, $D(P)$, in the manner explained in Chap. VII-8. In order to reinterpret the results of Chap. VII-8 more conveniently for our present purposes, we will modify its notations as follows.

As in Chap. VII-8, when P is any finite set of places of k, containing P_∞, we write G_P for the group of the ideles (z_v) of k such that $z_v = 1$ for all $v \in P$, and G_P' for the group of the ideles (z_v) such that $z_v = 1$ for $v \in P$, and $z_v \in r_v^\times$, i.e. $|z_v|_v = 1$, for v not in P. We will now write L_P for the free group generated by the places v not in P, that group being written multiplicatively; this may be identified in an obvious manner with the group $I(P)$ or $D(P)$ of Chap. VII-8, according to the characteristic of k. We write l_P for the morphism of G_P onto L_P, with the kernel G_P', given by $(z_v) \to \prod_{v \notin P} v^{r(v)}$ with $r(v) = \mathrm{ord}_v(z_v)$; moreover, for every $\xi \in k^\times$ such that $\xi \in r_v^\times$ for all finite places $v \in P$, we write $\mathrm{pr}(\xi) = \prod_{v \notin P} v^{\rho(v)}$ with $\rho(v) = \mathrm{ord}_v(\xi)$ for v not in P.

DEFINITION 1. *A subgroup J of L_P will be called a congruence group if one can find, for every $v \in P$, an open subgroup g_v of k_v^\times, contained in r_v^\times when v is finite, such that $\mathrm{pr}(\xi) \in J$ for every $\xi \in \bigcap (k^\times \cap g_v)$; the group $g = \prod_{v \in P} g_v$ will then be called a defining group for J.*

Clearly it would make no difference in this definition if the groups g_v were restricted to be of the form $1 + p_v^m$ with $m \geqslant 1$ for every finite $v \in P$.

PROPOSITION 15. *Notations being as above, call $\mathfrak{U}(P)$ the set of the open subgroups of k_A^\times containing k^\times and containing r_v^\times for all v not in P. Then, for each $U \in \mathfrak{U}(P)$, the formula $U \cap G_P = l_P^{-1}(J)$ defines a congruence subgroup $J = J(U, P)$ of L_P; a group $g = \prod g_v$, where the g_v are as in definition 1, is a defining group for J if and only if it is contained in U; U is the closure of $k^\times l_P^{-1}(J)$ in k_A^\times, and the canonical homomorphism of k_A^\times onto k_A^\times/U determines an isomorphism of L_P/J onto k_A^\times/U. Moreover, $U \to J(U, P)$ maps $\mathfrak{U}(P)$ bijectively onto the set of all congruence subgroups of L_P.*

Take $U \in \mathfrak{U}(P)$; call ω the canonical homomorphism of k_A^\times onto the discrete group $\Gamma = k_A^\times/U$; as the morphism of G_P into Γ induced by ω is trivial on G_P', it can be written as $\varphi \circ l_P$, where φ is a morphism of L_P into Γ; clearly the kernel of φ is J. By the corollary of prop. 17, Chap. VII-8, this implies that J is a congruence subgroup of L_P; then, by prop. 17, Chap. VII-8, ω is the unique extension of $\varphi \circ l_P$ to k_A^\times, trivial on k^\times, and it is trivial on g if g is a group of definition for J, so that $g \subset U$ when that is so. By prop. 15 of Chap. VII-8, $k^\times G_P$ is dense in k_A^\times; this implies that $\varphi \circ l_P$ maps G_P surjectively onto Γ, so that $\varphi(L_P) = \Gamma$, and also that $U \cap (k^\times G_P)$ is dense in U; this is the same as $k^\times \cdot (U \cap G_P)$, i.e. $k^\times l_P^{-1}(J)$. Conversely, let J be any congruence subgroup of L_P, and call φ the canonical homomorphism of L_P onto the discrete group $\Gamma = L_P/J$; again by prop. 17 of Chap. VII-8, $\varphi \circ l_P$ can be uniquely extended to a morphism ω of k_A^\times into Γ, trivial on k^\times; if then U is the kernel of ω, we have $U \in \mathfrak{U}(P)$ and $J = J(U, P)$. Finally, if the groups g_v are as in def. 1, and if $g = \prod g_v$, every $\xi \in \bigcap (k^\times \cap g_v)$ is in $g \times G_P$, so that, if $g \subset U$, the projection of ξ onto G_P is in $U \cap G_P$, and the image of that projection in L_P, which is the same as $\mathrm{pr}(\xi)$, is in J; thus g is then a defining group for J.

COROLLARY 1. *Notations being as in proposition 15, let P' be a finite set of places of k, containing P. Then, if J is any congruence subgroup of L_P, $J' = J \cap L_{P'}$ is a congruence subgroup of $L_{P'}$; if $J = J(U, P)$ with $U \in \mathfrak{U}(P)$, $J' = J(U, P')$.*

Here it is understood that $L_{P'}$ is to be regarded as a subgroup of L_P, in the obvious manner, for $P' \supset P$. Clearly, then, $\mathfrak{U}(P) \subset \mathfrak{U}(P')$. If now $U \in \mathfrak{U}(P)$ and $U \cap G_P = l_P^{-1}(J)$, it is obvious that $U \cap G_{P'} = l_{P'}^{-1}(J')$ with $J' = J \cap L_{P'}$; our corollary follows at once from this and proposition 15.

COROLLARY 2. *Let* P, P' *be two finite sets of places of* k, *containing* P_∞; *let* J, J' *be congruence subgroups of* L_P *and of* $L_{P'}$, *respectively. Then* $k^\times l_P^{-1}(J)$ *and* $k^\times l_{P'}^{-1}(J')$ *have the same closure* U *in* k_A^\times *if and only if there is a finite set* P'', *containing* P *and* P', *such that* $J \cap L_{P''} = J' \cap L_{P''}$; *when that is so, the same is true for all finite sets* P'' *containing* P *and* P', *and* U *is in* $\mathfrak{U}(P \cap P')$.

Call U, U' the closures of the two sets in question; then, by proposition 15, $J = J(U, P)$ and $J' = J(U', P')$. If $U = U'$, it follows at once from proposition 15 that U is in $\mathfrak{U}(P \cap P')$; therefore, by corollary 1, if $P'' \supset P \cup P'$, $J \cap L_{P''}$ and $J' \cap L_{P''}$ are both the same as $J(U, P'')$. On the other hand, if there is P'' and J'' such that $P'' \supset P \cup P'$ and $J'' = J \cap L_{P''} = = J' \cap L_{P''}$, corollary 1 gives $J'' = J(U, P'') = J(U', P'')$, hence $U = U'$ by proposition 15.

When two congruence groups J, J' are as in corollary 2, one says that they are *equivalent*. Since every open subgroup U of k_A^\times, containing k^\times, belongs to $\mathfrak{U}(P)$ when P is suitably chosen, it is now clear that there is a one-to-one correspondence between the set \mathfrak{U} of all such groups and the set of equivalence classes of congruence groups. Therefore the one-to-one correspondence between \mathfrak{U} and \mathfrak{R} (resp. $\mathfrak{R} \cup \mathfrak{R}_0$) mentioned above under (b) determines a similar correspondence between \mathfrak{R} (resp. $\mathfrak{R} \cup \mathfrak{R}_0$) and the equivalence classes of congruence groups. This will now be described more in detail.

To begin with, it is obvious, from proposition 15 and its corollaries, that, when an equivalence class of congruence groups is given, there is a smallest set P such that this class contains a congruence subgroup J of L_P; in fact, if U is the open subgroup of k_A^\times corresponding to that class, P consists of the infinite places, and of the finite places v such that r_v^\times is not contained in U; if we write $U_v = U \cap k_v^\times$ for all v, this is the same as to say that r_v^\times is not contained in U_v. Similarly, there is then a largest defining group for J; this is $\prod_{v \in P} g_v$, where $g_v = U_v$ for every infinite place, and $g_v = U_v \cap r_v^\times$ for every finite $v \in P$. When one considers only defining groups for which g_v is of the form $1 + p_v^m$ with $m \geqslant 1$ when v is finite, one must then take, for each such $v \in P$, the smallest integer $m(v) \geqslant 1$ such that $1 + p_v^{m(v)}$ is contained in U_v. If k is of characteristic $p > 1$, the divisor $\sum m(v) \cdot v$ is then called "the conductor" of U and of every congruence group equivalent to J. If k is of characteristic 0, one puts $m(v) = 0$ or 1, for each real place v of k, according as U_v is \mathbf{R}^\times or \mathbf{R}_+^\times; one puts $m(v) = 0$ for all imaginary places v of k; attaching then a symbol p_v, called an "infinite prime", to each infinite place v of k, one calls the symbol $\prod_{v \in P} p_v^{m(v)}$ "the conductor" of U, of J, and of the congruence groups equivalent to J.

In the case of characteristic $p > 1$, it is obvious that a congruence subgroup J of L corresponds to an open subgroup U of k_A^1 if and only if it consists of divisors of degree 0 when L_P is identified with the group $D(P)$ of divisors prime to P. From now on, this case will be excluded; in other words, when the characteristic is not 0, we consider exclusively open subgroups of k_A^\times of finite index in k_A^\times, abelian extensions of k of finite degree, and congruence groups which contain at least one divisor of degree $\neq 0$. This being understood, we can make use of prop. 14 of § 10 and its corollaries. In particular, if k' is the abelian extension of k corresponding to the open subgroup U of k_A^\times, corollary 4 of that proposition shows that U contains r_v^\times if and only if k'_w is unramified over k_v for all w above v, i.e. if and only if v does not occur in the discriminant \mathfrak{D} of k' over k. We will write Δ for the set consisting of the infinite places of k and of those occurring in the discriminant \mathfrak{D}; then there is a congruence subgroup J of L_P, corresponding to U, if and only if $P \supset \Delta$. As to the conductor of U, if we leave aside the infinite places, it is, in an obvious sense, $\sup_{\omega \in \gamma} (\mathfrak{f}(\omega))$ if notations are as in th. 9 of § 10; as to the infinite places, the proof of the corollary of th. 10, § 10, shows that such a place occurs in the conductor if and only if it is real and the places of k' lying above it are imaginary.

Before discussing the relation between the congruence groups associated with U and the Frobenius automorphisms, we introduce some definitions, valid for an arbitrary Galois extension k' of k of finite degree. Call \mathfrak{g} the Galois group of k' over k; let v be any place of k, and w a place of k' lying above v. By corollary 4 of th. 4, Chap. III-4, we can identify the Galois group \mathfrak{h} of k'_w over k_v with a subgroup of \mathfrak{g} by means of the restriction morphism of \mathfrak{h} into \mathfrak{g}. If v is a finite place, and k'_w is unramified over k_v, \mathfrak{h} is cyclic and generated by the Frobenius automorphism φ_w of k'_w over k_v; after \mathfrak{h} has been identified with its image in \mathfrak{g}, φ_w may be regarded as an element of \mathfrak{g}; this is called *the Frobenius automorphism of k' over k at w*. If w' is another place of k' above v, the same corollary shows that there is a k_v-linear isomorphism of k'_w onto $k'_{w'}$, determined by an automorphism σ of k' over k; then the Frobenius automorphism of k' over k at w' is $\sigma^{-1} \varphi_w \sigma$. Clearly φ_w is the identity if and only if v splits fully in k'. In particular, let k, k' be algebraic number-fields; let $\mathfrak{r}, \mathfrak{r}'$ be their maximal orders; let $\mathfrak{p}_v, \mathfrak{p}'_w$ be the prime ideals, in \mathfrak{r} and in \mathfrak{r}' respectively, corresponding to v and to w; then $\mathfrak{r}/\mathfrak{p}_v$, $\mathfrak{r}'/\mathfrak{p}'_w$ are finite fields, with $q = q_v$ and $q' = q'_w$ elements, respectively, and φ_w is the automorphism of k' over k which determines on $\mathfrak{r}'/\mathfrak{p}'_w$ the automorphism $x \to x^q$. This may also be defined as the automorphism φ of k' over k for which $\zeta^\varphi \equiv \zeta^q \ (\mathfrak{p}'_w)$ for every $\zeta \in \mathfrak{r}'$.

If, in addition to the above assumptions, we also assume k' to be abelian over k, i.e. \mathfrak{g} to be commutative, φ_w is the same for all the places w

above k; in this case, the only one with which we are concerned here, φ_w is called *the Frobenius automorphism of k' over k at v*; we will denote it by $(k'/k|v)$, or $(k'|v)$ when there is no risk of confusion. We may now reinterpret corollary 3 of prop. 14, § 10, as follows. As in that corollary, call U the open subgroup of k_A^\times associated with k', and identify the Galois group \mathfrak{g} of k' over k with k_A^\times/U by means of the canonical morphism. Take $P \supset \varDelta$, with \varDelta defined as above. The canonical homomorphism of k_A^\times onto $\mathfrak{g} = k_A^\times/U$ is trivial on r_v^\times for every v not in \varDelta, so that it induces on G_P a morphism of G_P into \mathfrak{g}, trivial on G'_P, which determines a morphism φ of $L_P = G_P/G'_P$ into \mathfrak{g}. Corollary 3 of prop. 14, § 10, says now that, for every v not in P, $\varphi(v)$ is the Frobenius automorphism $\varphi_v = (k'|v)$ of k' over k at v, as defined above. This morphism φ of L_P into \mathfrak{g}, defined for $P \supset \varDelta$, will be denoted by $\mathfrak{m} \to (k'/k|\mathfrak{m})$; one writes $(k'|\mathfrak{m})$ instead of $(k'/k|\mathfrak{m})$ when there is no risk of confusion, and calls this "the Artin symbol". It may be characterized as the morphism of L_P (or, what amounts to the same, of the group of ideals $I(P)$, or of the group of divisors $D(P)$, according to the characteristic) into \mathfrak{g} which maps every place of k, not in P, onto the Frobenius automorphism of k' over k at that place. In view of prop. 15, we have thus proved that this morphism is surjective and that its kernel $J = J(U, P)$ is a congruence subgroup of L_P. When one takes for P all the finite sets of places containing \varDelta, the kernels $J(U, P)$ make up an equivalence class of congruence groups; they are all contained in $J(k') = J(U, \varDelta)$.

The above results show also that a finite place v of k splits fully in k' if and only if it belongs to $J(k')$. It follows now from prop. 15 of Chap. VIII-5 that, if k'' is a separable extension of k contained in \bar{k}, and if almost all the places of k belonging to $J(k')$ split fully in k'', k'' is contained in k'. Obviously this implies that there are infinitely many places of k belonging to $J(k')$; it will be seen in § 12 that the same is true for all the cosets of $J(k')$ in L_\varDelta. The corollary of prop. 15, Chap. VIII-5, shows also that, if k'' is a Galois extension of k, it contains k' if and only if almost all the places of k which split fully in k'' are in $J(k')$. From this, it follows that, if k' and k'' are two abelian extensions of k contained in \bar{k}, k'' contains k' if and only if there is a set P for which $J(k'') \cap L_P$ is contained in $J(k') \cap L_P$; this may also be considered as a consequence of the results of § 9, combined with prop. 15 of this §. In particular, k' is uniquely determined by the equivalence class of congruence groups determined by $J(k')$; this, too, is an immediate consequence of the results of § 9 and of prop. 15 of this §. Traditionally, one says that k' is "the classfield" for that class of congruence groups or for any group belonging to that class.

The above characterization of the class of congruence groups for which k' is "the classfield" is based solely on the "Artin symbol"; another

one will now be derived from the fact that $U=k^\times N_{k'/k}(k_A'^\times)$. More generally, if we take for k' any extension of k of finite degree, th. 7 of § 9, and its corollaries, show that the group $U=k^\times N_{k'/k}(k_A'^\times)$ is the open subgroup of finite index of k_A^\times associated with the maximal abelian extension L of k, contained in k'. Take any finite set P of places of k, containing P_∞; for each $v\in P$, take an open subgroup g_v of k_v^\times, contained in r_v^\times when v is finite; put $g=\prod_{v\in P} g_v$, $U_g=k^\times g\,G_P'$ and $J_g=J(U_g,P)$; then, in the notation of def. 1, J_g is the subgroup of L_P consisting of the elements pr(ξ) for $\xi\in\bigcap (k^\times\cap g_v)$. As $U_g U$ contains G_P', it determines a congruence group $J=J(U_g U,P)$, given by $l_P^{-1}(J)=U_g U\cap G_P$. Call H_P the group of the ideles (z_w') of k' such that $z_w'=1$ for every place w of k' lying above a place $v\in P$, and H_P' the group of the ideles (z_w') of k' such that $z_w'=1$ when w lies above a place $v\in P$, and $|z_w'|_w=1$ otherwise; then $L_P'=H_P/H_P'$ is the free group generated by the places of k' which do not lie above P. As $N_{k'/k}$ maps H_P into G_P and H_P' into G_P', it determines a morphism \mathfrak{N} of L_P' into L_P, which is the same as the morphism $\mathfrak{N}_{k'/k}$ (resp. $\mathfrak{S}_{k'/k}$) of Chap. VIII-4 when L_P', L_P are interpreted as groups of ideals (resp. of divisors) of k' and of k. By prop. 15 of Chap. VII-8, $k'^\times H_P$ is dense in $k_A'^\times$, so that $k^\times N_{k'/k}(H_P)$ is dense in U. As U_g is open in k_A^\times, this implies that we have

$$U_g U=k^\times g\,G_P'\cdot N_{k'/k}(H_P).$$

From this, one concludes immediately that J is the subgroup of L_P generated by J_g and $\mathfrak{N}(L_P')$. Call k'' the classfield for the congruence group J; this is the abelian extension of k associated with the open subgroup $U_g U$ of k_A^\times, so that it is contained in the abelian extension L of k associated with U. Call n, n_0 the degrees of k' and of L, respectively, over k; it is now clear that the index of J in L_P, which is equal to that of $U_g U$ in k_A^\times and to the degree of k'' over k, is $\leq n_0$, and that it is equal to n_0 if and only if $U_g\subset U$, hence $k''=L$; this will be the case when P is taken large enough, and g small enough. We see at the same time that the index of J in L_P is always $\leq n$, and that it is equal to n if and only if k' is abelian over k and is the classfield for J. In other words, when a congruence subgroup J of L_P is given, an extension k' of k of finite degree is abelian and is the classfield for J if and only if J contains $\mathfrak{N}(L_P')$ and has an index in L_P equal to the degree of k' over k.

Finally, we can reinterpret corollary 5 of th. 7, § 9, as follows. As above, let k' be an extension of k of finite degree. Let M be an abelian extension of k contained in some extension of k', and call M' the compositum of M and k'. Assume that M is the classfield for a congruence subgroup J of L_P. Let v be a place of k, w a place of k' above v, u' a place of M' above

w, and u the place of M below u'; $M'_{u'}$ is the compositum of k'_w and M', hence of k'_w and M, hence of k'_w and M_u. If v is not in P, M_u is unramified over k_v; this implies that $M'_{u'}$ is then unramified over k'_w. Therefore M' is the classfield for some congruence subgroup J' of L'_P. Now let U, U' be the open subgroups of k_A^\times, $k_A'^\times$, respectively associated with M and with M'; by corollary 5 of th. 7, § 9, $U' = N_{k'/k}^{-1}(U)$. By prop. 15, J, J' can be defined by $l_P^{-1}(J) = U \cap G_P$ and by the similar formula for J', U'; therefore an element \mathfrak{m}' of L'_P is in J' if and only if it is the image of an element z' of H_P such that $z' \in U'$, i.e. $N_{k'/k}(z') \in U$; as $N_{k'/k}$ maps H_P into G_P, this is equivalent to $N_{k'/k}(z') \in U \cap G_P$, hence to $\mathfrak{N}(\mathfrak{m}') \in J$. Therefore we have $J' = \mathfrak{N}^{-1}(J)$.

As an illustration, we will now apply the above considerations to the case $k = \mathbf{Q}$, which has been treated from another point of view in § 4. Take $k' = \mathbf{Q}(\varepsilon)$, where ε is a primitive m-th root of 1; as before, identify its Galois group \mathfrak{g} with $(\mathbf{Z}/m\mathbf{Z})^\times$ by assigning to the automorphism $\varepsilon \to \varepsilon^x$, with $x \in \mathbf{Z}$, $(x, m) = 1$, the image of x in $(\mathbf{Z}/m\mathbf{Z})^\times$. As we have observed before, it is obvious that, for every rational prime p, not dividing m, and for every place w of k' above p, k'_w is unramified over \mathbf{Q}_p, and that the Frobenius automorphism of k' over \mathbf{Q} at p is the one given by $\varepsilon \to \varepsilon^p$, i.e. the image of p in $(\mathbf{Z}/m\mathbf{Z})^\times$. Consequently, only primes dividing m can occur in the discriminant of k' over \mathbf{Q}, and k' is the classfield for some congruence subgroup J of the group L_m of the fractional ideals of \mathbf{Q}, prime to m; L_m can be identified in an obvious manner with the group of the fractions $r = a/b$, where a, b are two integers > 0, both prime to m. Moreover, the Artin symbol $r \to (k'/\mathbf{Q}|r)$ is the morphism of L_m into $(\mathbf{Z}/m\mathbf{Z})^\times$ which maps every prime p, not dividing m, onto its image in $(\mathbf{Z}/m\mathbf{Z})^\times$; clearly this maps every integer $a > 0$, prime to m, onto its image in $(\mathbf{Z}/m\mathbf{Z})^\times$, and its kernel J consists of the elements a/b of L_m for which $a \equiv b \ (m)$. It can easily be verified that the "conductor" for this group J is 1 if $m = 1$ or 2, that it is $\mathfrak{p}_\infty(m/2)$ if m is even and $m/2$ is odd, and that it is $\mathfrak{p}_\infty m$ in all other cases. Except in the trivial cases $m = 1$ or 2, when $k' = \mathbf{Q}$, one may express this by saying that the conductor is $\mathfrak{p}_\infty m'$, where m' is the smallest integer such that $\mathbf{Q}(\varepsilon)$ is generated over \mathbf{Q} by a primitive m'-th root of 1. As we have seen, this implies that the primes occurring in the discriminant of $\mathbf{Q}(\varepsilon)$ over \mathbf{Q} are those which divide m'; it would be easy now to compute that discriminant itself, by means of th. 9 of § 10. It is also a consequence of what we have seen above that, if k is any algebraic number-field, and ε is again a primitive m-th root of 1, $k(\varepsilon)$ is the classfield for the congruence subgroup J' of the group L'_m of fractional ideals of k, prime to m, consisting of the fractional ideals \mathfrak{m} such that $\mathfrak{N}(\mathfrak{m}) \in J$, where J is as defined above.

§ 12. **"Coronidis loco".** The results of § 10 give the answer to a question which could not be settled in Chap. VII-5.

THEOREM 11. *Let ω be any non-trivial character of k_A^\times, trivial on k^\times. Then $L(1,\omega)\neq 0$.*

Except for the case $\omega^2=1$, this is contained in corollary 2 of th. 2, Chap. VII-5. Assume now that ω is of order 2; call U its kernel, which is an open subgroup of k_A^\times of index 2, containing k^\times. By corollary 2 of th. 7, § 9, there is a quadratic extension k' of k associated with U. By th. 10 of § 10, we have

$$\zeta_{k'}(s)=\zeta_k(s)L(s,\omega).$$

If k is of characteristic 0, by the corollary of th. 3, Chap. VII-6, both ζ_k and $\zeta_{k'}$ have a simple pole at $s=1$, and their residues there, whose values are given by that corollary, are >0. The same is true when k is of characteristic $p>1$, by th. 4 of Chap. VII-6. Therefore $L(1,\omega)>0$.

One should observe that the above proof can be extended in an obvious manner to any non-trivial character ω of k_A^\times of finite order, trivial on k^\times, by applying th. 10 of § 10 to the cyclic extension k' of k associated with the kernel U of ω; so far as the conclusion of theorem 11 is concerned, this adds nothing new to what has already been proved by a different method in corollary 2 of th. 2, Chap. VII-5, but it supplies some important relations between the class-numbers of k and k' and the values of the corresponding L-functions at $s=1$; more generally, th. 10 of § 10 shows at once that similar relations hold for all abelian extensions of k of finite degree. One should also note that, if ω_s, for $s\in\mathbf{C}$, has the same meaning as in Chap. VII, and if one replaces ω by $\omega_{it}\omega$ in theorem 11, one finds that $L(1+it,\omega)\neq 0$ for all $t\in\mathbf{R}$.

COROLLARY. *Let k_0 be an A-field contained in k; let V be a set of finite places of k, such that, for almost all the finite places v of k, not in V, the closure of k_0 in k_v is not k_v. Let ω be a non-trivial character of k_A^\times, trivial on k^\times, such that ω_v is unramified at all the places $v\in V$. Then the product*

$$q(k,V,\omega,s)=\prod_{v\in V}(1-\omega_v(\pi_v)q_v^{-s})^{-1}$$

is absolutely convergent for $\mathrm{Re}(s)>1$ and tends to a finite limit, other than 0, when s tends to 1.

For almost all v, by th. 1 of Chap. VIII-4, k_v is unramified over the closure $(k_0)_u$ of k_0 in k_v, so that its modular degree over $(k_0)_u$ is equal to its degree over the same field. In view of this, the assumption made above about V is identical with that made in corollary 3 of th. 2, Chap. VII-5. That being so, the proof of the latter corollary can be applied here;

when that is done, one sees that our assertion is an immediate consequence of theorem 11, combined with corollary 3 of prop. 1, Chap. VII-1.

THEOREM 12. *Let L be an A-field, k_0 an A-field contained in L, and α an automorphism of L over k_0. Then there are infinitely many places w of L such that L_w is unramified over the closure of k_0 in L_w and that the Frobenius automorphism of L_w over that closure induces α on L.*

Call k the subfield of L consisting of the elements of L, fixed under α; as $k_0 \subset k \subset L$, L has a finite degree d over k; by Galois theory, this implies that L is cyclic over k, its Galois group \mathfrak{g} over k being the one generated by α. For each place v of k, call u the place of k_0 which lies below v, and let w be any place of L above v; then the closure of k_0 in L_w is $(k_0)_u$. By th. 1 of Chap. VIII-4, there is a finite set P of places of k, containing P_∞, such that, when v is not in P, k_v is unramified over $(k_0)_u$, and L_w over k_v, hence also over $(k_0)_u$. Call then φ the Frobenius automorphism of L_w over $(k_0)_u$; as this generates the Galois group of L_w over $(k_0)_u$, it leaves no element of L_w fixed except those of $(k_0)_u$; therefore, if it induces α on L, we must have $k \subset (k_0)_u$, hence $k_v = (k_0)_u$, and then, in view of our definitions in § 11, α is the Frobenius automorphism of L over k at v. Call M_0 the set of the places v of k, not in P, such that $k_v \neq (k_0)_u$; for every place v of k, not in $P \cup M_0$, call φ_v the Frobenius automorphism of L over k at v; call M_1 the set of the places v of k, not in $P \cup M_0$, for which $\varphi_v = \alpha$, and call V the complement of $P \cup M_0 \cup M_1$ in the set of all places of k. Clearly the assertion in our theorem amounts to saying that M_1 is not a finite set, and M_1 is finite if and only if V has the property described in the corollary of th. 11. Assuming now that V has that property, we will derive a contradiction from this assumption. With our usual notations, call χ a character of \mathfrak{A} attached to the cyclic extension L of k; here, of course, \mathfrak{A} is the Galois group of k_{ab} over k, and L is regarded as a subfield of k_{ab}. Let \mathfrak{B} be the subgroup of \mathfrak{A} corresponding to L; then we may write $\mathfrak{g} = \mathfrak{A}/\mathfrak{B}$, and the group of the characters of \mathfrak{g} consists of the characters χ^i for $0 \leqslant i < d$. Put $\omega = \chi \circ \mathfrak{a}$; then, by corollary 3 of prop. 14, § 10, ω_v is unramified if and only if L_w is unramified over k_v, and then the Frobenius automorphism φ_v of L over k at v is the image of π_v in \mathfrak{g} under the morphism of k_A^\times onto \mathfrak{g} determined by \mathfrak{a}. This gives now, with the notation of the corollary of th. 11:

$$q(k, V, \omega^i, s) = \prod_{v \in V} (1 - \chi^i(\varphi_v) q_v^{-s})^{-1}.$$

For brevity, call this $q_i(s)$; we have now

$$\log q_i(s) = \sum_{v \in V} \sum_{n=1}^{+\infty} \chi^i(\varphi_v)^n q_v^{-ns}/n,$$

this being absolutely convergent for $\mathrm{Re}(s) > 1$. This gives:

$$\sum_{i=0}^{d-1} \chi^i(\alpha^{-1})\log q_i(s) =$$

$$= \sum_{v \in V} \left(\sum_{i=0}^{d-1} \chi^i(\alpha^{-1}\varphi_v) \right) q_v^{-s} + \sum_{v \in V} \sum_{n=2}^{+\infty} \sum_{i=0}^{d-1} \chi^i(\alpha^{-1}\varphi_v^n) q_v^{-ns}/n.$$

In the right-hand side, all the coefficients in the first series are 0, since $\varphi_v \neq \alpha$ for $v \in V$. On the other hand, $q_v \geq 2$ for all v, so that, for each v and for $\mathrm{Re}(s) > 1$, we have

$$\sum_{n=2}^{+\infty} |q_v^{-ns}|/n \leq \frac{1}{2} \sum_{n=2}^{+\infty} q_v^{-n} \leq q_v^{-2}.$$

Therefore the second series in the right-hand side of the above formula is majorized by $d\sum_v q_v^{-2}$, which is convergent by prop. 1 of Chap. VII-1. We have thus shown that the left-hand side remains bounded for $\mathrm{Re}(s) > 1$. On the other hand, the corollary of th. 11 shows that, for $1 \leq i < d$, $\log q_i(s)$ remains bounded when s tends to 1, and corollary 3 of th. 2, Chap. VII-5, shows that $\log q_0(s)$ does not. This is a contradiction.

COROLLARY. *Notations being as in definition 1 of § 11, let J be a congruence subgroup of L_P; if k is of characteristic $p > 1$, assume that J contains divisors of degree $\neq 0$. Then there are infinitely many places of k in every coset of J in L_P.*

In fact, let k' be the "classfield" for J, as explained in § 11; call \mathfrak{g} its Galois group over k. It has been shown in § 11 that the places v of k, in a given coset of J in L_P, are those places, not in P, where the Frobenius automorphism of k' over k is a given one. Our assertion is now a special case of theorem 12.

As an illustration for theorem 12, take $k_0 = \mathbf{Q}$, and take for L the field generated by a primitive m-th root of 1. Then our theorem says that, if a is any integer prime to m, there are infinitely many rational primes congruent to a modulo m. This is Dirichlet's "theorem of the arithmetic progression", and the proof given above for theorem 12 is directly modelled on Dirichlet's original proof for his theorem.

Finally, let ω, k and k' be again as in the proof of theorem 11, so that we have

$$\zeta_{k'}(s) = \zeta_k(s) L(s, \omega).$$

If k is of characteristic 0, we have also, by the corollary of th. 10, § 10:

$$Z_{k'}(s) = \pi^\rho Z_k(s) \Lambda(s, \omega),$$

where ρ is as explained in that corollary. Now write that the functions in these formulas satisfy the functional equations contained in theorems 3 and 4 of Chap. VII-6 and theorems 5 and 6 of Chap. VII-7. Writing that the exponential factors must be the same in the functional equations for both sides, one gets nothing new; the relation obtained in this manner is an immediate consequence of th. 9 of § 9. Writing that the constant factors are the same on both sides, one gets $\kappa\omega(b)=1$, with κ and b defined as in theorems 5 and 6 of Chap. VII-7. This will now be applied to a special case. Assume that we have taken for ω a character of k_A^\times of order 2, trivial on $k^\times \Omega(P_\infty)$, or, what amounts to the same, trivial on k^\times, on k_v^\times whenever v is an infinite place, and on r_v^\times whenever v is a finite place. According to prop. 14 of Chap. VII-7, we have then $\kappa_v=1$ for all v, hence $\kappa=1$, and the idele b is the same as the differential idele a. Therefore, for every such character ω, we have $\omega(a)=1$. Here, if k is an algebraic number-field, a may be assumed to have been chosen as in prop. 12 of Chap. VIII-4, i.e. so that id(a) is the different \mathfrak{d} of k over \mathbf{Q}; if k is of characteristic $p>1$, we know, by the definition of a differential idele in Chap. VII-2, that $\mathfrak{c}=$ div(a) is a divisor belonging to the canonical class. On the other hand, the conditions imposed on ω amount to saying that it is trivial on $k^\times(k_A^\times)^2 \Omega(P_\infty)$; therefore a is in that group. As $k_A^\times/k^\times \Omega(P_\infty)$ may be identified with the group $I(k)/P(k)$ of the ideal-classes of k, if k is an algebraic number-field, and with the group $D(k)/P(k)$ of the divisor-classes of k if k is of characteristic $p>1$, we have thus proved the following theorem (due to Hecke in the case of algebraic number-fields):

THEOREM 13. *If k is an algebraic number-field, there is an ideal-class of k whose square is the class defined by the different of k over \mathbf{Q}. If k is of characteristic $p>1$, there is a divisor-class of k whose square is the canonical class of k.*

Notes to the text

(The places in the text to which these notes belong have been marked by a * in the margin.)

P. 1: Cf. E. Witt, Hamb. Abhandl. 8 (1931), 413.

P. 27: The analogy in the text can be pursued much further. Let K and V be as in definition 1; call two norms N, N' on V equivalent if N'/N is constant on V. Then the quotient of the set of all K-norms on V by this equivalence relation can be identified with the so-called "building" associated by F. Bruhat and J. Tits (cf. Publ. Math. IHES, n° 41, 1971) with the group $\mathrm{Aut}(V)$, i.e. with $GL(n, K)$ if $V = K^n$; this corresponds to the "Riemannian symmetric space" associated with $GL(n, K)$ for $K = \mathbf{R}$, \mathbf{C} or \mathbf{H} in the classical theory. An "apartment" of that building consists of the points determined by norms of the form given by proposition 3 for a fixed decomposition $V = V_1 + \cdots + V_n$ of V. The "buildings" associated with the other "classical groups" over K can also be interpreted by means of norms in the spaces on which these groups operate.

P. 74: The proof of theorem 4 given in the text is the one due to G. Fujisaki (J. Fac. Sc. Tokyo (I) VII (1958), 567–604). It is in this proof that the "Minkowski argument" (which appears here in the form of lemma 1, Chap. II-4) plays a decisive role, just as it did at the corresponding place in the classical theory.

P. 101: For a treatment (due to C. Chevalley) of the topic of "linear compacity", cf. Chapter II, §§ 27–33, of S. Lefschetz, *Algebraic Topology*, A. M. S. 1942. In a locally linearly compact vector-space V over a (discretely topologized) field K, one can attach, to each linearly compact open subspace W, an integer $d(W)$ so that, if $W \supset W'$, $d(W) - d(W')$ is the dimension of W/W' over K; this takes the place of the Haar measure in the theory of locally compact groups.

P. 122: The proof given here is Tate's (cf. J. Tate, *Thesis*, Princeton 1950 = Chapter XV of Cassels-Fröhlich, *Algebraic Number Theory*, Acad. Press 1967).

P. 125: The proof given here, based on lemma 7, is the classical one, due to Hadamard (Bull. Soc. Math. 24 (1896), 199–220), with the improvements due to F. Mertens (Sitz.-ber. Ak. Wiss., Wien (Math.-nat. Kl.), 107 (1898), 1429–1434).

P. 126: In fact, it will be seen (cf. proof of th. 11, Chap. XIII-12) that, if $\omega^2 = 1$, $\omega \neq 1$, there is a quadratic extension k' of k such that

$$p(k, P, \omega, s) = p(k', P', s) p(k, P, s)^{-1}$$

where P' is the set of places of k' above P; in substance, this is equivalent to the "law of quadratic reciprocity" for k. As both factors in the right-hand side have a simple pole at $s = 1$, this proves the assertion. That proof, however, can be replaced by a simple function-theoretic argument, as follows. Note first that, for $\omega^2 = 1$, the product

$$p_1(s) = p(k, P, \omega, s) p(k, P, s)$$

is a product of factors respectively equal to

$$(1 - q_v^{-s})^{-2} = \sum_{n=0}^{\infty} (n+1) q_v^{-ns}$$

or to

$$(1 - q_v^{-2s})^{-1} = \sum_{n=0}^{\infty} q_v^{-2ns}$$

according as $\lambda(v)$ is 1 or -1. Expanding this into a Dirichlet series, we get for $p_1(s)$ a series with coefficients in \mathbf{R}_+ which diverges for $s = 0$. By an elementary lemma, originally due to Landau (cf. e.g. E.C. Titchmarsh, *The Theory of Functions* (2nd ed.), Oxford 1939, §9.2) the function defined by such a series must have a singular point on \mathbf{R}_+. On the other hand, in view of our results in §§6–7, $p_1(s)$ would be holomorphic in the whole plane if $p(k, P, \omega, s)$ was 0 at $s = 1$. Cf. also the remark at the end of the proof of th. 11, Chap. XIII-12, and the Notes to p. 288.

P. 152: The theorem expressed by formula (11) is due to J. Herbrand (J. de Math. (IX) 10 (1931), 481–498); hence the name we have given to "the Herbrand distribution".

P. 165: This argument is incomplete. Before applying prop. 2 to C/C', Z, M, one should first observe that M, regarded as a (C/C')-module, is both faithful and simple. For any $z \in Z$, the mapping $m \to zm$ is an endomorphism of M as a (C/C')-module, hence also of M as a C-module, hence of the form $m \to \xi m$ with $\xi \in K$; therefore Z is isomorphic to K, and C/C' is an algebra over K in the sense of §1 (this was tacitly assumed in the text). The proof proceeds then as before.

P. 178: Cf. R. Brauer, Math. Zeit. 28 (1928), 677–696.

P. 202: An alternative proof (communicated by A. Dress) is as follows. Call $N = n^2$ the dimension of A over k; take α as in the text; identify $\text{End}_k(A)$ with $M_N(k)$ by means of the basis α. Then prop. 3 of Chap. IX-1 defines an isomorphism F of $A \otimes A^0$ onto $M_N(k)$. As $\alpha \otimes \alpha$ is a basis of

$A \otimes A^0$ over k, F determines, for almost all v, an isomorphism F_v of the r_v-lattice λ_v generated in $A_v \otimes A_v^0$ by $\alpha \otimes \alpha$ onto $M_N(r_v)$; when that is so, by th. 1 of Chap. X-1, λ_v is a maximal compact subring of $A_v \otimes A_v^0$; as easily seen, this implies that α_v is a maximal compact subring of A_v. By th. 1 of Chap. X-1, there is then a division algebra D over k_v, an integer v and an isomorphism φ of $M_v(D)$ onto A_v such that φ maps $M_v(R)$ onto α_v, R being the maximal compact subring in D. Let π be a prime element of k_v; using prop. 5 of Chap. I-4, one sees easily that $R/\pi R$ and $M_v(R)/\pi M_v(R)$ are simple rings (i.e. that they have no non-trivial two-sided ideals) if and only if $D = k_v$. Consequently, A is unramified at v if and only if the ring $\alpha_v/\pi \alpha_v$ is simple; but it must be so if v is as above, as one sees at once by using the isomorphism F_v and the fact that the ring $M_N(r_v)/\pi M_N(r_v)$ is simple.

P. 206: Cf. M. Eichler, Math. Zeit. 43 (1938), 481–494.

P. 208: This statement is obviously false if K is of characteristic $p > 1$; for instance, it contradicts the results of Chap. XII-3 if those of Chap. II-3 are taken into account. If K is of characteristic 0, the statement is correct.

P. 241: The proof of the transfer theorem given here is the one due to C. Chevalley (J. Math. Soc. Japan 3 (1951), 36–44). For another proof, cf. Appendix I in this volume.

P. 256: Cf. H. Hasse, Math. Ann. 107 (1933), 731–760.

P. 262: The content of proposition 8 may be expressed by saying that, in the duality between $k_A^\times/(k_A^\times)^n$ and itself defined by the Hilbert symbol (cf. prop. 7), the image of k^\times in that group (which is a discrete subgroup with compact factor-group) is self-dual, i.e. that it is the group "associated by duality" with itself in the sense of Chap. II-5.

P. 273: Cf. C. Chevalley, loc. cit. (in the Note to p. 241).

P. 288: Cf. above, Note to p. 126.

P. 288: Of course the same argument applies to $\omega = 1$; in other words, $\zeta_k(1 + it) \neq 0$ for $t \in \mathbf{R}$, $t \neq 0$. As first shown by Hadamard for $k = \mathbf{Q}$ (loc. cit., Note to p. 125), this fact is essentially equivalent to the "prime number theorem" (more precisely, the "prime ideal theorem") for k.

P. 291: This proof (originally arising from a suggestion by J.-P. Serre) is taken from J. V. Armitage, Invent. Math. 2 (1967), 238–246.

Appendix I

The transfer theorem

1. As in Chap. IX-3, take an arbitrary field K and an extension K' of K of finite degree n, contained in K_{sep}; write \mathfrak{G}, \mathfrak{G}' for the Galois groups of K_{sep} over K and over K', respectively. Call t the transfer homomorphism of $\mathfrak{G}/\mathfrak{G}^{(1)}$ into $\mathfrak{G}'/\mathfrak{G}'^{(1)}$; as explained in Chap. XII-5, this may be defined by means of any full set $\{\sigma_1, \dots, \sigma_n\}$ of representatives of the cosets $\sigma \mathfrak{G}'$ of \mathfrak{G}' in \mathfrak{G}.

Let f' be any factor-set of K' (cf. Chap. IX-3, def. 4). For any ρ, σ, τ in \mathfrak{G}, and for $1 \leq i \leq n$, we can write $\rho \sigma_i$, $\sigma \sigma_i$, $\tau \sigma_i$ uniquely in the form

$$(1) \qquad \rho \sigma_i = \sigma_j \alpha_i, \qquad \sigma \sigma_i = \sigma_k \beta_i, \qquad \tau \sigma_i = \sigma_l \gamma_i,$$

with $1 \leq j, k, l \leq n$ and with $\alpha_i, \beta_i, \gamma_i$ in \mathfrak{G}'. Then the formula

$$(\rho, \sigma, \tau) \to f(\rho, \sigma, \tau) = \prod_{i=1}^{n} f'(\alpha_i, \beta_i, \gamma_i)^{\sigma_i^{-1}}$$

defines a factor-set f of K; we will write $f = v(f')$. If z' is a covariant mapping of $\mathfrak{G}' \times \mathfrak{G}'$ into K_{sep}^{\times}, we can define quite similarly a covariant mapping $z = v(z')$ of $\mathfrak{G} \times \mathfrak{G}$ into K_{sep}^{\times}; then, if f' is the coboundary of z', $v(f')$ is the coboundary of $v(z')$. Therefore v maps coboundaries into coboundaries and determines a morphism, for which we also write v, of factor-classes of K' into factor-classes of K. If, for each i, we replace σ_i by $\sigma_i \lambda_i$ with $\lambda_i \in \mathfrak{G}'$, then, for a given f', $v(f')$ is modified by the coboundary of the covariant mapping

$$(\rho, \sigma) \to \prod_i [f'(\alpha_i, \lambda_j^{-1} \alpha_i, \lambda_k^{-1} \beta_i) \, f'(\alpha_i, \beta_i, \lambda_k^{-1} \beta_i)^{-1}]^{\sigma_i^{-1}},$$

where j, k, α_i, β_i are as in (1). This shows that the morphism v for factor-classes does not depend upon the choice of the σ_i.

2. Now let notations be as in Chap. IX-4; instead of $\{\chi, \theta\}$, however, we will write $\{\chi, \theta\}_K$; and we write $\{\chi', \theta'\}_{K'}$ for the similarly defined symbol over K'.

LEMMA A. *Let χ' be a character of \mathfrak{G}'; then, for all $\theta \in K^{\times}$:*

$$\{\chi' \circ t, \theta\}_K = v(\{\chi', \theta\}_{K'}).$$

As in Chap. IX-4, write $\chi' = e \circ \Phi'$, where Φ' is a mapping of \mathfrak{G}' into the interval $[0, 1[$ on \mathbf{R}; Φ' is constant on cosets modulo $\mathfrak{G}'^{(1)}$. Then

$\chi' \circ t = \mathbf{e} \circ \Phi$ with $\Phi = \Phi' \circ t$; if ρ and the α_i are as in (1), this gives (by definition of the transfer) $\Phi(\rho) = \Phi'(\prod_i \alpha_i)$. In the formula to be proved, both sides are defined as the classes of certain factor-sets; one has to show that those factor-sets differ only by the coboundary of some co-variant mapping z. For any ρ, σ in \mathfrak{G}, define the α_i, β_i as in (1), and put $z(\rho, \sigma) = \theta^N$ where N is the integer

$$N = \sum_i \Phi'(\beta_i \alpha_i^{-1}) - \Phi'(\prod_i \beta_i \alpha_i^{-1}).$$

It is trivial to verify that z is then a covariant mapping with the required property.

LEMMA B. *Let χ be a character of \mathfrak{G}, and χ' its restriction to \mathfrak{G}'. Then, for all $\theta' \in K'^{\times}$, we have*

$$\{\chi, N_{K'/K}(\theta')\}_K = v(\{\chi', \theta'\}_{K'}).$$

The proof is similar to that of lemma A. Write $\chi = \mathbf{e} \circ \Phi$; both sides of the formula to be proved are defined as the classes of certain factor-sets; one verifies that the latter differ by the coboundary of the covariant mapping z given, for all ρ, σ, by the formulas

$$z(\rho, \sigma) = \prod_i (\theta'^{\sigma_i^{-1}})^{N_i},$$

$$N_i = \Phi(\sigma \rho^{-1}) - \Phi(\beta_i \alpha_i^{-1}) + \Phi(\sigma_j) - \Phi(\sigma_k),$$

where j, k, α_i, β_i are given by (1), so that the N_i are integers.

3. Now we take for K a commutative p-field. In view of the definition of the canonical morphism in Chap. XII-2, the local "transfer theorem", i.e. theorem 6 of Chap. XII-5, is equivalent to the following statement:

THEOREM. *Let K, K' be as in theorem 8 of Chapter XII-5; then, for all $\chi' \in X_{K'}$ and all $\theta \in K^{\times}$, we have*

$$(\chi' \circ t, \theta)_K = (\chi', \theta)_{K'}.$$

Consider the symbol η defined in Chap. XII-2; let η' be the corresponding symbol for K'. In view of lemma A, the theorem will be proved if we show that, for any factor-class c' of K', we have $\eta[v(c')] = \eta'(c')$. By th. 1 of Chap. XII-2, we may write c' in the form $\{\chi', \theta'\}_{K'}$ with an unramified character χ' of \mathfrak{G}' and some $\theta' \in K'$. Then χ' is attached to a cyclic extension $K'(\mu)$ of K' generated by a root μ of 1 of order prime to p, and it is the restriction to \mathfrak{G}' of a suitably chosen character χ of \mathfrak{G} attached to the cyclic unramified extension $K(\mu)$ of K. Our conclusion follows now at once from lemma B, combined with th. 2 of Chap. XII-2.

4. In order to deduce the global transfer theorem (theorem 8 of Chap. XIII-9) from the local one, we first observe the following. Let notations be as in Chap. XIII-10; let k' be an extension of k of finite degree, contained in k_{sep}. For any place v of k, and any place w of k' lying above v, let \mathfrak{A}'_w be the Galois group of $k'_{w,ab}$ over k'_w, and ρ'_w the restriction morphism of \mathfrak{A}'_w into the Galois group \mathfrak{A}' of k'_{ab} over k'. Call t, t_w the transfer homomorphisms of \mathfrak{A} into \mathfrak{A}' and of \mathfrak{A}_v into \mathfrak{A}'_w, respectively. Then we have

$$t \circ \rho_v = \prod_{w/v} (\rho'_w \circ t_w),$$

the product being taken over all the places w of k' lying above v; the proof of this is easy (and purely group-theoretical) and will be left as an exercise to the reader. This being granted, the global transfer theorem is an immediate consequence of the local theorem and of the definitions.

Appendix II

W-groups for local fields

1. For the formulation of Shafarevitch's theorem and related results, it is convenient to introduce modified Galois groups, to be called W-groups, as follows. Let K be a commutative p-field; as in Chap. XII-2, let $K_0 = K(\mathfrak{M})$ be the subfield of K_{sep} generated over K by the set \mathfrak{M} of all roots of 1 of order prime to p in K_{sep}. Let \mathfrak{K} be a Galois extension of K between K_0 and K_{sep}; let \mathfrak{G}, \mathfrak{G}_0 be the Galois groups of \mathfrak{K} over K and over K_0, respectively. Let φ be the restriction to \mathfrak{K} of a Frobenius automorphism of K_{sep} over K. We put

$$\mathfrak{W} = \bigcup_{n \in \mathbf{Z}} \varphi^n \mathfrak{G}_0$$

and give to \mathfrak{W} the topology determined by a fundamental system of neighborhoods of the identity in \mathfrak{G}_0 (e.g., by all open subgroups of \mathfrak{G}_0). This makes \mathfrak{W} into a locally compact group with the maximal compact subgroup \mathfrak{G}_0; $\mathfrak{W}/\mathfrak{G}_0$ is discrete and isomorphic to \mathbf{Z}. With this topology, \mathfrak{W} will be called the W-group of \mathfrak{K} over K; it has an obvious injective morphism δ into \mathfrak{G}, which maps it onto a dense subgroup of \mathfrak{G}.

Call q the module of K; the Frobenius automorphism φ determines on \mathfrak{M} the bijective mapping $\mu \to \mu^\varphi = \mu^q$, and φ^n determines on \mathfrak{M}, for every $n \in \mathbf{Z}$, a bijection which we write as $\mu \to \mu^Q$ with $Q = q^n$. Then \mathfrak{W} may be described as consisting of those automorphisms ω of \mathfrak{K} over K which determine on \mathfrak{M} a bijection of the form $\mu \to \mu^\omega = \mu^Q$ with $Q = q^n$, $n \in \mathbf{Z}$; when ω and Q are such, we will write $|\omega|_{\mathfrak{W}} = Q^{-1}$ and call $|\omega|_{\mathfrak{W}}$ the *module* of ω in \mathfrak{W}. Clearly $\omega \to |\omega|_{\mathfrak{W}}$ is a morphism of \mathfrak{W} into \mathbf{R}_+^\times with the compact kernel \mathfrak{G}_0, and it maps \mathfrak{W} onto the subgroup of \mathbf{R}_+^\times generated by q.

2. If \mathfrak{K}' is any Galois extension of K between K_0 and \mathfrak{K}, and Γ is the Galois group of \mathfrak{K} over \mathfrak{K}', we may clearly identify the W-group of \mathfrak{K}' over K with \mathfrak{W}/Γ. On the other hand, let K' be any finite extension of K between K and \mathfrak{K}; let \mathfrak{G}' be the Galois group of \mathfrak{K} over K', and \mathfrak{W}' its W-group over K'; clearly we have $\mathfrak{W}' = \delta^{-1}(\mathfrak{G}')$. As \mathfrak{G}' and its cosets in \mathfrak{G} are open in \mathfrak{G}, \mathfrak{W}' is open in \mathfrak{W} and has a finite index, equal to that of \mathfrak{G}' in \mathfrak{G} and to the degree of K' over K. If K' is a Galois extension of K, we can identify its Galois group over K with $\mathfrak{W}/\mathfrak{W}'$ as well as with $\mathfrak{G}/\mathfrak{G}'$. Conversely, let \mathfrak{W}' be any open subgroup of \mathfrak{W} of finite index in \mathfrak{W}.

Then $\mathfrak{G}_0 \cap \mathfrak{W}'$ is open in \mathfrak{G}_0 and therefore belongs, in the sense of Galois theory, to some finite extension $K_0(\xi)$ of K_0, contained in \mathfrak{R}. Let L be a finite Galois extension of K between $K(\xi)$ and \mathfrak{R}. Let φ' be in \mathfrak{W}' and not in \mathfrak{G}_0; replacing φ' by φ'^{-1} if necessary, we may assume that $|\varphi'|_{\mathfrak{m}} = q^n$ with $n > 0$. Take an integer $v > 0$ such that φ'^v induces the identity on L; call K'' the compositum of L and of the unramified extension K_{nv} of degree nv of K in K_0, and let \mathfrak{W}'' be the W-group of \mathfrak{R} over K''. Take any $\omega \in \mathfrak{W}''$; as ω induces the identity on K_{nv}, we have $|\omega|_{\mathfrak{m}} = q^{nvi}$ with some $i \in \mathbf{Z}$. Then $\omega \varphi'^{-vi}$ induces the identity on K_0 and on L, hence on $K_0(\xi)$, so that it is in \mathfrak{W}'. Thus \mathfrak{W}'' is contained in \mathfrak{W}'. As we have seen that the Galois group of K'' over K may be identified with $\mathfrak{W}/\mathfrak{W}''$, this shows that \mathfrak{W}' belongs to some field K' between K and K'', and, more precisely, that it is the W-group of \mathfrak{R} over K'. Thus we see that W-groups have the same formal properties as Galois groups.

In particular, a cyclic extension L of K of degree n corresponds to an open subgroup \mathfrak{W}' of \mathfrak{W} of index n whose factor-group is cyclic and may be identified with the Galois group of L over K, and conversely. If χ is a character of \mathfrak{G} attached to L, it determines a character $\chi \circ \delta$ of \mathfrak{W}, also of order n; conversely, a character of \mathfrak{W} is of the form $\chi \circ \delta$ if and only if it is of finite order. We will frequently (by abuse of notation) make no distinction between a character χ of \mathfrak{G} and the corresponding character of \mathfrak{W}.

3. In applying the above concepts, the field \mathfrak{R} will mostly be taken of the form L_{ab}, where L is a finite Galois extension of K. In particular, we will always denote by W_K the W-group of K_{ab} over K. It follows at once from prop. 7 and corollary 2 of th. 3, Chap. XII-3, that the image $\delta(W_K)$ of W_K in the Galois group \mathfrak{A} of K_{ab} over K is the same as the image $\mathfrak{a}(K^\times)$ of K^\times in \mathfrak{A} under the canonical morphism \mathfrak{a}. Consequently, there is a canonical isomorphism \mathfrak{w}_K of K^\times onto W_K such that $\mathfrak{a} = \delta \circ \mathfrak{w}_K$. Moreover, it follows from the same results that $|\mathfrak{w}_K(\theta)|_{W_K} = |\theta|_K$ for all $\theta \in K^\times$.

Let for instance L be cyclic of degree n over K; as L is contained in K_{ab}, it corresponds to an open subgroup Γ of W_K, of index n, and we may identify W_K/Γ with the Galois group \mathfrak{g} of L over K; every character of \mathfrak{g} may be regarded as a character of W_K, trivial on Γ. If χ is such a character of order n, i.e. if it is attached to L (in the sense of Chap. IX-4), then, by the definition of the canonical morphisms \mathfrak{a} and \mathfrak{w}_K, $\chi[\mathfrak{w}_K(\theta)]$, for any $\theta \in K^\times$, is the Hasse invariant $h(A) = (\chi, \theta)_K$ of the cyclic algebra $A = [L/K; \chi, \theta]$ over K.

4. Let K' be any extension of K of finite degree; we assume that K_{sep} is contained in K'_{sep}. Let \mathfrak{R}, \mathfrak{R}' be Galois extensions of K and of K', respectively, such that $K_0 \subset \mathfrak{R} \subset \mathfrak{R}' \subset K'_{sep}$. Let \mathfrak{W}, \mathfrak{W}' be the W-groups of \mathfrak{R} over K and of \mathfrak{R}' over K', respectively. Then, just as for ordinary

Galois groups, there is a restriction morphism of \mathfrak{W}' into \mathfrak{W}, which we again denote by ρ; obviously $|\rho(\omega')|_{\mathfrak{W}} = |\omega'|_{\mathfrak{W}'}$ for all $\omega' \in \mathfrak{W}'$. Such is the case, for instance, if $\mathfrak{K} = K_{ab}$, $\mathfrak{K}' = K'_{ab}$; it is then an immediate consequence of th. 2, Chap. XII-2 (just as in corollary 1 of that theorem) that
$$\rho \circ \mathfrak{w}_{K'} = \mathfrak{w}_K \circ N_{K'/K}.$$

5. On the other hand, let \mathfrak{K}, \mathfrak{K}' be two Galois extensions of K such that $K_0 \subset \mathfrak{K} \subset \mathfrak{K}' \subset K_{sep}$; let \mathfrak{W}, \mathfrak{W}' be their W-groups over K, and let Γ be the Galois group of \mathfrak{K}' over \mathfrak{K}. Then we can identify \mathfrak{W} with \mathfrak{W}'/Γ, and the canonical morphism of \mathfrak{W}' onto \mathfrak{W} preserves the module. Thus \mathfrak{K} is abelian over K if and only if Γ contains the closure of the commutator-group of \mathfrak{W}'.

Now take any finite extension K' of K, contained in K_{sep}; let \mathfrak{K}' be any Galois extension of K between K'_{ab} and K_{sep}, e. g. K_{sep} itself. Call Ω, Ω' the W-groups of \mathfrak{K}' over K and over K', respectively; write Ω^c, Ω'^c for the closures of their commutator-groups; as Ω' is an open subgroup of finite index of Ω, we may introduce, just as in Chap. XII-5, the transfer homomorphism t of Ω/Ω^c into Ω'/Ω'^c. As K_{ab}, K'_{ab} are respectively the maximal abelian extensions of K and of K', contained in \mathfrak{K}', the Galois groups of \mathfrak{K}' over K_{ab} and over K'_{ab} are Ω^c and Ω'^c, respectively, and we may identify W_K with Ω/Ω^c and $W_{K'}$ with Ω'/Ω'^c, so that t maps W_K into $W_{K'}$. Combining now the transfer theorem (cf. Chap. XII-5 and Appendix I) with our definitions for the W-groups, one sees at once that the theorem in question may be expressed by the formula
$$t \circ \mathfrak{w}_K = \mathfrak{w}_{K'} \circ j,$$
where j is the natural injection of K^\times into K'^\times. Clearly this implies that t is injective and maps W_K onto $\mathfrak{w}_{K'}(K^\times)$.

Appendix III

Shafarevitch's theorem

This theorem gives the structure of the W-group of L_{ab} over K whenever K is a commutative p-field and L a finite Galois extension of K. We begin by supplementing the results of Chapter IX with some additional observations.

1. Let assumptions and notations be as in Chap. IX, so that K is an arbitrary field, \mathfrak{G} the Galois group of K_{sep} over K, and all algebras over K are understood to be as stated in Chap. IX-1. Let A be a central simple algebra of dimension n^2 over K. Let L be an extension of K of degree n, and f a K-linear isomorphism of L into A. Call V the vectorspace of dimension n over L, with the same underlying space as A, defined by $(\xi, x) \to xf(\xi)$ for $\xi \in L$, $x \in A$. For every $a \in A$, the mapping $x \to ax$ is an endomorphism $F(a)$ of V; F is then a representation of A into $\mathrm{End}_L(V)$, and, by corollary 5 of prop. 3, Chap. IX-1, its L-linear extension F_L to A_L is an isomorphism of A_L onto $\mathrm{End}_L(V)$. Let $z \in A$ be such that $zf(\xi) = f(\xi)z$ for all $\xi \in L$; then $x \to xz$ is in $\mathrm{End}_L(V)$ and commutes with $F(a)$ for all $a \in A$; therefore it is in the center of $\mathrm{End}_L(V)$, i.e. of the form $x \to xf(\zeta)$ with some $\zeta \in L$, so that $z = f(\zeta)$. In other words, $f(L)$ is its own "commutant" in A, and $f(L^\times)$ its own centralizer in A^\times. Let now f' be another embedding of L into A; let V', F' be to f' what V, F are to f. As noted in Chap. IX-2, it follows from prop. 4, Chap. IX-1 that there is an isomorphism Y of V onto V' such that $F' = Y^{-1}FY$. This means that Y is a bijection of A onto A such that $Y(xf(\xi)) = Y(x)f'(\xi)$ and $Y(ax) = aY(x)$ for all $\xi \in L$ and all x, a in A. Take $x = 1_A$ and put $b = Y(1_A)$; then we see that $b \in A^\times$ and that $f' = b^{-1}fb$. In other words, two embeddings f, f' can differ only by an inner automorphism of A. In particular, let \mathfrak{g} be the group of all automorphisms of L over K; then, for every $\alpha \in \mathfrak{g}$, there is $b_\alpha \in A^\times$ such that $f(\xi^\alpha) = b_\alpha^{-1}f(\xi)b_\alpha$ for all $\xi \in L$; consequently, the normalizer N of $f(L^\times)$ in A^\times is given by $N = \bigcup_\alpha b_\alpha f(L^\times)$, and $N/f(L^\times)$ can be identified with \mathfrak{g}. For any α, β in \mathfrak{g}, $b_{\alpha\beta}^{-1}b_\alpha b_\beta$ commutes with L^\times, so that we can write $b_\alpha b_\beta = b_{\alpha\beta}\lambda(\alpha, \beta)$ with $\lambda(\alpha, \beta) \in L^\times$. Moreover, the b_α are linearly independent over L in V; for otherwise, taking a maximal subset $\{b_\lambda\}$ of linearly independent ones among them, we could write, for any b_α not in that set, $b_\alpha = \sum b_\lambda f(\xi_\lambda)$; then, writing that $f(\eta)b_\alpha = b_\alpha f(\eta^\alpha)$ for all $\eta \in L$, we get a contradiction.

2. In particular, assume that L is a Galois extension of K, so that \mathfrak{g} is its Galois group; at the same time, simplify notations by identifying L with $f(L)$ by means of f. Then the b_α make up a basis of V over L, so that A consists of the elements $\sum b_\alpha \xi_\alpha$ with $\xi_\alpha \in L$ for all α. Clearly A is completely defined as an algebra by the multiplication laws:

$$(1) \qquad\qquad b_\alpha b_\beta = b_{\alpha\beta}\,\lambda(\alpha,\beta), \qquad \xi\,b_\alpha = b_\alpha\,\xi^\alpha$$

for all α, β in \mathfrak{g} and all $\xi \in L$. Moreover, writing that $(b_\alpha b_\beta)\,b_\gamma$ is the same as $b_\alpha(b_\beta b_\gamma)$, one gets

$$(2) \qquad\qquad \lambda(\alpha\beta,\gamma)\,\lambda(\alpha,\beta)^\gamma = \lambda(\alpha, \beta\gamma)\,\lambda(\beta,\gamma).$$

Conversely, let L be a Galois extension of K of degree n, contained in K_{sep}; \mathfrak{G} being as before, call \mathfrak{H} the Galois group of K_{sep} over L, so that the Galois group of L over K is $\mathfrak{g}=\mathfrak{G}/\mathfrak{H}$. For any $\rho\in\mathfrak{G}$, write ρ^* for the image of ρ in $\mathfrak{g}=\mathfrak{G}/\mathfrak{H}$. For any mapping λ of $\mathfrak{g}\times\mathfrak{g}$ into L^\times, we define an \mathfrak{H}-regular covariant mapping f of $\mathfrak{G}\times\mathfrak{G}\times\mathfrak{G}$ into K_{sep} by

$$(3) \qquad (\rho,\sigma,\tau) \to f(\rho,\sigma,\tau) = \lambda(\tau^*\sigma^{*-1}, \sigma^*\rho^{*-1})^{\rho^*};$$

this is a factor-set if (and only if) λ satisfies (2). It is now easily verified that the algebra A defined by means of f by Brauer's construction (as described in the proof of lemma 4, Chap. IX-3) is precisely as above if we call b_α the element of A given (in terms of that construction) by the covariant mapping $(\rho,\sigma)\to\delta_{\rho^*,\alpha\sigma^*}$.

As in §1, consider the normalizer $N=\bigcup b_\alpha L^\times$ of L^\times in A^\times; write N^c for its commutator-group, and τ for the transfer homomorphism of N/N^c into L^\times. As the definition of τ is invariant with respect to all inner automorphisms of N, and as such automorphisms determine the identity on N/N^c, τ must map N/N^c into the subgroup of the elements of L^\times which are invariant under such automorphisms, i.e. into K^\times. On the other hand, regarding τ as a morphism of N into L^\times, and calculating it (according to definition) by means of the representatives b_α of the cosets of L^\times in N, one sees at once that, on L^\times, τ coincides with $N_{L/K}$.

3. Assumptions being as in §2, let K' be a field between K and L, corresponding to a subgroup \mathfrak{g}' of \mathfrak{g}. One verifies at once that an element of A commutes with all elements of $1_A \cdot K'$ if and only if it is of the form $\sum b_\alpha \xi_\alpha$, with $\xi_\alpha \in L$ and $\xi_\alpha = 0$ unless $\alpha\in\mathfrak{g}'$. Clearly these elements make up a subring A' of A (the "commutant" of K' in A) which is the algebra over K' defined by means of K', L and the restriction of λ to $\mathfrak{g}'\times\mathfrak{g}'$ just as A was defined above by means of K, L, λ; in particular, it is a central simple algebra over K'.

4. Let K, L, A be again as in §2; consider the case where \mathfrak{g} is cyclic; if α is a generator of \mathfrak{g}, we have $\mathfrak{g}=\{1,\alpha,\dots,\alpha^{n-1}\}$. For $\beta=\alpha^i$, we have

$b_\alpha^{-i} \xi b_\alpha^i = \xi^\beta$, so that we may take $b_\beta = b_\alpha^i$ for $0 \le i \le n-1$. If N and τ are as in § 2, we also see at once that $\tau(b_\alpha) = b_\alpha^n$; similarly, for any $a \in b_\alpha L^\times$, we may take $1, a, \ldots, a^{n-1}$ as the representatives of the cosets of L^\times in N, and see thus that $\tau(a) = a^n$, hence $a^n \in K^\times$. In particular, if we put $\theta = b_\alpha^n$, θ is in K^\times; it is clear that A is then no other than the cyclic algebra defined in prop. 11 of Chap. IX-4, i.e. the algebra $[L/K; \chi, \theta]$ if χ is the character of g given by $\chi(\alpha) = e(1/n)$.

Under those same assumptions, we have, for every $\xi \in L^\times$, $\xi^{-1} b_\alpha^{-1} \xi b_\alpha = \xi^{\alpha-1}$; therefore the image U of L^\times under $\xi \to \xi^{\alpha-1}$ is contained in N^c. Conversely, the image of b_α in N/U commutes with the image of L^\times in N/U; as these images generate N/U, N/U is commutative, so that $U \supset N^c$. Therefore, in this case, N^c is the same as U, i.e. (by Hilbert's theorem) the same as the kernel of the morphism $N_{L/K}$ of L^\times into K^\times.

5. As in § 2, let K be any field, and L a Galois extension of K of degree n, with the Galois group g.

LEMMA A. *Let φ be a morphism of a group G onto g; let H be its kernel. Let ω be a morphism of H into L^\times; assume that we have, for all $g \in G$ and all $h \in H$:*

(4) $$\omega(g^{-1} h g) = \omega(h)^{\varphi(g)}.$$

Then there is a central simple algebra A of dimension n^2 over K, containing L, such that ω can be extended to a morphism ω^ of G into A^\times satisfying*

$$\omega^*(g^{-1}) \xi \omega^*(g) = \xi^{\varphi(g)}$$

for all $g \in G$ and all $\xi \in L$. Moreover, these conditions determine A and ω^ uniquely, up to isomorphism; and $\omega^*(G) L^\times$ is then the normalizer of L^\times in A^\times.*

For each $\alpha \in$ g, choose $g_\alpha \in G$ such that $\varphi(g_\alpha) = \alpha$. For any α, β in g, we can write $g_\alpha g_\beta = g_{\alpha\beta} h(\alpha, \beta)$ with $h(\alpha, \beta) \in H$. Writing that $(g_\alpha g_\beta) g_\gamma$ is the same as $g_\alpha(g_\beta g_\gamma)$, we get

$$h(\alpha \beta, \gamma) \cdot g_\gamma^{-1} h(\alpha, \beta) g_\gamma = h(\alpha, \beta \gamma) h(\beta, \gamma).$$

Putting $\lambda(\alpha, \beta) = \omega[h(\alpha, \beta)]$, we see now, in view of (4), that λ satisfies (2), so that we can construct an algebra $A = \sum b_\alpha L$ with the multiplication laws (1). It is then obvious that the formulas $\omega^*(g_\alpha h) = b_\alpha \omega(h)$, for all $\alpha \in$ g, $h \in H$, define a morphism ω^* with the required properties. If A', ω'^* have the same properties, then, putting $b'_\alpha = \omega'^*(g_\alpha)$, we see that the b'_α satisfy relations similar to (1); from the results of § 2, it follows then that they are a basis for A' over L and that A', ω'^* differ from A, ω^* only by an isomorphism of A onto A'.

6. Now consider the following situation. Let K, L, \mathfrak{g} be as before; let L' be a Galois extension of K, containing L and contained in K_{sep}, of finite degree d over L. Call Γ, \varDelta the Galois groups of L' over K and over L, respectively, so that $\mathfrak{g} = \Gamma/\varDelta$. Let G be a group, φ' a morphism of G onto Γ, H' the kernel of φ', and ω' a morphism of H' into L'^{\times}; we assume that these data satisfy (4) when they are substituted there for G, φ, H, ω respectively, so that we can apply lemma A to them. This determines an algebra A' of dimension $n^2 d^2$ over K. Call ψ the canonical morphism of Γ onto $\mathfrak{g} = \Gamma/\varDelta$; put $\varphi = \psi \circ \varphi'$ and $H = \varphi'^{-1}(\varDelta)$; φ is a morphism of G onto \mathfrak{g} with the kernel H. To simplify notations, assume that H' is commutative, and let H^c be the commutator-group of H; then we can define (as in Chap. XII-5) the transfer homomorphism t of H/H^c into H', and regard it as a morphism of H into H'; we have $t(g^{-1}hg) = g^{-1}t(h)g$ for all $g \in G$ and $h \in H$. Now put $\omega = \omega' \circ t$. We have, for all $h \in H$, $h \in G$:

$$\omega(g^{-1}hg) = \omega(h)^{\varphi'(g)};$$

for $g \in H$, this implies that $\omega(h)$ is invariant under $\varphi'(H) = \varDelta$, so that it is in L^{\times} and that we may replace φ' by φ in the above formula. Therefore we can apply lemma A to G, H, φ, ω; this defines an algebra A of dimension n^2 over K. The following lemma and its proof are due to Artin and Tate (E. Artin and J. Tate, *Classfield theory*, Harvard 1961, Chap. XIII-3, th. 6, p. 188):

LEMMA B. *Let A, A' be as above; then, in the Brauer group $B(K)$, we have* $\text{Cl}(A) = \text{Cl}(A')^d$ *with* $d = [L':L]$.

For each $\xi \in \Gamma$, choose $g_{\xi} \in G$ such that $\varphi'(g_{\xi}) = \xi$; for all ξ, η in Γ, put

$$h'(\xi, \eta) = g_{\xi\eta}^{-1} g_{\xi} g_{\eta}; \qquad \lambda'(\xi, \eta) = \omega'[h'(\xi, \eta)].$$

As in the proof of lemma A, $\text{Cl}(A')$ is determined by λ', or, in the language of Chap. IX-3, by the factor-set f' of K determined in terms of λ' by the formula similar to (3). On the other hand, the definition of the transfer gives, for any $h \in H$:

$$t(h) = \prod_{\theta \in \varDelta} (g_{\varphi'(h)\theta}^{-1} \cdot h \cdot g_{\theta}).$$

Choose a full set M of representatives of the cosets $\xi\varDelta$ of \varDelta in Γ; for any $\xi \in \Gamma$, call $\mu(\xi)$ the representative in M of the coset $\xi\varDelta$. The elements g_{μ}, for $\mu \in M$, make up a full set of representatives of the cosets of H in G, so that we may use them, as in the proof of lemma A, to construct a factor-set defining $\text{Cl}(A)$; this is done as follows. Take any two elements ξ, η of Γ; put $\alpha = \mu(\xi)$, $\beta = \mu(\eta)$, $\gamma = \mu(\xi\eta)$, $\delta = \gamma^{-1}\alpha\beta$; put

$$h(\xi, \eta) = g_{\gamma}^{-1} g_{\alpha} g_{\beta}, \qquad \lambda(\xi, \eta) = \omega[h(\xi, \eta)];$$

as these are constant on cosets of Δ in Γ, λ may be regarded as a mapping of $\mathfrak{g} \times \mathfrak{g}$ into L^\times, and a factor-set f defining $\text{Cl}(A)$ is given in terms of λ by (3). By the definition of ω, we have

$$\lambda(\xi, \eta) = \omega' \left[\prod_{\theta \in \Delta} g_{\delta\theta}^{-1} g_\gamma^{-1} g_\alpha g_\beta g_\theta \right].$$

For any $\theta \in \Delta$, put

$$\theta' = \beta \theta \eta^{-1}, \qquad \theta'' = \alpha \theta' \xi^{-1} = \alpha \beta \theta (\xi \eta)^{-1}.$$

When θ runs through Δ, so do θ', θ'' and $\delta\theta$. In G, we have the following (easily verified) group-theoretical identity:

$$g_{\delta\theta}^{-1} g_\gamma^{-1} g_\alpha g_\beta g_\theta = h'(\gamma, \delta\theta)^{-1} h'(\theta'', \xi\eta) h'(\xi, \eta)$$

$$\cdot g_\eta^{-1} [h'(\theta'', \xi)^{-1} h'(\alpha, \theta')] g_\eta \cdot h'(\theta', \eta)^{-1} h'(\beta, \theta).$$

For every $\xi \in \Gamma$, put

$$c(\xi) = \omega' \left[\prod_{\theta \in \Delta} h'(\mu(\xi), \theta) h'(\theta, \xi)^{-1} \right].$$

Then, taking into account the fact that ω' satisfies (4), we get

$$\lambda(\xi, \eta) = c(\xi)^\eta c(\eta) c(\xi\eta)^{-1} \lambda'(\xi, \eta)^d.$$

This proves the lemma; in fact, if f, f' are as above, ff'^{-d} is the coboundary of $(\rho, \sigma) \to c(\sigma \rho^{-1})^\rho$.

7. From now on, we will take for K a commutative p-field. Also, if G is any topological group (e.g. a W-group), we will denote by G^c its topological commutator-group, i.e. the closure of its commutator-group in the algebraic sense.

Let A be any central simple algebra over K; if its dimension over K is n^2, we can write it as $M_d(D)$, where D is a division algebra of dimension $(n/d)^2$ over K; then $\text{Cl}(A)$ is the same as $\text{Cl}(D)$, and this, as shown in th. 1 of Chap. XII-2 and its corollaries, is of order n/d in the Brauer group $B(K)$; in other words, the Hasse invariant $h(A)$ is a root of 1 of order n/d, and it is of order n if and only if A is a division algebra. By corollary 2 of th. 2, Chap. XII-2, combined with corollary 3 of th. 3, Chap. IX-3, this implies that every separable extension L of K of degree n can be embedded in A; in view of § 1 above, this embedding is unique, up to an inner automorphism of A, so that we can apply to K, L and A all the results of that §. In particular, if L is a Galois extension of K with the Galois group \mathfrak{g}, and if N is the normalizer of L^\times in A^\times, the inner automorphisms $x \to a^{-1} x a$, for $a \in N$, induce \mathfrak{g} on L, and we can thus identify N/L^\times with \mathfrak{g}.

8. A straightforward application of lemma A gives now:

THEOREM I. *Let K be a commutative p-field, L a Galois extension of K of degree n; let W, W_L be the W-groups of L_{ab} over K and over L, respectively. Then there is a central simple algebra A of dimension n^2 over K, an embedding of L into A, and an isomorphism \mathfrak{w} of the normalizer N of L^\times in A^\times onto W, such that the restriction of \mathfrak{w} to L^\times is the canonical isomorphism \mathfrak{w}_L of L^\times onto W_L, and that, for every $a \in N$, the automorphism induced on L by $x \to a^{-1} x a$ is the restriction of $\mathfrak{w}(a)$ to L.*

In fact, in lemma A, substitute W, W_L for G, H; for φ, substitute the canonical morphism of W onto W/W_L when W/W_L is identified with the Galois group of L over K (cf. Appendix II, § 2); for ω, substitute \mathfrak{w}_L^{-1}. Then (4) follows at once (by "transport of structure") from the fact that \mathfrak{w}_L is "canonically" attached to the pair (L, L_{ab}). Under these circumstances, it is obvious that the morphism ω^* of lemma A is an isomorphism of W onto N; its inverse \mathfrak{w} has then the required properties.

COROLLARY 1. *In theorem I, A, L and \mathfrak{w} are uniquely characterized (up to an isomorphism) by the properties stated there.*

Also this is part of lemma A. The algebra A, with a given embedding of L into A, will be called *the canonical algebra* for the pair (K, L); \mathfrak{w} will be called *the canonical isomorphism* of N onto W.

COROLLARY 2. *With A and \mathfrak{w} as above, write $v_{A/K}$ for the reduced norm in A over K. Then, for all $a \in N$:*

$$|v_{A/K}(a)|_K = |\mathfrak{w}(a)|_W.$$

In fact, both sides define morphisms of N into \mathbf{R}_+^\times; as such, they must be equal if they coincide on a subgroup of N of finite index, e.g. on L^\times. On L^\times, $v_{A/K}$ coincides with $N_{L/K}$ (cf. the proof of lemma 4, Chap. IX-3), and \mathfrak{w} with \mathfrak{w}_L; therefore, for $a = \xi \in L^\times$, the left-hand side is $|\xi|_L$, and the right-hand side is $|\mathfrak{w}_L(\xi)|_{W_L}$; in view of Appendix II, § 3, this proves our assertion.

COROLLARY 3. *With A, L, N as above, the transfer homomorphism τ of N/N^c into L^\times is injective and maps N/N^c onto K^\times; if it is regarded as a morphism of N onto K^\times, then, for every $a \in N$, $\mathfrak{w}_K[\tau(a)]$ is the restriction to K_{ab} of the automorphism $\mathfrak{w}(a)$ of L_{ab} over K.*

This follows at once from the transfer theorem, as reformulated for W-groups at the end of Appendix II, § 5, when this is combined with the above results.

9. With the same notations as in § 8, the structure of W will now be completely determined by Shafarevitch's theorem (I.R. Shafarevitch, C. R. Ac. Sc. URSS 53 (1946), 15–16):

THEOREM II. *Let K, L, n be as in theorem I. Then the canonical algebra for (K, L) is the division algebra with the Hasse invariant $\mathbf{e}(1/n)$ over K.*

We will write $(K; L)$ for the Hasse invariant $h(A)$ of A; we have to prove $(K; L) = \mathbf{e}(1/n)$, and this will imply that A is a division algebra. The proof will consist of three steps:

(a) Let K be as above; let L, L' be as in § 6; call W', W'', $W_{L'}$ the W-groups of L'_{ab} over K, L, L', respectively; take for ω' the inverse of the canonical isomorphism $\mathfrak{w}_{L'}$ of L'^\times onto $W_{L'}$; let t be the transfer homomorphism of W''/W''^c into $W_{L'}$. As in Appendix II, § 5, we can identify W''/W''^c with W_L. If then we call ω the inverse of the canonical isomorphism \mathfrak{w}_L of L^\times onto $W_L = W''/W''^c$, the transfer theorem, as restated in Appendix II, § 5, gives $\omega = \omega' \circ t$. We are therefore exactly in the situation described in § 6, and lemma B gives $(K; L) = (K; L')^d$ with $d = [L':L]$.

(b) Let K, L, n be as in theorem I; let K' be any cyclic extension of K of degree n, e.g. the unramified one. Call L' the compositum of L and K' in K_{sep}; put $K_1 = K' \cap L$ and $d = [L:K_1]$. Then L' is of degree d both over L and over K'; consequently, by (a), $(K; L)$ and $(K; K')$ are both equal to $(K; L')^d$. In particular, we see that $(K; L)$ depends only upon n, and that it is enough to prove our theorem in the cyclic case.

(c) Take L cyclic over K; take notations as in § 4 above; with those notations, A is the cyclic algebra $[L/K; \chi, \theta]$ with $\theta = \tau(b_\alpha)$. In view of Appendix II, § 3, we have $h(A) = \chi[\mathfrak{w}_K(\theta)]$. By corollary 3 of theorem I, § 8, $\mathfrak{w}_K(\theta)$ is the restriction to K_{ab} of the automorphism $\mathfrak{w}(b_\alpha)$ of L_{ab} over K; by theorem I, the restriction to L of the latter automorphism, and therefore also of the former one, is the one induced on L by the automorphism $x \to b_\alpha^{-1} x b_\alpha$ of A, which is α. This gives $h(A) = \chi(\alpha) = \mathbf{e}(1/n)$, which completes the proof of Shafarevitch's theorem.

Appendix IV

The Herbrand distribution

1. We begin by stating some general facts about the Herbrand distributions, as defined in Chap. VIII-3. Let again K be a commutative p-field.

LEMMA A. *Let* \mathfrak{R}, \mathfrak{R}' *be two Galois extensions of* K, *finite or not, such that* $K \subset \mathfrak{R} \subset \mathfrak{R}'$. *Let* \mathfrak{G}, \mathfrak{G}', \mathfrak{H} *be the Galois groups of* \mathfrak{R} *over* K, *of* \mathfrak{R}' *over* K *and of* \mathfrak{R}' *over* \mathfrak{R}, *respectively; let* Φ *be the canonical morphism of* \mathfrak{G}' *onto* $\mathfrak{G} = \mathfrak{G}'/\mathfrak{H}$. *Let* H, H' *be the Herbrand distributions on* \mathfrak{G} *and on* \mathfrak{G}', *respectively. Then, for every locally constant function* f *on* \mathfrak{G}, *we have* $H(f) = H'(f \circ \Phi)$.

This is obvious. We may express the conclusion by saying that H is the image (more precisely, the "direct image") of H' under Φ.

LEMMA B. *Let* \mathfrak{R} *be a Galois extension of* K, *finite or not. Let* K' *be an extension of* K *of finite degree, contained in* \mathfrak{R}, *with the order of rami-fication* e *and the differential exponent* d *over* K. *Let* \mathfrak{G}, \mathfrak{G}' *be the Galois groups of* \mathfrak{R} *over* K *and over* K', *respectively; let* H, H' *be the Herbrand distributions on* \mathfrak{G} *and on* \mathfrak{G}'. *Then, for every locally constant function* f *on* \mathfrak{G}, *equal to* 0 *outside* \mathfrak{G}', *we have* $H'(f) = eH(f) - df(\varepsilon)$, *where* ε *is the identity in* \mathfrak{G}.

This is also obvious. It may be expressed by saying that H' coincides with eH on open and compact subsets of \mathfrak{G}', disjoint from ε, or more briefly that it coincides with eH on \mathfrak{G}' outside ε; this fact, together with the trivial condition $H'(1) = 0$, determines H' completely in terms of H.

2. Now let assumptions and notations be as in § 1 of Appendix II. Let H be the Herbrand distribution on \mathfrak{G}; as has been shown in the proof of lemma 3, Chap. XII-4, it is 0 outside \mathfrak{G}_0. More precisely:

LEMMA C. *Let* K, K_0, \mathfrak{R}, \mathfrak{G}, \mathfrak{G}_0 *be as in* § 1 *of Appendix* II. *Then the support of the Herbrand distribution* H *on* \mathfrak{G} *is* \mathfrak{G}_0.

Take any $\lambda \in \mathfrak{G}_0$, other than the identity; take any open subgroup \mathfrak{G}' of \mathfrak{G}, not containing λ; then, by Chap. VIII-3, we have $H(\mathfrak{G}''\lambda) < 0$ for all open subgroups \mathfrak{G}'' of \mathfrak{G}'. Now take $\lambda \in \mathfrak{G} - \mathfrak{G}_0$; then there is a root μ of 1 of order prime to p such that $\mu^\lambda \neq \mu$; call \mathfrak{G}' the open subgroup of \mathfrak{G} corresponding to $K(\mu)$. Then λ is not in \mathfrak{G}', and, by Chap. VIII-3, we

have $H(\mathfrak{G}''\lambda')=0$ whenever $\lambda'\in\mathfrak{G}'\lambda$ and \mathfrak{G}'' is an open subgroup of \mathfrak{G}', so that H is 0 on $\mathfrak{G}'\lambda$.

As noted above, it also follows from the definition of H that $H(\mathfrak{G})=0$, so that, in view of lemma C, $H(\mathfrak{G}_0)=0$.

Let now \mathfrak{W} be the W-group of \mathfrak{R} over K. Clearly there is a unique distribution \mathbf{H} on \mathfrak{W} which coincides with H on \mathfrak{G}_0 and is 0 outside \mathfrak{G}_0. This will be called *the Herbrand distribution* on \mathfrak{W}. As explained in Chap. VIII-3, we extend it to a linear form, also denoted by \mathbf{H}, on the space of locally constant functions on \mathfrak{W}.

3. Now we will apply theorems I and II of Appendix III, §§ 8–9. As in those theorems, we take a Galois extension L of K, of finite degree n; we call W, W_L the W-groups of L_{ab} over K and over L, respectively. We call A the canonical algebra for (K, L), N the normalizer of L^\times in A^\times, and \mathfrak{w} the canonical isomorphism of N onto W. We use the isomorphism \mathfrak{w}^{-1} of W onto N to transport to N the Herbrand distribution on W, and denote again by \mathbf{H} this distribution on N. Our purpose is to give an explicit formula for \mathbf{H} on N.

As before, we write $\nu_{A/K}$ for the reduced norm in A over K; moreover, we put $\|x\|=|\nu_{A/K}(x)|_K$ for every $x\in A$. In view of corollary 2 of theorem I, Appendix III, § 8, and of lemma C above, the support of \mathbf{H} on N is the compact subgroup N_0 of N determined by $\|a\|=1$, i.e. the kernel of the morphism $a\to\|a\|$ of N into \mathbf{R}_+^\times. As noted above, we have $\mathbf{H}(1)=0$.

Let da be the Haar measure on N, normalized so that the measure of N_0 is 1. The following theorem, in substance, is due to J. Tate and Shankar Sen (J. Ind. Math. Soc. 27 (1964), 197–202):

THEOREM. *For any locally constant function f on N, we have:*

(1)
$$\mathbf{H}(f)=-\int_{N_0}[f(a)-f(1_A)]\cdot\|1_A-a\|^{-1}\,da.$$

As both sides of (1) are 0 for $f=1$, it is enough to prove it for the case $f(1_A)=0$; this will be assumed from now on. The proof will consist of several steps:

(a) Take the "abelian case" where $L=K$, $n=1$, $A=K$, $N=K^\times$, $N_0=R^\times$; as usual, we write R for the maximal compact subring of K, and P for its maximal ideal. Clearly it is enough to verify (1) when f is the characteristic function of any set X of the form $X=(1+P^\nu)\xi$ with $0\le\operatorname{ord}(1-\xi)<\nu$; then $\mathbf{H}(X)$ is given by theorem 5 of Chap. XII-4. At the same time, the integrand in (1) is 0 outside X and has on X the constant value q^ρ with $\rho=\operatorname{ord}(1-\xi)$. As $1+P^\nu$ has the index $q^{\nu-1}(q-1)$ in R^\times, this proves (1) in this case.

(b) Take now the general case, and take a field K' between K and L, corresponding to a subgroup \mathfrak{g}' of the Galois group \mathfrak{g} of L over K; put

$n' = [L:K']$. Then, so far as A, L, N, K and K' are concerned, we are in the situation considered in §3 of Appendix III; if we write $N = \bigcup b_\alpha L^\times$ and $A = \sum b_\alpha L$ as there, we have seen in that § that the "commutant" algebra of K' in A is $A' = \sum b_\lambda L$, where the sum is taken over all $\lambda \in \mathfrak{g}'$; this is a central simple algebra of dimension n'^2 over K', and the normalizer of L^\times in A'^\times is the subgroup $N' = \bigcup b_\lambda L^\times$ of N, the union being taken again over all $\lambda \in \mathfrak{g}'$. In view of our observations in §2 of Appendix II, it is clear that the canonical isomorphism \mathfrak{w} of N onto W maps N' onto the W-group W' of L_{ab} over K'; corollary 1 of theorem I, Appendix III, §8, shows now that A' is the canonical algebra for (K', L), and that the canonical isomorphism of N' onto W' is the restriction \mathfrak{w}' of \mathfrak{w} to N'. Consequently, if \mathbf{H}' is the Herbrand distribution on N', lemma B shows that, on N' and outside 1_A, \mathbf{H}' coincides with $e\mathbf{H}$, where e is the order of ramification of K' over K. Now call (1') the formula, similar to (1), with \mathbf{H}', A', N' substituted for \mathbf{H}, A, N. For any f, equal to 0 outside N' (and at 1_A, as assumed above), call $\mathbf{H}_1(f)$, $\mathbf{H}'_1(f)$ the right-hand sides of (1) and of (1'), respectively; it will be shown now that $\mathbf{H}'_1(f) = e\mathbf{H}_1(f)$.

Take any $x' \in A'^\times$; by corollary 1 of prop. 6, Chap. IX-2, and corollary 3 of th. 3, Chap. I-2, the automorphism $y' \to x'y'$ of the additive group of A' has the module

$$\mathrm{mod}_{A'}(x') = |v_{A'/K'}(x')|_{K'}^{n'}.$$

Similarly, the module of $y \to x'y$ in A is

$$\mathrm{mod}_A(x') = |v_{A/K}(x')|_K^n.$$

But we may also regard A as a left vector-space over the division algebra A'; as the dimensions of A and A' over K are n^2 and $n'^2 d$ with $d = [K':K] = n/n'$, A has the dimension d over A'. By corollary 2 of th. 3, Chap. I-2, we have then $\mathrm{mod}_A(x') = \mathrm{mod}_{A'}(x')^d$. This gives

$$|v_{A/K}(x')|_K = |v_{A'/K'}(x')|_{K'}.$$

Therefore the integrands in $\mathbf{H}_1(f)$ and $\mathbf{H}'_1(f)$ are the same. Put now $N'_0 = N' \cap N_0$. If K_0 is as before (cf. §2), \mathfrak{w} maps N_0 onto the Galois group \mathfrak{G}_0 of L_{ab} over K_0, and similarly it maps N'_0 onto the Galois group of L_{ab} over the compositum $K'_0 = K'K_0$; therefore the index of N'_0 in N_0 is equal to the degree of K'_0 over K_0, which is the same as that of K' over $K' \cap K_0$; this is e, by corollary 4 of th. 7, Chap. I-4. Consequently, if $d'a$ is the Haar measure on N', normalized so that the measure of N'_0 is 1, we have $d'a = e \cdot da$ on N'. This gives $\mathbf{H}'_1(f) = e\mathbf{H}_1(f)$, as we had asserted.

(c) In particular, apply (b) to the case $K' = L$. In view of (a), it shows that (1) holds whenever f is 0 outside L^\times (and at 1_A).

(d) To prove (1), it is enough to show that both sides coincide on each coset of L^x in N; we have found in (c) that they do so on L^x itself; we still have to verify that they coincide on all other cosets. In other words, let $b_\alpha L^x$ be any coset of L^x in N, other than L^x; we have to show that (1) holds whenever f is 0 outside $b_\alpha L^x$. Let \mathfrak{g}' be the cyclic subgroup of \mathfrak{g} generated by α; let K' be the field between K and L, corresponding to \mathfrak{g}'; apply to K' what has been proved above in (b). We see thus that it is enough to verify (1') for f equal to 0 outside $b_\alpha L^x$. Writing now K, N, \mathfrak{g} instead of K', N', \mathfrak{g}', we see that our theorem will be proved if we verify (1) under the additional assumptions that \mathfrak{g} is cyclic of order $n > 1$, generated by α. and that f is 0 outside the coset $b_\alpha L^x$.

(e) That being now assumed, we are once more in the situation described in §4 of Appendix III. Let notations be the same as there; N^c is then the same as the kernel U of the morphism $N_{L/K}$ of L^x into K^x; as we have $|\xi|_L = |N_{L/K}(\xi)|_K$ for all $\xi \in L^x$, U is compact. By corollary 3 of theorem I, Appendix III, §8, the transfer τ of N into L^x has the kernel $N^c = U$ and maps N onto K^x; the same corollary shows also that, if we identify N with W by means of \mathfrak{w}, and K^x with $W_K = W/W^c$ by means of \mathfrak{w}_K, τ becomes the canonical morphism of W onto $W/W^{c\cdot}$ therefore we can apply lemma A of §1 above, and conclude that the direct image under τ of the Herbrand measure \mathbf{H} on W is the Herbrand measure \mathbf{H}_K on K^x, as given by (a) above. In other words, for any locally constant function F on K^x, equal to 0 at 1, we have

$$(2) \qquad \mathbf{H}(F \circ \tau) = \mathbf{H}_K(F) = -\int_{R^x} F(x) \cdot |1 - x|_K^{-1} d^x x,$$

with the Haar measure $d^x x$ on K normalized so that the measure of R^x is 1. Clearly a locally constant function on N can be written as $F \circ \tau$ if and only if it is constant on the cosets of U.

We have to prove (1) for those functions f on N which are 0 outside $b_\alpha L^x$. Observe now that both sides of (1) are clearly invariant under all inner automorphisms of N, and in particular under any automorphism $a \to \xi a \xi^{-1}$ with $\xi \in L^x$. For $a \in b_\alpha L^x$, we have $a^{-1} \xi a = \xi^\alpha$, hence $\xi a \xi^{-1} = a u$ with $u = \xi^{\alpha - 1}$. By Hilbert's theorem, the kernel U of $N_{L/K}$ is the group consisting of the elements $u = \xi^{\alpha - 1}$ for $\xi \in L^x$; consequently, on the coset $b_\alpha L^x$, the inner automorphisms $a \to \xi a \xi^{-1}$, for $\xi \in L^x$, induce the same mappings as the translations $a \to a u$ for $u \in U$. Therefore, if f is 0 outside $b_\alpha L^x$, both sides of (1) remain unchanged, for any $u \in U$, when one replaces f by the function $a \to f(a u)$; hence they are still unchanged if we replace f by $a \to \bar{f}(a)$, where $\bar{f}(a)$ is the mean value of $u \to f(a u)$ on U for the Haar measure on U. Thus our theorem will be proved if we verify (1) for such a function \bar{f}, i.e. for one which is 0 outside $b_\alpha L^x$ and

constant under the translations $a \to a u$. From now on, let f be such a function; as we have seen, it can be written as $f = F \circ \tau$.

As before, put $\theta = \tau(b_\alpha)$. As τ coincides with $N_{L/K}$ on L^\times, it maps the cosets $b_\alpha^i L^\times$ of L^\times in N, for $0 \le i < n$, onto the cosets $\theta^i N_{L/K}(L^\times)$ of $N_{L/K}(L^\times)$ in K^\times, respectively; as it maps N onto K^\times, and as its kernel U is contained in L^\times, K^\times is the disjoint union of these n cosets (a result which is substantially contained in corollary 2 of th. 3, Chap. XII-3). In particular, a function $f = F \circ \tau$ is 0 outside $b_\alpha L^\times$ if and only if F is 0 outside $\theta N_{L/K}(L^\times)$. To complete our proof, we have to compare the right-hand sides of (1) and of (2) for such a pair of functions f, F. By corollaries 2 and 3 of theorem I, Appendix III, § 8, we have $|\tau(a)|_K = \|a\|$ for all $a \in N$; therefore $\tau^{-1}(R^\times) = N_0$, and the direct image of the measure da in (1) is the measure $d^\times x$ in (2). Now take any $a \in b_\alpha L^\times$; in § 4 of Appendix III, we have seen that $\tau(a) = a^n \in K^\times$, and that the a^i, for $0 \le i < n$, may be taken as representatives of the cosets of L^\times in N and therefore also as a basis of A over L; consequently, if $\{\eta_0, \ldots, \eta_{n-1}\}$ is a basis of L over K, the elements $a^i \eta_j$, for $0 \le i, j < n$, make up a basis of A over K. In order to evaluate the integrand of (1) for $a \in b_\alpha L^\times$, put $\mu = \|1_A - a\|$. Then the automorphism $z \to (1_A - a)z$ of the additive group of A has the module μ^n. On the other hand, this module may also be expressed by means of corollary 3 of th. 3, Chap. I-2, in terms of the determinant of the linear substitution determined by that automorphism on the basis $\{a^i \eta_j\}$; this determinant is easily seen to have the value $(1 - x)^n$, with $x = a^n = \tau(a)$. This gives $\mu = |1 - x|_K$. Therefore, for $x = \tau(a)$, the integrands in the right-hand sides of (1) and of (2) are the same. This concludes the proof.

Index of definitions

(This index contains all concepts and terms whose definition is given or recalled in the text, even if this is not done by way of a formal definition. A reference such as IV-1 (62) means Chapter IV, § 1, page 62; IV-1, d. 1 (59) means Chapter IV, § 1, definition 1, page 59; P & N (XII) means Prerequisites and Notations, page XII).

Index of definitions

factor-set, IX-3, d. 4 (175); cyclic —, IX-4
 (181).
faithful, IX-1 (162).
field of constants, IV-4 (77).
finite: — field, I-1 (1); — place, III-1, d. 3
 (44).
Fourier transform, VII-2 (105).
fractional ideal, V-3, d. 4 (85).
Frobenius automorphism, I-4, d. 5 (20),
 XII-2 (221), XIII-1 (247), XIII-11
 (284—285).
fully: — ramified, I-4, d. 4 (15); to split —,
 VIII-5 (158).
fundamental set, V-4 (89).

genus, VI, d. 1 (100).

Hasse invariant, XII-2 (221, 224), XIII-3
 (252).
Herbrand distribution, VIII-3 (152), XII-4
 (239).
hermitian form, X-3 (198).
Hilbert symbol, XIII-5 (262).
Hom, X-1 (188—189).
homomorphism, P & N (XIV); transfer —,
 XII-5 (241).
ᕼ-regular, IX-3, d. 2 (171).

id, V-3 (87).
ideal: fractional —, V-3, d. 4 (85).
ideal-class, V-3 (87).
idele, idele group, IV-3 (71); differential —,
 VII-2, d. 4 (113).
image (of a Haar measure), II-4 (36).
imaginary (place), III-1, d. 3 (44).
infinite (place), III-1, d. 3 (44).
integral (over a ring), I-4 (22).
invariant (Hasse —), cf. Hasse.
inverse (algebra), IX-1 (164).
isomorphism, P & N (XV).

k-lattice, V-2, d. 3 (83).
K-lattice, II-2, d. 2 (28).
K-norm, II-1, d. 1 (24).

lattice: dual —, II-5 (39); k- —, V-2, d. 3
 (83); K- —, II-2, d. 2 (28); normal —,
 X-1 (193), XI-4 (212); Q- —, V-1, d. 1
 (80); R- —, II-4, d. 3 (35).
left order, X-1 (192), XI-4 (212).
L-function, VII-7 (133).
lie (to — above, below), III-1, d. 4 (45).
local field, I-4 (20).
locally constant, VII-2 (106).

maximal order, V-2 (82), X-1 (191), XI-4
 (211).
mod, I-2 (3—4).
modular degree, I-4, d. 4 (15).
module: (of an automorphism) I-2 (3); (of
 a p-field) I-4, d. 3 (13); (of an idele)
 IV-3 (73).
monic, P & N (XIV).
morphism, P & N (XV); canonical —,
 XII-1 (215), XII-2 (224), XIII-1 (245);
 natural —, IX-3 (171); restriction —,
 IX-3 (173).
multiplicative character, XIII-2 (250).

N, III-3 (53).
ᕉ, V-3, d. 5 (88), VIII-4 (155).
natural: — embedding, VIII-4 (154);
 — morphism, IX-3 (171).
norm, III-3 (53), V-3, d. 5 (88), VIII-4 (155);
 K- —, II-1, d. 1 (24); reduced —, IX-2
 (169); regular —, III-3 (53).
normal lattice, X-I (193), XI-4 (212).
N-orthogonal, II-1 (25—26).
numerator, V-3 (87).

ord, I-4, d. 3 (13), II-5, d. 4 (41).
order, P & N (XV), V-1, d. 2 (81), X-1 (191
 —192), XI-4 (211—212); — (of a char-
 acter), II-5, d. 4 (41); — of ramification,
 I-4, d. 4 (15); left, right —, X-1 (192),
 XI-4 (212); maximal —, V-2 (82), X-1
 (191), XI-4 (211).
orthogonal: N- —, II-1 (25—26).
orthonormal, X-3 (198).

p-adic, I-3 (11).
pairing (canonical —), XII-1 (215), XII-2
 (224), XIII-1 (245).
p-field, I-3, d. 2 (12).
place, III-1, d. 2 (43—44); finite, imaginary,
 infinite, real —, III-1, d. 3 (44).
Poisson summation formula, VII-2 (106).
polynomial: — mapping, III-3 (52); Eisen-
 stein —, VIII-2 (147); prime —, III-1
 (46).
power-series (formal —), I-4 (20).
prime: — element, I-4, d. 3 (13); — poly-
 nomial, III-1 (46); — to P, VII-8 (136);
 mutually —, V-3 (86).
primitive root of 1, P & N (XV).
principal: — divisor, VI (97); — ideal, V-3
 (87); — quasicharacter, VII-3, d. 6
 (115).
proper embedding, III-2 (50).

Springer-Verlag
and the Environment

We at Springer-Verlag firmly believe that an international science publisher has a special obligation to the environment, and our corporate policies consistently reflect this conviction.

We also expect our business partners – paper mills, printers, packaging manufacturers, etc. – to commit themselves to using environmentally friendly materials and production processes.

The paper in this book is made from low- or no-chlorine pulp and is acid free, in conformance with international standards for paper permanency.